SEVENTH EDITION

Historical Geology

Evolution of Earth and Life Through Time

Reed Wicander
Central Michigan University

James S. Monroe
Emeritus, Central Michigan University

BROOKS/COLE
CENGAGE Learning

Australia • Brazil • Japan • Korea • Mexico • Singapore • Spain • United Kingdom • United States

BROOKS/COLE
CENGAGE Learning·

Historical Geology: Evolution of Earth and Life Through Time, Seventh Edition
Reed Wicander, James S. Monroe

Publisher: Yolanda Cossio

Senior Acquisitions Sponsoring Editor: Aileen Berg

Developmental Editor: Liana Sarkisian

Assistant Editor: Shannon Elderon

Editorial Assistant: Margaux Cameron

Media Editor: Alexandria Brady

Marketing Manager: Jack Cooney

Marketing Communications Manager: Darlene Macanan

Content Project Manager: Carol Samet

Art Director: Pam Galbreath

Manufacturing Planner: Becky Cross

Rights Acquisitions Specialist: Tom McDonough

Production and Composition Service: Jennifer Bonnar, Lachina Publishing Services

Photo Researcher: Wendy Granger

Text Researcher: Pablo D'Stair

Copy Editor: Maggie Jarpey

Illustrators: Accurate Art, Precision Graphics, Rolin Graphics, Pre-Press PMG, Lachina Publishing Services

Cover and Text Designer: Lisa Buckley

Cover Image: Eryk Jaegermann/Design Pics/ Corbis

For product information and technology assistance, contact us at **Cengage Learning Customer & Sales Support, 1-800-354-9706**

For permission to use material from this text or product, submit all requests online at **www.cengage.com/permissions**
Further permissions questions can be e-mailed to **permissionrequest@cengage.com**

Library of Congress Control Number: 2012932126

Student Edition:

ISBN-13: 978-1-111-98729-9

ISBN-10: 1-111-98729-7

Loose-leaf Edition:

ISBN-13: 978-1-111-99057-2

ISBN-10: 1-111-99057-3

Brooks/Cole
20 Davis Drive
Belmont, CA 94002-3098
USA

Cengage Learning is a leading provider of customized learning solutions with office locations around the globe, including Singapore, the United Kingdom, Australia, Mexico, Brazil, and Japan. Locate your local office at **www.cengage .com/global**

Cengage Learning products are represented in Canada by Nelson Education, Ltd.

To learn more about Brooks/Cole, visit **www.cengage.com/brookscole**

Purchase any of our products at your local college store or at our preferred online store **www.CengageBrain.com**

Printed in the United States of America
1 2 3 4 5 6 7 16 15 14 13 12

About the Authors

Reed Wicander

REED WICANDER is a geology professor at Central Michigan University, where he teaches physical geology, historical geology, prehistoric life, and invertebrate paleontology. He has coauthored numerous geology textbooks with James S. Monroe. His main research interests involve various aspects of Paleozoic palynology, specifically the study of acritarchs, a subject on which he has published many papers. He is a past president of the American Association of Stratigraphic Palynologists, a former chairman of the Acritarch Subcommission of the Commission Internationale de Microflore du Paléozoique, and former councillor of the International Federation of Palynological Societies.

Courtesy of Melanie G. Wicander

James S. Monroe

JAMES S. MONROE is professor emeritus of geology at Central Michigan University, where he has taught physical geology, historical geology, prehistoric life, and stratigraphy and sedimentology since 1975. He has coauthored numerous textbooks with Reed Wicander and has interests in Cenozoic geology and geologic education. Now retired, he continues to teach geology classes for Osher Lifelong Learning Institute, an affiliate of California State University, Chico, and leads field trips to areas of geologic interest.

James S. Monroe

Brief Contents

Contents

CHAPTER 4
Geologic Time Concepts and Principles   62

CHAPTER 5
Rocks, Fossils, and Time
Making Sense of the Geologic Record   81

CHAPTER 9
Precambrian Earth and Life History The Proterozoic Eon   168

CHAPTER 10
Early Paleozoic Earth History   190

CHAPTER 11
Late Paleozoic Earth History   210

Smithsonian Institution, Transparency No. 86-13471A

CHAPTER 12
Paleozoic Life History
Invertebrates   233

Courtesy of Ken Higgs, Department of Geology, University College Cork, Ireland

CHAPTER 13
Paleozoic Life History
Vertebrates and Plants   253

© 2009 Lars Oppermann/Flickr/Getty Images

CHAPTER 14
Mesozoic Earth History   274

CHAPTER 15

Life of the Mesozoic Era   297

CHAPTER 16

Cenozoic Earth History The Paleogene and Neogene Periods   320

CHAPTER 17

Cenozoic Earth History The Quaternary Period   344

© Carl Buell

CHAPTER 18
Life of the Cenozoic Era 365

Courtesy of Persis Sturgis

CHAPTER 19
Primate and Human Evolution 388

Epilogue 403

Preface

Earth is a dynamic planet that has changed continuously during its 4.6 billion years of existence. The size, shape, and geographic distribution of the continents and ocean basins have changed through time, as have the atmosphere and biota. As scientists and concerned citizens, we have become increasingly aware of how fragile our planet is and, more importantly, how interdependent all of its various systems and subsystems are.

We have also learned that we cannot continually pollute our environment and that our natural resources are limited and, in most cases, nonrenewable. Furthermore, we are coming to realize how central geology is to our everyday lives. For these and other reasons, geology is one of the most important college or university courses a student can take.

Historical geologists are concerned with all aspects of Earth and its life history. They seek to determine what events occurred during the past, place those events into an orderly chronological sequence, and provide conceptual frameworks for explaining such events. Equally important is using the lessons learned from the geologic past to understand and place in context some of the global issues facing the world today, such as depletion of natural resources, global climate warming, and decreasing biodiversity. Thus, what makes historical geology both fascinating and relevant is that, like the dynamic Earth it seeks to understand, it is an exciting and ever-changing science in which new discoveries and insights are continually being made.

Historical Geology: Evolution of Earth and Life Through Time, seventh edition, is designed for a one-semester geology course and is written with students in mind. One of the problems with any introductory science course is that students are overwhelmed by the amount of material that must be learned. Furthermore, most of the material does not seem to be linked by any unifying theme and does not always appear to be relevant to their lives. This book, however, is written to address that problem in that it shows, in its easy-to-read style, that historical geology is an exciting science, and one that is increasingly relevant in today's world.

The goal of this book is to provide students with an understanding of the principles of historical geology and how these principles are applied in unraveling Earth's history. It is our intent to present the geologic and biologic history of Earth, not as a set of encyclopedic facts to memorize, but rather as a continuum of interrelated events reflecting the underlying geologic and biologic principles and processes that have shaped our planet and life upon

it. Instead of emphasizing individual, and seemingly unrelated, events, we seek to understand the underlying causes of why things happened the way they did and how all of Earth's systems and subsystems are interrelated. Using this approach, students will gain a better understanding of how everything fits together and why events occurred in a particular sequence.

Because of the nature of the science, all historical geology textbooks share some broad similarities. Most begin with several chapters on concepts and principles, followed by a chronological discussion of Earth and life history. In this respect, we have not departed from convention. We have, however, attempted to place greater emphasis on basic concepts and principles, their historical development, and their importance in deciphering Earth history: in other words, how do we know what we know? By approaching Earth history in this manner, students come to understand Earth's history as part of a dynamic and complex integrated system, and not as a series of isolated and unrelated events.

Features in the Seventh Edition

Just as Earth is dynamic and evolving, so too is *Historical Geology: Evolution of Earth and Life Through Time*. The seventh edition has undergone significant and considerable rewriting and updating, resulting in a book that is still easy to read yet has a high level of current information. The new edition features many new photographs and figures, as well as numerous new *Perspectives*, all of which are designed to help students maximize their learning and understanding of their planet's history. Drawing on the comments and suggestions of reviewers and users of the sixth edition, we have expanded and retained many of the features that were successful in the previous edition.

- The *Chapter Objectives* outline at the beginning of each chapter has been retained to alert students to the key points that the chapter will address.
- Chapter 2 (Mineral and Rocks), which provides the necessary background for those students who are unfamiliar with minerals, rocks, and the rock cycle, has been retained, as well as Appendix C on Mineral Identification.

- Chapter content has been extensively updated and rewritten to (1) help clarify concepts, (2) emphasize underlying processes, and (3) make the material more exploratory.
- An added emphasis has been placed on global climate warming throughout the text, with both new material added and previous sections rewritten and updated.
- Twelve of the seventeen *Perspectives* are new.
- New, bold, and dramatic photos have been added to open each chapter. In addition, numerous new images have been added to the text of each chapter, and many figures have been updated.
- The format of about 10 multiple-choice questions and 10 short-essay questions has been retained in the *Review Questions* section at the end of each chapter. A significant number of the questions have been rewritten or are new. Answers to all of the multiple-choice questions are provided at the back of the book.
- The *Epilogue* has been reformatted, updated, and expanded. Several new figures have also been added to help the reader visualize and better understand the topics covered. The *Epilogue* is designed to tie together current issues with the historical perspective of geology presented in the previous 19 chapters.

It is our strong belief that the major rewriting and updating done in the text, as well as the addition of new and dramatic photos and figures, have greatly improved the seventh edition of *Historical Geology: Evolution of Earth and Life Through Time*. We think that these changes and enhancements make the textbook easier to read and comprehend, as well as a more effective teaching tool that engages students in the learning process, thereby fostering a better understanding of the material and how it relates to Earth today.

Text Organization

As in the previous editions, we develop three major themes in this textbook that are essential to the interpretation and appreciation of historical geology. These themes are introduced early and reinforced throughout the book. The themes are *plate tectonics* (Chapter 3), a unifying theory for interpreting much of Earth's physical history and, to a large extent, its biologic history; *time* (Chapter 4), the dimension that sets historical geology apart from most of the other sciences; and *evolutionary theory* (Chapter 7), the explanation for inferred relationships among living and fossil organisms. Additionally, we have emphasized the intimate interrelationship existing between physical and biologic events, and the fact that Earth is a complex, dynamic, and evolving planet whose history is best studied by using a systems approach.

This book was written for a one-semester course in historical geology to serve both majors and non-majors in geology and in the Earth sciences. We have organized

Historical Geology: Evolution of Earth and Life Through Time, seventh edition, into the following informal categories:

- Chapter 1 reviews the principles and concepts of geology and the three themes this book emphasizes. The text is written at an appropriate level for those students taking historical geology with no prerequisites, but the instructor may have to spend more time expanding some of the concepts and terminology discussed in Chapter 1.
- Chapter 2, "Minerals and Rocks," can be used to introduce those students who have not had an introductory geology course to minerals, rocks, and the rock cycle, or as a review for those students that have had such a course.
- Chapter 3 explores plate tectonics, which is the first major theme of this book. Particular emphasis is placed on the evidence substantiating plate tectonic theory, why this theory is one of the cornerstones of geology, and why plate tectonic theory serves as a unifying paradigm in explaining many apparently unrelated geologic phenomena.
- The second major theme of this book, the concepts and principles of geologic time, is examined in Chapter 4. In addition, a new *Perspective*, "The Anthropocene: A New Geologic Epoch?" has been added to show how humans' impact on Earth might show up in the geologic record, possibly necessitating a new geologic epoch: the Anthropocene.
- Chapter 5 expands on the theme of geologic time by integrating it with rocks and fossils.
- Depositional environments are sometimes covered rather superficially (perhaps with little more than a summary table) in some historical geology textbooks. However, Chapter 6, "Sedimentary Rocks: The Archives of Earth History," is completely devoted to this topic; it contains sufficient detail to be meaningful while avoiding an overly detailed discussion more appropriate for advanced courses.
- The third major theme of this book, organic evolution, is examined in Chapter 7. In this chapter, the theory of evolution is covered, as well as its supporting evidence.
- Precambrian time—fully 88 percent of all geologic time—is sometimes considered in a single chapter in other historical geology textbooks. However, in this book, Chapter 8 is devoted to the geologic and biologic histories of the Hadean and Archean Eon, whereas Chapter 9 covers the same topics for the Proterozoic Eon.
- Chapters 10 through 19 constitute our chronological treatment of the Phanerozoic geologic and biologic history of Earth. These chapters are arranged so that the geologic history of an era is followed by a discussion of the biologic history of that era. We think that

this format allows easier integration of life history with geologic history.

- An *Epilogue* summarizes the major topics and themes of this book, with an updated and added emphasis on global climate warming.

In these chapters, there is an integration of the three themes of this book, as well as an emphasis on the underlying principles of geology and how they have helped us decipher Earth's history. We have found that presenting the material in the order just discussed works well for most students. We know, however, that many instructors prefer an entirely different order of topics, depending on the emphasis in their course. We have, therefore, written this book so that instructors can present the chapters in whatever order suits the needs of a particular course.

Chapter Organization

All chapters have the same organizational format as follows:

- Each chapter opens with a photograph relating to the chapter material, many of which are new and dramatic, followed by an *Outline* of the topics covered, and a list of *Chapter Objectives* that alerts students to the learning-outcome objectives of the chapter.

- An *Introduction* follows that is intended to stimulate interest in the chapter and show how the chapter material fits into the larger geologic perspective. Many of the *Introductions* have been rewritten and updated in this edition.

- The text is written in a clear, informal style, making it easy for students to comprehend.

- Numerous color diagrams and photographs complement the text and provide a visual representation of the concepts and information presented.

- All of the chapters, except Chapter 2, contain a *Perspective*, most of which are new to this edition, presenting a brief discussion of an interesting aspect of historical geology or geologic research pertinent to that chapter.

- Each of the chapters on geologic history in the second half of this book contains a final section on mineral resources characteristic of that time period. These sections provide applied economic material of interest to students.

- The end-of-chapter materials begin with a concise review of important concepts and ideas in the *Summary*.

- The *Important Terms*, which are printed in boldface type in the chapter text, are listed at the end of each chapter for easy review, along with the page number where that term is first defined. A full *Glossary* of important terms appears at the end of the text.

- The *Review Questions* are another important feature of this book. They include about 10 multiple-choice questions with answers, as well as about 10 short-essay questions. Many of both types of questions are new in each chapter of the seventh edition.

- The global paleogeographic maps that illustrate in stunning relief the geography of the world during various time periods have been retained in this edition. These maps enable students to visualize what the world looked like during the time period being studied and thus add a visualization dimension to the text material.

- As in the previous editions, end-of-chapter summary tables are provided for the chapters on geologic and biologic history. These tables are designed to give an overall perspective of the geologic and biologic events that occurred during a particular time interval and to show how these events are interrelated. The emphasis in these tables is on the geologic evolution of North America. Global tectonic events and sea-level changes are also incorporated into these tables to provide global insights and perspective.

Ancillary Materials

We are pleased to offer a full suite of text and multimedia products to accompany the seventh edition of *Historical Geology: Evolution of Earth and Life Through Time.*

For Instructors

Online Instructor's Manual with Test Bank This comprehensive manual is designed to help instructors prepare for lectures. It contains lecture outlines, teaching suggestions and tips, and an expanded list of references, plus a bank of test questions. (ISBN: 978-1-111-83108-0)

ExamView® Computerized Testing Create, deliver, and customize tests and study guides (both print and online) in minutes with this easy-to-use assessment and tutorial system. ExamView offers both a *Quick Test Wizard* and an *Online Test Wizard* that guide you step-by-step through the process of creating tests, while its unique capabilities allow you to see the test you are creating on the screen exactly as it will print or display online. You can build tests of up to 250 questions using up to 12 question types. Using *ExamView*'s complete word processing capabilities, you can enter an unlimited number of new questions or edit existing questions. (ISBN: 978-1-111-83117-2)

PowerLecture™: A One-Stop Microsoft® PowerPoint® Tool PowerLecture instructor resources are a collection of book-specific lecture and class tools on DVD. The fastest and easiest way to build powerful, customized media-rich lectures, PowerLecture includes chapter-specific PowerPoint presentations, images, animations and video, an instructor manual, a test bank, and videos. PowerLecture media-teaching tools are an effective way to enhance the educational experience. (ISBN: 978-1-111-98742-8)

For Students

Geology CourseMate The more you study, the better the results. Make the most of your study time by accessing everything you need to succeed in one place. Geology CourseMate includes:

- An interactive eBook with highlighting, note taking, and an interactive glossary
- Interactive learning tools, including:
  - Quizzes
  - Flashcards
  - Animations
  - . . . and more!

(ISBN-10: 1-111-83112-2 | ISBN-13: 978-1-111-83112-7)

Acknowledgments

As the authors, we are, of course, responsible for the organization, style, and accuracy of the text, and any mistakes, omissions, or errors are our responsibility. The finished product is the culmination of many years of work during which we received numerous comments and advice from many geologists who reviewed parts of the text. We wish to express our sincere appreciation to the reviewers who reviewed the sixth edition and made many helpful and useful comments that led to the improvements seen in this seventh edition.

Paul Buchanan, *Kilgore College*

Donald Burt, *Arizona State University*

Ntungwa Maasha, *College of Coastal Georgia*

Billy Morris, *Georgia Highlands College*

We would like to thank the reviewers of the first five editions. Their comments and suggestions resulted in many improvements and pedagogical innovations, making this book a success for students and instructors alike.

Paul Belasky, *Ohlone College*

Allen I. Benimof, *College of Staten Island*

Stuart Birnbaum, *University of Texas, San Antonio*

Claudia Bolze, *Tulsa Community College*

Thomas W. Broadhead, *University of Tennessee at Knoxville*

Mark J. Camp, *University of Toledo*

James L. Carew, *College of Charleston*

James F. Coble, *Tidewater Community College*

Mitchell W. Colgan, *College of Charleston*

Bruce Corliss, *Duke University*

William C. Cornell, *University of Texas at El Paso*

John F. Cottrell, *Monroe Community College*

Rex E. Crick, *University of Texas at Arlington*

Michael Dalman, *Blinn College*

Chris Dewey, *Mississippi State University*

Dean A. Dunn, *University of Southern Mississippi*

Mike Farabee, *Estrella Mountain Community College*

Chad Ferguson, *Bucknell University*

Robert Fillmore, *Western State College of Colorado*

Michael Fix, *University of Missouri, St. Louis*

Richard Fluegeman, Jr., *Ball State University*

Annabelle Foos, *University of Akron*

Sarah Fowell, *University of Alaska, Fairbanks*

Heather Gallacher, *Cleveland State University*

Susan Goldstein, *University of Georgia*

Richard Gottfrief, *Frederick Community College*

Joseph C. Gould, *University of South Florida*

Bryan Gregor, *Wright State University*

Thor A. Hansen, *Western Washington University*

Richard D. Harnell, *Monroe Community College*

Rob Houston, *Northwestern Michigan College*

Paul D. Howell, *University of Kentucky*

Amanda Julson, *Blinn College*

Jonathan D. Karr, *North Carolina State University*

Lanna Kopachena, *Eastfield College (Dallas County Community College District)*

William W. Korth, *Buffalo State College*

Niranjala Kottachchi, *Fresno City College*

R. L. Langenheim, Jr., *University of Illinois at Urbana-Champaign*

Steve LoDuca, *Eastern Michigan University*

Ntungwa Maasha, *College of Coastal Georgia*

L. Lynn Marquez, *Millersville University*

George F. Maxey, *University of North Texas*

Michael McKinney, *University of Tennessee at Knoxville*

Glen K. Merrill, *University of Houston-Downtown*

Arthur Mirsky, *Indiana University, Purdue University at Indianapolis*

Pamela Nelson, *Glendale Community College*

Christine O'Leary, *Wallace State Community College*

N. S. Parate, *HCC-NE College, Pinemont Center*

William C. Parker, *Florida State University*

Mark W. Presley, *Eastfield College (Dallas County Community College District)*

Anne Raymond, *Texas A&M University*

G. J. Retallack, *University of Oregon*

Mark Rich, *University of Georgia*

Scott Ritter, *Brigham Young University*

Gary D. Rosenberg, *Indiana University, Purdue University at Indianapolis*

Barbara L. Ruf, *University of Georgia*

Michael Rygel, *SUNY, Potsdam*

W. Bruce Saunder, *Bryn Mawr College*

Steven Schimmrich, *SUNY, Ulster County Community College*

William A. Smith, *Charleston Southern University*

Ronald D. Stieglitz, *University of Wisconsin, Green Bay*

Michelle Stoklosa, *Boise State University*

Edmund Stump, *Arizona State University*

David Sunderlin, *Lafayette College*

Donald Swif, *Old Dominion University*

Carol M. Tang, *Arizona State University*

Jane L. Teranes, *Scripps Institution of Oceanography*

Michael J. Tevesz, *Cleveland State University*

Matthew S. Tomaso, *Montclair State University*

Art Troell, *San Antonio College*

Robert A. Vargo, *California University of Pennsylvania*

Shawn Willsey, *College of Southern Idaho*

Mark A. Wilson, *College of Wooster*

We also wish to thank Kathy Benison, Richard V. Dietrich (Professor Emeritus), David J. Matty, Jane M. Matty, Wayne E. Moore (Professor Emeritus), and Sven Morgan of the Geology Department of Central Michigan University, and Bruce M. C. Pape (Emeritus) of the Geography Department of Central Michigan University, as well as Eric Johnson (Hartwick College, New York) and Stephen D. Stahl (St. Bonaventure, New York) for providing us with photographs and answering our questions concerning various topics. We are also grateful for the generosity of the various agencies and individuals from many countries that provided photographs.

Special thanks must go to Yolanda Cossio, publisher for the Earth Sciences discipline of Cengage Learning, who initiated this seventh edition, to Aileen Berg, senior acquisitions sponsoring editor, and to our developmental editor, Liana Sarkisian, who not only kept us on task but encouraged us throughout its revision. We would also like to thank Pam Galbreath for the fresh design, as well as acknowledging Carol Samet, Cengage Learning Content Project Manager, for all her help. We also thank Maggie Jarpey for her copyediting skills. We appreciate her help in improving our manuscript. We thank Pablo D'Stair, text permissions manager, for securing the necessary permissions, and Wendy Granger for her invaluable help in locating appropriate photos.

Because historical geology is largely a visual science, we extend thanks to the artists at Precision Graphics, Graphic World, Pre-Press PMG, and Lachina Publishing Services, who were responsible for much of the art program. We also thank the artists at Magellan Geographix, who rendered many of the maps, and Dr. Ron Blakey, who allowed us to use his global paleogeographic maps.

As always, our families were very patient and encouraging when most of our spare time and energy were devoted to this book. We thank them for their continued support and understanding during the writing and revisions of these seven editions.

Reed Wicander

James S. Monroe

The Dynamic and Evolving Earth

True-color satellite image of Asia (partly in shadow), the Arctic ice cap, and the Sun. The complex interactions between the atmosphere, biosphere, hydrosphere, lithosphere, and Earth's interior have resulted in a dynamically changing planet since its origin 4.6 billion years ago.

Planet Observer/Getty Images

At the end of this chapter, you will have learned that

▶ Earth is a complex, dynamic planet that has continually evolved since its origin some 4.6 billion years ago.

▶ Earth can be viewed as an integrated system of interconnected components that interact and affect one another in various ways.

▶ Theories are based on the scientific method and can be tested by observation and/or experiment.

▶ The universe is thought to have originated approximately 14 billion years ago with a Big Bang, and the solar system and planets evolved from a turbulent, rotating cloud of material surrounding the embryonic Sun.

▶ Earth consists of three concentric layers—core, mantle, and crust—and this orderly division resulted during Earth's early history.

▶ Plate tectonics is the unifying theory of geology, and it revolutionized the science.

▶ The theory of organic evolution provides the conceptual framework for understanding the evolution of Earth's fauna and flora.

▶ An appreciation of geologic time and the principle of uniformitarianism are central to understanding the evolution of Earth and its biota.

▶ Geology is an integral part of our lives.

Introduction

A major benefit of the space age has been the ability to look back from space and view our planet in its entirety. We can see not only the beauty of our planet, but also its fragility. But what about its long and frequently turbulent history? Unfortunately, we can't travel back in time to witness it firsthand, but we *can* tell the story of Earth's 4.6-billion-year existence, because that history is preserved in its geologic record.

In this book, you will learn how to decipher that history from the clues Earth provides. But before we learn how to do this, let's take a sneak preview of our imaginary full-length feature film *The History of Earth*.

In this movie, we would see a planet undergoing remarkable change as continents moved about its surface. As a result of these movements, ocean basins would open and close, and mountain ranges would form along continental margins, or where continents collided with each other. Oceanic and atmospheric circulation patterns would shift in response to the moving continents, sometimes causing massive ice sheets to form, grow, and then melt away. At other times, extensive swamps or vast interior deserts would appear.

We would also witness the first living cells evolving from a primordial organic soup, sometime between 4.6 and 3.5 billion years ago. Somewhere around 1.2 billion years ago, cells with a nucleus had evolved, and not long thereafter multicelled soft-bodied animals made their appearance in the world's oceans and were followed, in relatively short order, by animals with skeletons, and then by animals with backbones.

Until about 450 million years ago, Earth's landscape was essentially barren. At that time, however, the landscape came to life as plants and animals moved from their home in the water to take up residency on land. Viewed from above, Earth's landmasses took on new hues and colors as different life-forms began inhabiting the terrestrial environment. From that point on, Earth would never be the same, as plants, insects, amphibians, reptiles, birds, and mammals made the land their home. Near the end of our film, humans evolve, and we see how their activities greatly impact the global ecosystem. It seems only fitting that the movie's final image is of Earth as a shimmering blue-green oasis in the black void of space, with a voiceover saying, "To be continued."

Every good movie has a theme, and the major theme of *The History of Earth* is that Earth is a complex, dynamic planet that has changed continuously since its origin some 4.6 billion years ago. Because of its epic nature, three interrelated subthemes run throughout *The History of Earth*. The first is that Earth's outermost part is composed of a series of moving plates (*plate tectonics*) whose interactions have affected the planet's physical and biological history. The second is that Earth's biota has evolved or changed throughout its history (*organic evolution*). The third is that the physical and biological changes that occurred did so over long periods of time (*geologic* or *deep time*). These three interrelated subthemes are central to our understanding and appreciation of our planet's history.

By viewing Earth as a whole—that is, thinking of it as a *system*—we not only see how its various components are interconnected, but we can also better appreciate its complex and dynamic nature. The system concept makes it easier for us to study a multifaceted subject, such as Earth, because it divides the whole into smaller components that we can easily understand, without losing sight of how the separate components fit together as a whole. In the same way, you can think of this book as a large, panoramic landscape painting. Each chapter fills in the details of the landscape, thereby enhancing the overall enjoyment and understanding of the entire painting.

A **system** is a combination of related parts that interact in an organized manner. We can thus consider Earth as a system of interconnected components that interact and affect each other in many different ways. The principal subsystems of Earth are the *atmosphere, biosphere, hydrosphere, lithosphere, mantle,* and *core* (▶ Figure 1.1). The complex

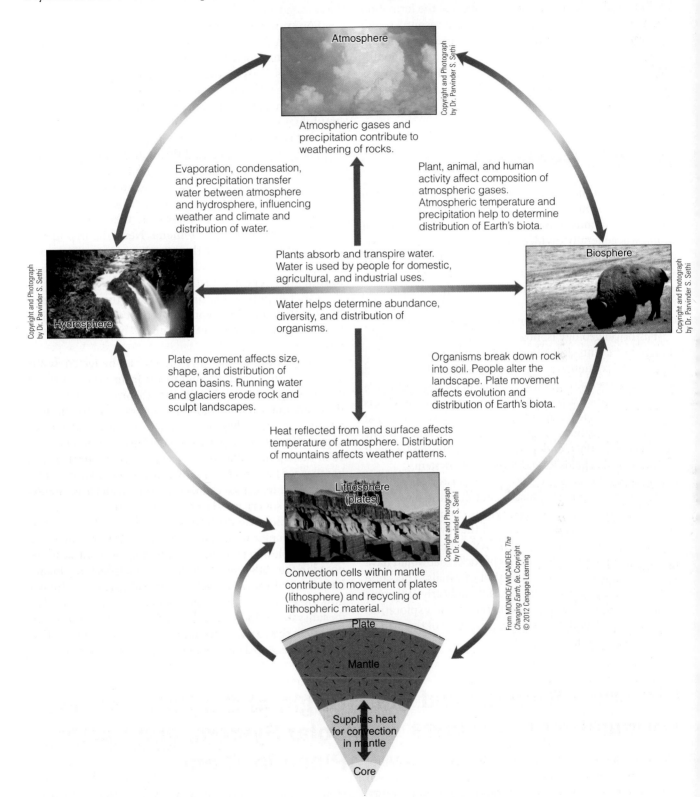

Atmospheric gases and precipitation contribute to weathering of rocks.

Evaporation, condensation, and precipitation transfer water between atmosphere and hydrosphere, influencing weather and climate and distribution of water.

Plant, animal, and human activity affect composition of atmospheric gases. Atmospheric temperature and precipitation help to determine distribution of Earth's biota.

Plants absorb and transpire water. Water is used by people for domestic, agricultural, and industrial uses.

Water helps determine abundance, diversity, and distribution of organisms.

Plate movement affects size, shape, and distribution of ocean basins. Running water and glaciers erode rock and sculpt landscapes.

Organisms break down rock into soil. People alter the landscape. Plate movement affects evolution and distribution of Earth's biota.

Heat reflected from land surface affects temperature of atmosphere. Distribution of mountains affects weather patterns.

Convection cells within mantle contribute to movement of plates (lithosphere) and recycling of lithospheric material.

Supplies heat for convection in mantle

▶ **Figure 1.1 Subsystems of Earth** The atmosphere, hydrosphere, biosphere, lithosphere, mantle, and core are all subsystems of Earth. This simplified diagram shows how these subsystems interact, with some examples of how materials and energy are cycled throughout the Earth system. The interactions between these subsystems make Earth a dynamic planet that has evolved and changed since its origin 4.6 billion years ago.

interactions among these subsystems result in a dynamically changing planet in which matter and energy are continuously recycled into different forms. For example, the movement of plates has profoundly affected the formation of landscapes, the distribution of mineral resources, and atmospheric and oceanic circulation patterns, which in turn have affected global climate changes. Examined in this manner, the continuous evolution of Earth and its life is not a series of isolated and unrelated events but a dynamic interplay between its various subsystems.

What Is Geology?

Geology, from the Greek *geo* and *logos,* is defined as the study of Earth, but now must also include the study of the planets and moons in our solar system. The discipline of geology is generally divided into two broad areas: physical geology and historical geology. *Physical geology* is the study of Earth materials, such as minerals and rocks, as well as the processes operating within Earth and on its surface. *Historical geology* examines the origin and evolution of Earth, its continents, oceans, atmosphere, and life.

Historical geology is, however, more than just a recitation of past events. It is the study of a dynamic planet that has changed continuously during the past 4.6 billion years. In addition to determining what occurred in the past, geologists are also concerned with explaining how and why past events happened. It is one thing to observe in the fossil record that dinosaurs went extinct but quite another to ask how and why they became extinct, and, perhaps more importantly, what implications that holds for today's global ecosystem.

Not only do the basic principles of historical geology aid in interpreting Earth's history, but they also have practical applications. For example, William Smith, an English surveyor and engineer, recognized that by studying the sequences of rocks and the fossils they contained, he could predict the kinds and thicknesses of rocks that would have to be excavated in the construction of canals. The same principles Smith used in the late 18th and early 19th centuries are still used today in mineral and oil exploration and also in interpreting the geologic history of the planets and moons of our solar system.

Historical Geology and the Formulation of Theories

The term **theory** has various meanings and is frequently misunderstood and consequently misused. In colloquial usage, it means a speculative or conjectural view of something—hence the widespread belief that scientific theories are little more than unsubstantiated wild guesses. In scientific usage, however, a theory is a coherent explanation for one or several related natural phenomena supported by a large body of objective evidence. From a theory, scientists derive predictive statements that can be tested by observations and/or experiments so that their validity can be assessed.

For example, one prediction of plate tectonic theory is that oceanic crust is youngest near spreading ridges, and becomes progressively older with increasing distance from ridges. This prediction has been verified by observations (see Chapter 3). Likewise, according to the theory of evolution, fish should appear in the fossil record before amphibians, followed by reptiles, mammals, and birds—and that is indeed the case (see Chapter 7).

Theories are formulated through the process known as the **scientific method**. This method is an orderly, logical approach that involves gathering and analyzing facts or data about the problem under consideration. Tentative explanations, or **hypotheses**, are then formulated to explain the observed phenomena. Next, the hypotheses are tested to see if what they predicted actually occurs in a given situation. Finally, if one of the hypotheses is found, after repeated tests, to explain the phenomena, then that hypothesis is proposed as a theory. Remember, however, that in science, even a theory is still subject to further testing and refinement as new data become available.

The fact that a scientific theory can be tested and is subject to such testing separates it from other forms of human inquiry. Because scientific theories can be tested, they have the potential for being supported or even proved wrong. Accordingly, science must proceed without any appeal to beliefs or supernatural explanations, not because such beliefs or explanations are necessarily untrue, but because we have no way to investigate them. For this reason, science makes no claim about the existence or nonexistence of a supernatural or spiritual realm.

Each scientific discipline has certain theories that are of particular importance. For example, the theory of organic evolution revolutionized biology when it was proposed in the 19th century. In geology, plate tectonic theory has changed the way geologists view Earth. Geologists now view Earth from a global perspective in which all of its subsystems and cycles are interconnected, and Earth history is seen as a continuum of interrelated events that are part of a global pattern of change.

Origin of the Universe and Solar System, and Earth's Place in Them

How did the universe begin? What has been its history? What is its eventual fate, or is it infinite? These are just some of the basic questions people have asked and wondered about since they first looked into the nighttime sky and saw the vastness of the universe beyond Earth.

Origin of the Universe—Did It Begin with a Big Bang?

Most scientists think that the universe originated about 14 billion years ago in what is popularly called the **Big Bang**. The Big Bang is a model for the evolution of the universe in which a dense, hot state was followed by expansion, cooling, and a less dense state.

According to modern *cosmology* (the study of the origin, evolution, and nature of the universe), the universe has no edge and therefore no center. Thus, when the universe began, all matter and energy were compressed into an infinitely small high-temperature and high-density state in which both time and space were set at zero. Therefore, there is no "before the Big Bang" but only what occurred after it. As demonstrated by Einstein's theory of relativity, space and time are unalterably linked to form a space–time continuum—that is, without space, there can be no time.

How do we know that the Big Bang took place approximately 14 billion years ago? Why couldn't the universe have always existed as we know it today? Two fundamental phenomena indicate that the Big Bang occurred. First, the universe is expanding, and second, it is permeated by background radiation.

When astronomers look beyond our own solar system, they observe that everywhere in the universe galaxies are moving away from each other at tremendous speeds. Edwin Hubble first recognized this phenomenon in 1929. By measuring the optical spectra of distant galaxies, Hubble noted that the velocity at which a galaxy moves away from Earth increases proportionally to its distance from Earth. He observed that the spectral lines (wavelengths of light) of the galaxies are shifted toward the red end of the spectrum; that is, the lines are shifted toward longer wavelengths. Galaxies receding from each other at tremendous speeds would produce such a *redshift*. This is an example of the *Doppler effect,* which is a change in the frequency of a sound, light, or other wave caused by movement of the wave's source relative to the observer (▶Figure 1.2).

One way to envision how velocity increases with increasing distance is by reference to the popular analogy of a rising loaf of raisin bread in which the raisins are uniformly distributed throughout the loaf (▶Figure 1.3). As the dough rises, the raisins are uniformly pushed away from each other at velocities directly proportional to the distance between any two raisins. The farther away a given raisin is to begin with, the farther it must move to maintain the regular spacing during the expansion, and hence the greater its velocity must be.

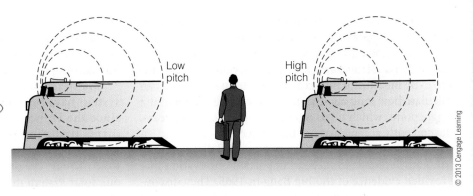

▶ Figure 1.2 **The Doppler Effect** The sound waves of an approaching whistle are slightly compressed so that the individual hears a shorter-wavelength, higher-pitched sound. As the whistle passes and recedes from the individual, the sound waves are slightly spread out, and a longer-wavelength, lower-pitched sound is heard.

© 2013 Cengage Learning

▶ Figure 1.3 **The Expanding Universe** The motion of raisins in a rising loaf of raisin bread illustrates the relationship that exists between distance and speed and is analogous to an expanding universe. In this diagram, adjacent raisins are located 2 cm apart before the loaf rises. After one hour, any raisin is now 4 cm away from its nearest neighbor and 8 cm away from the next raisin over, and so on. Therefore, from the perspective of any raisin, its nearest neighbor has moved away from it at a speed of 2 cm per hour, and the next raisin over has moved away from it at a speed of 4 cm per hour. In the same way that raisins move apart in a rising loaf of bread, galaxies are receding from each other at a rate proportional to the distance between them.

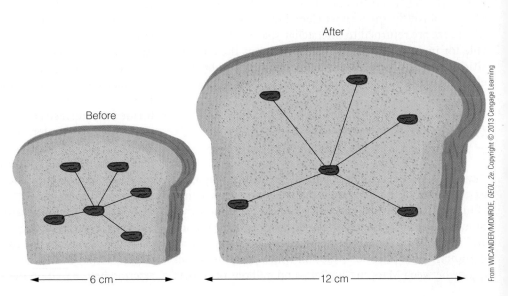

Before

6 cm

After

12 cm

In the same way that raisins move apart in a rising loaf of bread, galaxies are receding from each other at a rate proportional to the distance between them, which is exactly what astronomers see when they observe the universe. By measuring this expansion rate, astronomers can calculate how long ago the galaxies were all together at a single point, which turns out to be about 14 billion years, the currently accepted age of the universe.

Arno Penzias and Robert Wilson of Bell Telephone Laboratories made the second important observation that provided evidence of the Big Bang in 1965. They discovered that there is a pervasive background radiation of 2.7 Kelvin (K) above absolute zero (absolute zero equals –273°C; 2.7 K = –270.3°C) everywhere in the universe. This background radiation is thought to be the fading afterglow of the Big Bang.

Currently, cosmologists cannot say what it was like at time zero of the Big Bang because they do not understand the physics of matter and energy under such extreme conditions. However, it is thought that during the first second following the Big Bang, the four basic forces—(1) *gravity* (the attraction of one body toward another), (2) *electromagnetic force* (the force associated with electric and magnetic fields; responsible for atomic structure, chemical reactions, and the attractive and repulsive forces associated with electrical charge and magnetism), (3) *strong nuclear force* (the binding of protons and neutrons together), and (4) *weak nuclear force* (the force responsible for the breakdown of an atom's nucleus, producing radioactive decay)—separated, and the universe experienced enormous expansion.

As the universe continued expanding and cooling, stars and galaxies began to form and the chemical makeup of the universe changed. Initially, the universe was 100 percent hydrogen and helium, whereas today it is 98 percent hydrogen and helium, and 2 percent all other elements by weight.

How did such a change in the universe's composition occur? Throughout their life cycle, stars undergo many nuclear reactions in which lighter elements are converted into heavier elements by nuclear fusion. When a star dies, often explosively, the heavier elements that were formed in its core are returned to interstellar space and are available for inclusion in new stars. In this way, the composition of the universe is gradually enhanced in heavier elements.

Our Solar System—Its Origin and Evolution

Our solar system, which is part of the Milky Way Galaxy, consists of the Sun, eight planets, one dwarf planet (Pluto), 101 known moons or satellites (although this number keeps changing with the discovery of new moons and satellites surrounding the Jovian planets), a tremendous number of asteroids—most of which orbit the Sun in a zone between Mars and Jupiter—and millions of comets

and meteoroids, as well as interplanetary dust and gases (▶Figure 1.4). Any theory formulated to explain the origin and evolution of our solar system must, therefore, take into account its various features and characteristics.

Many scientific theories for the origin of the solar system have been proposed, modified, and discarded since the French scientist and philosopher René Descartes first proposed, in 1644, that the solar system formed from a gigantic whirlpool within a universal fluid. Today, the **solar nebula theory** for the origin of our solar system involves the condensation and collapse of interstellar material in a spiral arm of the Milky Way Galaxy (▶Figure 1.5).

The collapse of this cloud of gases and small grains into a counterclockwise-rotating disk concentrated about 90 percent of the material in the central part of the disk and formed an embryonic Sun, around which swirled a rotating cloud of material called a *solar nebula*. Within this solar nebula were localized eddies in which gases and solid particles condensed. During the condensation process, gaseous, liquid, and solid particles began accreting into ever-larger masses called *planetesimals*, which collided and grew in size and mass until they eventually became planets.

The composition and evolutionary history of the planets are a consequence, in part, of their distance from the Sun (see Perspective). The **terrestrial planets**—Mercury, Venus, Earth, and Mars—so named because they are similar to *terra*, Latin for "earth," are all small and composed of rock and metallic elements that condensed at the high temperatures of the inner nebula. The **Jovian planets**—Jupiter, Saturn, Uranus, and Neptune—so named because they resemble Jupiter (the Roman god was also named *Jove*) all have small, rocky cores compared to their overall size, and are composed mostly of hydrogen, helium, ammonia, and methane, which condense at low temperatures.

While the planets were accreting, material that had been pulled into the center of the nebula also condensed, collapsed, and was heated to several million degrees by gravitational compression. The result was the birth of a star, our Sun.

During the early accretionary phase of the solar system's history, collisions between various bodies were common, as indicated by the craters on many planets and moons. Asteroids probably formed as planetesimals in a localized eddy between what eventually became Mars and Jupiter in much the same way that other planetesimals formed the terrestrial planets. The tremendous gravitational field of Jupiter, however, prevented this material from ever accreting into a planet. Comets, which are interplanetary bodies composed of loosely bound rocky and icy material, are thought to have condensed near the orbits of Uranus and Neptune.

The solar nebula theory of the formation of the solar system thus accounts for most of the characteristics of the planets and their moons, the differences in composition between the terrestrial and Jovian planets, and the presence of the asteroid belt. Based on the available data, the solar nebula

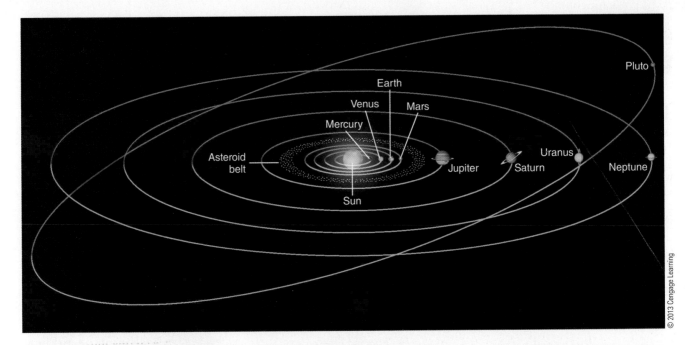

▶ **Figure 1.4 Diagrammatic Representation of the Solar System** This representation of the solar system shows the planets and their orbits around the Sun. On August 24, 2006, the International Astronomical Union downgraded Pluto from a planet to a dwarf planet. A dwarf planet has the same characteristics as a planet, except that it does not clear the neighborhood around its orbit of debris. Pluto orbits among the icy debris of the Kuiper Belt and therefore does not meet the criteria for a true planet.

(a) A huge rotating cloud of gas contracts and flattens

(b) to form a disk of gas and dust with the Sun forming in the center,

(c) and eddies gathering up material to form planets.

▶ **Figure 1.5 Solar Nebula Theory** According to the currently accepted theory for the origin of our solar system, the planets and the Sun formed from a rotating cloud of gas.

theory best explains the features of the solar system and provides a logical explanation for its evolutionary history.

Earth—Its Place in Our Solar System

Some 4.6 billion years ago, various planetesimals in our solar system gathered enough material together to form Earth and the other planets. Scientists think that this early Earth was probably cool, of generally uniform composition and density throughout, and composed mostly of silicates (compounds consisting of silicon and oxygen), iron and magnesium oxides, and smaller amounts of all the other chemical elements (▶ Figure 1.6a). Subsequently,

when the combination of meteorite impacts, gravitational compression, and heat from radioactive decay increased the temperature of Earth enough to melt iron and nickel, this homogeneous composition disappeared (Figure 1.6b) and was replaced by a series of concentric layers of differing composition and density, resulting in a differentiated planet (Figure 1.6c).

This differentiation into a layered planet is probably the most significant event in Earth's history. Not only did it lead to the formation of a crust and eventually continents, but it also was probably responsible for the emission of gases from the interior that eventually led to the formation of the oceans and atmosphere (see Chapter 8).

The Terrestrial and Jovian Planets

The planets of our solar system are divided into two major groups that are quite different, indicating that the two underwent very different evolutionary histories. The four inner planets—Mercury, Venus, Earth, and Mars—are the terrestrial planets; they are small and dense (composed of a metallic core and silicate mantle-crust), ranging from no atmosphere (Mercury) to an oppressively thick one (Venus). The outer four planets (Pluto is now considered a dwarf planet)—Jupiter, Saturn, Uranus, and Neptune—are the Jovian planets; they are large, ringed, low-density planets composed of hydrogen compounds, metals, and rocks, and surrounded by thick atmospheres.

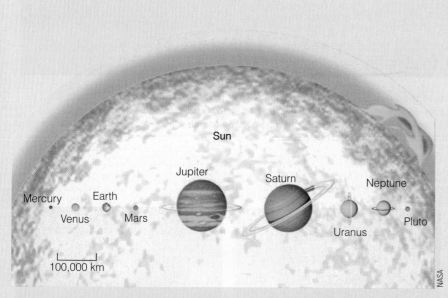

100,000 km

NASA

Figure 1 The relative sizes of the planets and the Sun. (Distances between planets are not to scale.)

NASA

Figure 2 Mercury has a heavily cratered surface that has changed very little since its early history. Because Mercury is so small, its gravitational attraction is insufficient to retain atmospheric gases; any atmosphere that it may have held when it formed probably escaped into space quickly.

NASA

Figure 3 Venus is surrounded by an oppressively thick atmosphere that completely obscures its surface. However, radar images from orbiting spacecraft reveal a wide variety of terrains, including volcanic features, folded mountain ranges, and a complex network of faults.

NASA

Figure 4 Earth is unique among our solar system's planets in that it has a hospitable atmosphere, oceans of water, and a variety of climates, as well as supporting life.

Figure 5 The Moon is one-fourth the diameter of Earth, has a low density relative to the terrestrial planets, and is extremely dry. Its surface is divided into low-lying, dark-colored plains and light-colored highlands that are heavily cratered, attesting to a period of massive meteorite bombardment in our solar system more than 4 billion years ago. The hypothesis that best accounts for the origin of the Moon has a giant planetesimal, the size of Mars or larger, crashing into Earth 4.6 to 4.4 billion years ago, causing the ejection of a large quantity of hot material that cooled and formed the Moon.

Figure 6 Mars has a thin atmosphere, little water, and distinct seasons. Its southern hemisphere is heavily cratered like the surfaces of Mercury and the Moon. The northern hemisphere has large, smooth plains, fewer craters, and evidence of extensive volcanism. The largest volcano in the solar system is found in the northern hemisphere. Huge canyons, the largest of which would stretch from San Francisco to New York, also appear in the northern half of the planet.

Figure 8 Saturn's most conspicuous feature is its ring system, consisting of thousands of rippling, spiraling bands of countless particles. The width of Saturn's rings would just reach from Earth to the Moon.

Figure 7 Jupiter is the largest of the Jovian planets. With its moons, rings, strong magnetic field, and intense radiation belts, Jupiter is the most complex and varied planet in our solar system. Jupiter's cloudy and violent atmosphere is divided into a series of different colored bands and a variety of spots (the Great Red Spot being one) that interact in incredibly complex motions.

Figure 9 Uranus is the only planet that lies on its side; that is, its axis of rotation nearly parallels the plane in which the planets revolve around the Sun. Some scientists think that a collision with an Earth-sized body early in its history may have knocked Uranus on its side. Like the other Jovian planets, Uranus has a ring system, albeit a faint one.

Figure 10 Neptune is a dynamic, stormy planet with an atmosphere similar to those of the other Jovian planets. Winds of up to 2,000 km/hr blow over the planet, creating tremendous storms, the largest of which, the Great Dark Spot, seen in the center, is nearly as big as Earth and is similar to the Great Red Spot on Jupiter.

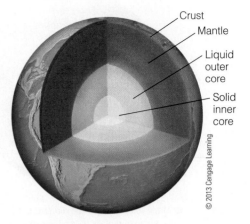

(a) Early Earth probably had a uniform composition and density throughout.

(b) The temperature of early Earth reached the melting point of iron and nickel, which, being denser than silicate minerals, settled to Earth's center. At the same time, the lighter silicates flowed upward to form the mantle and the crust.

(c) In this way, a differentiated Earth formed, consisting of a dense iron–nickel core, an iron-rich silicate mantle, and a silicate crust with continents and ocean basins.

Why Is Earth a Dynamic and Evolving Planet?

Earth is a dynamic planet that has continuously changed during its 4.6-billion-year existence. The size, shape, and geographic distribution of continents and ocean basins have changed through time; the composition of the atmosphere has evolved; and life-forms existing today differ from those that lived during the past. Mountains and hills have been worn away by erosion, and the forces of wind, water, and ice have sculpted a diversity of landscapes. Volcanic eruptions and earthquakes reveal an active interior, and folded and fractured rocks are testimony to the tremendous power of Earth's internal forces.

Earth consists of three concentric layers: the core, the mantle, and the crust (▶ Figure 1.7). This orderly division results from density differences between the layers as a function of variations in composition, temperature, and pressure.

The **core** has a calculated density of 10–13 grams per cubic centimeter (g/cm^3) and occupies about 16 percent of Earth's total volume. Seismic (earthquake) data indicate that the core consists of a small, solid inner region and a larger, apparently liquid outer portion. Both are thought to consist largely of iron and a small amount of nickel.

The **mantle** surrounds the core and comprises about 83 percent of Earth's volume. It is less dense than the core (3.3–5.7 g/cm^3) and is thought to be composed largely of *peridotite,* a dark, dense igneous rock containing abundant iron and magnesium (see Figure 2.10). The mantle is divided into three distinct zones based on physical characteristics. The lower mantle is solid and forms most of the volume of Earth's interior. The **asthenosphere** surrounds the lower mantle. It has the same composition as

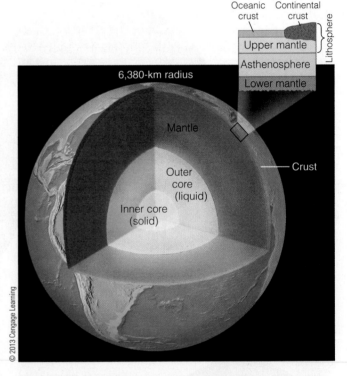

▶ Figure 1.7 **Cross Section of Earth Illustrating the Core, Mantle, and Crust** The enlarged portion shows the relationship between the lithosphere (composed of the continental crust, oceanic crust, and upper mantle) and the underlying asthenosphere and lower mantle.

the lower mantle but behaves plastically and flows slowly. Partial melting within the asthenosphere generates *magma* (molten material), some of which rises to the surface because it is less dense than the rock from which it was derived. The upper mantle surrounds the asthenosphere. The solid upper mantle and the overlying crust constitute

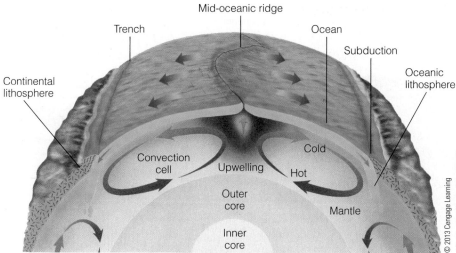

© 2013 Cengage Learning

▶ **Figure 1.8** **Movement of Earth's Plates** Earth's plates are thought to move partially as a result of underlying mantle convection cells in which warm material from deep within Earth rises toward the surface, cools, and then, upon losing heat, descends back into the interior, as shown in this diagrammatic cross section.

the **lithosphere**, which is broken into numerous individual pieces called **plates** that move over the asthenosphere, partially as a result of underlying *convection cells* (▶ Figure 1.8). Interactions of these plates are responsible for such phenomena as earthquakes, volcanic eruptions, and the formation of mountain ranges and ocean basins.

The **crust**, Earth's outermost layer, consists of two types. *Continental crust* is thick (20–90 km), has an average density of 2.7 g/cm³, and contains considerable silicon and aluminum. *Oceanic crust* is thin (5–10 km), denser than continental crust (3.0 g/cm³), and is composed of the dark igneous rocks *basalt* and *gabbro* (see Figures 2.11a and 2.11b).

Plate Tectonic Theory

The recognition that the lithosphere is divided into rigid plates that move over the asthenosphere (▶ Figure 1.9) forms the foundation of **plate tectonic theory**, a unifying theory that will be discussed in greater detail in Chapter 3. Zones of volcanic activity, earthquakes, or both mark most plate boundaries. Along these boundaries, plates separate (diverge), collide (converge), or slide sideways past each other (▶ Figure 1.10).

The acceptance of plate tectonic theory is recognized as a major milestone in the geologic sciences, comparable to

© 2013 Cengage Learning

| Ridge axis Divergent boundary | Transform fault Transform boundary | Subduction zone Convergent boundary | Zones of extension within continents | Uncertain plate boundary |

▶ **Figure 1.9** **Earth's Plates** Earth's lithosphere is divided into rigid plates of various sizes that move over the asthenosphere.

Continental–
continental
convergent
plate
boundary

Continental–
oceanic
convergent
plate
boundary

Trench

Oceanic–
oceanic
convergent
plate
boundary

Mid-oceanic
ridge

Divergent
plate
boundary

Divergent
plate
boundary

Transform
plate
boundary

© 2013 Cengage Learning

Upwelling

Asthenosphere

Upwelling

Lithosphere

▶ **Figure 1.10 Relationship Between Lithosphere, Asthenosphere, and Plate Boundaries** An idealized block diagram illustrating the relationship between the lithosphere and the underlying asthenosphere, and the three principal types of plate boundaries: divergent, convergent, and transform.

the revolution that Darwin's theory of evolution caused in biology. Plate tectonic theory has provided the framework for interpreting the composition, structure, and internal processes of Earth on a global scale. The theory has led to the realization that the continents and ocean basins are part of a lithosphere-atmosphere-hydrosphere system that evolved together with Earth's interior.

A revolutionary concept when it was proposed in the 1960s, plate tectonic theory has had far-reaching consequences in all fields of geology because it provides the basis for relating many seemingly unrelated phenomena. Besides being responsible for the major features of Earth's crust, plate movements also affect the formation and occurrence of Earth's natural resources, as well as influencing the distribution and evolution of the world's biota.

The impact of plate tectonic theory has been particularly notable in the interpretation of Earth's history. For example, the Appalachian Mountains in eastern North America and the mountain ranges of Greenland, Scotland, Norway, and Sweden are not the result of unrelated mountain-building episodes but, rather, are part of a larger mountain-building event that involved the closing of an ancient Atlantic Ocean and the formation of the supercontinent Pangaea approximately 251 million years ago (see Chapter 11).

Organic Evolution and the History of Life

Plate tectonic theory provides us with a model for understanding the internal workings of Earth and their effect on Earth's surface. The theory of **organic evolution** (whose central thesis is that all present-day organisms are related, and

that they have descended with modifications from organisms that lived during the past) provides the conceptual framework for understanding the history of life. Together, the theories of plate tectonics and organic evolution have changed the way we view our planet, and we should not be surprised at the intimate association between them. Although this relationship between plate tectonic processes and the evolution of life is incredibly complex, paleontological data provide indisputable evidence of the influence of plate movement on the distribution of organisms.

The publication in 1859 of Darwin's *On the Origin of Species by Means of Natural Selection* revolutionized biology and marked the beginning of modern evolutionary biology. With its publication, most naturalists recognized that evolution provided a unifying theory that explained an otherwise encyclopedic collection of biologic facts.

When Darwin proposed his theory of organic evolution, he cited a wealth of supporting evidence, including the way organisms are classified, embryology, comparative anatomy, the geographic distribution of organisms, and, to a limited extent, the fossil record. Furthermore, Darwin proposed that *natural selection,* which results in the survival to reproductive age of those organisms best adapted to their environment, is the mechanism that accounts for evolution.

Perhaps the most compelling evidence in favor of evolution can be found in the fossil record. Just as the geologic record allows geologists to interpret physical events and conditions in the geologic past, **fossils,** which are the remains or traces of once-living organisms, not only provide evidence that evolution has occurred, but also demonstrate that Earth has a history extending beyond that recorded by humans. The succession of fossils in the rock record provides geologists with a means for dating rocks and allowed for a *relative geologic time scale* to be constructed in the 1800s.

Geologic Time and Uniformitarianism

An appreciation of the immensity of geologic time is central to understanding the evolution of Earth and its biota. Indeed, time is one of the main aspects that sets geology apart from the other sciences except astronomy. Most people have difficulty comprehending geologic time because they tend to think in terms of the human perspective—seconds, hours, days, and years. Ancient history is what occurred hundreds or even thousands of years ago. When geologists talk of "ancient geologic history," however, they are referring to events that happened hundreds of millions, or even billions, of years ago.

It is equally important to remember that Earth goes through cycles of much longer duration than the human perspective of time. Although they may have disastrous effects on the human species, global warming and cooling are part of a larger cycle that has resulted in numerous glacial advances and retreats during the past 2.6 million years. Because of their geologic perspective on time and the interrelationships of Earth's various systems, subsystems, and cycles, geologists are in a unique position to make valuable contributions to such current environmental debates as those involving global warming and rising sea levels.

The **geologic time scale** subdivides geologic time into a hierarchy of increasingly shorter time intervals, with each time subdivision having a specific name. The geologic time scale resulted from the work of many 19th-century geologists who pieced together information from numerous rock exposures and constructed a chronology based on changes in Earth's biota through time. Subsequently, with the discovery of radioactivity in 1895 and the development of various radiometric-dating techniques, geologists have been able to assign numerical ages (also known as *absolute ages*) in years to the subdivisions of the geologic time scale (▶ Figure 1.11).

One of the cornerstones of geology is the **principle of uniformitarianism**, which is based on the premise that present-day processes have operated throughout geologic time. Therefore, in order to understand and interpret geologic events from evidence preserved in rocks, we must first understand present-day processes and their results. In fact, uniformitarianism fits in completely with the system approach we are following for the study of Earth.

Uniformitarianism is a powerful principle that allows us to use present-day processes as the basis for interpreting the past and for predicting potential future events. We should keep in mind, however, that uniformitarianism does not exclude sudden or catastrophic events such as volcanic eruptions, earthquakes, tsunami, landslides, or floods. These are processes that shape our modern world, and some geologists view Earth history as a series of such short-term or punctuated events. This view is certainly in keeping with the modern principle of uniformitarianism,

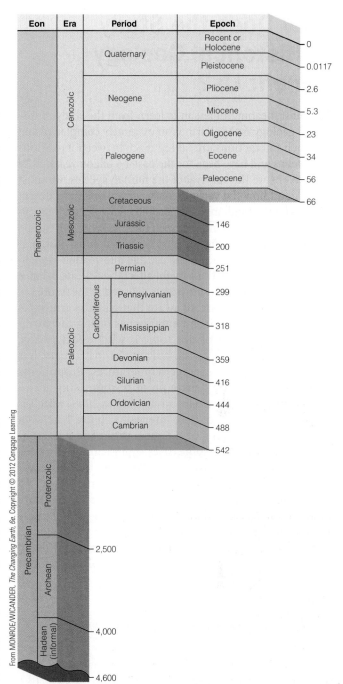

▶ Figure 1.11 **The Geologic Time Scale** The numbers to the right of the columns are the ages in millions of years before the present. Dates are from the 2009 International Stratigraphic Chart. © 2009 by the International Commission of Stratigraphy.

From MONROE/WICANDER, *The Changing Earth, 6e*. Copyright © 2012 Cengage Learning

which does not require that the rates and intensities of geologic processes be constant through time.

Therefore, we can conclude that even though the rates and intensities of geologic processes have varied during the past, the physical and chemical laws of nature have remained the same. Thus, although Earth is in a dynamic state of change, and has been ever since it formed, the processes that shaped its past are the same ones operating today.

How Does the Study of Historical Geology Benefit Us?

The most meaningful lesson to learn from the study of historical geology is that Earth is an extremely complex planet in which interactions are taking place between its various subsystems and have been for the past 4.6 billion years. If we want to ensure the survival of the human species, we must understand how the various subsystems work and interact with each other and, more importantly, how our actions can affect the delicate balance between these systems. We can do this, in part, by studying what has happened in the past, particularly on the global scale, and use that information to try to determine how our actions might affect the delicate balance between Earth's various subsystems now and in the future.

For example, we are all aware that global warming is one of the major environmental issues facing the world today. At the center of this topic is what effect, if any, human activity, such as the burning of fossil fuels, has had on rising global average temperatures. Proponents on both sides of the debate point to increasing mean global temperature changes from 1880 to 2008, which is a historical, or human, time interval (▶Figure 1.12). On the other hand, mean global temperatures and precipitation have changed, in some cases significantly, on a geologic time scale (▶Figure 1.13). We must therefore ask the question of whether our historical perspective and the possible role humans have played in this temperature increase is reflective of an impending long-term climate change, or if it is merely a short-term event when viewed from a geologic perspective. In any case, continued global warming will most certainly affect Earth's biota, just as it has done in the past.

The study of geology also goes beyond learning numerous facts about Earth. We don't just study geology—we *live* it. Geology is an integral part of our lives. Our standard of living depends directly on our consumption of natural resources, most of which formed millions and billions of years ago. However, the way we consume natural resources and interact with the environment, as individuals and as a society, also determines our ability to pass on this standard of living to the next generation.

As you study the various topics and historical accounts covered in this book, keep in mind the themes discussed

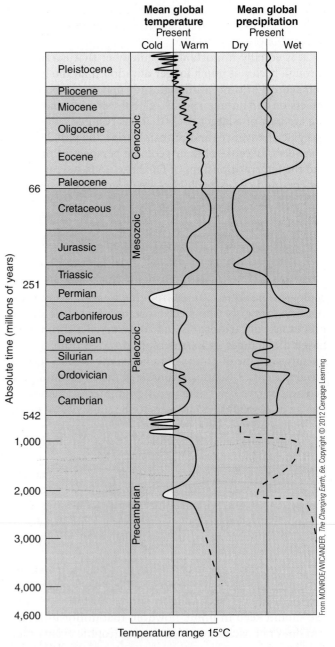

▶ Figure 1.13 **Mean Global Temperature and Precipitation Changes Throughout Earth History** There have been several episodes of global cooling throughout Earth history, notably at the end of the Precambrian (Proterozoic Eon), the end of the Ordovician Period, the Late Carboniferous to Permian periods, and a series of glacial and interglacial episodes during the Pleistocene Epoch.

▶ Figure 1.12 **Mean Global Temperature Changes from 1880 to 2008** The zero line represents the average from 1951 to 1980. The plus and minus values represent deviations (in C°) from the average for the annual mean (in black) and 5-year mean (in red). Note the dramatic increase in temperature from approximately 1970 to 2008.

in this chapter and how, like the parts of a system, they are interrelated and responsible for the 4.6-billion-year history of Earth. View each chapter's topic in the context of how it fits in the whole Earth system, and remember that Earth's history is a continuum and the result of interaction among its various subsystems. By relating each chapter's topic to its place in the Earth system, you will gain a greater appreciation of why geology is so integral to our lives.

It is also important to remember that historical geology is not a dry history of Earth but a vibrant, dynamic science in which what we see today is based on what went on before. As Stephen Jay Gould states in his book *Wonderful Life: The Burgess Shale and the Nature of History* (1989), "Play the tape of life again starting with the Burgess Shale, and a different set of survivors—not including vertebrates this time—would grace our planet today."

Summary

▶ Earth can be viewed as a system of interconnected components that interact with and affect each other. The principal subsystems of Earth are the atmosphere, hydrosphere, biosphere, lithosphere, mantle, and core.

▶ Earth is a continually changing and dynamic planet because of the interaction among its various subsystems and cycles.

▶ Geology, the study of Earth, is divided into (1) physical geology, which is the study of Earth materials and the processes that operate both within Earth and on its surface; and (2) historical geology, which examines the origin and evolution of Earth, its continents, oceans, atmosphere, and life.

▶ The scientific method is an orderly, logical approach that involves gathering and analyzing facts about a particular phenomenon, formulating hypotheses to explain the phenomenon, testing the hypotheses, and finally proposing a theory. A theory is a testable explanation for some natural phenomenon that has a large body of supporting evidence.

▶ The universe began, in what is popularly called the Big Bang, approximately 14 billion years ago. Astronomers have deduced this age by observing that celestial objects are moving away from each other in an ever-expanding universe. Furthermore, the universe has a pervasive background radiation of 2.7 K above absolute zero (2.7 K = −270.3°C), which is thought to be the faint afterglow of the Big Bang.

▶ About 4.6 billion years ago, our solar system formed from a rotating cloud of interstellar matter. As this cloud condensed, it eventually collapsed under the influence of gravity and flattened into a counterclockwise rotating disk. Within this rotating disk, the Sun, planets, and moons formed from the turbulent eddies of nebular gases and solids.

▶ Earth formed from a swirling eddy of nebular material 4.6 billion years ago, accreting as a solid body and soon thereafter differentiating into a layered planet.

▶ Earth's outermost layer is the crust, which is divided into continental and oceanic portions. The crust and underlying solid upper mantle, together known as the lithosphere, overlie the asthenosphere, a zone that behaves plastically and flows slowly. The asthenosphere is underlain by the solid lower mantle. Earth's core consists of an outer liquid portion and an inner solid portion.

▶ The lithosphere is divided into a series of plates that diverge, converge, and slide sideways past one another.

▶ Plate tectonic theory provides a unifying explanation for many geologic features and events. The interaction between plates is responsible for volcanic eruptions, earthquakes, the formation of mountain ranges and ocean basins, and the recycling of rock materials.

▶ The central thesis of the theory of organic evolution is that all living organisms evolved (descended with modifications) from organisms that existed in the past.

▶ Time sets geology apart from the other sciences except astronomy, and an appreciation of the immensity of geologic time is central to understanding Earth's evolution. The geologic time scale is the calendar geologists use to date past events.

▶ The principle of uniformitarianism is basic to the interpretation of Earth history. This principle holds that the laws of nature have been constant through time and that the same processes operating today have operated in the past, although not necessarily at the same rates.

▶ Geology is an integral part of our lives. Our standard of living depends directly on our consumption of natural resources, which formed millions and billions of years ago.

▶ An appreciation of Earth history and how Earth's various subsystems interact is crucial in understanding the role that human activities might have in such current environmental issues as global warming and sea-level changes.

Important Terms

asthenosphere, p. 10
Big Bang, p. 5
core, p. 10
crust, p. 11
fossil, p. 12
geologic time scale, p. 13
geology, p. 4

hypothesis, p. 4
Jovian planets, p. 6
lithosphere, p. 11
mantle, p. 10
organic evolution, p. 12
plate, p. 11
plate tectonic theory, p. 11

principle of uniformitarianism, p. 13
scientific method, p. 4
solar nebula theory, p. 6
system, p. 3
terrestrial planets, p. 6
theory, p. 4

Review Questions

1. The development of the geologic time scale was based primarily on
   a. _____ absolute dating techniques; b. _____ sedimentary structures; c. _____ plate movements; d. _____ changes in Earth's biota through time; e. _____ all of the previous answers.
2. The layer of Earth that is composed predominantly of iron with a small amount of nickel is the
   a. _____ crust; b. _____ mantle; c. _____ asthenosphere; d. ✓ core; e. _____ lithosphere.
3. The premise that present-day processes have operated throughout geologic time is the principle of
   a. _____ organic evolution; b. _____ plate tectonics; c. ✓ uniformitarianism; d. _____ geologic time; e. _____ scientific deduction.
4. Which of the following statements about a scientific theory is *not* true?
   a. _____ It is an explanation for some natural phenomenon; b. _____ It is a conjecture or guess; c. _____ It has a large body of supporting evidence; d. _____ It is testable; e. _____ Predictive statements can be derived from it.
5. The study of Earth materials, as well as the processes operating within Earth and on its surface, is
   a. _____ geomorphology; b. _____ historical geology; c. _____ cosmology; d. ✓ physical geology; e. _____ seismology.
6. A combination of related parts interacting in an organized fashion is
   a. _____ a cycle; b. _____ a theory; c. _____ uniformitarianism; d. _____ a hypothesis; e. _____ a system.
7. That all living organisms are the descendants of different life-forms that existed in the past is the central claim of
   a. _____ the principle of fossil succession; b. _____ the principle of uniformitarianism; c. _____ plate tectonic theory; d. ✓ theory of organic evolution; e. _____ none of the previous answers.
8. Plates are composed of
   a. _____ the crust and upper mantle; b. _____ the asthenosphere and upper mantle; c. _____ the crust and asthenosphere; d. _____ continental and oceanic crust only; e. _____ the core and mantle.
9. Two observations that led scientists to conclude that the Big Bang occurred approximately 14 billion years ago are
   a. _____ a steady-state universe and background radiation of 2.7 K above absolute zero; b. _____ a steady-state universe and opaque background radiation; c. _____ an expanding universe and opaque background radiation; d. ✓ an expanding universe and background radiation of 2.7 K above absolute zero; e. _____ a shrinking universe and opaque background radiation.
10. The movement of plates is thought to result from
    a. _____ density differences between the inner and outer core; b. _____ rotation of the mantle around the core; c. _____ gravitational forces; d. _____ the Coriolis effect; e. _____ convection cells.
11. Why is viewing Earth as a system a good way to study it? Are humans a part of the Earth system? If so, what role, if any, do we play in Earth's evolution?
12. Describe how you would use the scientific method to formulate a hypothesis explaining the similarity of mountain ranges on the east coast of North America and those in England, Scotland, and the Scandinavian countries. How would you test your hypothesis?
13. What are the three concentric layers of Earth, and what are their characteristics?
14. Explain how the principle of uniformitarianism allows for catastrophic events.
15. Why is plate tectonic theory a unifying theory of geology, and how does it fit into a systems approach to the study of Earth?
16. What is the Big Bang? What evidence do we have that the universe began approximately 14 billion years ago?
17. Why is Earth considered a dynamic planet?
18. Why is it important to have a basic knowledge and understanding of historical geology?
19. How does the solar nebula theory account for the formation of our solar system, its features, and evolutionary history?
20. Discuss why an accurate geologic time scale is particularly important for geologists in examining changes in global temperature during the past, and how an understanding of geologic time is crucial to the current debate on global warming and its consequences.

Minerals and Rocks

2

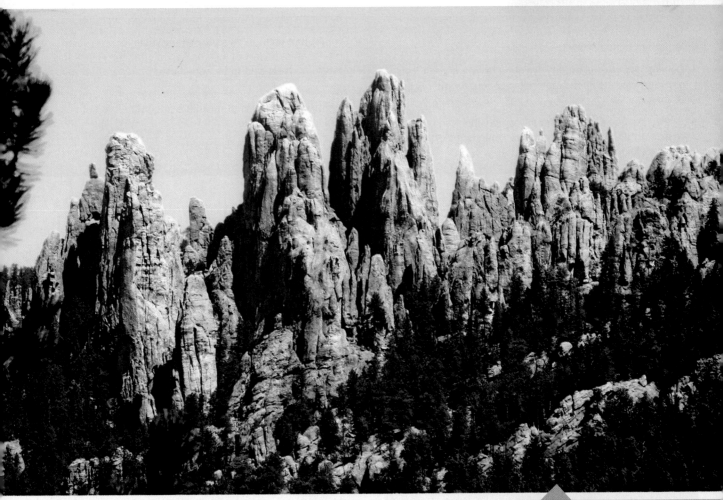

This exposure of the Harney Peak Granite, called The Needles, is in Custer State Park in the Black Hills of South Dakota. It is part of a large body of granite that cooled and crystallized from magma about 12 km below Earth's surface 1.7 billion years ago. The granite, composed mostly of the minerals quartz, potassium feldspar, and plagioclase feldspar, is now well exposed following uplift and deep erosion. The 18-m-high images of presidents Washington, Jefferson, Lincoln, and Theodore Roosevelt are carved in the Harney Peak Granite at Mount Rushmore National Memorial, which is also in the Black Hills.

Sue Monroe

At the end of this chapter, you will have learned that

▶ Chemical elements are composed of atoms, all of the same kind, whereas compounds form when different atoms bond together. Most minerals are compounds, which are characterized as naturally occurring, inorganic, crystalline solids.

▶ Of the 3,500 or so minerals known, only a few, perhaps two dozen, are common in rocks, but many others are found in small quantities in rocks, and some of these are important natural resources.

▶ Cooling and crystallization of magma or lava and the consolidation of pyroclastic materials account for the origin of igneous rocks.

▶ Geologists use mineral content (composition) and textures to classify plutonic rocks (intrusive igneous rocks) and volcanic rocks (extrusive igneous rocks).

▶ Mechanical and chemical weathering of rocks yields sediment that is transported, deposited, and then lithified to form detrital and chemical sedimentary rocks.

▶ Texture and composition are the criteria geologists use to classify sedimentary rocks.

▶ Any type of rock may be altered by heat, pressure, fluids, or any combination of these to form metamorphic rocks.

▶ One feature used to classify metamorphic rocks is foliation—that is, a platy or layered aspect, but some lack this feature and are said to be nonfoliated.

▶ The fact that Earth materials are continually recycled and that the three groups of rocks are interrelated is summarized in the rock cycle.

Introduction

Throughout most of Earth's history there were no human observers to witness and record events, so we rely on the **geologic record**, that is, the record of physical and biological events preserved in rocks. Accordingly, we must be familiar with the kinds of rocks in Earth's crust, what they are composed of, how they form, and what they tell us about prehistoric events. As a starting point, we can say that rocks are composed of **minerals**, which are naturally occurring, inorganic, crystalline solids, with a narrowly defined chemical composition and distinctive physical properties (▶ Figure 2.1a). So **rocks** are made up of one or more minerals (Figure 2.1b).

The definition of the term *mineral* is lengthy, but the only part that is probably unfamiliar is *crystalline solid*. Of course, a solid is any object that retains its shape and volume unless deformed by a force. A crystalline solid is one in which its atoms are arranged in a specific three-dimensional framework, unlike the arrangement of atoms in liquids and gases. Ice, for example, is a crystalline solid that meets all the criteria for the definition of the term *mineral*, and so do quartz, diamond, emerald, and many others.

Some minerals and rocks have served as religious symbols or talismans, or have been worn, carried, applied externally, or ingested for their presumed mystical or

Figure 2.1 Minerals and Rocks

(a) Quartz crystals. Quartz is a very common mineral in many rocks, such as granite and sandstone. In fact, most of the world's sand is composed of quartz.

(b) Minerals in granite. This igneous rock is made up mostly of three minerals—quartz, potassium feldspar, and plagioclase feldspar—but it may also contain small amounts of biotite, muscovite, and hornblende.

curative powers. There is no evidence that minerals or rocks have any such powers, but many are important economically, accounting for complex political and economic ties between nations. For example, the United States must import all the manganese, aluminum ore, and cobalt that it uses. Iron ore, on the other hand, is found in abundance in the Great Lakes region of the United States and Canada. Even some rather common minerals are important: quartz, which makes up most sand, is used to make abrasives, glass, and optical instruments; clay minerals are used for ceramics; gypsum is needed to manufacture wallboard; and phosphate minerals are used for fertilizers and animal-feed supplements.

In addition to their economic importance, another good reason to study minerals is that they are the building blocks of rocks. Granite, for instance, is made up of specific percentages of quartz, potassium feldspars, and plagioclase feldspars (Figure 2.1b), whereas limestone consists of the mineral calcite. Furthermore, some minerals are attractive and eagerly sought by mineral collectors and for museum displays, and others are known as *gemstones*—that is, precious and semiprecious minerals used for decorative purposes, especially jewelry.

From our discussion so far, we know how minerals are defined, and that minerals are the basic components of rocks. Now let's delve deeper into what minerals are made of by considering matter, atoms, elements, and bonding.

Matter and Its Composition

Matter is anything that has mass and occupies space, so it exists as solids, liquids, gases, and plasma. The last is an ionized gas, as in neon lights and the Sun, but here we are interested mostly in solids, because by definition minerals are solids and rocks are composed of minerals. Liquids and gases are also important, though, because some features we see in the geologic record resulted from moving liquids (running water) and gases (wind).

Elements and Atoms

Matter consists of chemical **elements**, each of which is composed of tiny particles called **atoms**, the smallest units of matter that retain the characteristics of an element. Atoms have a compact nucleus made up of one or more *protons*—particles with a positive electrical charge—and electrically neutral *neutrons*. Negatively charged *electrons* rapidly orbit the nucleus at specific distances in one or more *electron shells* (▶Figure 2.2).

The number of protons in an atom's nucleus determines its **atomic number**; carbon has six protons, so its atomic number is 6 (Figure 2.2). An atom's **atomic mass number**, in contrast, is found by adding the number of protons and neutrons in the nucleus (electrons have negligible mass). However, the number of neutrons in the nucleus of an element might vary. Carbon atoms (with six protons) have six, seven, or eight neutrons, making three *isotopes*, or different forms, of carbon (▶Figure 2.3). Some elements have only one isotope, but many have several.

Bonding and Compounds

The process of atoms joining to other atoms is known as **bonding**. Should atoms of two or more elements bond, the resulting substance is a **compound**. Thus, gaseous oxygen is an element, whereas ice, made up of hydrogen and oxygen (H_2O), is a compound. Most minerals are compounds, but there are a few important exceptions.

▶ Figure 2.2 **Shell Model for Atoms** The shell model for several atoms and their electron configurations. A circle represents the nucleus of each atom, but remember that atomic nuclei are made up of protons and neutrons as shown in Figure 2.3.

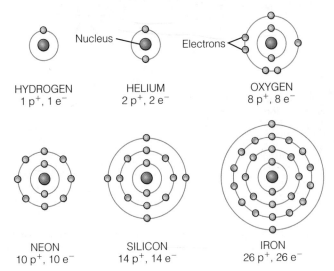

HYDROGEN
$1\,p^+, 1\,e^-$

HELIUM
$2\,p^+, 2\,e^-$

OXYGEN
$8\,p^+, 8\,e^-$

NEON
$10\,p^+, 10\,e^-$

SILICON
$14\,p^+, 14\,e^-$

IRON
$26\,p^+, 26\,e^-$

Element	Symbol	Atomic Number	Distribution of Electrons			
			First Shell	Second Shell	Third Shell	Fourth Shell
Hydrogen	H	1	1	—	—	—
Helium	He	2	2	—	—	—
Carbon	C	6	2	4	—	—
Oxygen	O	8	2	6	—	—
Neon	Ne	10	2	8	—	—
Sodium	Na	11	2	8	1	—
Magnesium	Mg	12	2	8	2	—
Aluminum	Al	13	2	8	3	—
Silicon	Si	14	2	8	4	—
Phosphorus	P	15	2	8	5	—
Sulfur	S	16	2	8	6	—
Chlorine	Cl	17	2	8	7	—
Potassium	K	19	2	8	8	1
Calcium	Ca	20	2	8	8	2
Iron	Fe	26	2	8	14	2

Nucleus

6 p
6 n

^{12}C (Carbon 12)

6 p
7 n

^{13}C (Carbon 13)

6 p
8 n

^{14}C (Carbon 14)

▶ Figure 2.3 **Isotopes of Carbon**
The element carbon has six protons (p) in its nucleus so its atomic number is 6, but its atomic mass number may be 12, 13, or 14 depending on how many neutrons (n) are in the nucleus.

Except for hydrogen, with one proton and one electron, the innermost electron shell of an atom has no more than two electrons, and the outermost shell contains no more than eight; these outer ones are those involved in chemical bonding. *Ionic* and *covalent bonding* are the most important types in minerals, although some useful properties of some minerals result from metallic bonding and van der Waals bonds or forces. A few elements, known as *noble gases*, have complete outer shells with eight electrons, so they rarely react with other elements to form compounds.

One way for the noble gas configuration of eight outer electrons to be attained is by transfer of one or more electrons from one atom to another. A good example is sodium (Na) and chlorine (Cl); sodium has only one electron in its outer shell, whereas chlorine has seven. Sodium loses its outer electron, leaving the next shell with eight (▶ Figure 2.4a). But

now sodium has one fewer electron (negative charge) than it has protons (positive charge), so it is a positively charged *ion*, symbolized Na^{+1}.

The electron lost by sodium goes into chlorine's outer shell, which had seven to begin with, so now it has eight (Figure 2.4a). In this case, though, chlorine has one too many negative charges and is thus an ion symbolized Cl^{-1}. An attractive force exists between the Na^{+1} and Cl^{-1} ions, so an *ionic bond* forms between them, yielding the mineral halite (NaCl).

Covalent bonds result when the electron shells of adjacent atoms overlap and they share electrons. A carbon atom in diamond shares all four of its outer electrons with a neighbor to produce the stable noble gas configuration (Figure 2.4b). Among the most common minerals, the silicates, silicon forms partly covalent and partly ionic bonds with oxygen.

▶ Figure 2.4 **Ionic and Covalent Bonding**

electron transfer

Sodium atom
11 p$^+$
10 e$^-$

Chlorine atom
17 p$^+$
17 e$^-$

Sodium ion
11 p$^+$
10 e$^-$

Chlorine ion
17 p$^+$
18 e$^-$

+   −

(a) In ionic bonding, the electron in the outermost electron shell of sodium is transferred to the outermost electron shell of chlorine. Once the transfer has taken place, the positively charged sodium ion and negatively charged chlorine ion attract one another.

(b) Covalent bonds form when adjacent atoms share electrons, as in these carbon atoms.

Minerals—The Building Blocks of Rocks

A mineral's composition is shown by a chemical formula, a shorthand way of indicating how many atoms of different kinds it contains. For example, quartz (SiO_2) is made up of one silicon atom for every two oxygen atoms, whereas the formula for orthoclase is $KAlSi_3O_8$. A few minerals known as *native elements*, such as gold (Au) and diamond (C), consist of only one element and accordingly are not compounds.

Recall our definition of the term *mineral*. *Naturally occurring* means not made by humans, and *inorganic* reminds us that animal and vegetable matter are not minerals. Nevertheless, corals and clams and some other organisms build shells of the mineral calcite ($CaCO_3$) or silica (SiO_2). By definition, minerals are **crystalline solids** in which their atoms are arranged in a specific three-dimensional framework. Ideally, minerals grow and form perfect crystals with planar surfaces (crystal faces), sharp corners, and straight edges (▸Figure 2.5). If numerous minerals grow in proximity, as in cooling lava, they do not develop well-formed crystals.

Some minerals have very specific chemical compositions, but others have a range of compositions because one element can substitute for another if the ions of the two elements have the same electrical charge and are about the same size. Iron and magnesium meet these criteria and thus substitute for one another in olivine [$(Fe,Mg)_2SiO_4$], which may have magnesium, iron, or both. Calcium (Ca) and sodium (Na) substitute for one another in the plagioclase feldspars, which vary from calcium-rich ($CaAl_2Si_2O_8$) to sodium-rich ($NaAlSi_3O_8$) varieties.

Composition and structure control the characteristic physical properties of minerals. These properties are particularly useful for mineral identification (see Appendix C).

How Many Minerals Are There?

Geologists define several mineral groups, each of which is composed of minerals sharing the same negatively charged ion or ion group (Table 2.1). More than 3,500 minerals are known, but only about two dozen are particularly common, although many others are important resources.

Silicate Minerals Given that oxygen (62.6 percent) and silicon (21.2 percent) account for nearly 84 percent of all atoms in Earth's crust, you might expect these elements to be common in minerals. And, indeed, they are. In fact, a combination of silicon and oxygen is called *silica*, and the minerals made up of silica are **silicates**. Silicates account for about one-third of all known minerals and make up perhaps 95 percent of Earth's crust.

All silicates are composed of a basic building block called the *silica tetrahedron*, consisting of one silicon atom

TABLE 2.1 Some of the Mineral Groups Recognized by Geologists

Mineral Group	Negatively Charged Ion or Ion Group	Examples	Composition
Carbonate	$(CO_3)^{-2}$	Calcite	$CaCO_3$
		Dolomite	$CaMg(CO_3)_2$
Halide	Cl^{-1} F^{-1}	Halite	$NaCl$
		Fluorite	CaF_2
Native element	—	Gold	Au
		Silver	Ag
		Diamond	C
		Graphite	C
Phosphate	$(PO_4)^{-3}$	Apatite	$Ca_5(PO_4)_3$ (F, Cl)
Oxide	O^{-2}	Hematite	Fe_2O_3
		Magnetite	Fe_3O_4
Silicate	$(SiO_4)^{-4}$	Quartz	SiO_2
		Potassium feldspar	$KAlSi_3O_8$
		Olivine	$(Mg,Fe)_2SiO_4$
Sulfate	$(SO_4)^{-2}$	Anhydrite	$CaSO_4$
		Gypsum	$CaSO_4 \cdot 2H_2O$
Sulfide	S^{-2}	Galena	PbS
		Pyrite	FeS_2

© 2013 Cengage Learning

▸ Figure 2.5 **A Variety of Mineral Crystal Shapes**

(a) Cubic crystals are typical of the minerals halite and galena.

(b) Pyritohedron crystals such as those of pyrite have 12 sides.

(c) Diamond has octahedral, or eight-sided, crystals.

(d) A prism terminated by a pyramid is found in quartz.

© 2013 Cengage Learning

(a) Ferromagnesian silicates.

(b) Nonferromagnesian silicates.

surrounded by four oxygen atoms. These tetrahedra exist in minerals as isolated units bonded to other elements, or they may be arranged in single chains, double chains, sheets, or complex three-dimensional networks, thus accounting for the incredible diversity of silicate minerals.

Among the silicates, geologists recognize *ferromagnesian silicates*, which contain iron (Fe), magnesium (Mg), or both (▶ Figure 2.6a). They tend to be dark colored and denser than the *nonferromagnesian silicates*, which, of course, lack these elements (Figure 2.6b).

Other Mineral Groups Representatives of the several other mineral groups in Table 2.1 may be important resources, but most of them are not common constituents of rocks. For example, the oxides have oxygen combined with some other element as in hematite (Fe_2O_3) and magnetite (Fe_3O_4), which are iron ores. In the sulfates, an element is combined with the complex sulfate molecule (SO_4)$^{-2}$ as in gypsum ($CaSO_4 \cdot 2H_2O$), much of which is used for drywall or wallboard.

One group of minerals that deserve special attention are the **carbonate minerals**, which include those with the carbonate ion (CO_3)$^{-2}$ as in the mineral calcite ($CaCO_3$). Many other carbonate minerals are known, but only dolomite [$CaMg(CO_3)_2$] is very common. Calcite and dolomite are important because they make up the sedimentary rocks limestone and dolostone, respectively, both of which are widespread at or near Earth's surface.

Rock-Forming Minerals and the Rock Cycle

Geologists have identified hundreds of minerals in rocks, but only a few are common enough to be designated as **rock-forming minerals**—that is, minerals that are essential for the identification and classification of rocks. As you might expect from our previous discussion, most

rock-forming minerals are silicates (Figure 2.6), but the carbonate minerals calcite and dolomite are also important. In most cases, minerals present in small amounts, called *accessory minerals*, can be ignored.

The **rock cycle** is a pictorial representation of events leading to the origin, destruction and/or changes, and reformation of rocks as a consequence of Earth's internal and surface processes (▶ Figure 2.7). In short, it is one manifestation of the interaction among Earth's systems.

Furthermore, the rock cycle shows that the three major rock groups—igneous, sedimentary, and metamorphic—are interrelated; that is, any rock type can be derived from the others. Figure 2.7 shows that the ideal cycle involves the events depicted on the circle leading from magma to igneous rocks and so on. Notice also that the circle has several internal arrows indicating interruptions in the cycle.

Igneous Rocks

The ultimate source of all **igneous rocks** is **magma**, molten rock below Earth's surface. The term **lava** is used to describe magma that reaches the surface. Igneous rocks also result from magma that erupts explosively, thereby forming particulate matter known as **pyroclastic materials**, such as volcanic ash, which become consolidated. Geologists characterize the igneous rocks that formed at the surface as **volcanic rocks** or **extrusive igneous rocks**, and those that formed below the surface as **plutonic rocks** or **intrusive igneous rocks**. The plutonic or intrusive rocks are found in several types of *plutons* (intrusive igneous bodies) shown in ▶ Figure 2.8.

Texture and Composition

Several igneous rock textures tell us something about how the rocks formed in the first place. For instance, rapid cooling in a lava flow results in a fine-grained, or *aphanitic* texture, in

▶ **Figure 2.7** **The Rock Cycle** This cycle shows the interrelationships among Earth's internal and surface processes and how the three major rock groups are related. An ideal cycle includes the events on the outer margin of the cycle, but interruptions indicated by internal arrows are common.

which individual minerals are too small to see without magnification (▶Figure 2.9a). In contrast, a coarse-grained, or *phaneritic* texture is the outcome of comparatively slow cooling that takes place in plutons (Figure 2.9b). Thus, these two textures are usually sufficient to determine whether an igneous rock is volcanic or plutonic. However, magma in a shallow pluton may crystallize and have an aphanitic texture, whereas the interior of a thick lava flow may be phaneritic.

Some igneous rocks, though, have a combination of markedly different-sized minerals—a so-called *porphyritic texture*;

the large minerals are *phenocrysts*, whereas the smaller ones constitute the rock's *groundmass* (Figure 2.9c). A porphyritic texture might indicate a two-stage cooling history

▶ **Figure 2.8** **Block Diagram Showing the Origin of Extrusive Igneous (Volcanic) Rocks and Intrusive Igneous (Plutonic) Rocks** Notice that some of the plutons cut across the layering in the intruded rock (country rock) and are said to be *discordant*, but others are parallel to the layering and are *concordant*.

Igneous Rocks **23**

▶ Figure 2.9 **Igneous Rock Textures** Texture is one criterion used to classify igneous rocks, and it tells something about the history of these rocks.

(a) Rapid cooling as in a lava flow results in many small minerals and an aphanitic, or fine-grained, texture.

(b) Slower cooling in plutons yields a phaneritic, or coarse-grained, texture.

(c) A porphyritic texture indicates a complex cooling history.

(d) Obsidian has a glassy texture because magma cooled too quickly for mineral crystals to form.

(e) Gases expand in cooling lava and yield a vesicular texture.

(f) Microscopic view of a fragmental texture. The colorless, angular objects are pieces of volcanic glass measuring up to 2 mm.

in which magma began cooling below the surface and then was expelled onto the surface, where cooling continued. The resulting igneous rocks are characterized as *porphyry*.

A *glassy texture* results from cooling so rapidly that the atoms in lava have too little time to form the three-dimensional framework of minerals. As a result, the natural glass *obsidian* forms (Figure 2.9d). Cooling lava might have a large content of trapped water vapor and other gases that form small holes or cavities called *vesicles*; rocks with numerous vesicles are *vesicular* (Figure 2.9e). And finally, a *pyroclastic* or *fragmental texture* characterizes igneous rocks composed of pyroclastic materials (Figure 2.9f).

With few exceptions, the primary constituent of magma is silica, but the silica content varies enough for us to recognize magmas characterized as *felsic* (>65 percent silica), *intermediate* (53–65 percent silica), *mafic* (45–52 percent silica), and *ultramafic* (<45 percent silica). Felsic magma also contains considerable sodium, potassium, and aluminum, but little calcium, iron, and magnesium. In contrast, mafic magma has proportionately more calcium, iron, and magnesium. Intermediate magma, of course, has a composition between those of felsic and mafic magmas.

Classifying Igneous Rocks

Texture and composition are the criteria that geologists use to classify most igneous rocks. Notice in ▶ Figure 2.10

that all of the rocks shown, except peridotite, are pairs; each member of a pair has the same composition but a different texture. Thus, basalt (aphanitic) and gabbro (phaneritic) have the same minerals, and because they contain a large proportion of ferromagnesian silicates they tend to be dark colored (▶ Figures 2.11a and 2.11b). Likewise, andesite (aphanitic) and diorite (phaneritic) have the same composition but differ in texture (Figures 2.10 and 2.11c, 2.11d). And lastly, rhyolite (aphanitic) and granite (phaneritic) also are made up of the same minerals but have different textures. In addition, rhyolite and granite are composed mostly of nonferromagnesian silicates, so they tend to be light colored (Figures 2.10 and 2.11e, 2.11f). We can infer from their textures that basalt, andesite, and rhyolite are volcanic rocks, whereas gabbro, diorite, and granite are plutonic. Notice also from Figure 2.10 that composition is related to the type of magma from which the rocks cooled.

For a few igneous rocks, texture is the primary criterion for classification (▶ Figure 2.12a). *Tuff* is composed of volcanic ash, a designation for pyroclastic materials measuring less than 2 mm (Figure 2.12b). Some ash deposits are so hot when they form that the particles fuse together, forming *welded tuff*. Consolidation of larger pyroclastic materials, such as *lapilli* (2–64 mm) and blocks and bombs*

---

*Blocks are angular, whereas bombs are smooth and commonly shaped like a teardrop.

▶ **Figure 2.10 Classification of Igneous Rocks** This diagram shows the proportions of the main minerals and the textures of common igneous rocks. For example, an aphanitic (fine-grained) rock of mostly calcium-rich plagioclase and pyroxene is basalt.

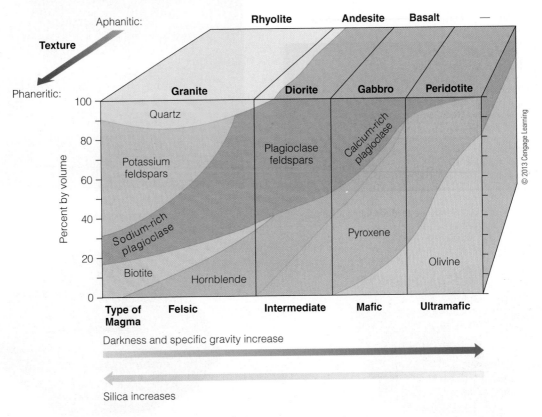

▶ **Figure 2.11 The Common Igneous Rocks** Can you tell from the images which rocks are volcanic and which are plutonic?

(a) Basalt.

(b) Gabbro.

(c) Andesite.

(d) Diorite.

(e) Rhyolite.

(f) Granite.

Composition	Felsic ◄———————► Mafic	
Texture	Vesicular	Pumice	Scoria
	Glassy	Obsidian	
	Pyroclastic or Fragmental	◄———— Volcanic Breccia ————►	
		Tuff/welded tuff	

© 2013 Cengage Learning

(a) Vesicular, glassy, and pyroclastic are textures found in some volcanic rocks.

Courtesy of David J. Matty

(b) Tuff is composed of pyroclastic materials such as those in Figure 2.9f.

Sue Monroe

(c) Pumice is glassy and vesicular; its density is so low that it floats.

Sue Monroe

(d) Scoria is also vesicular, but it is darker, heavier, and more crystalline than pumice.

(>64 mm), yields *volcanic breccia. Obsidian* and *pumice* are varieties of volcanic glass. The former looks like red, brown, or black glass (Figure 2.9d), whereas the latter has a frothy appearance, because it has numerous vesicles. *Scoria* also has numerous vesicles (Figure 2.12d), but it is more crystalline than pumice (Figure 2.12c).

Sedimentary Rocks

Notice in the rock cycle (Figure 2.7) that weathering processes disintegrate or decompose rocks at or near the surface. Accordingly, rocks are broken into smaller particles of gravel (>2 mm), sand (1/16–2 mm), silt (1/256–1/16 mm), and clay (<1256 mm), and some minerals may be dissolved—that is, taken into solution. In either case, these solid or dissolved materials are commonly transported elsewhere and deposited as *sediment*, perhaps as sand on a beach or as minerals extracted from solution by inorganic chemical processes or by the activities of organisms. Any sedimentary deposit may be lithified—that is, transformed into **sedimentary rock**, which by definition is any rock made up of sediment.

Sediment Transport, Deposition, and Lithification

Sediment transport involves removing sediment from its source area and carrying it elsewhere. Running water is the most effective method of transport, but glaciers, wind, and waves are also important in some areas. In any case, transported sediment must eventually be deposited, as in a stream channel, on a beach, or on the seafloor. Sediment also accumulates as minerals that precipitate from solution or that organisms extract from solution. In all cases, sediment is an unconsolidated aggregate of solid particles, such as sand or mud.

Lithification is the geologic phenomenon of converting sediment into sedimentary rock. As sediment accumulates, *compaction*, resulting from pressure exerted by the weight of overlying sediments, reduces the amount of pore space (the void space between particles) and thus the volume of the deposit (▶ Figure 2.13a). Compaction alone is usually sufficient for lithification of mud (particles measuring less than 1/16 mm). But for gravel and sand, *cementation* is necessary, which involves the precipitation of minerals within pores that effectively binds the sediment

▶ **Figure 2.13 Lithification of Detrital Sediment**

Sediment

Gravel > 2 mm   Compaction

Sand 2 mm–1/16 mm   Compaction

Silt 1/16 mm–1/256 mm   Compaction

Clay < 1/256 mm   Compaction

© 2013 Cengage Learning

(a) Compaction of gravel, sand, silt, and clay. Notice that little compaction takes place in gravel and sand, but the volume of silt and clay deposits is considerably reduced by compaction.

Bruce Railsback, Department of Geology, University of Georgia

(b) Cementation of sand to form sandstone. This sandstone is made up mostly of quartz (Q) and the cement is calcite (C).

together (Figure 2.13b). The most common cements are calcium carbonate ($CaCO_3$), silica (SiO_2), and, in a distant third place, iron oxides and hydroxides such as hematite (Fe_2O_3) and limonite [$FeO(OH)$], respectively.

Classification of Sedimentary Rocks

Detrital sedimentary rocks are made up of detritus—that is, the solid particles such as gravel, sand, and mud derived by the breakdown of preexisting rocks. In contrast, **chemical sedimentary rocks** are made up of minerals derived from materials in solution and extracted by either inorganic chemical processes or by the activities of organisms. The subcategory called *biochemical sedimentary rocks* is so named because of the importance of organisms in their origin.

Detrital Sedimentary Rocks All detrital sedimentary rocks are composed of fragments or particles known as *clasts* that are derived from preexisting rocks of any kind, and they are defined primarily by the size of their constituent clasts (Table 2.2). *Conglomerate* and *sedimentary breccia* are both made up of *gravel* (detrital particles measuring more than 2 mm). The only difference between them is that conglomerate (▶ Figure 2.14a) has rounded clasts, as opposed to the angular ones in sedimentary breccia (Figure 2.14b). Particle rounding results from wear and abrasion during transport.

The term *sandstone* applies to any detrital sedimentary rock composed of sand (particles measuring 1/16–2 mm) (Table 2.2; Figure 2.14c). *Quartz sandstone* is the most common and, as its name implies, is made up mostly of quartz. Another fairly common type of sandstone is *arkose*, which in addition to quartz has at least 25 percent feldspar minerals.

All detrital sedimentary rocks composed of particles measuring less than 1/16 mm are collectively called *mudrocks* (Table 2.2; Figure 2.14c). Among the mudrocks we distinguish between *siltstone*, composed of silt-sized particles, *mudstone*, a mixture of silt- and clay-sized particles; and *claystone*, which is composed mostly of particles

TABLE 2.2 Classification of Detrital Sedimentary Rocks

Sediment Name and Size	Features	Rock Name
Gravel >2 mm	Rounded gravel Angular gravel	Conglomerate Sedimentary breccia
Sand 1/16–2 mm	Mostly quartz >25% feldspars	Quartz sandstone Arkose
Silt 1/256–1/16 mm	>50% silt-sized particles Mixture of silt and clay	Siltstone ⎤ Mudstone* ⎟ —— Mudrocks
Clay <1/256 mm	>50% clay-sized particles	Claystone* ⎦

*These rocks are called *shale* if they are fissile, that is, they split along closely spaced parallel planes.

© 2013 Cengage Learning

(a) Outcrop of conglomerate in Death Valley, California. The largest clast measures about 30 cm long.

(b) Sedimentary breccia in Death Valley, California. Notice the angular gravel-sized particles. The largest clast is about 12 cm across.

(c) Mudrock overlain by sandstone at Mesa Verde National Park in Colorado.

(d) This mudrock in Tennessee is fissile, so it is called shale.

measuring less than 1/256 mm. However, some mudstones and claystones are further designated as *shale* if they are *fissile*—meaning that they break along closely spaced parallel planes (Figure 2.14d, Table 2.2).

Chemical Sedimentary Rocks Chemical sedimentary rocks form when minerals are extracted from solution by inorganic chemical processes or by organisms. Some of the resulting rocks have a *crystalline texture*—that is, they are made up of an interlocking mosaic of mineral crystals. But others have a *clastic texture*—for example, an accumulation of broken shells on a beach, if lithified, becomes a type of limestone.

Limestone (▶ Figure 2.15) and *dolostone* are the most common rocks in this category, and geologists refer to both

as **carbonate rocks** because they are composed of carbonate minerals: calcite and dolomite, respectively (Table 2.3). Actually, most dolomite was originally limestone that was altered when magnesium replaced some of the calcium in calcite. Furthermore, most limestone is also biochemical, because organisms are so important in its origin.

The **evaporites**, such as *rock salt* and *rock gypsum*, form by inorganic chemical precipitation of minerals from solutions made concentrated by evaporation (Table 2.3; ▶ Figure 2.16a). If you were to evaporate seawater, for instance, you would end up with a small amount of rock gypsum and considerably more rock salt. Several other evaporites would also form but only in very small amounts.

Chert, a dense, hard rock, consists of silica, some of which forms as oval to spherical masses within other rocks

Sue Monroe

Sue Monroe

(a) Limestone with numerous fossil shells is called fossiliferous limestone.

(c) Coquina is limestone made up entirely of broken shells. Compare with the fossiliferous limestone in (a).

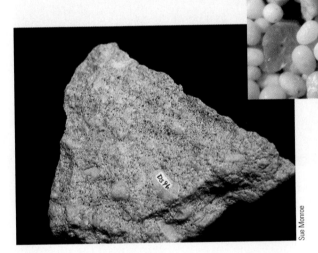

Stan Celestian/Glendale Community College

Sue Monroe

(b) This oolitic limestone is made up partly of ooids (see inset), which are rather spherical grains of calcium carbonate.

by inorganic chemical precipitation. Other chert consists of layers of microscopic shells of silica-secreting organisms and is thus a biochemical sedimentary rock (Table 2.3; Figure 2.16c).

A biochemical sedimentary rock of great economic importance is *coal*, which consists of partially altered, compressed remains of land plants. Coal forms in oxygen-deficient swamps and bogs where the decomposition process is interrupted and the accumulating vegetation alters first to *peat* and, if buried and compressed, changes to coal (Table 2.3; Figure 2.16b).

TABLE 2.3 **Classification of Chemical and Biochemical Sedimentary Rocks**

Chemical Sedimentary Rocks

Texture	Composition	Rock Name	
Crystalline	Calcite ($CaCO_3$)	Limestone ⎤	Carbonates
Crystalline	Dolomite [$CaMg(CO_3)_2$]	Dolostone ⎦	
Crystalline	Gypsum ($CaSO_4 \cdot 2H_2O$)	Rock gypsum ⎤	Evaporites
Crystalline	Halite (NaCl)	Rock salt ⎦	

Biochemical Sedimentary Rocks

Texture	Composition	Rock Name
Clastic	Calcium carbonate ($CaCO_3$) shells	Limestone (various types such as chalk and coquina)
Usually crystalline	Altered microscopic shells of silicon dioxide (SiO_2)	Chert
–	Mostly carbon from altered plant remains	Coal

(a) Rock gypsum. When deeply buried, gypsum ($CaSO_4 \cdot 2H_2O$) loses its water and is converted to anhydrite ($CaSO_4$).

(b) Bituminous coal is the most common type of coal used for fuel.

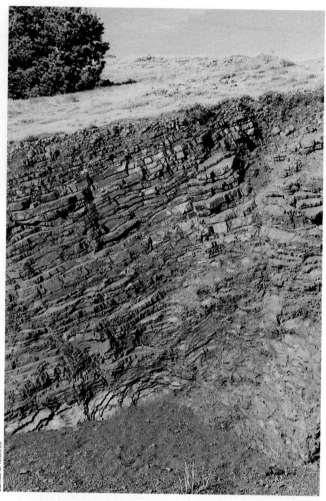

(c) Bedded chert exposed in Marin County, California. Most of the layers are about 5 cm thick.

Metamorphic Rocks

All sedimentary and igneous rocks form under specific conditions of temperature, pressure, and fluid activity. However, should they be subjected to different conditions, they may be altered in the solid state—that is, they become **metamorphic rocks**. Although sedimentary rocks, followed by igneous rocks, are the most susceptible to metamorphism, any type of rock, including metamorphic rocks, can be changed (metamorphosed) under the right conditions. Metamorphic changes may be *compositional* (new minerals form) or *textural* (minerals become aligned) or both, thus forming a foliated texture (▶ Figure 2.17). Some of these changes are minor, so the parent rock is easily recognized, but changes may be so great that it can be very difficult or impossible to identify the parent rock.

What Causes Metamorphism?

Heat, pressure, and chemical fluids—the agents responsible for metamorphism—may act singly or in any combination, and the time during which they are effective varies considerably. For example, a lava flow may bake the underlying rocks for a short time but otherwise has little effect, whereas the rocks adjacent to plutons, especially batholiths (Figure 2.8), may be altered over a long period and for a great distance.

Heat is important because it increases the rates of chemical reactions that may yield minerals different from those in the parent rock. Sources of heat include plutons, especially the larger ones such as stocks and batholiths, and deep burial, because Earth's temperature increases with depth. Deep burial also subjects rocks to *lithostatic pressure* resulting from the weight of the overlying rocks. Under these conditions minerals in a rock become more

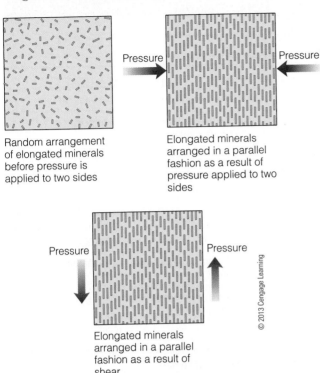

Random arrangement of elongated minerals before pressure is applied to two sides

Elongated minerals arranged in a parallel fashion as a result of pressure applied to two sides

Elongated minerals arranged in a parallel fashion as a result of shear

© 2013 Cengage Learning

(a) When rocks are subjected to differential pressure, the mineral grains are typically arranged in a parallel fashion, producing a foliated texture.

Reed Wicander

(b) Photomicrograph of a metamorphic rock with a foliated texture showing the parallel arrangement of mineral grains.

closely packed and may *recrystallize*—that is, form smaller and denser minerals. Lithostatic pressure operates with the same intensity in all directions, but *differential pressure* exerts force more intensely from one direction, as occurs at convergent plate boundaries.

In any region where metamorphism takes place, water is present in varying amounts and may contain ions in solution that enhance metamorphism by increasing the rate of chemical reactions. Accordingly, *fluid activity* is also an important metamorphic agent.

Contact metamorphism takes place when heat and chemical fluids from an igneous body alter adjacent rocks. The rocks in contact with a batholith may be heated to nearly 1,000°C, and, of course, fluids from the magma also bring about changes. The degree of metamorphism decreases with increasing distance from a pluton until the surrounding rocks are unaffected.

Most metamorphic rocks result from **regional metamorphism**, which takes place over large but elongated areas as the result of tremendous pressure, elevated temperatures, and fluid activity. This kind of metamorphism is most obvious along convergent plate boundaries where the rocks are intensely deformed during convergence and subduction. It also takes place at divergent plate boundaries, though usually at shallower depths, and here only high temperature and fluid activity are important.

Dynamic metamorphism is much more restricted in its extent, being confined to zones adjacent to *faults* (fractures along which rocks have moved) where high levels of differential pressure develop.

Metamorphic Rock Classification

Many metamorphic rocks, especially those subjected to intense differential pressure, have their platy and elongate minerals aligned in a parallel fashion, giving them a *foliated texture* (Figure 2.17). In contrast, some metamorphic rocks do not develop any discernable orientation of their minerals. Instead, they consist of a mosaic of roughly equidimensional minerals and have a *nonfoliated texture*.

Foliated Metamorphic Rocks *Slate*, a very fine-grained rock, results from low-grade metamorphism of mudrocks or, more rarely, volcanic ash (Table 2.4; ▶Figure 2.18a). *Phyllite* is coarser-grained than slate, but the minerals are still too small to see without magnification. Actually, the change from some kind of mudrock (shale perhaps) to slate and then to phyllite is a continuum, and so is the origin of *schist*. In schist, though, the elongate and platy minerals are clearly visible and they impart a *schistosity*, or *schistose foliation*, to the rock (Table 2.4; Figure 2.18b).

Gneiss, with its alternating dark and light bands of minerals, is one of the more attractive metamorphic rocks (Table 2.4; Figure 2.18c). Quartz and feldspars are most common in the light-colored bands, and biotite and hornblende are found in the dark-colored bands. Most gneiss probably forms from regional metamorphism of clay-rich sedimentary rocks, but metamorphism of granite and other metamorphic rocks also yields gneiss.

In some areas of regional metamorphism, exposures of "mixed rocks" having both igneous and metamorphic characteristics are present. These rocks, called *migmatites*, consist of streaks, or lenses of granite intermixed with ferromagnesian mineral-rich metamorphic rocks (Figure 2.18d). Where do you think these rocks would be in the rock cycle?

TABLE 2.4 Classification of Common Metamorphic Rocks

Texture	Metamorphic Rock	Typical Minerals	Metamorphic Grade	Characteristics of Rocks	Parent Rock
Foliated	Slate	Clays, micas, chlorite	Low	Fine-grained, splits easily into flat pieces	Mudrocks, volcanic ash
	Phyllite	Fine-grained quartz, micas, chlorite	Low to medium	Fine-grained, glossy or lustrous sheen	Mudrocks
	Schist	Micas, chlorite, quartz, talc, hornblende, garnet, staurolite, graphite	Low to high	Distinct foliation, minerals visible	Mudrocks, carbonates, mafic igneous rocks
	Gneiss	Quartz, feldspars, hornblende, micas	High	Segregated light and dark bands visible	Mudrocks, sandstones, felsic igneous rocks
	Amphibolite	Hornblende, plagioclase	Medium to high	Dark-colored, weakly foliated	Mafic igneous rocks
	Migmatite	Quartz, feldspars, hornblende, micas	High	Streaks or lenses of granite intermixed with gneiss	Felsic igneous rocks mixed with metamorphic rocks
Nonfoliated	Marble	Calcite, dolomite	Low to high	Interlocking grains of calcite or dolomite, reacts with HCl	Limestone or dolostone
	Quartzite	Quartz	Medium to high	Interlocking quartz grains, hard, dense	Quartz sandstone
	Greenstone	Chlorite, epidote, hornblende	Low to high	Fine-grained, green	Mafic igneous rocks
	Hornfels	Micas, garnet, andalusite, cordierite, quartz	Low to medium	Fine-grained, equidimensional grains, hard, dense	Mudrocks
	Anthracite	Carbon	High	Black, lustrous, subconcoidal fracture	Coal

© 2013 Cengage Learning

Nonfoliated Metamorphic Rocks Nonfoliated metamorphic rocks may form by either contact or regional metamorphism, but in all cases they lack the platy and elongate minerals found in foliated rocks. *Marble*, a well-known metamorphic rock composed of calcite or dolomite, results from metamorphism of limestone or dolostone (Table 2.4; ▶ Figure 2.19a). Some marble is attractive and has been a favorite with sculptors and for building material for centuries. Metamorphism of quartz sandstone yields *quartzite*, a hard, compact rock (Table 2.4; Figure 2.19b).

The name *greenstone* applies to any compact, dark green, altered mafic igneous rock that formed under metamorphic conditions. Minerals such as chlorite and epidote give it its green color. Several varieties of *hornfels* are known, most of which formed when clay-rich sedimentary rocks were altered during contact metamorphism.

Anthracite is a black, lustrous, hard coal with mostly carbon and a low amount of volatile matter. It forms from other types of coal.

Plate Tectonics and the Rock Cycle

Notice in the rock cycle (Figure 2.7) that interactions among Earth systems are responsible for the origin and alternation of rocks. The atmosphere, hydrosphere, and biosphere acting on Earth materials account for weathering, transport, and deposition, whereas Earth's internal heat is responsible for melting and contributes to metamorphism. Plate tectonics also plays an important role in recycling Earth materials.

Sediment along continental margins may become lithified and be incorporated into a moving plate along with underlying oceanic crust. Where plates collide at convergent plate boundaries, heat and pressure commonly lead to metamorphism, igneous activity, and the origin of new rocks. In addition, some of the rocks in a subducted plate are deformed and incorporated into an evolving mountain system that in turn is weathered and eroded, yielding sediment to begin yet another cycle.

▶ Figure 2.18 **Foliated Metamorphic Rocks**

(a) Hand specimen of slate.

(b) Schist.

(c) Gneiss.

(d) Migmatites. A migmatite boulder in the Rocky Mountain National Park, near Estes Park, Colorado. Migmatites consist of high-grade metamorphic rock intermixed with streaks, or lenses, of granite.

▶ Figure 2.19 **Nonfoliated Metamorphic Rocks** What were these rocks before they were metamorphosed?

(a) Marble.

(b) Quartzite.

Summary

▶ Elements are composed of atoms, which have a nucleus of protons and neutrons around which electrons orbit in electron shells.

▶ The number of protons in a nucleus determines the atomic number of an element; the atomic mass number is the number of protons plus neutrons in the nucleus.

▶ Atoms bond by transferring electrons from one atom to another (ionic bond) or by sharing electrons (covalent bond). Most minerals are compounds of two or more elements bonded together.

▶ By far the most common minerals are the silicates (composed of at least silicon and oxygen), but carbonate minerals (containing the CO_3 ion) are prevalent in some rocks.

▶ The two broad groups of igneous rocks, plutonic (or intrusive) and volcanic (or extrusive) are classified by composition and texture. However, for a few volcanic rocks texture is the main consideration.

▶ Lithification involving compaction and cementation is the process whereby sediment is transformed into sedimentary rock.

▶ Sedimentary rocks are also grouped into two broad categories: detrital (made up of solid particles of preexisting rocks) and chemical/biochemical (composed of minerals derived by inorganic chemical processes or the activities of organisms).

▶ Metamorphic rocks result from compositional and/or textural transformation of other rocks by heat, pressure, and fluid activity. Most metamorphism is regional, occurring deep within the crust over large areas, but some, called contact metamorphism, takes place adjacent to hot igneous rocks.

▶ Metamorphism imparts a foliated texture to many rocks (parallel alignment of minerals), but some rocks have a mosaic of equidimensional minerals and are nonfoliated.

▶ Metamorphic rocks are classified largely by their textures, but composition is a consideration for some.

▶ Plate tectonics, which is driven by Earth's internal heat coupled with surface processes such as weathering, erosion, and deposition, accounts for the recycling of Earth materials in the rock cycle.

Important Terms

atom, p. 19
atomic mass number, p. 19
atomic number, p. 19
bonding, p. 19
carbonate mineral, p. 22
carbonate rock, p. 28
chemical sedimentary rock, p. 27
compound, p. 19
contact metamorphism, p. 31
crystalline solid, p. 21
detrital sedimentary rock, p. 27

dynamic metamorphism, p. 31
element, p. 19
evaporite, p. 28
extrusive igneous rock, p. 22
geologic record, p. 18
igneous rock, p. 22
intrusive igneous rock, p. 22
lava, p. 22
lithification, p. 26
magma, p. 22
metamorphic rock, p. 30

mineral, p. 18
plutonic rock, p. 22
pyroclastic materials p. 22
regional metamorphism, p. 31
rock, p. 18
rock cycle, p. 22
rock-forming mineral, p. 22
sedimentary rock, p. 26
silicate, p. 21
volcanic rock, p. 22

Review Questions

1. Magma that cools and crystallizes beneath the surface forms _____ rocks.
   a. _____ sedimentary; b. _____ plutonic; c. _____ extrusive; d. _____ metamorphic; e. _____ extraterrestrial.

2. The most common rock-forming minerals other than silicates are the
   a. _____ sulfides; b. _____ native elements; c. _____ phosphates; d. _____ carbonates; e. _____ oxides.

3. Evidence in rocks of past physical and biological events is known as the
   a. _____ geologic record; b. _____ plutonic history; c. _____ planetary chronicle; d. _____ story of evolution; e. _____ plate tectonic narrative.

4. An example of an aphanitic volcanic rock is
   a. _____ conglomerate; b. _____ phyllite; c. _____ granite; d. _____ calcite; e. _____ basalt.

5. Naturally occurring, solid substances with all of their atoms arranged in a specific three-dimensional framework are said to be
   a. _____ natural glass; b. _____ magma; c. _____ crystalline; d. _____ detrital; e. _____ biochemical.

6. An atom with 8 protons and 14 neutrons has an atomic mass number of
   a. _____ 22; b. _____ 6; c. _____ 14; d. _____ 8; e. _____ 112.

7. Many limestones are made up of minerals that were extracted from seawater by organisms, so they are called _____ sedimentary rocks.
   a. _____ porphyritic; b. _____ biochemical; c. _____ plutonic; d. _____ foliated; e. _____ silicate.

8. The type of bonding in which electrons are shared by atoms is called _____ bonding.
   a. _____ van der Waals; b. _____ ionic; c. _____ crystalline; d. _____ covalent; e. _____ evaporite.

9. Metamorphism that occurs over a large but usually elongate area is called _____ metamorphism.
a. _____ contact; b. _____ dynamic; c. _____ lithification;
d. _____ compound; e. _____ regional.

10. Lithification is a process whereby
a. _____ lava cools and forms a phaneritic texture;
b. _____ sediment is converted to sedimentary rock;
c. _____ mudstone is changed to slate; d. _____ heat, pressure, and chemical fluids alter rocks; e. _____ rocks and minerals decompose.

11. If diamond were perfectly cleaved, it would yield geometric figures like the ones shown in the figure below, but this mineral is fluorite. From the image alone you should be able to tell that these specimens are not diamond. How? Look up in Appendix C other mineral properties of diamond and fluorite that you can use to tell them apart.

12. What are rock-forming minerals, and why do geologists distinguish between them and accessory minerals?

13. Examine the rocks in Figures 2.14a and 2.14b. Why are the gravel clasts in (a) rounded whereas those in (b) are angular?

14. What are plutons, and how are they distinguished from one another?

15. Ice is a mineral, but water vapor and liquid water are not. Why?

16. How is it possible for some chemical elements to substitute for one another in the plagioclase feldspars?

17. Refer to the the rock cycle (Figure 2.7) and explain how the interactions of Earth systems can explain how any one of three main rock groups can be derived from any of the others.

18. Compare contact and regional metamorphism, and give an example of a rock that forms under each condition.

19. What are the two groups of silicate minerals, and how do they differ from one another?

20. Why are volcanic rocks usually aphanitic whereas most plutonic rocks are phaneritic?

Sue Monroe

3

Plate Tectonics
A Unifying Theory

The Mand River winds through the Zagros Mountains in western Iran. The Zagros Mountains consist of a series of anticlines and synclines, which are rocks that were folded by compressive forces as a result of the collision of the Arabian plate with the Eurasian plate.

USGS EROS Data Center/Landsat/NASA

At the end of this chapter, you will have learned that

▶ Plate tectonics is the unifying theory of geology and revolutionized how geologists study Earth.

▶ The hypothesis of continental drift was based on considerable geologic, paleontologic, and climatologic evidence.

▶ Earth has a magnetic field in which the magnetic poles are in proximity to the geographic poles, and the strength of the magnetic field is strongest at the poles and weakest at the equator.

▶ The hypothesis of seafloor spreading accounts for continental movement, and thermal convection cells are thought to be the mechanism for plate movement.

▶ The three types of plate boundaries are divergent, convergent, and transform. Along these boundaries new plates are formed, consumed, or slide past one another.

▶ Interaction along plate boundaries accounts for most of Earth's earthquake and volcanic activity.

▶ The rate of movement and direction of motion of plates can be calculated in several ways, all of which agree and which indicate that the plates move at different average velocities.

▶ Mountain building is a consequence of compressive forces related to plate movement.

▶ Plate movement affects not only the distribution of the world's biota and its evolution, but it also affects the formation and distribution of some natural resources.

Introduction

Imagine it is the day after Christmas, December 26, 2004, and you are vacationing on a beautiful beach in Thailand. You look up from the book you're reading to see the sea suddenly retreat from the shoreline, exposing a vast expanse of seafloor that moments before was underwater and teeming with exotic and colorful fish. It is hard to believe that within minutes of this unusual event, a powerful *tsunami* (a large sea wave that is usually produced by an earthquake but can also result from submarine landslides and volcanic eruptions) will sweep over your resort and everything in its path for several kilometers inland. Within hours, the coasts of Indonesia, Sri Lanka, India, Thailand, Somalia, Myanmar, Malaysia, and the Maldives will be inundated by the deadliest tsunami in history. More than 220,000 people will die, and the region will incur billions of dollars in damage.

Now go forward to October 25, 2010, when Mount Merapi in Central Java, Indonesia, began a series of violent eruptions that lasted into late November. Huge quantities of ash, lava, and ash flows were emitted during these eruptions, and although warnings were issued, at least 350 people were killed by the eruptions, and more than 350,000 were evacuated from areas around the volcano.

Almost five months later, on March 11, 2011, a catastrophic 9.0-magnitude earthquake struck Japan, killing more than 13,000 people and leaving thousands injured and homeless. This devastating earthquake also generated a tsunami that sent walls of water crashing into the northeastern shores of the island, causing further damage and casualties.

What do these three tragic events have in common? They are part of the dynamic interactions involving Earth's plates. When two plates come together along a convergent plate boundary, one plate is pushed or pulled under the other plate, triggering large earthquakes such as the one that shook Japan in 2011. If conditions are right, earthquakes can also produce a devastating tsunami, such as occurred in Indonesia in 2004, and Japan in 2011.

As the descending plate continues to move downward along a convergent plate boundary and is assimilated into Earth's interior, magma is generated. Being less dense than the surrounding material, the magma rises toward the surface, where it may erupt as a volcano. It therefore should not be surprising that the distribution of volcanoes and earthquakes closely follows plate boundaries.

If you're like most people, you probably have only a vague notion of what plate tectonic theory is. Yet plate tectonics affects all of us. Volcanic eruptions, earthquakes, and tsunami are the result of interactions between plates. Global weather patterns and oceanic currents are caused, in part, by the configuration of the continents and ocean basins. The formation and distribution of many natural resources are related to plate movement, and thus have an impact on the economic well-being and political decisions of nations. It is therefore important to understand this unifying theory, not only because it impacts us as individuals and citizens of nation-states, but also because it ties together many aspects of geology and Earth history that you will be studying.

Early Ideas About Continental Drift

The idea that Earth's past geography was different from today's is not new. The earliest maps showing the east coast of South America and the west coast of Africa probably provided people with the first evidence that continents may have once been joined, then broken apart and moved to their present locations.

During the late 19th century, Austrian geologist Edward Suess noted the similarities between the Late Paleozoic plant fossils of India, Australia, South Africa, and South America, as well as evidence of glaciation in the rock sequences of these continents. The plant fossils comprise a unique flora that occurs in the coal layers just above the glacial deposits of these southern continents. This flora is very different from the contemporaneous coal swamp flora of the northern continents and is collectively known as the **Glossopteris flora**, after its most conspicuous genus (▶ Figure 3.1).

Suess also proposed the name *Gondwanaland* (or **Gondwana**, as we will use here) for a supercontinent composed of the aforementioned southern continents. Abundant fossils of the *Glossopteris* flora are found in coal beds in Gondwana, a province in India. Suess thought these southern continents were at one time connected by land bridges over which plants and animals migrated. Thus, in his view, the similarities of fossils on these continents were due to the appearance and disappearance of the connecting land bridges.

In 1910, an American geologist, Frank Taylor, published his views concerning continental movement. He explained the formation of mountain ranges as a result of the lateral movement of continents, and he envisioned the present-day continents as parts of larger polar continents that had broken apart and migrated toward the equator after Earth's rotation was slowed by gigantic tidal forces. According to Taylor, these tidal forces were generated when Earth captured the Moon approximately 100 million years ago.

Although we now know that Taylor's mechanism is incorrect, one of his most significant contributions was his suggestion that the Mid-Atlantic Ridge, discovered by the 1872–1876 British HMS *Challenger* expeditions, might mark the site along which an ancient continent broke apart to form the present-day Atlantic Ocean.

Alfred Wegener and the Continental Drift Hypothesis

Alfred Wegener, a German meteorologist (▶ Figure 3.2), is generally credited with developing the hypothesis of **continental drift**. In his monumental book, *The Origin of Continents and Oceans* (first published in 1915), Wegener proposed that all landmasses were originally united in a single supercontinent that he named **Pangaea**, from the Greek meaning "all land." Wegener portrayed his grand concept of continental movement in a series of maps showing the breakup of Pangaea and the movement of the various continents to their present-day locations. Wegener amassed a tremendous amount of geologic, paleontologic, and climatologic evidence in support of continental drift; however, the initial reaction of scientists to his then-heretical ideas can best be described as mixed.

Nevertheless, Alexander du Toit, a South African geologist, and one of Wegener's more ardent supporters, further developed Wegener's arguments and gathered more geologic and paleontologic evidence in support of his continental drift hypothesis. In 1937, du Toit published *Our Wandering Continents*, in which he contrasted the glacial deposits of Gondwana with coal deposits of the same age found in the continents of the Northern Hemisphere. To resolve this apparent climatologic paradox, du Toit moved the Gondwana continents to the South Pole and brought the northern continents together such that the coal deposits were located at the equator. He named this northern landmass **Laurasia**, and it consisted of present-day North America, Greenland, Europe, and Asia (except for India, which was part of Gondwana).

▶ Figure 3.1 **Fossil *Glossopteris* Leaves** Plant fossils, such as these *Glossopteris* leaves from the Upper Permian Dunedoo Formation in Australia, are found on all five of the Gondwana continents. Their presence on continents with widely varying climates today is evidence that the continents were at one time connected. The distribution of the plants at that time was in the same climatic latitudinal belt.

▶ Figure 3.2 **Alfred Wegener** Alfred Wegener, a German meteorologist, proposed the continental drift hypothesis in 1912 based on a tremendous amount of geologic, paleontologic, and climatologic evidence. He is shown here waiting out the Arctic winter in an expedition hut in Greenland.

What Is the Evidence for Continental Drift?

What, then, was the evidence that Wegener, du Toit, and others used to support the hypothesis of continental drift? It included the fit of the shorelines of continents, the appearance of the same rock sequences and mountain ranges of the same age on continents now widely separated, the matching of glacial deposits and paleoclimatic zones, and the similarities of many extinct plant and animal groups whose fossil remains are found today on widely separated continents. Wegener and his supporters argued that this vast amount of evidence from a variety of sources surely indicated that the continents must have been close together in the past.

Continental Fit

Wegener, like some before him, was impressed by the close resemblance between the coastlines of continents on opposite sides of the Atlantic Ocean, particularly South America and Africa. He cited these similarities as partial evidence that the continents were at one time joined together as a supercontinent that subsequently split apart. His critics pointed out, however, that the configuration of coastlines results from erosional and depositional processes, and therefore is continuously being modified.

A more realistic approach is to fit the continents together along the continental slope, where erosion would be minimal. Sir Edward Bullard, an English geophysicist, and two associates, showed in 1965 that the best fit between continents occurs at a depth of 2,000 m, confirming the close fit between continents when they are reassembled to form Pangaea (▶ Figure 3.3).

Similarity of Rock Sequences and Mountain Ranges

If the continents were at one time joined, then the rocks and mountain ranges of the same age in adjoining locations on the opposite continents should closely match. Such is the case for the Gondwana continents (▶ Figure 3.4). Marine, nonmarine, and glacial rock sequences of Pennsylvanian to Triassic age are almost identical on all five Gondwana continents, strongly indicating that they were joined at one time.

Furthermore, the trends of several major mountain ranges also support the hypothesis of continental drift. These mountain ranges seemingly end at the coastline of one continent, only to apparently continue on another continent across the ocean. The folded Appalachian Mountains of North America, for example, trend northeastward through the eastern United States and Canada and terminate abruptly at the Newfoundland coastline. Mountain ranges of the same age and deformational style are found

▶ **Figure 3.3 Continental Fit** When continents are placed together based on their outlines, the best fit isn't along their present-day coastlines, but rather along the continental slope at a depth of about 2,000 m, where erosion would be minimal.

in eastern Greenland, Ireland, Great Britain, and Norway. When the continents are positioned next to each other as they were during the Paleozoic Era, they form an essentially continuous mountain range (see Chapters 10 and 11).

Glacial Evidence

During the Late Paleozoic Era, massive glaciers covered large continental areas of the Southern Hemisphere (see Figures 11.1 and 11.2). Evidence for this glaciation includes layers of *till* (sediments deposited by glaciers) and glacial *striations* (scratch marks) in the bedrock beneath the till (▶ Figure 3.5). Fossils and sedimentary rocks of the same age from the Northern Hemisphere, however, give no indication of glaciation.

If the continents did not move during the past, one would have to explain how glaciers moved from the oceans onto land, and how large-scale continental glaciers formed near the equator. But, if the continents are reassembled as a single landmass with South Africa located at the South Pole, the direction of movement of Late Paleozoic continental glaciers makes sense (Figure 3.5a).

Fossil Evidence

Some of the more compelling evidence for continental drift comes from the fossil record (▶ Figure 3.6). For example, fossils of the *Glossopteris* flora (which include the seed fern *Glossopteris*, as well as many other distinctive and easily identifiable plants) are found in equivalent Pennsylvanian- and Permian-age coal deposits on all five

Similarity of Rock Sequences on the Gondwana Continents Sequences of marine, nonmarine, and glacial rocks of Pennsylvanian (UC) to Triassic (TR) age are nearly the same on all five Gondwana continents (South America, Africa, India, Australia, and Antarctica). These continents are widely separated today and have different environments and climates ranging from tropical to polar. Thus, the rocks forming on each continent are very different. When the continents were all joined together in the past, however, the environments of adjacent continents were similar and the rocks forming in those areas were similar. The range indicated by G in each column is the age range of the *Glossopteris* flora.

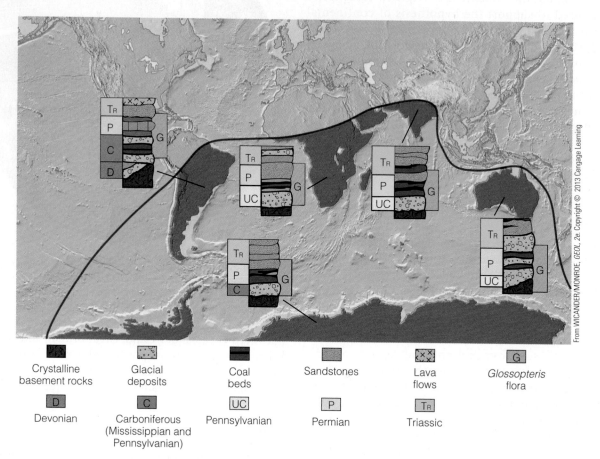

From WICANDER/MONROE. *GEOL*, 2e. Copyright © 2013 Cengage Learning

Crystalline basement rocks	Glacial deposits	Coal beds	Sandstones	Lava flows	*Glossopteris* flora
Devonian	Carboniferous (Mississippian and Pennsylvanian)	Pennsylvanian	Permian	Triassic	

Gondwana continents. However, the present-day climates of South America, Africa, India, Australia, and Antarctica range from tropical to polar, and are much too diverse to support the type of plants comprising the *Glossopteris* flora. Wegener therefore reasoned that these continents must once have been joined so that these widely separated localities were all in the same latitudinal climatic belt (Figure 3.6).

Du Toit also provided paleontologic support for continental drift by noting that fossils of the freshwater reptile *Mesosaurus* are found in Permian-age rocks in certain regions of Brazil and South Africa and nowhere else in the world (Figure 3.6). Because the physiologies of freshwater and marine animals are completely different, it is hard to imagine how a freshwater reptile could have swum across the Atlantic Ocean and found a freshwater environment nearly identical to its former habitat. Moreover, if *Mesosaurus* could have swum across the ocean, its fossil remains should be widely dispersed. It is more logical to assume that *Mesosaurus* lived in lakes in what are now adjacent areas of South America and Africa but were once united into a single continent.

Notwithstanding all of the empirical evidence presented by Wegener, and later by du Toit and others, most geologists simply refused to entertain the idea that continents might have moved in the past. The geologists were not necessarily being obstinate about accepting new ideas; rather, they found the evidence for continental drift inadequate and unconvincing. In part, this was because no one could provide a suitable mechanism to explain how continents could move over Earth's surface.

Interest in continental drift waned until studies of Earth's magnetic field and oceanographic research, conducted during the 1950s, showed that the present-day ocean basins are not as old as continents, but are geologically young features that resulted from the breakup of Pangea.

Earth's Magnetic Field

Magnetism is a physical phenomenon resulting from the spin of electrons in some solids—particularly those of iron—and moving electricity. A **magnetic field** is an area

iStockPhoto/Jan Rysavy

Courtesy of Scott Katz

(a) When the Gondwana continents are placed together so that South Africa is located at the South Pole, the glacial movements indicated by striations (red arrows) found on rock outcrops on each continent make sense. In this situation, the glacier (white area) is located in a polar climate and has moved radially outward from its thick central area toward its periphery.

(b) Glacial striations (scratch marks) on an outcrop of Permian-age bedrock exposed at Hallet's Cove, Australia, indicate the general direction of glacial movement more than 200 million years ago. As a glacier moves over a continent's surface, it grinds and scratches the underlying rock. The glacial striations that are preserved on a rock's surface thus provide evidence of the direction (red arrows) the glacier moved at that time.

▶ Figure 3.6 **Fossil Evidence Supporting Continental Drift** Some of the animals and plants whose fossils are found today on the widely separated continents of South America, Africa, India, Australia, and Antarctica. During the Late Paleozoic Era, these continents were joined together to form Gondwana, the southern landmass of Pangaea. Plants of the *Glossopteris* flora are found on all five continents, which today have widely different climates; however, during the Pennsylvanian and Permian periods, they were all located in the same general climatic belt. *Mesosaurus* is a freshwater reptile whose fossils are found only in similar nonmarine Permian-age rocks in Brazil and South Africa. *Cynognathus* and *Lystrosaurus* are land reptiles that lived during the Early Triassic Period. Fossils of *Cynognathus* are found in South America and Africa, whereas fossils of *Lystrosaurus* have been recovered

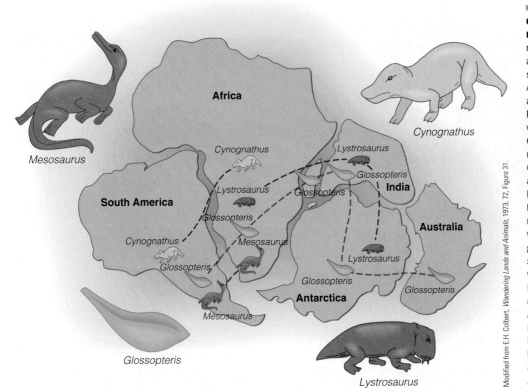

Modified from E.H. Colbert, *Wandering Lands and Animals*, 1973, 72, Figure 31.

from Africa, India, and Antarctica. It is hard to imagine how a freshwater reptile and land-dwelling reptiles could have swum across the wide oceans that presently separate these continents. It is more logical to assume that the continents were once connected.

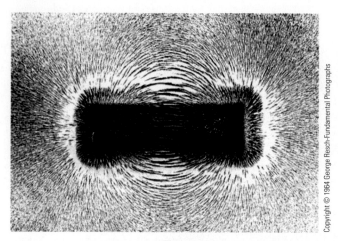

Copyright © 1964 George Resch-Fundamental Photographs

▶ **Figure 3.7 Magnetic Field** Iron filings align along the lines of magnetic force radiating from a bar magnet.

in which magnetic substances such as iron are affected by lines of magnetic force emanating from Earth (▶Figure 3.7). The magnetic field shown in Figure 3.7 is *dipolar*, meaning that it possesses two unlike magnetic poles, referred to as the *north and south poles*.

A useful analogy is to think of Earth as a giant dipole magnet in which the magnetic poles are in proximity to the geographic poles (▶Figure 3.8). This arrangement means that the strength of the magnetic field is not constant, but varies. Notice in Figure 3.8 that the lines of magnetic force around Earth parallel its surface only near the equator, just as the iron filings do around a bar magnet (Figure 3.7). As the lines of force approach the poles, they are oriented at increasingly larger angles with respect to the surface, and the strength of the magnetic field increases; it is strongest at the poles and weakest at the equator.

Another important aspect of the magnetic field is that the magnetic poles, where the lines of force leave and enter Earth, do not coincide with the geographic (rotational) poles. Currently, an 11.5-degree angle exists between the two (Figure 3.8). Studies of Earth's magnetic field show that the locations of the magnetic poles vary slightly over time, but that they still correspond closely, on average, with the locations of the geographic poles.

When magma cools, the magnetic iron-bearing minerals align themselves with Earth's magnetic field, recording both its direction and strength. The temperature at which iron-bearing minerals gain their magnetization is called the **Curie point**. As long as the rock is not subsequently heated above the Curie point, it will preserve that *remanent magnetism* (paleomagnetism). Thus, an ancient lava flow provides a record of the orientation and strength of Earth's magnetic field at the time the lava flow cooled.

▶ Figure 3.8 **Earth's Magnetic Field**

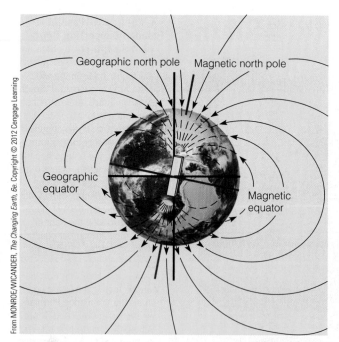

From MONROE/WICANDER, The Changing Earth, 6e. Copyright © 2012 Cengage Learning

(a) Earth's magnetic field has lines of force just like those of a bar magnet.

From MONROE/WICANDER, The Changing Earth, 6e. Copyright © 2012 Cengage Learning

(b) The strength of the magnetic field changes from the magnetic equator to the magnetic poles. This change in strength causes a dip needle (a magnetic needle that is balanced on the tip of a support so that it can freely move vertically) to be parallel to Earth's surface only at the magnetic equator, where the strength of the north and south magnetic poles are equally balanced. Its inclination, or dip, with respect to Earth's surface increases as it moves toward the magnetic poles until it is at 90 degrees, or perpendicular, to Earth's surface at the magnetic poles.

Magnetic Reversals and Seafloor Spreading

Geologists refer to Earth's present magnetic field as being *normal*—that is, with the north and south magnetic poles located approximately at the north and south geographic poles, respectively. At various times in the geologic past, however, Earth's magnetic field has completely reversed; that is, the magnetic north and south poles reversed positions, so that the magnetic north pole became the magnetic south pole, and the magnetic south pole became the magnetic north pole. During such a reversal, the magnetic field weakens until it temporarily disappears. When the magnetic field returns, the magnetic poles have reversed their position. The existence of such **magnetic reversals** was discovered by dating and determining the orientation of the remanent magnetism in lava flows on land (▶ Figure 3.9). Although the cause of magnetic reversals is still uncertain, their occurrence in the geologic record is well documented.

As a result of the oceanographic research conducted in the 1950s, Harry Hess of Princeton University proposed, in a 1962 landmark paper, the theory of **seafloor spreading** to account for continental movement. He suggested that continents do not move through oceanic crust as do ships plowing through sea ice, but rather that the continents and oceanic crust move together as a single unit. Thus, the theory of seafloor spreading answered a major objection of the opponents of continental drift—namely, how could continents move through oceanic crust? As we now know, continents move with the oceanic crust as part of a lithospheric system.

As a mechanism to drive this system, Hess revived the idea (first proposed in the late 1920s by the British geologist Arthur Holmes) of a heat transfer system—or **thermal convection**

(a)
South magnetic pole (normal position)   North magnetic pole (normal position)

(b)
North magnetic pole (reversed)   South magnetic pole (reversed)

(c)
South magnetic pole (normal)   North magnetic pole (normal)

(d)
North magnetic pole (reversed)   South magnetic pole (reversed)

Reprinted with permission from A. Cox, *Science*, Vol. 163, pp. 237–245, January 17, 1969. Copyright © 1969 American Association for the Advancement of Science.

▶ Figure 3.9 **Magnetic Reversals** During the time period shown (a–d), volcanic eruptions produced a succession of overlapping lava flows. At the time of these volcanic eruptions, Earth's magnetic field completely reversed—that is, the north magnetic pole moved to the south geographic pole, and the south magnetic pole moved to the north geographic pole. Thus, the end of the needle of a magnetic compass that today would point to the North Pole would point to the South Pole if the magnetic field should again reverse. We know that Earth's magnetic field has reversed numerous times in the past because when lava flows cool below the Curie point, magnetic minerals within the flow orient themselves parallel to the magnetic field at the time. They thus record whether the magnetic field was normal or reversed at that time. The white arrows in this diagram show the direction of the north magnetic pole for each individual lava flow, thus confirming that Earth's magnetic field has reversed during the past.

cells—within the mantle as a mechanism to move the plates. According to Hess, hot magma rises from the mantle,

Modified from Kious and Tiling, USGS and Hyndman & Hyndman, *Natural Hazards and Disasters*, Brooks/Cole, 2006, p. 15, Fig. 2.6b.

▶ **Figure 3.10 Magnetic Anomalies and Sea-floor Spreading** The sequence of magnetic anomalies preserved within the oceanic crust is both parallel to and symmetric around oceanic ridges. Basaltic lava intruding into an oceanic ridge today and spreading laterally away from the ridge records Earth's current magnetic field or polarity (considered by convention to be normal). Basaltic intrusions 3, 9, and 15 million years ago record Earth's reversed magnetic field at those times. This schematic diagram shows how the solidified basalt moves away from the oceanic ridge (or spreading ridge), carrying with it the magnetic anomalies that are preserved in the oceanic crust. Magnetic anomalies are magnetic readings that are either higher (positive magnetic anomalies) or lower (negative magnetic anomalies) than Earth's current magnetic field strength. The magnetic anomalies are recorded by a magnetometer, which measures the strength of the magnetic field.

intrudes along fractures defining oceanic ridges, and thus forms new crust. Cold crust is subducted back into the mantle at oceanic trenches, where it is heated and recycled, thus completing a thermal convection cell (see Figure 1.8).

How could Hess's hypothesis be confirmed? Magnetic surveys of the oceanic crust revealed a pattern of striped **magnetic anomalies** (deviations from the average strength of Earth's present-day magnetic field) in the rocks that are both parallel to and symmetric around the oceanic ridges (▶Figure 3.10). A positive magnetic anomaly results when Earth's magnetic field at the time of oceanic crust formation along an oceanic ridge summit was the same as today, thus yielding a stronger than normal (positive) magnetic signal. A negative magnetic anomaly results when Earth's magnetic field at the time of oceanic crust formation was reversed, therefore yielding a weaker than normal (negative) magnetic signal.

Thus, as new oceanic crust forms at oceanic ridge summits and records Earth's magnetic field at the time, the previously formed crust moves laterally away from the ridge. These magnetic stripes therefore represent times of normal and reversed polarity at oceanic ridges (where upwelling magma forms new oceanic crust), and they conclusively confirm Hess's theory of seafloor spreading.

One of the consequences of the seafloor spreading theory is its confirmation that ocean basins are geologically young features whose openings and closings are partially responsible for continental movement (▶Figure 3.11). Radiometric dating reveals that the oldest oceanic crust is somewhat less than 180 million years old, whereas the oldest continental crust is 3.96 billion years old. Although

geologists do not universally accept the idea of thermal convection cells as the sole driving mechanism for plate movement, most accept that plates are created at oceanic ridges and destroyed at deep-sea trenches, regardless of the driving mechanism involved.

Plate Tectonics: A Unifying Theory

Plate tectonic theory is based on a simple model of Earth. The rigid lithosphere, composed of both oceanic and continental crust, as well as the underlying upper mantle, consists of numerous variable-sized pieces called *plates* (▶Figure 3.12). There are seven major plates (Eurasian, Indian-Australian, Antarctic, North American, South American, Pacific, and African), and numerous smaller ones ranging from only a few tens to several hundreds of kilometers in width (for example, the Nazca plate). Plates also vary in thickness; those composed of upper mantle and continental crust are as much as 250 km thick, whereas those of upper mantle and oceanic crust are up to 100 km thick.

The lithosphere overlies the hotter and weaker semiplastic asthenosphere. It is thought that movement resulting from some type of heat-transfer system within the asthenosphere causes the overlying plates to move. As plates move over the asthenosphere, they separate, mostly at oceanic ridges; in other areas, such as at oceanic trenches, they collide and are subducted back into the mantle.

▶ **Figure 3.11 Age of the World's Ocean Basins** The age of the world's ocean basins has been determined from magnetic anomalies preserved in oceanic crust. The red colors adjacent to the oceanic ridges are the youngest oceanic crust. Moving laterally away from the ridges, the red colors grade to yellow at 48 million years ago, to green at 68 million years ago, and to dark blue some 155 million years ago. The darkest blue color is adjacent to the continental margins and is just somewhat less than 180 million years old.

An easy way to visualize plate movement is to think of a conveyor belt moving luggage from an airplane's cargo hold to a baggage cart. The conveyor belt represents convection currents within the mantle, and the luggage represents Earth's lithospheric plates. The luggage is moved along by the conveyor belt until it is dumped into the baggage cart in the same way that plates are moved by convection cells until they are subducted into Earth's interior.

Although this analogy allows you to visualize how the mechanism of plate movement takes place, remember that it is limited as follows: whereas all of the luggage is dumped into the cart, plates consist of both continental and oceanic lithosphere, which have different densities, and only the oceanic lithosphere is subducted into Earth's interior.

Plate tectonic theory is presently accepted because the evidence for it is overwhelming, and because it ties together many seemingly unrelated geologic features and events and shows how they are interrelated. Consequently, geologists now view many geologic processes from the global perspective of plate tectonic theory, in which interactions along plate margins are responsible for such phenomena as mountain building, earthquakes, and volcanism.

The Three Types of Plate Boundaries

Because it appears that plate tectonics has operated since at least the Proterozoic Eon (see Chapter 9), it is important that we understand how plates move and interact with each other and how ancient plate boundaries are recognized. After all, the movement of plates has profoundly affected the geologic and biologic history of this planet.

Geologists recognize three major types of plate boundaries: *divergent, convergent,* and *transform* (Table 3.1 on page 48). Along these boundaries, new plates are formed, are consumed, or slide laterally past one another. To understand the implications of plate interactions in terms of how they have affected Earth's history, geologists must study present-day plate boundaries (see Perspective).

Divergent Boundaries

Divergent plate boundaries, or *spreading ridges,* occur where plates are separating and new oceanic lithosphere is forming. Thus, divergent boundaries are places where the crust is extended, thinned, and fractured as magma—derived from the partial melting of the mantle—rises to the surface. The magma is almost entirely basaltic and intrudes into vertical fractures to form dikes and pillow lava flows (▶ Figure 3.13). As successive injections of magma cool and solidify, they form new oceanic crust and record the intensity and orientation of Earth's magnetic field (Figure 3.10).

Divergent boundaries most commonly occur along the crests of oceanic ridges (for example, the Mid-Atlantic Ridge). Oceanic ridges are thus characterized by rugged topography with high relief resulting from displacement of rocks along large fractures, shallow-depth earthquakes, high heat flow, and basaltic flows or pillow lavas.

Divergent boundaries are also present under continents during the early stages of continental breakup. When

Perspective

Plate Boundaries, Earthquakes, and Tsunami

In the afternoon of January 12, 2010, a magnitude 7.0 earthquake struck the island nation of Haiti. According to official estimates, 222,570 people died, at least 300,000 were injured, and more than 285,000 residences and businesses were destroyed or severely damaged (Figure 1). Widespread devastation occurred in the capital of Port-au-Prince and elsewhere throughout the region, exacerbated by an almost total collapse of the vital infrastructure needed to respond to such a disaster, including medical, transportation, and communications systems.

A little more than a year later, on February 23, 2011, a 6.3-magnitude earthquake struck Christchurch, New Zealand. In addition to the tremendous damage to buildings and infrastructure, 181 people were killed in this earthquake, making it the second deadliest earthquake to strike New Zealand (Figure 2).

Less than a month following the New Zealand earthquake, a 9.0-magnitude earthquake and tsunami struck Japan on March 11, 2011, causing more than 13,000 deaths and tremendous property damage, including severe damage to a nuclear power plant in the northeastern part of the island. Within minutes after the earthquake, walls of water, some as high as 37 m, inundated low-lying areas along the Japanese coast and extended as far as 10 km inland, sweeping aside boats, vehicles, and structures, as if they were toys (Figure 3).

What do these tragic events have in common besides the terrible loss of lives and massive property damage? They are all associated with Earth's plates and the interactions taking place along their boundaries. For example, the Haiti earthquake in 2010 resulted from movement along a strike-slip fault zone (a *strike-slip fault* is one involving horizontal movement of rock) that is part of the boundary separating the southwestward-moving North American plate from the eastward-moving Caribbean plate to the

Figure 1 People walking over the rubble of destroyed buildings in downtown Port-au-Prince, Haiti, following the 7.0-magnitude earthquake that struck the island nation on January 12, 2010, and left more than 222,500 dead.

Figure 2 Some of the damage done to buildings by the violent ground shaking experienced in the heart of Christchurch, New Zealand, as a result of the 6.3-magnitude earthquake that struck February 23, 2011.

south. This earthquake also caused a localized tsunami that struck a small fishing village and swept at least three people out to sea.

The New Zealand earthquake also resulted from movement along a strike-slip fault, but in this case there was an additional component of vertical movement. Some geologists think that this particular earthquake was actually an aftershock of the 7.1-magnitude Canterbury earthquake that struck the south island of New Zealand on September 4, 2010. Although both earthquakes occurred on previously unknown faults, the faults and subsequent earthquakes are both associated with the regional plate boundary deformation taking place between the Pacific and Australian plates in the central South Island of New Zealand (Figure 3.12).

Lastly, the 2011 Japanese earthquake and resulting tsunami were caused by the Pacific plate subducting beneath the North American plate in the region of northern Japan. Here, the Pacific plate is moving approximately westward with respect to the North American and Eurasian plates and plunges beneath Japan along the Japan Trench, which marks the surface expression of an oceanic–oceanic convergent plate boundary (Figure 3.16a).

Figure 3 Aerial view of Minamisanriku, Miyagi Prefecture, in northeastern Japan, showing houses clogged with debris and a large boat that had been swept inland as a result of the tsunami.

magma wells up beneath a continent, the crust is initially elevated, stretched, and thinned, producing fractures, faults, rift valleys, and volcanic activity (▶Figure 3.14a). As magma intrudes into faults and fractures, it solidifies or flows out onto the surface as lava flows, the latter often covering the rift valley floor (Figure 3.14b). The East

Ridge axis   Subduction zone   Hot spot   Direction of movement

▶ **Figure 3.12 Earth's Plates** A world map showing Earth's plates, their boundaries, their relative motion, average rates of movement in centimeters per year, and hot spots.

Type	Example	Landforms	Volcanism
Divergent			
Oceanic	Mid-Atlantic Ridge	Mid-oceanic ridge with axial rift valley	Basalt
Continental	East African Rift Valley	Rift valley	Basalt and rhyolite, no andesite
Convergent			
Oceanic–oceanic	Aleutian Islands	Volcanic island arc, offshore oceanic trench	Andesite
Oceanic–continental	Andes	Offshore oceanic trench, volcanic mountain chain, mountain belt	Andesite
Continental–continental	Himalayas	Mountain belt	Minor
Transform	San Andreas Fault	Fault valley	Minor

TABLE 3.1 Types of Plate Boundaries

African Rift Valley is an excellent example of continental breakup at this stage (▶ Figure 3.15).

As spreading proceeds, some rift valleys continue to lengthen and deepen until the continental crust eventually breaks and a narrow linear sea is formed, separating two continental blocks (Figure 3.14c). The Red Sea, separating the Arabian Peninsula from Africa (Figure 3.15), and the Gulf of California, which separates Baja California from mainland Mexico, are good examples of this more advanced stage of rifting.

As a newly created narrow sea continues to enlarge, it may eventually become an expansive ocean basin such as the Atlantic Ocean basin is today, separating North and South America from Europe and Africa by thousands of kilometers (Figure 3.14d). The Mid-Atlantic Ridge is the boundary between these diverging plates; the American plates are moving westward, and the Eurasian and African plates are moving eastward.

An Example of Ancient Rifting

What features in the geologic record can geologists use to recognize ancient rifting? Associated with regions of continental rifting are faults, dikes, sills, lava flows, and thick sedimentary sequences within rift valleys, all features that are preserved in the geologic record. The Triassic fault-block basins of the eastern United States are a good example of ancient continental rifting (see Figure 14.7). These fault-block basins mark the zone of rifting that occurred when North America split apart from Africa (see Chapter 14). The basins contain thousands of meters of continental sediment and are riddled with dikes and sills.

Pillow lavas (Figure 3.13), in association with deep-sea sediment, are also evidence of ancient rifting. The presence of pillow lavas marks the formation of a spreading ridge in a narrow linear sea (Figures 3.14 and 3.15). A narrow linear sea forms when the continental crust in the rift valley finally breaks apart and the area is flooded with seawater. Magma, intruding into the sea along this newly formed spreading ridge, solidifies as pillow lavas, which are preserved in the geologic record, along with the sediment that was deposited on them.

Image courtesy of Submarine Ring of Fire 2002 Exploration, NOAA-OE

▶ **Figure 3.13** **Pillow Lavas** Pillow lavas forming along the Mid-Atlantic Ridge. Their distinctive bulbous shape is the result of underwater eruption.

Convergent Boundaries

Whereas new crust forms at divergent plate boundaries, older crust must be destroyed and recycled in order for the entire surface area of Earth to remain the same. Otherwise, we would have an expanding Earth. Such plate destruction occurs at **convergent plate boundaries**, where two plates collide and the leading edge of one plate descends beneath the margin of the other plate by a process known as *subduction*. As the subducting plate moves into the asthenosphere, it is heated and eventually incorporated into the mantle.

Deformation, volcanism, mountain building, metamorphism, earthquake activity, and deposits of valuable minerals characterize convergent boundaries. Three types of convergent plate boundaries are recognized: *oceanic–oceanic, oceanic–continental, and continental–continental.*

Oceanic–Oceanic Boundaries When two oceanic plates converge, one is subducted beneath the other along an **oceanic–oceanic plate boundary** (▶ Figure 3.16a). The subducting plate bends downward to form the outer wall of an oceanic trench. A *subduction complex*, composed of wedge-shaped slices of highly folded and faulted

Volcanic activity

Magma

Continental crust

(a) Rising magma beneath a continent pushes the crust up, producing numerous fractures, faults, rift valleys, and volcanic activity.

Rift valley

(b) As the crust is stretched and thinned, rift valleys develop, and lava flows onto the valley floors, such as seen today in the East African Rift Valley.

Coastal mountain range

Narrow fault-bounded sea

(c) Continued spreading further separates the continent until it splits apart and a narrow seaway develops. The Red Sea, which separates the Arabian Peninsula from Africa, is a good example of this stage of development.

Continental "seaboard" (coastal mountains gone)

Wide ocean

(d) As spreading continues, an oceanic ridge system forms, and an ocean basin develops and grows. The Mid-Atlantic Ridge illustrates this stage in a divergent plate boundary's history.

© 2013 Cengage Learning

▶ Figure 3.15 **East African Rift Valley and the Red Sea—Present-Day Examples of Divergent Plate Boundaries** The East African Rift Valley and the Red Sea represent different stages in the history of a divergent plate boundary. The East African Rift Valley is being formed by the separation of eastern Africa from the rest of the continent along a divergent plate boundary. The Red Sea represents a more advanced stage of rifting, in which two continental blocks (Africa and the Arabian Peninsula) are separated by a narrow sea.

marine sediments and oceanic lithosphere scraped off the descending plate, forms along the inner wall of the oceanic trench. As the subducting plate descends into the mantle, it is heated and partially melted, generating magma that is commonly of andesitic composition. This magma is less dense than the surrounding mantle rocks and rises to the surface of the nonsubducted plate to form a curved chain of volcanic islands called a *volcanic island arc* (any plane

intersecting a sphere makes an arc). This arc is nearly parallel to the oceanic trench and is separated from it by a distance of up to several hundred kilometers—the distance depending on the angle of dip of the subducting plate (Figure 3.16a).

In those areas where the rate of subduction is faster than the forward movement of the overriding plate, the lithosphere on the landward side of the volcanic island arc may be subjected to tensional stress and stretched and thinned, resulting in the formation of a *back-arc basin*. This back-arc basin may grow by spreading if magma breaks through the thin crust and forms new oceanic crust (Figure 3.16a). A good example of a back-arc basin associated with an oceanic–oceanic plate boundary is the Sea of Japan between the Asian continent and the islands of Japan (see Figure 16.7).

Most present-day active volcanic island arcs are in the Pacific Ocean basin and include the Aleutian Islands, the Kermadec–Tonga arc, and the Japanese (Figure 3.16a) and Philippine Islands. The Scotia and Antillean (Caribbean) island arcs are present in the Atlantic Ocean basin.

Oceanic–Continental Boundaries When an oceanic and a continental plate converge, the denser oceanic plate is subducted under the continental plate along an **oceanic–continental plate boundary** (Figure 3.16b). Just as at oceanic–oceanic plate boundaries, the descending oceanic plate forms the outer wall of an oceanic trench.

The magma generated by subduction rises beneath the continent and either crystallizes as large igneous bodies (called *plutons*) before reaching the surface, or erupts at the surface to produce a chain of andesitic volcanoes, also called a *volcanic arc*.

An excellent example of an oceanic–continental plate boundary is the Pacific coast of South America where the oceanic Nazca plate is currently being subducted under South America (Figure 3.16b; see also Figure 16.8). The Peru-Chile Trench marks the site of subduction, and the Andes Mountains are the resulting volcanic mountain chain on the nonsubducting plate. This particular example demonstrates the effect that plate tectonics has on our lives. For instance, earthquakes are commonly associated with subduction zones, and the western side of South America is the site of frequent and devastating earthquakes.

Continental–Continental Boundaries Two continents approaching each other are initially separated by an ocean floor that is being subducted under one continent. The edge of that continent displays the features characteristic of oceanic–continental convergence along an active continental plate margin. As the ocean floor continues to be subducted, the two continents come closer together until they eventually collide. Because continental lithosphere, which consists of continental crust and the upper mantle, is less dense than oceanic lithosphere (oceanic crust and upper mantle), it cannot sink into the asthenosphere. Although one continent may partly slide under the other,

▶ Figure 3.16 **Three Types of Convergent Plate Boundaries**

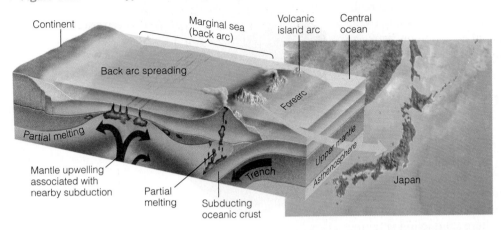

Continent
Marginal sea (back arc)
Volcanic island arc
Central ocean
Back arc spreading
Forearc
Partial melting
Upper mantle
Asthenosphere
Japan
Mantle upwelling associated with nearby subduction
Partial melting
Trench
Subducting oceanic crust

(a) Oceanic–oceanic plate boundary. An oceanic trench forms where one oceanic plate is subducted beneath another. On the nonsubducted plate, a volcanic island arc forms from the rising magma generated from the subducting plate. The Japanese Islands are a volcanic island arc resulting from the subduction of one oceanic plate beneath another oceanic plate.

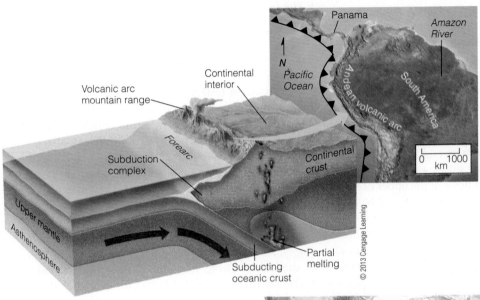

Volcanic arc mountain range
Continental interior
Panama
Amazon River
N
Pacific Ocean
Andean volcanic arc
South America
Forearc
Subduction complex
Continental crust
Upper mantle
Asthenosphere
0   1000 km
Subducting oceanic crust
Partial melting

(b) Oceanic–continental plate boundary. When an oceanic plate is subducted beneath a continental plate, an andesitic volcanic mountain range is formed on the continental plate as a result of rising magma. The Andes Mountains in Peru are one of the best examples of continuing mountain building at an oceanic–continental convergent plate boundary.

© 2013 Cengage Learning

Secondary rifting caused by collision
Mountain range
Himalaya Mts.
India
Indian Ocean
Rift faulting, Tibetan plateau
N
Continental crust
Continental crust
Thrust zone
Upper mantle
Asthenosphere
0   1000 km
Oceanic crust
Little or no partial melting

(c) Continental–continental plate boundary. When two continental plates converge, neither is subducted because of their great thickness and low and equal densities. As the two continental plates collide, a mountain range is formed in the interior of a new and larger continent. The Himalayas in central Asia resulted from the collision between India and Asia approximately 40 to 50 million years ago.

it cannot be pulled or pushed down into a subduction zone (Figure 3.16c).

When two continents collide, they are welded together along a zone marking the former site of subduction. At this **continental–continental plate boundary**, an interior mountain belt is formed consisting of deformed sediments and sedimentary rocks, igneous intrusions, metamorphic rocks, and fragments of oceanic crust. In addition, the entire region is subjected to numerous earthquakes. The Himalayas in central Asia, the world's youngest and highest mountain system, resulted from the collision between India and Asia that began about 40 to 50 million years ago and is still continuing (Figure 3.16c; see also Figure 16.6).

Recognizing Ancient Convergent Plate Boundaries

How can former subduction zones be recognized in the geologic record? Igneous rocks provide one such clue. The magma that erupted at the surface, forming island arc volcanoes and continental volcanoes, is of andesitic composition. Another clue is the zone of intensely deformed rocks between the deep-sea trench where subduction is taking place and the area of igneous activity. Here, sediments and submarine rocks are folded, faulted, and metamorphosed into a chaotic mixture of rocks termed a *mélange*.

During subduction, pieces of oceanic lithosphere are sometimes incorporated into the mélange and accreted onto the edge of the continent. Such slices of oceanic crust and upper mantle are called *ophiolites* (▶ Figure 3.17). Detailed studies reveal that an ideal ophiolite consists of a layer of deep-sea sediments that includes graywackes (poorly sorted sandstones containing abundant feldspars and rock fragments, usually in a clay-rich matrix), black shales, and cherts, deposited on pillow lava and sheet lava flows of the upper oceanic crust. A sheeted dike complex consisting of vertical basaltic dikes, followed by massive gabbro, and layered gabbro, form the rest of the oceanic crust. Beneath the gabbro is peridotite, which probably represents the upper mantle. The presence of ophiolites in an outcrop or drilling core is a key feature in recognizing plate convergence along a subduction zone.

Elongate belts of folded and faulted marine sedimentary rocks, andesites, and ophiolites are found in the Appalachians, Alps, Himalayas, and Andes. The combination of such features is significant evidence that these mountain ranges resulted from deformation along convergent plate boundaries.

Transform Boundaries

The third type of plate boundary is a **transform plate boundary**. These mostly occur along fractures in the seafloor known as *transform faults* where plates slide laterally past one another, roughly parallel to the direction of plate movement. Although lithosphere is neither created nor destroyed along a transform boundary, the movement between plates results in a zone of intensely shattered rock and numerous shallow-depth earthquakes.

© 2013 Cengage Learning

▶ Figure 3.17 **Ophiolites** Ophiolites are sequences of rock on land consisting of deep-sea sediments, oceanic crust, and upper mantle. Ophiolites are one feature used to recognize ancient convergent plate boundaries.

- Deep-sea sediments
- Pillow lava and sheet lava
- Sheeted dikes
- Massive gabbro
- Layered gabbro
- Upper mantle peridotite

Transform faults "transform," or change, one type of motion between plates into another type of motion. Most commonly, transform faults connect two oceanic ridge segments (▶ Figure 3.18), but they can also connect ridges to trenches, and trenches to trenches. Although the majority of transform faults are in oceanic crust and are marked by distinct fracture zones, they may also extend into continents.

One of the best-known transform faults is the San Andreas Fault in California. It separates the Pacific plate from the North American plate and connects spreading ridges in the Gulf of California with the Juan de Fuca and Pacific plates off the coast of northern California (▶ Figure 3.19).

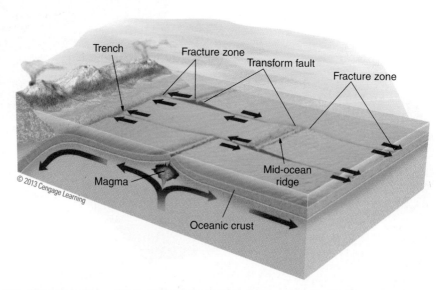

© 2013 Cengage Learning

Trench · Fracture zone · Transform fault · Fracture zone · Mid-ocean ridge · Magma · Oceanic crust

▶ Figure 3.18 **Transform Plate Boundaries** Horizontal movement between plates occurs along transform faults. Extensions of transform faults on the seafloor form fracture zones. Most transform faults connect two oceanic ridge segments. Note that relative motion between the plates only occurs between the two ridges.

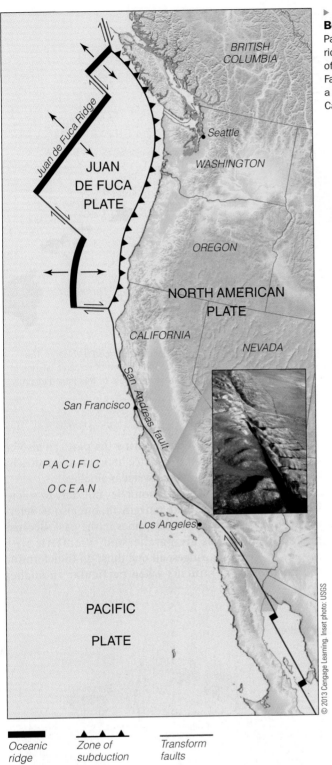

▶ Figure 3.19 **The San Andreas Fault—A Transform Plate Boundary** The San Andreas Fault is a transform fault separating the Pacific plate from the North American plate. It connects the spreading ridges in the Gulf of California with the Juan de Fuca and Pacific plates off the coast of northern California. Movement along the San Andreas Fault has caused numerous earthquakes. The insert photograph shows a segment of the San Andreas Fault as it cuts through the Carrizo Plain, California.

▬	▲▲▲	───
Oceanic ridge	*Zone of subduction*	*Transform faults*

Many of the earthquakes that affect California are the result of movement along this fault (see Chapter 16).

Unfortunately, transform faults generally do not leave any characteristic or diagnostic features except for the obvious displacement of the rocks with which they are associated. This displacement is usually large, on the order of tens to hundreds of kilometers. Such large displacements in ancient rocks can sometimes be related to transform fault systems.

Hot Spots and Mantle Plumes

Before leaving the topic of plate boundaries, we should briefly mention **hot spots**, which are intraplate features found beneath both oceanic and continental plates. Hot spots are locations on Earth's surface where stationary columns of magma, originating deep within the mantle (*mantle plume*), have slowly risen to the surface and formed volcanoes (▶Figure 3.20).

Because the mantle plumes apparently remain stationary (although some evidence suggests that they might not) within the mantle while the plates move over them, the resulting hot spots leave a trail of extinct and progressively older volcanoes (*aseismic ridges*) that record the movement of the plate. Some examples of aseismic ridges and hot spots are the Emperor Seamount–Hawaiian Island chain and Yellowstone National Park in Wyoming.

Mantle plumes and hot spots help geologists explain some of the geologic activity occurring within plates as opposed to activity occurring at or near plate boundaries. In addition, if mantle plumes are essentially fixed with respect to Earth's rotational axis, they can be used to determine not only the direction of plate movement, but also the rate of movement. They can also provide reference points for determining paleolatitude, an important tool when reconstructing the location of continents in the geologic past.

How Are Plate Movement and Motion Determined?

How fast and in what direction are Earth's plates moving? Do they all move at the same rate? Rates of plate movement can be calculated in several ways. The least accurate method is to determine the age of the sediments immediately above any portion of the oceanic crust and then divide the distance from the spreading ridge by that age. Such calculations give an average rate of movement.

A more accurate method of determining both the average rate of movement and relative motion is by dating the magnetic anomalies in the crust of the seafloor (Figure 3.11). The distance from an oceanic ridge axis to any magnetic anomaly indicates the width of new seafloor that formed during that time interval. For example, if the distance between the present-day Mid-Atlantic Ridge

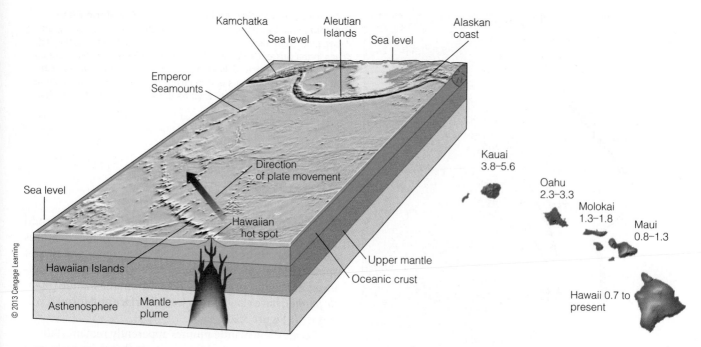

Kamchatka   Aleutian
Sea level   Islands   Sea level   Alaskan coast

Emperor Seamounts

Direction of plate movement

Sea level

Hawaiian hot spot

Upper mantle

Hawaiian Islands

Oceanic crust

Asthenosphere   Mantle plume

Kauai 3.8–5.6

Oahu 2.3–3.3

Molokai 1.3–1.8

Maui 0.8–1.3

Hawaii 0.7 to present

© 2013 Cengage Learning

▶ **Figure 3.20 Hot Spots** A hot spot is the location where a stationary mantle plume has risen to the surface and formed a volcano. The Emperor Seamount–Hawaiian Island chain formed as a result of the Pacific plate moving over a mantle plume, and the line of volcanic islands in this chain traces the direction of plate movement. The island of Hawaii and Loihi Seamount are the only current hot spots of this island chain. The numbers indicate the age of the islands in millions of years.

and Anomaly 31 is 2,010 km, and Anomaly 31 formed 67 million years ago (▶ Figure 3.21), then the average rate of movement during the past 67 million years has been 3 cm per year (2,010 km, which equals 201 million cm divided by 67 million years; 201,000,000 cm/67,000,000 years = 3 cm/year). Thus, for a given interval of time, the wider the strip of seafloor, the faster the plate has moved. In this way, not only can the present average rate of movement and relative motion be determined (Figure 3.12), but

the average rate of movement during the past can also be calculated by dividing the distance between anomalies by the amount of time elapsed between anomalies.

Geologists use magnetic anomalies not only to calculate the average rate of plate movement, but also to determine plate positions at various times in the past. Because magnetic anomalies are parallel and symmetric with respect to spreading ridges, all one must do to determine the position of continents when particular anomalies

▶ **Figure 3.21 Reconstructing Plate Positions Using Magnetic Anomalies**

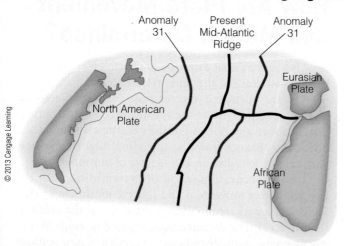

Anomaly 31   Present Mid-Atlantic Ridge   Anomaly 31

North American Plate

Eurasian Plate

African Plate

© 2013 Cengage Learning

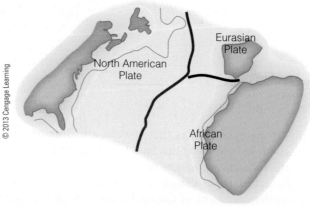

Eurasian Plate

North American Plate

African Plate

© 2013 Cengage Learning

(b) The Atlantic Ocean 67 million years ago. Anomaly 31 marks the plate boundary 67 million years ago. By moving the anomalies back together, along with the plates they are on, we can reconstruct the former positions of the continents.

(a) The present North Atlantic, showing the Mid-Atlantic Ridge and Magnetic Anomaly 31, which formed 67 million years ago.

formed is to move the anomalies back to the spreading ridge, which will also move the continents with them (Figure 3.21). Because subduction destroys oceanic crust and the magnetic record that it carries, we therefore have an excellent record of plate movements since the breakup of Pangaea, but not as good an understanding of plate movement before that time.

The average rate of movement, as well as the relative motion between any two plates, can also be determined by satellite-laser ranging techniques. Laser beams from a station on one plate are bounced off a satellite (in geosynchronous orbit) and returned to a station on a different plate. As the plates move away from each other, the laser beam takes more time to go from the sending station to the stationary satellite and back to the receiving station. This difference in elapsed time is used to calculate the rate of movement and the relative motion between plates.

Plate motions derived from magnetic reversals and satellite-laser ranging techniques give only the relative motion of one plate with respect to another. Hot spots, however, allow geologists to determine absolute motion because they provide an apparently fixed reference point from which the rate and direction of plate movement can be measured. The previously mentioned Emperor Seamount–Hawaiian Island chain (Figure 3.20) formed as a result of movement over a hot spot. Thus, the line of the volcanic islands traces the direction of plate movement, and dating the volcanoes enables geologists to determine the rate of movement.

▶ **Figure 3.22 Thermal Convection Cells as the Driving Force of Plate Movement** A cutaway view of Earth shows that the lithosphere glides horizontally across the asthenosphere. Heat from the core, supplemented by heat produced from radioactive decay, drives huge mantle convection cells that move the lithosphere. Spreading ridges mark the location of ascending limbs of the convection cells, and oceanic trenches are the surface expression of subduction, where plates descend into Earth's interior.

The Driving Mechanism of Plate Tectonics

A major obstacle to the acceptance of the continental drift hypothesis was the lack of a driving mechanism to explain continental movement. When it was shown that continents and ocean floors moved together, not separately, and that new crust formed at spreading ridges by rising magma, most geologists accepted some type of convective heat system (convection cells) as the basic process responsible for plate motion. However, the question of what exactly drives the plates remains.

Most of the heat from Earth's interior results from the decay of radioactive elements, such as uranium (see Chapter 4), in the core and lower mantle. The most efficient way for this heat to escape Earth's interior is through some type of slow convection system. Heat from the core, supplemented by heat generated from radioactive decay, thus drives large mantle convection cells (▶Figure 3.22). In this manner, hot rock from the interior rises toward the surface, loses heat to the overlying lithosphere, becomes denser as it cools, and then sinks back into the interior, where it is heated, and the process repeats itself. This type of convective heat system is analogous to a pot of stew cooking on a stove (▶Figure 3.23).

In this mantle convection cell model, spreading ridges mark the ascending limbs of adjacent convection cells, and trenches are present where convection cells descend back into Earth's interior. The convection cells therefore determine the location of spreading ridges and trenches, with the lithosphere lying above the thermal convection cells.

▶ **Figure 3.23** **Convection in a Pot of Stew** Heat from the stove is applied to the base of the stew pot, causing the stew to heat up. As heat rises through the stew, pieces of the stew are carried to the surface, where the heat is dissipated and the pieces of stew cool, then sink back to the bottom of the pot. The bubbling seen at the surface of the stew is the result of convection cells churning the stew. In the same manner, heat from the decay of radioactive elements produces convection cells within Earth's interior.

Thus, each plate corresponds to a single convection cell that moves as a result of the convective movement of the cell itself (Figure 3.22).

Although most geologists agree that Earth's internal heat plays an important role in plate movement, two other processes, referred to as *slab-pull* and *ridge-push*, might also help facilitate the movement of plates (▶ Figure 3.24). Both mechanisms are gravity driven but still depend on thermal differences within Earth.

In slab-pull, the subducting cold slab of lithosphere, being denser than the surrounding warmer asthenosphere, pulls the rest of the plate along as it descends into the asthenosphere (Figure 3.24). As the lithosphere moves

downward, there is a corresponding upward flow back into the spreading ridge.

Operating in conjunction with slab-pull is the ridge-push mechanism. As a result of rising magma, the oceanic ridges are higher than the surrounding oceanic crust. It is thought that gravity pushes the oceanic lithosphere away from the higher spreading ridges and toward the trenches, where it is subducted back into Earth's interior (Figure 3.24).

Recent research suggests another possible driver of plate movement, namely, mantle plumes. Recall that mantle plumes are somewhat stationary columns of magma originating deep within the mantle and slowly

▶ **Figure 3.24** **Plate Movement Resulting from Gravity-Driven Mechanisms** Plate movement is also thought to occur, at least partially, from gravity-driven slab-pull or ridge-push mechanisms. In slab-pull, the edge of the subducting plate descends into the interior, and the rest of the plate is pulled downward. In ridge-push, rising magma pushes the oceanic ridges higher than the rest of the oceanic crust. Gravity thus pushes the oceanic lithosphere away from the ridges and toward the trenches.

rising to the surface, where they form either a subsurface mushroom-shaped plume head or erupt at the surface as a volcano (hot spot) (Figure 3.20). According to this research, the subsurface plume head might aid plate movement by spreading outward faster than the overlying plates, thus dragging them along, or the hot plume head might reduce the friction beneath the plates, allowing for easier subduction in slab-pull.

Currently, geologists are fairly certain that some type of convective system is involved in plate movement, but the extent to which other mechanisms, such as slab-pull, ridge-push, and mantle plumes are involved is still unresolved. However, the fact that plates have moved in the past and are still moving today has been proven beyond a doubt. And although a comprehensive theory of plate movement has not yet been developed, more and more of the pieces are falling into place as geologists learn more about Earth's interior.

Plate Tectonics and Mountain Building

How are plate tectonics and mountain building, also known as *orogenesis*, related? An **orogeny** is an episode of intense rock deformation, or mountain building, that results from compressive forces generated by colliding plates. As one plate is subducted under another plate, sedimentary and volcanic rocks are folded and faulted along the plate margin while the more deeply buried rocks are subjected to regional metamorphism. Magma generated within the mantle either rises to the surface to erupt as andesitic volcanoes or cools and crystallizes beneath the surface, forming intrusive igneous bodies. Typically, most orogenies occur along either oceanic–continental or continental–continental plate boundaries.

As we discussed earlier in this chapter, ophiolites are evidence of ancient convergent plate boundaries. The slivers of ophiolites found in the interiors of such mountain ranges as the Alps, Himalayas, and Urals mark the sites of former subduction zones. The relationship between mountain building and the opening and closing of ocean basins is called the *Wilson cycle* in honor of the Canadian geologist J. Tuzo Wilson, who first suggested that an ancient ocean had closed to form the Appalachian Mountains and then reopened and widened to form the present Atlantic Ocean.

Although much of the geology of continents can be described in terms of a succession of Wilson cycles, the movement of microplates and microcontinents such as Avalonia (see Chapter 10), and their accretion at the margin of continents, must also be considered when dealing with the tectonic history of a continent. These small, accreted lithospheric blocks, also known as *terranes*, differ completely in their fossil content, stratigraphy, structural trends, and paleomagnetic properties from the rocks of the surrounding mountain system. To date, most terranes in North America have been identified in mountains of the Pacific coast region, but a number of such terranes are also present in other mountain systems, such as the Appalachians.

Plate Tectonics and the Distribution of Life

Plate tectonic theory is as revolutionary and far-reaching in its implications for geology as the theory of evolution was for biology when it was proposed. Interestingly, it was the fossil evidence that convinced Wegener, Suess, and du Toit, as well as many other geologists, of the correctness of the continental drift hypothesis. Together, the theories of plate tectonics and evolution have changed the way we view our planet, and we should not be surprised at the intimate association between them. Although the relationship between plate tectonic processes and the evolution of life is incredibly complex, paleontologic data provide convincing evidence of the influence of plate movement on the distribution of organisms.

The present distribution of plants and animals is not random but rather is controlled largely by climate and geographic barriers. The world's biota occupy *biotic provinces*, which are regions characterized by a distinctive assemblage of plants and animals. Organisms within a province have similar ecologic requirements, and the boundaries separating provinces are, therefore, natural ecologic breaks. Climatic or geographic barriers are the most common province boundaries, and these are mostly controlled by plate movement.

Because adjacent biotic provinces usually have less than 20 percent of their species in common, global diversity is a direct reflection of the number of provinces; the more provinces there are, the greater the global diversity. When continents break up, for example, the opportunity for new biotic provinces to form increases, with a resultant increase in diversity. Just the opposite occurs when continents come together. Plate tectonics thus plays an important role in the distribution of organisms and their evolutionary history.

The marine invertebrates presently found on opposite sides of the Isthmus of Panama provide an excellent example of what happens when a once-uniform fauna is separated by the formation of a geographic barrier. Prior to the rise of this land connection between North and South America, a homogeneous population of bottom-dwelling invertebrates inhabited the shallow seas of the area. After the rise of the Isthmus of Panama by subduction of the Pacific plate approximately 5 million years ago, the original population was divided into two populations—one in the Caribbean Sea, and the other in the Pacific Ocean. In response to the changing environment, new species evolved on opposite sides of the isthmus (▸Figure 3.25).

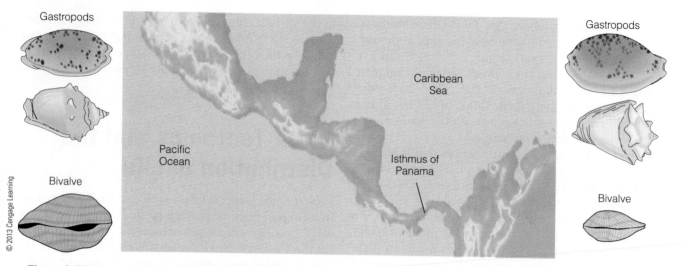

Gastropods

Bivalve

© 2013 Cengage Learning

Caribbean Sea

Pacific Ocean

Isthmus of Panama

Gastropods

Bivalve

▶ **Figure 3.25** **Plate Tectonics and the Distribution of Organisms** The Isthmus of Panama forms a barrier that divides a once-uniform fauna of molluscs. Thus, species have evolved differently on opposite sides of the isthmus, reflecting the different conditions caused by the creation of this geographic barrier. Each pair of gastropod and bivalve belong to the same genus, but are a different species.

The formation of the Isthmus of Panama also influenced the evolution of the North and South American mammalian faunas (see Figure 18.19). During most of the Cenozoic Era, South America was an island continent, and its mammalian fauna evolved in isolation from the rest of the world's faunas. When North and South America were connected by the Isthmus of Panama, migrants from North America replaced most of the indigenous South American mammals. Surprisingly, only a few South American mammal groups migrated northward.

Plate Tectonics and the Distribution of Natural Resources

In addition to being responsible for the major features of Earth's crust and influencing the distribution and evolution of the world's biota, plate movement also affects the formation and distribution of some natural resources. Consequently, geologists are using plate tectonic theory in their search for petroleum and mineral deposits and in explaining the occurrence of these natural resources.

Although significant concentrations of petroleum occur in many areas of the world, more than 50 percent of all proven reserves are in the Persian Gulf region. It should therefore not be surprising that many of the conflicts in the Middle East have shared as their underlying cause the desire to control these vast amounts of petroleum. Most people, however, are not aware of *why* there is so much oil in this region of the world. The answer lies in the paleogeography and plate movement of this region during the Mesozoic and Cenozoic eras.

During the Mesozoic Era, the Persian Gulf area was a broad marine shelf. This passive continental margin lay near the equator where countless microorganisms lived in the surface waters. The remains of these organisms accumulated with the bottom sediments and were buried, beginning the complex process of petroleum formation.

During the Cenozoic Era, convergent plate movement in the Persian Gulf area resulted in these sediments being subducted, which generated heat that broke down the organic molecules of the microorganisms buried with the sediments, leading to the generation of petroleum. The continued collisions in the area folded the rocks, creating traps for the petroleum to accumulate such that this is now a major oil-producing region.

Many metallic mineral deposits such as copper, gold, lead, silver, tin, and zinc are related to igneous and associated hydrothermal (hot water) activity. So it is not surprising that a close relationship exists between plate boundaries and the occurrence of these valuable deposits.

The magma generated by partial melting of a subducting plate rises toward the surface, and as it cools, it precipitates and concentrates various metallic ores. Many of the world's major metallic ore deposits, such as the porphyry copper deposits of western North and South America (▶ Figure 3.26a), are associated with convergent plate boundaries. The majority of the copper deposits in the Andes and southwestern United States were formed less than 60 million years ago when oceanic plates were subducted under the North and South American plates. The rising magma and associated hydrothermal fluids carried minute amounts of copper, which were originally widely disseminated but eventually became concentrated in the cracks and fractures of the surrounding andesites. These low-grade copper deposits contain from 0.2 to

▶ Figure 3.26 **Copper Deposits and Convergent Plate Boundaries**

(a) Valuable copper deposits are located along the west coasts of North and South America in association with convergent plate boundaries. The rising magma and associated hydrothermal activity resulting from subduction carried small amounts of copper, which became trapped and concentrated in the surrounding rocks through time.

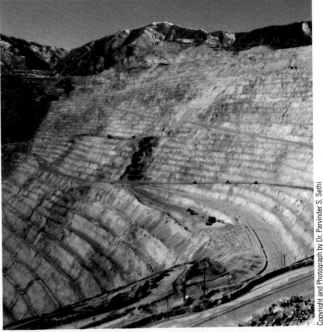

(b) Bingham Mine, near Salt Lake City, Utah, is a huge, open-pit copper mine with reserves estimated at 1.7 billion tons. More than 400,000 tons of rock are removed for processing each day. Note the small specks toward the middle of the photograph that are the 12-foot-high dump trucks!

2 percent copper and are extracted from large open-pit mines (Figure 3.26b).

It is becoming increasingly clear that if we are to keep up with the continuing demands of a global industrialized society, the application of plate tectonic theory to the origin and distribution of natural resources is essential. We will discuss natural resources and their formation as part of the geologic history of Earth in later chapters in this book.

Summary

- ▶ The concept of continental movement is not new. The earliest maps showing the similarity between the east coast of South America and the west coast of Africa provided the first evidence that continents may once have been united and subsequently separated from each other.

- ▶ Alfred Wegener is generally credited with developing the hypothesis of continental drift. He provided abundant geologic and paleontologic evidence to show that the continents were once united in one supercontinent, which he named Pangaea. Unfortunately, Wegener could not explain how the continents moved, and most geologists ignored his ideas.

- ▶ Seafloor spreading was confirmed by the discovery of magnetic anomalies in the ocean crust that were both parallel to and symmetric around ocean ridges, indicating that new oceanic crust must have formed as the seafloor was spreading. The pattern of oceanic magnetic anomalies matched the pattern of magnetic reversals already known from continental lava flows and showed that Earth's magnetic field has reversed itself numerous times during the past.

- ▶ Radiometric dating reveals the oldest oceanic crust is less than 180 million years old, whereas the oldest continental crust is 3.96 billion years old. Clearly, the ocean basins are recent geologic features.

- ▶ Plate tectonic theory became widely accepted by the 1970s because the evidence overwhelmingly supports it and because it provides geologists with a powerful theory for explaining such phenomena as volcanism, earthquake activity, mountain building, global climatic changes, the distribution of the world's biota, and the distribution of some mineral resources.

- ▶ Geologists recognize three types of plate boundaries: divergent boundaries, where plates move away from each other; convergent boundaries, where two plates collide;

and transform boundaries, where two plates slide past each other.

- ▶ Ancient plate boundaries can be recognized by their associated rock assemblages and geologic structures. For divergent boundaries, these may include rift valleys with thick sedimentary sequences and numerous dikes and sills. For convergent boundaries, ophiolites and andesitic rocks are two characteristic features. Transform plate boundaries generally do not leave any characteristic or diagnostic features in the geologic record.

- ▶ The average rate of movement and relative motion of plates can be calculated in several ways. The results of these different methods all agree and indicate that the plates move at different average velocities.

- ▶ The absolute motion of plates can be determined by the movement of plates over mantle plumes. A mantle plume is an apparently stationary column of magma that rises to the surface from deep within the mantle and forms either a subsurface mushroom-shaped plume head, or erupts at the surface as a volcano.

- ▶ Although a comprehensive theory of plate movement has yet to be developed, geologists think that some type of convective system is involved in plate movement.

- ▶ The formation of most mountain systems is the result of plate movement, typically along convergent plate boundaries.

- ▶ The relationship between plate tectonic processes and the evolution of life is complex. The distribution of plants and animals is not random but rather is controlled mostly by climate and geographic barriers, which are influenced, to a great extent, by the movement of plates.

- ▶ A close relationship exists between the formation of petroleum, as well as some mineral deposits, and plate boundaries. Furthermore, the formation and distribution of some natural resources are related to plate movements.

Important Terms

continental–continental plate boundary, p. 51
continental drift, p. 38
convergent plate boundary, p. 48
Curie point, p. 42
divergent plate boundary, p. 45
Glossopteris flora, p. 38
Gondwana, p. 38

hot spot, p. 53
Laurasia, p. 38
magnetic anomaly, p. 44
magnetic field, p. 40
magnetic reversal, p. 43
magnetism, p. 40
oceanic–continental plate boundary, p. 50
oceanic–oceanic plate boundary, p. 48

orogeny, p. 57
Pangaea, p. 38
plate tectonic theory, p. 44
seafloor spreading, p. 43
thermal convection cell, p. 43
transform fault, p. 52
transform plate boundary, p. 52

Review Questions

1. Along what type of plate boundary does subduction occur?
   a. _____ Divergent; b. _____ Transform; c. _____ Convergent; d. _____ Intraplate; e. _____ Answers a and c.
2. The man credited with developing the continental drift hypothesis is
   a. _____ Wilson; b. _____ Hess; c. _____ Vine; d. _____ Wegener; e . _____ du Toit.
3. Iron-bearing minerals in magma gain their magnetism and align themselves with the magnetic field when they cool through the
   a. _____ Curie point; b. _____ magnetic anomaly point; c. _____ thermal convection point; d. _____ hot spot point; e. _____ isostatic point.
4. Divergent plate boundaries are areas where
   a. _____ two plates are joined together; b. _____ plates move over a mantle plume; c. _____ two plates move toward each other; d. _____ two plates slide past each other; e. _____ two plates move away from each other.

5. The northern part of Pangaea, consisting of present-day North America, Greenland, Europe, and Asia (except for India), is called
   a. _____ Gondwana; b. _____ Laurentia; c. _____ Atlantis; d. _____ Laurasia; e. _____ Pacifica.
6. Magnetic surveys of the ocean basins indicate that
   a. _____ the oceanic crust is youngest adjacent to mid-oceanic ridges; b. _____ the oceanic crust is oldest adjacent to mid-oceanic ridges; c. _____ the oceanic crust is youngest adjacent to the continents; d. _____ the oceanic crust is the same age everywhere; e. _____ answers b and c.
7. The most common biotic province boundaries are
   a. _____ geographic barriers; b. _____ biologic barriers; c. _____ climatic barriers; d. _____ answers a and b; e. _____ answers a and c.
8. The driving mechanism of plate movement is thought to be largely the result of
   a. _____ isostasy; b. _____ Earth's rotation; c. _____ thermal convection cells; d. _____ magnetism; e. _____ polar wandering.

9. The Himalayas formed along what type of plate boundary? a. _____ Continental–continental convergent; b. _____ Oceanic–oceanic convergent; c. _____ Oceanic–continental convergent; d. _____ Continental–continental divergent; e. _____ Oceanic–continental transform.

10. Hot spots and aseismic ridges can be used to determine the a. _____ location of divergent plate boundaries; b. _____ absolute motion of plates; c. _____ location of magnetic anomalies in the oceanic crust; d. __✓__ relative motion of plates; e. _____ location of convergent plate boundaries.

11. Explain why such natural disasters as volcanic eruptions and earthquakes are associated with divergent and convergent plate boundaries.

12. If the movement along the San Andreas Fault, which separates the Pacific plate from the North American plate, averages 5.5 cm per year, how long will it take before Los Angeles is opposite San Francisco?

13. Why was the continental drift hypothesis proposed by Wegener rejected by so many geologists for so long?

14. Explain why global diversity increases with an increase in the number of biotic provinces. How does plate movement affect the number of biotic provinces?

15. What can hot spots tell us about the absolute direction of plate movement?

16. What evidence convinced Wegener and others that continents must have moved in the past and at one time formed a supercontinent?

17. You've been selected to be part of the first astronaut team to go to Mars. While your two fellow crew members descend to the Martian surface, you'll be staying in the command module and circling the Red Planet. As part of the geologic investigation of Mars, one of the crew members will be mapping the geology around the landing site and deciphering the geologic history of the area. Your job will be to observe and photograph the planet's surface and try to determine whether Mars had an active plate tectonic regime in the past and whether there is current plate movement. What features would you look for, and what evidence might reveal current or previous plate activity?

18. Why is some type of thermal convection system thought to be the major force driving plate movement?

19. How did the theory of seafloor spreading, proposed by Harry Hess in 1962, overcome the objections of those opposed to the continental drift hypothesis?

20. Using the ages (the numbers represent ages in millions of years) for each of the Hawaiian Islands, as well as the scale given in the accompanying figure, calculate the average rate of movement per year for the Pacific plate since each island formed. Is the average rate of movement the same for each island? Would you expect it to be? Explain why it may not be and why there are different ages for some of the islands.

4

Geologic Time
Concepts and Principles

The Grand Canyon, Arizona. Major John Wesley Powell led two expeditions down the Colorado River and through the canyon in 1869 and 1871. He was struck by the seemingly limitless time represented by the rocks exposed in the canyon walls and by the recognition that these rock layers, like the pages in a book, contain the geologic history of this region.

Copyright and Photograph by Dr. Parvinder S. Sethi

At the end of this chapter, you will have learned that

▶ The concept of geologic time, its measurement, and Earth's age, has changed through human history.

▶ The principle of uniformitarianism is fundamental to geology and to an understanding of Earth history.

▶ The fundamental principles of relative dating provide a means to interpret geologic history.

▶ The discovery of radioactivity provided geologists with a clock that could determine the ages of ancient rocks and validate the supposition that Earth was very old.

▶ Absolute dating methods are used to date geologic events in terms of years before present, and the most accurate radiometric dates are obtained from igneous rocks.

▶ Geologic time and the methods used to measure it are important elements in the study of climate change.

Introduction

In 1869, Major John Wesley Powell, a Civil War veteran who had lost his right arm in the battle of Shiloh, led a group of hardy explorers down the uncharted Colorado River through the Grand Canyon. With no maps or other information, Powell and his group ran the many rapids of the Colorado River in fragile wooden boats, hastily recording what they saw. Powell wrote in his diary that "all about me are interesting geologic records. The book is open and I read as I run."

Probably no one has contributed as much to the understanding of the Grand Canyon as Major Powell. In recognition of his contributions, the Powell Memorial was erected on the South Rim of the Grand Canyon in 1969 to commemorate the 100th anniversary of this history-making first expedition.

Most tourists today, like Powell and his fellow explorers in 1869, are astonished by the seemingly limitless time represented by the rocks exposed in the walls of the Grand Canyon. For most visitors, viewing a 1.5-kilometer-deep cut into Earth's crust is the only encounter they'll ever have with the magnitude of geologic time. When standing on the rim and looking down into the Grand Canyon, we are really looking far back in time, all the way back to the early history of our planet. In fact, more than 1 billion years of history are preserved in the rocks of the Grand Canyon, and reading what is preserved in those rocks is, just as Powell noted in his diary more than 100 years ago, like reading the pages in a history book.

In reading this "book"—the rock layers of the Grand Canyon—we learn that this area underwent episodes of mountain building as well as periods of advancing and retreating shallow seas. How do we know this? The answer lies in applying the principles of relative dating to the rocks we see exposed in the canyon walls, and also in recognizing that present-day processes have operated throughout Earth history. In this chapter, you will learn what those principles are and how they can be used to interpret and understand the geologic history of an area.

Vast periods of time set geology apart from most of the other sciences, and an appreciation of the immensity of geologic time is fundamental to understanding the physical and biological history of our planet. In fact, understanding and accepting the magnitude of geologic time is one of the major contributions geology has made to the sciences.

Why is the study of geologic time important? One reason is that it is important to realize that Earth has gone through periods of warmer and colder conditions in the past, and an understanding of what caused these might be helpful in the current debate over global climate change. Secondly, one of the most valuable lessons you will learn in this chapter is how to reason and apply some of the fundamental geologic principles to solve basic geologic problems. The logic used in applying these principles to interpret the geologic history of an area involves basic reasoning skills that can be applied in almost any profession or discipline.

How Is Geologic Time Measured?

In some respects, time is defined by the methods used to measure it. Geologists use two different frames of reference when discussing geologic time. **Relative dating** is placing geologic events in a sequential order as determined from their positions in the geologic record. Relative dating will not tell us how long ago a particular event took place, only that one event preceded another.

The various principles used to determine relative dating were discovered hundreds of years ago, and since then they have been used to construct the *relative geologic time scale* (▶Figure 4.1). Furthermore, these principles are still widely used by geologists today, especially in reconstructing the geologic history of the terrestrial planets and their moons.

Absolute dating (also called *numerical dating*) provides specific dates for rock units or events expressed in years before the present. *Radiometric dating* is the most common method of obtaining absolute ages. Dates are calculated

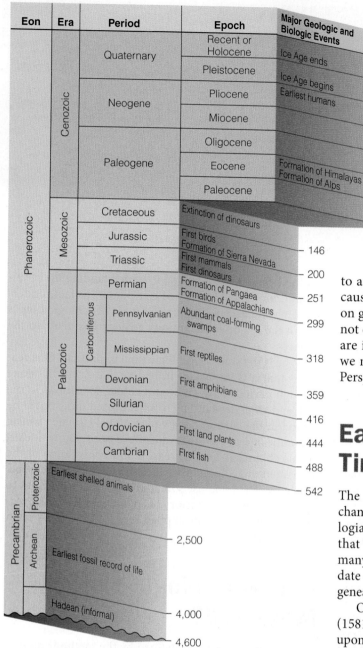

Eon	Era	Period	Epoch	Major Geologic and Biologic Events	Millions of Years Ago
Phanerozoic	Cenozoic	Quaternary	Recent or Holocene	Ice Age ends	0
			Pleistocene	Ice Age begins	0.0117
		Neogene	Pliocene	Earliest humans	2.6
			Miocene		5.3
		Paleogene	Oligocene		23
			Eocene	Formation of Himalayas Formation of Alps	34
			Paleocene		56
	Mesozoic	Cretaceous		Extinction of dinosaurs	66
		Jurassic		First birds Formation of Sierra Nevada	146
		Triassic		First mammals First dinosaurs	200
	Paleozoic	Permian		Formation of Pangaea Formation of Appalachians	251
		Carboniferous	Pennsylvanian	Abundant coal-forming swamps	299
			Mississippian	First reptiles	318
		Devonian		First amphibians	359
		Silurian			416
		Ordovician		First land plants	444
		Cambrian		First fish	488
				Earliest shelled animals	542
Precambrian	Proterozoic				2,500
	Archean			Earliest fossil record of life	
				Hadean (informal)	4,000
					4,600

▶ Figure 4.1 **The Geologic Time Scale**
Some of the major geologic and biologic events are indicated for the various eras, periods, and epochs.

Dates are from the 2009 International Stratigraphic Chart, © 2009 by the International Commission on Stratigraphy.

to accurately determine past climatic changes and their causes has important implications for the current debate on global warming and its effects on humans. In addition, not only does global warming affect humans, but humans are impacting the planet in a myriad of ways, such that we might be responsible for a new geologic epoch (see Perspective).

Early Concepts of Geologic Time and Earth's Age

The concept of geologic time and its measurement have changed throughout human history. Early Christian theologians were largely responsible for formulating the idea that time is linear rather than circular. This prompted many religious scholars and clerics to try to establish the date of creation by analyzing historical records and the genealogies found in Scripture.

One of the most influential scholars was James Ussher (1581–1665), Archbishop of Armagh, Ireland, who, based upon Old Testament genealogy, asserted that God created Earth on Sunday, October 23, 4004 BC. In 1701, an authorized version of the Bible made this date accepted Church doctrine. For nearly a century thereafter, it was considered heresy to assume that Earth and all its features were more than about 6,000 years old. Thus, the idea of a very young Earth provided the basis for most Western chronologies of Earth history prior to the 18th century.

During the 18th and 19th centuries, several attempts were made to determine Earth's age on the basis of scientific evidence rather than revelation. The French zoologist Georges Louis de Buffon (1707–1788) assumed Earth gradually cooled to its present condition from a molten beginning. To simulate this history, he melted iron balls of various diameters and allowed them to cool to the surrounding temperature. By extrapolating their cooling rate to a ball the size of Earth, he determined that Earth was at least 75,000 years old. Although this age was much older

from the natural rates of decay of various radioactive elements present in trace amounts in some rocks. It was not until the discovery of radioactivity near the end of the 19th century that absolute ages could be accurately applied to the relative geologic time scale.

Today, the geologic time scale is really a dual scale—a relative scale based on rock sequences with radiometric dates expressed as years before the present (Figure 4.1).

Advances and refinements in absolute dating techniques during the 20th century have changed the way we view Earth in terms of when events occurred in the past and the rates of geologic change through time. The ability

Perspective

The Anthropocene: A New Geologic Epoch?

The term *anthropocene* was coined on the spur of the moment by the Nobel Prize–winning Dutch atmospheric chemist Paul Crutzen at a scientific conference in 2000. Shortly thereafter, the term appeared in print in a 2000 article by Paul Crutzen and Eugene Stoermer. This was followed by a 2002 paper in the journal *Nature* in which Crutzen discussed the idea of an Anthropocene Epoch, based on the increasing impact humans are having on the planet. Since then, the term has caught on both in the scientific community and among the general public.

Interestingly, as far back as 1873, Antonio Stoppani, an Italian geologist, proposed that humans had introduced a new geologic era that he called the anthropozoic, and over the years, other terms followed. Why, then, has Crutzen's term—which comes from the Latin *antropos*, meaning "man," or "human," and the geologic epoch suffix *cene*, as in Paleocene or Holocene—gained so much favor, whereas others have been largely ignored? Perhaps it is because of an increasing realization that humans are altering the environment, and at an alarmingly rapid rate.

Just because a term is widely used in the scientific literature doesn't mean that it will be formally accepted as the name for a new geologic epoch. For this to happen, many criteria and formal steps must be taken and met. Because the geologic time scale is fundamental to the work geologists do, and is the framework on which Earth history is based, adding another epoch is not to be taken lightly.

Although *anthropocene* is an informal and imprecisely defined term, a proposal was presented to the Stratigraphy Commission of the Geological Society of London in 2008 to make the Anthropocene a formal geologic epoch. A significant majority thought that the proposal had merit and should be examined further. Since then, various working groups that are part of the International Commission on Stratigraphy (ICS), have been convened to determine if the proposed Anthropocene Epoch will join the other seven epochs of the Cenozoic Era. Eventually, votes must

Figure 1 Although cities have changed the landscape by expropriating the land surface on which they are built, using large amounts of energy and natural resources, as well as contributing to pollution, they will leave little, if any, direct evidence of their existence in the future geologic record. Dubai in the United Arab Emirates, which houses some 2 million people, has the world's tallest building, and it depends on air conditioning and desalinated seawater, both of which involve high energy consumption of fossil fuels, as well as contributing to atmospheric pollution.

be taken, not only by the ICS, but by its parent organization, the International Union of Geological Sciences, as well. One thing is clear: the work needed to satisfy the various organizations that will determine whether the Anthropocene is incorporated into the geologic time scale will take many years to complete, and there is no guarantee it will be successful.

Subdivisions of the geologic time scale are based on the recognition of distinctive events that are preserved in the geologic record, such as changes in fossil assemblages, which represent changes in the Earth's biota. For example, the end of the Permian Period is marked by a global mass-extinction event, as is the end of the Cretaceous Period. Even the boundaries of epochs, which are typically, geologically speaking, short in duration, are based on distinctive and recognizable geologic events. So, what is the evidence for the boundary marking the start of the Anthropogene Epoch—that is, at what point does human impact to the environment show up in the geologic record? Obviously, the geologic record of the present doesn't

yet exist, but what evidence of our impact is being imparted to the sediments that will make up the future geologic record?

Humans have certainly impacted the planet in a number of ways. For one, the population has increased dramatically in the past several centuries, rising to 7 billion in 2011, and projected to reach 9 billion people by 2045. Such a rise has placed great demands on the planet to feed, house, and clothe so many people. For example, there has been a dramatic increase in the burning of fossil fuels since the Industrial Revolution in the late 18th century, as well as extensive land clearing for agriculture. These actions have led to substantial increases in soil erosion, greenhouse gases, acid rain, climate warming, rising sea levels, and the introduction of many toxic substances into the environment. But how will these activities be preserved in the geologic record, and are they of such consequence that they will stand out above the background "noise" of the geologic record?

Some of the most plainly visible evidence of human activities, such as cities (Figure 1) and factories and power plants (Figure 2), will leave barely any traces in the geologic record, or none at all, because they occur on land that is, for the most part, the site of erosion rather than deposition. However, changes in the pollen record, beginning with the rise of farming, will be visible as a result of the sharp increase in the pollen of corn, wheat, soy, and rice as opposed to the more diverse pollen record of rain forests, prairies, and other plant biomes.

Probably the most significant change, from a geologic perspective, will be the change in the composition of the atmosphere, especially in carbon dioxide levels, which can be detected in the air trapped in polar ice. Changes in the climate can also be deduced by the latitudinal shifts of some plants and animals as they extend their geographic range closer to the poles. Furthermore, increasing carbon dioxide levels help acidify the world's oceans, which can cause environmental stress such that corals no longer construct reefs, and this would also show up in the fossil record (Figure 3).

If we have indeed entered a new epoch, when did it begin? On this point, there is still disagreement. Some scientists favor the Anthropocene Epoch as having begun approximately 8,000 years ago with the rise of farming and the subsequent land clearing, soil erosion, and global spread and increase in the human population. Others, including Paul Crutzen, consider the new epoch as having begun with the Industrial Revolution, with its increasing levels of carbon dioxide and methane resulting from the exploitation of coal, oil, and natural gas to fuel a burgeoning population and industrialized society.

As mentioned earlier, much work remains to be done before formalizing the Anthropocene Epoch. For now, however, humans and their global impact remain firmly part of the Holocene Epoch.

Figure 2 Large coal-fired power plants, like this one in England, have contributed to the sizable surge in carbon dioxide emissions into the atmosphere, which has exacerbated climate warming. Although cities, factories, and power plants will probably not leave any direct evidence of their existence in the geologic record, the emissions from the burning of coal, oil, and natural gas are likely to be discernible in atmospheric gases trapped in glacial ice and will probably be indirectly noticeable by changes in the fossil record.

Figure 3 This bleached, white coral (*Acropora* sp.) is evidence of increasing ocean water temperatures caused by global warming. High water temperatures kill the individual coral animals, leaving behind dead, white-colored reefs, such as seen here.

than that derived from Scripture, it was still vastly younger than we now know our planet to be.

Other scholars were equally ingenious in attempting to calculate Earth's age. For example, if deposition rates could be determined for various sediments, geologists reasoned that they could calculate how long it would take to deposit any rock layer. They could then extrapolate how old Earth was from the total thickness of sedimentary rock in its crust. Rates of deposition vary, however, even for the same type of rock. Furthermore, it is impossible to estimate how much rock has been removed by erosion, or how much a rock sequence has been reduced by compaction. As a result of these variables, estimates of Earth's age ranged from younger than 1 million years to older than 2 billion years.

The Granger Collection, New York

▶ Figure 4.2 **James Hutton** James Hutton, a Scottish geologist, originated the principle of uniformitarianism and, through his writings, profoundly influenced the course of geologic thinking.

Another method involved ocean salinity. John Joly, a 19th-century Irish geologist, measured the amount of salt currently in the world's streams. He then calculated that it would have taken at least 90 million years for the oceans to reach their present salinity level, assuming Earth's oceans were originally freshwater. This was still much younger than Earth's now accepted age of 4.6 billion years. This was due, in part, because Joly had no way to calculate how much salt has been recycled, or the amount of salt that is stored in continental salt deposits and seafloor clay deposits.

Besides trying to determine Earth's age, the naturalists of the 18th and 19th centuries were also formulating some of the fundamental geologic principles that are used in deciphering Earth history. From the evidence preserved in the geologic record, it was clear to them that Earth is very old and that geologic processes have operated over long periods of time.

James Hutton and the Recognition of Geologic Time

Many consider the Scottish geologist James Hutton (1726–1797) to be the founder of modern geology (▶Figure 4.2). His detailed studies and observations of rock exposures and present-day geologic processes formed the foundation for the **principle of uniformitarianism**, the concept that the same processes seen today have operated throughout geologic time.

Hutton viewed Earth history as cyclical—that is, continents are worn down by erosion, the eroded sediment is deposited in the sea, and uplift of the seafloor creates new continents, thus completing a cycle. He thought the mechanism for uplift was thermal expansion from Earth's hot interior.

Hutton also recognized the importance of unconformities in his cyclical view of Earth history. At Siccar Point, Scotland, he observed steeply inclined metamorphic rocks that had been eroded and covered by flat-lying younger rocks (▶Figure 4.3). It was clear to him that severe upheavals had tilted the lower rocks and formed mountains. These were then worn away and covered by younger, flat-lying

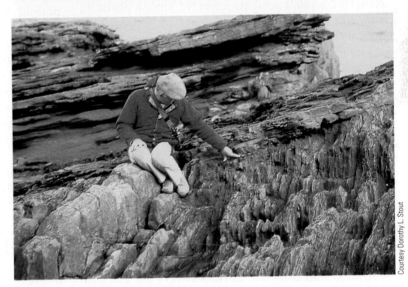

Courtesy Dorothy L. Stout

▶ Figure 4.3 **Angular Unconformity at Siccar Point, Scotland** It was at this location in 1788 that James Hutton first realized the significance of unconformities in interpreting Earth history.

rocks. The erosion surface meant that there was a gap in the geologic record, and the rocks above and below this surface provided evidence that both mountain building and erosion had occurred. Although he did not use the word *unconformity*, Hutton was the first to understand and explain the significance of such gaps in the geologic record. You will learn more about the different kinds of unconformities in the next chapter.

Hutton was also instrumental in establishing the concept that geologic processes had vast amounts of time in which to operate. Because Hutton relied on known processes to account for Earth history, he concluded that Earth must be very old. He did not, however, estimate how old Earth was, or how long it took to complete a cycle of erosion, deposition, and uplift. He merely allowed that "we find no vestige of a beginning, and no prospect of an end," which was in keeping with a cyclical view of Earth history.

In 1830, Charles Lyell published a landmark book, *Principles of Geology*, in which he championed Hutton's concept of uniformitarianism. Instead of relying on catastrophic events to explain various Earth features, Lyell recognized that imperceptible changes brought about by present-day processes could, over long periods of time, have tremendous cumulative effects. Through his writings, Lyell firmly established uniformitarianism as the guiding principle of geology. Furthermore, the recognition of vastly long periods of time was also necessary for, and instrumental in, the acceptance of Darwin's 1859 theory of evolution (see Chapter 7).

Lord Kelvin and a Crisis in Geology

After establishing that present-day processes have operated over vast periods of time, geologists were nevertheless nearly forced to accept a very young age for Earth. Lord Kelvin (1824–1907), a highly respected English physicist, claimed in a paper written in 1866 to have destroyed the uniformitarian foundation on which Huttonian–Lyellian geology was based. Starting with the generally accepted belief that Earth was originally molten, Kelvin assumed that it has gradually been losing heat and that, by measuring this heat loss, he could determine its age.

Kelvin knew from deep mines in Europe that Earth's temperature increases with depth, and he reasoned that Earth is losing heat from its interior. By knowing the size of Earth, the melting temperature of rocks, and the rate of heat loss, Kelvin calculated the age at which Earth was entirely molten. From these calculations, he concluded that Earth could not be older than 400 million years or younger than 20 million years. This wide discrepancy in age reflected uncertainties in average temperature increases with depth and the various melting points of Earth's constituent materials.

After establishing that Earth was very old and that present-day processes operating over long periods of time account for geologic features, geologists were in a quandary. If they accepted Kelvin's dates, they would have to abandon the concept of seemingly limitless time that was the underpinning of uniformitarian geology, and one of the foundations of Darwinian evolution, and squeeze events into a shorter time frame.

Kelvin's reasoning and calculations were sound, but his basic premises were false, thereby invalidating his conclusions. Kelvin was unaware that Earth has an internal heat source—radioactivity—that has allowed it to maintain a fairly constant temperature through time.* His 40-year campaign for a young Earth ended with the discovery of radioactivity near the end of the 19th century and the insight in 1905 that natural radioactive decay can be used in many cases to date how long ago a rock formed. His "unassailable calculations" were no longer valid, and his proof for a geologically young Earth collapsed.

Although the discovery of radioactivity destroyed Kelvin's arguments, it provided geologists with a clock that could measure Earth's age and validate what geologists had long thought—namely, that Earth was indeed very old!

Modern View of Uniformitarianism

Geologists today assume that the principles, or laws, of nature are constant but that rates and intensities of change have varied through time. For example, glaciation has been more prevalent during the last 2.6 million years than in the previous 300 million years. Because rates and intensities of geologic processes have varied through time, some geologists prefer to use the term *actualism* instead of *uniformitarianism* to remove the idea of "uniformity" from the concept. Most geologists, though, still use the term *uniformitarianism* because it indicates that even though rates and intensities of change have varied in the past, the laws of nature have remained the same.

Uniformitarianism is a powerful concept that allows us, through analogy and inductive reasoning, to use present-day processes as the basis for interpreting the past and for predicting potential future events. It also does not eliminate occasional, sudden, short-term events such as volcanic eruptions, earthquakes, floods, or even meteorite impacts as forces that shape our modern world. In fact, some geologists view Earth history as a series of such short-term, or punctuated, events, and this view is certainly in keeping with the modern principle of uniformitarianism.

Earth is in a state of dynamic change and has been since it formed. Although rates of change may have varied in the past, natural laws governing the processes have not.

Relative Dating Methods

Before the development of radiometric dating techniques, geologists had no reliable means of absolute dating and therefore depended solely on relative dating methods. Recall

---

*Actually, Earth's temperature has decreased through time because the original amount of radioactive materials has been decreasing and therefore is not supplying as much heat. However, the temperature is decreasing at a rate considerably slower than would be required to lend any credence to Kelvin's calculations (see p. 161, Chapter 8).

Copyright and Photograph by Dr. Parvinder S. Sethi

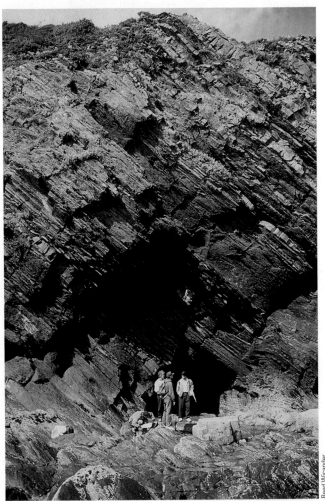

Reed Wicander

(a) The sedimentary rocks of Bryce Canyon National Park, Utah, illustrate three of the six fundamental principles of relative dating. These rocks were originally deposited horizontally in a variety of continental environments (principle of original horizontality). The oldest rocks are at the bottom of this highly dissected landscape, and the youngest rocks are at the top, forming the rims (principle of superposition). The exposed rock layers extend laterally in all directions for some distance (principle of lateral continuity).

(b) These shales and limestones of the Postolonnec Formation, at Postolonnec Beach, Crozon Peninsula, France, were originally deposited horizontally but have been significantly tilted since their formation.

that relative dating places events in sequential order, but does not tell us how long ago an event took place. Although the principles of relative dating may now seem self-evident, their discovery was an important scientific achievement because they provided geologists with a means to interpret geologic history and develop a *relative geologic time scale*.

Fundamental Principles of Relative Dating

Six fundamental geologic principles are used in relative dating: superposition, original horizontality, lateral continuity, cross-cutting relationships (all discussed in this chapter), and inclusions and fossil succession (discussed in Chapter 5).

The 17th century was an important time in the development of geology as a science because of the widely circulated writings of the Danish anatomist Nicolas Steno (1638–1686). Steno observed that when streams flood, they spread out across their floodplains and deposit layers of sediment that bury organisms dwelling on the floodplain. Subsequent floods produce new layers of sediments that are deposited, or superposed, over previous deposits. When lithified, these layers of sediment become sedimentary rock.

Thus, in an undisturbed succession of sedimentary rock layers, the oldest layer is at the bottom and the youngest layer is at the top. This **principle of superposition** is the basis for relative-age determinations of strata and their contained fossils (▶ Figure 4.4a and chapter opening photograph).

Steno also observed that because sedimentary particles settle from water under the influence of gravity, sediment is deposited in essentially horizontal layers, thus illustrating the **principle of original horizontality** (Figure 4.4a and chapter opening photograph). Therefore, a sequence of sedimentary rock layers that is steeply inclined from the horizontal must have been tilted after deposition and lithification (Figure 4.4b).

Steno's third principle, the **principle of lateral continuity**, states that sediment extends laterally in all directions until it thins and pinches out, or terminates, against the edge of the depositional basin (Figure 4.4a).

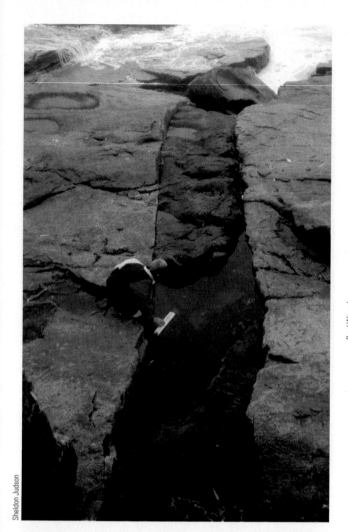

(b) A small fault (arrows show direction of movement) cuts across, and thus displaces, tilted sedimentary beds along Templin Highway in Castaic, California. The fault is therefore younger than the youngest beds that are displaced.

(a) A dark gabbro dike cuts across granite in Acadia National Park, Maine. The dike is younger than the granite it intrudes.

James Hutton is credited with formulating the **principle of cross-cutting relationships**. He recognized that an igneous intrusion or a fault must be younger than the rocks it intrudes or displaces (▶Figure 4.5).

The two other relative dating principles, the *principle of inclusions*, and the *principle of fossil succession* are discussed in the next chapter. The application of these six relative dating principles provided geologists with the means for correlating rock units (see Chapter 5) and deciphering the geologic history of an area. It was not until the discovery of radioactivity, however, that geologists would be able to determine how long ago geologic events occurred.

Absolute Dating Methods

Although most of the isotopes of the 92 naturally occurring elements are stable, some are radioactive and therefore spontaneously decay to other, more stable isotopes of elements, releasing energy in the process. The discovery in

1903 by Pierre and Marie Curie that radioactive decay produces heat meant that geologists finally had a mechanism for explaining Earth's internal heat that did not rely on residual cooling from a molten origin. Furthermore, geologists now had a powerful tool to date geologic events accurately and to verify the long time periods postulated by Hutton and Lyell.

Atoms, Elements, and Isotopes

As discussed in Chapter 2, all matter is made up of chemical *elements*, each composed of extremely small particles called *atoms*. The *nucleus* of an atom is composed of *protons* (positively charged particles) and *neutrons* (neutral particles) with *electrons* (negatively charged particles) encircling it (see Figure 2.2). The number of protons defines an element's *atomic number* and helps determine its properties and characteristics.

The combined number of protons and neutrons in an atom is its *atomic mass number*. However, not all atoms of the same element have the same number of neutrons in their nuclei. These variable forms of the same element are called *isotopes* (see Figure 2.3). Most isotopes are stable, but some are unstable and spontaneously decay to a more stable form. It is the decay rate of unstable isotopes that geologists measure to determine the absolute age of rocks.

Radioactive Decay and Half-Lives

Radioactive decay is the process whereby an unstable atomic nucleus is spontaneously transformed into an

▶ Figure 4.6 **Three Types of Radioactive Decay**

Parent nucleus | Alpha particle | Daughter nucleus | Changes in atomic number and atomic mass number

Atomic number = –2
Atomic mass number = –4

(a) Alpha decay, in which an unstable parent nucleus emits two protons and two neutrons.

● Proton   ◐ Neutron   ○ Electron

Parent nucleus | Beta particle | Daughter nucleus

Atomic number = +1
Atomic mass number = 0

(b) Beta decay, in which an electron is emitted from a neutron in the nucleus.

● Proton   ◐ Neutron   ○ Electron

Parent nucleus | Daughter nucleus

Atomic number = –1
Atomic mass number = 0

(c) Electron capture, in which a proton captures an electron and is thereby converted to a neutron.

© 2013 Cengage Learning

● Proton   ◐ Neutron   ○ Electron

atomic nucleus of a different element. Three types of radioactive decay are recognized, all of which result in a change of atomic structure (▶Figure 4.6).

In *alpha decay,* two protons and two neutrons are emitted from the nucleus, resulting in the loss of two atomic numbers and four atomic mass numbers. In *beta decay*, a fast-moving electron is emitted from a neutron in the nucleus, changing that neutron to a proton and consequently increasing the atomic number by one, with no resultant atomic mass number change. *Electron capture decay* is when a proton captures an electron from an electron shell and thereby converts to a neutron, resulting in a loss of one atomic number but not changing the atomic mass number.

Some elements undergo only one decay step in the conversion from an unstable form to a stable form. For example, rubidium 87 decays to strontium 87 by a single beta emission, and potassium 40 decays to argon 40 by a single electron capture. Other radioactive elements undergo several decay steps. Uranium 235 decays to lead 207 by seven alpha steps and six beta steps, whereas uranium 238 decays to lead 206 by eight alpha and six beta steps (▶Figure 4.7).

When discussing decay rates, it is convenient to refer to them in terms of half-lives. The **half-life** of a radioactive element is the time it takes for one-half of the atoms of the original unstable *parent element* to decay to atoms of a new, stable *daughter element*. The half-life of a given radioactive element is constant and can be precisely measured. Half-lives of various radioactive elements range from less than a billionth of a second to 49 billion years.

Radioactive decay occurs at a geometric rate rather than a linear rate. Therefore, a graph of the decay rate produces a curve rather than a straight line (▶Figure 4.8). For

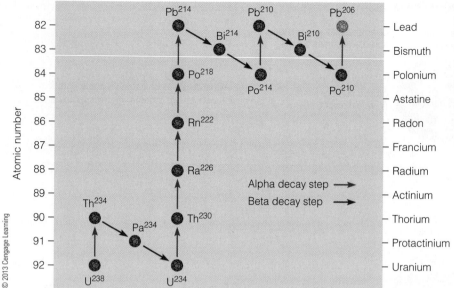

▶ **Figure 4.7** **Radioactive Decay Series for Uranium 238 and Lead 206** Radioactive uranium 238 decays to its stable daughter product, lead 206, by eight alpha and six beta decay steps. A number of different isotopes are produced as intermediate steps in the decay series.

example, an element with *1,000,000* parent atoms will have *500,000* parent atoms and *500,000* daughter atoms after one half-life. After two half-lives, it will have *250,000* parent atoms (one-half of the previous parent atoms, which is equivalent to one-fourth of the original parent atoms) and *750,000* daughter atoms. After three half-lives, it will have *125,000* parent atoms (one-half of the previous parent atoms, or one-eighth of the original parent atoms) and *875,000* daughter atoms, and so on, until the number of parent atoms remaining is so few that they cannot be accurately measured by present-day instruments.

By measuring the parent–daughter ratio and knowing the half-life of the parent (which has been determined in the laboratory), geologists can calculate the age of a sample that contains the radioactive element. The parent–daughter ratio is usually determined by a *mass spectrometer,* an instrument that measures the proportions of elements of different masses.

Sources of Uncertainty

The most accurate radiometric dates are obtained from igneous rocks. As magma cools and begins to crystallize, radioactive parent atoms are separated from previously formed daughter atoms. Because they are the right size, some radioactive parent atoms are incorporated into the crystal

▶ **Figure 4.8** **Uniform, Linear Change Compared to Geometric Radioactive Decay**

(a) Uniform, linear change is characteristic of many familiar processes. In this example, water is being added to a glass at a constant rate.

(b) A geometric radioactive decay curve, in which each time unit represents one half-life, and each half-life is the time it takes for half of the parent element to decay to the daughter element.

▶ Figure 4.9 **Crystallization of Magma Containing Radioactive Parent and Stable Daughter Atoms**

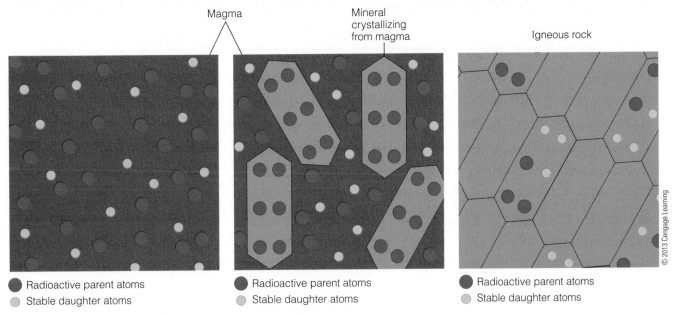

Magma

Mineral
crystallizing
from magma

Igneous rock

© 2013 Cengage Learning

● Radioactive parent atoms
● Stable daughter atoms

● Radioactive parent atoms
● Stable daughter atoms

● Radioactive parent atoms
● Stable daughter atoms

(a) Magma contains both radioactive parent atoms and stable daughter atoms. The radioactive parent atoms are larger than the stable daughter atoms.

(b) As magma cools and begins to crystallize, some of the radioactive parent atoms are incorporated into certain minerals because they are the right size and can fit into the crystal structure. In this example, only the larger radioactive parent atoms fit into the crystal structure. Therefore, at the time of crystallization, minerals in which the radioactive parent atoms can fit into the crystal structure will contain 100 percent radioactive parent atoms and 0 percent stable daughter atoms.

(c) After one half-life, 50 percent of the radioactive parent atoms will have decayed to stable daughter atoms, such that those minerals that had radioactive parent atoms in their crystal structure will now have 50 percent radioactive parent atoms and 50 percent stable daughter atoms.

structure of certain minerals. The stable daughter atoms, though, are a different size from the radioactive parent atoms and consequently cannot fit into the crystal structure of the same mineral as the parent atoms. Therefore, a mineral crystallizing in a cooling magma will contain radioactive parent atoms but no stable daughter atoms (▶ Figure 4.9). Thus, the time that is being measured is the time of crystallization of the mineral that contains the radioactive atoms and not the time of formation of the radioactive atoms.

Except in unusual circumstances, sedimentary rocks cannot be radiometrically dated because one would be measuring the age of a particular mineral rather than the time that it was deposited as a sedimentary particle. However, the greenish mineral glauconite is precipitated in certain marine environments, thus forming when the sedimentary rock forms. Glauconite contains radioactive potassium 40, which decays to argon 40 (Table 4.1); thus, a radiometric date can be obtained as to the time of the sedimentary rock's origin. Being a gas, however, the daughter product, argon, can easily escape from a mineral; therefore, any date obtained from the potassium 40 and argon 40 pair must be considered a minimum age.

To obtain accurate radiometric dates, geologists must be sure that they are dealing with a *closed system*, meaning that neither parent nor daughter atoms have been added

or removed from the system since crystallization, and that the ratio between them results only from radioactive decay. Otherwise, an inaccurate date will result. If daughter atoms have leaked out of the mineral being analyzed, the calculated age will be too young; if parent atoms have been removed, the calculated age will be too old.

Leakage may take place if the rock is heated or subjected to intense pressure as can sometimes occur during metamorphism. If this happens, some of the parent or daughter atoms may be driven from the mineral being analyzed, resulting in an inaccurate age determination. If the daughter product were completely removed, then one would be measuring the time since metamorphism (a useful measurement itself) and not the time since crystallization of the mineral (▶ Figure 4.10).

Because heat and pressure affect the parent–daughter ratio, metamorphic rocks are difficult to date accurately. Remember, that although the resulting parent–daughter ratio of the sample being analyzed may have been affected by heat, the decay rate of the parent element remains constant regardless of any physical or chemical changes.

To obtain an accurate radiometric date, geologists must make sure that the sample is fresh and unweathered and that it has not been subjected to high temperature or

TABLE 4.1 Five of the Principal Long-Lived Radioactive Isotope Pairs Used in Radiometric Dating

Isotopes		Half-Life of Parent (Years)	Effective Dating Range (Years)	Minerals and Rocks That Can Be Dated	
Parent	**Daughter**				
Uranium 238	Lead 206	4.5 billion	10 million to 4.6 billion	Zircon Uraninite	
Uranium 235	Lead 207	704 million			
Thorium 232	Lead 208	14 billion			
Rubidium 87	Strontium 87	48.8 billion	10 million to 4.6 billion	Muscovite Biotite Potassium feldspar Whole metamorphic or igneous rock	
Potassium 40	Argon 40	1.3 billion	100,000 to 4.6 billion	Glauconite Muscovite Biotite	Hornblende Whole volcanic rock

© 2013 Cengage Learning

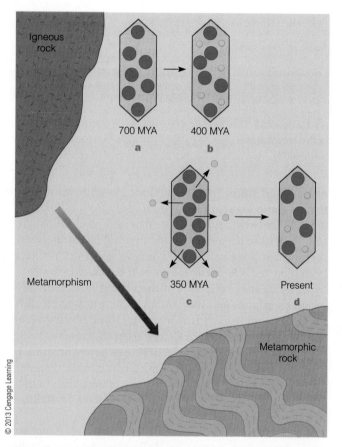

▶ Figure 4.10 **Effects of Metamorphism on Radiometric Dating**
The effect of metamorphism in driving out daughter atoms from a mineral that crystallized 700 million years ago (MYA). The mineral is shown immediately after crystallization (a), then at 400 MYA (b), when some of the parent atoms had decayed to daughter atoms. Metamorphism at 350 MYA (c) drives the daughter atoms out of the mineral into the surrounding rock. (d) If the rock has remained a closed chemical system throughout its history, dating the mineral today yields the time of metamorphism, whereas dating the whole rock provides the time of its crystallization, 700 MYA.

intense pressures after crystallization. Furthermore, it is sometimes possible to cross-check the radiometric date obtained by measuring the parent–daughter ratio of two different radioactive elements in the same mineral.

For example, naturally occurring uranium consists of both uranium 235 and uranium 238 isotopes. Through various decay steps, uranium 235 decays to lead 207, whereas uranium 238 decays to lead 206 (Figure 4.7). If the minerals that contain both uranium isotopes have remained closed systems, the ages obtained from each parent–daughter ratio should agree closely. If they do, they are said to be *concordant,* thus reflecting the time of crystallization of the magma. If the ages do not closely agree, then they are said to be *discordant,* and other samples must be taken and ratios measured to see which, if either, date is correct.

Recent advances and the development of new techniques and instruments for measuring various isotope ratios have enabled geologists not only to analyze increasingly smaller samples, but to analyze samples with a greater precision than ever before. Presently, the measurement error for many radiometric dates is typically less than 0.5 percent of the age and, in some cases, is even better than 0.1 percent. Thus, for a rock 540 million years old (near the beginning of the Cambrian Period), the possible error could range from nearly 2.7 million years to as low as less than 540,000 years.

Long-Lived Radioactive Isotope Pairs

Table 4.1 shows the five common, long-lived parent–daughter isotope pairs used in radiometric dating. Long-lived pairs have half-lives of millions or billions of years. All of these were present when Earth formed and are still present in measurable quantities. Other, shorter-lived radioactive isotope pairs have decayed to the point that only small quantities near the limit of detection remain.

The most commonly used isotope pairs are the uranium–lead and thorium–lead series, which are used principally to date ancient igneous intrusives, lunar samples, and some meteorites. The rubidium–strontium pair is also used for very old samples and has been effective in dating the oldest rocks on Earth as well as meteorites.

The potassium–argon method is typically used for dating fine-grained volcanic rocks from which individual crystals cannot be separated; hence, the whole rock is analyzed. Because argon is a gas, great care must be taken to ensure that the sample has not been subjected to heat, which would allow argon to escape; such a sample would yield an age that is too young.

Other Radioactive Isotope Pairs

Three other radioactive isotope pairs bear mentioning because, although they are rare, they are useful in certain types of situations. They are argon–argon, samarium–neodymium, and thorium–uranium.

In *argon–argon dating*, the ratio of argon 30 to argon 40 is used. Just as with the potassium–argon method, fine-grained volcanic rocks are the most common rocks used in this dating method, but it can also be effectively used with metamorphic rocks. Developed in the 1960s, and further refined during the early 1980s, the argon–argon method has a reliable dating range of 2,000 to 4.6 billion years.

The *samarium–neodymium dating* method uses the ratio of samarium 147 to neodymium 143, both of which are rare earth elements. Samarium 147 decays to neodymium 143 by a single alpha decay step, and has a half-life of 106 billion years. It is thus very useful in dating extremely old igneous and metamorphic rocks as well as meteorites. The samarium–neodymium method has been very helpful in determining the composition and evolution of Earth's mantle.

The *uranium–thorium dating* method is frequently used as a cross-check for carbon 14 dates, and, with a half-life of 75,000 years, is reliable back to approximately 500,000 years. In this method, an age is calculated by determining the degree to which equilibrium has been established between the radioactive parent isotope uranium 234 and the radioactive isotope thorium 230.

This technique is based on the fact that uranium is very slightly soluble in natural waters at, or near, Earth's surface, whereas thorium is not. Thus, carbonate minerals precipitating from surface or near-surface water would typically not contain thorium. However, the trace amounts of uranium in the precipitated carbonates would, as part of the radioactive decay process, produce radioactive thorium 230 and eventually achieve equilibrium with the uranium 234. What is being measured, then, is the degree to which equilibrium has been restored between thorium 230 and uranium 234. Because thorium 230 has a half-life of 75,000 years, the upper limit using this method is approximately 500,000 years.

▶ **Figure 4.11** **Fission-Track Dating** Each fission track (about 16 microns [= 16/1,000 mm] long) in this apatite crystal is the result of the radioactive decay of a uranium atom. The apatite crystal, which has been etched with hydrofluoric acid to make the fission tracks visible, comes from one of the dikes at Shiprock, New Mexico, and has a calculated age of 27 million years.

Fission-Track Dating

The emission of atomic particles resulting from the spontaneous decay of uranium within a mineral damages its crystal structure. The damage appears as microscopic linear tracks that are visible only after the mineral has been etched with hydrofluoric acid, an acid so powerful that its vapors can destroy one's sense of smell without careful handling. The age of the sample is determined from the number of fission tracks present and the amount of uranium the sample contains: the older the sample, the greater the number of tracks (▶ Figure 4.11).

Fission-track dating is of particular interest to archaeologists and geologists because the technique can be used to date samples ranging from only a few hundred to hundreds of millions of years old. It is most useful for dating samples between approximately 40,000 and 1.5 million years ago, a period for which other dating techniques are not always suitable. One of the problems in fission-track dating occurs when the rocks have later been subjected to high temperatures. If this happens, the damaged crystal structures are repaired by annealing, and consequently the tracks disappear. In such instances, the calculated age will be younger than the actual age.

Radiocarbon and Tree-Ring Dating Methods

Carbon is an important element in nature and is one of the basic elements found in all forms of life. It has three isotopes; two of these, carbon 12 and 13, are stable, whereas carbon 14 is radioactive (see Figure 2.3). Carbon 14 has a half-life of 5,730 years plus or minus 30 years. The **carbon-14 dating** technique is based on the ratio of carbon 14 to carbon 12 and is generally used to date once-living material.

The short half-life of carbon 14 makes this dating technique practical only for specimens younger than about 70,000 years. Consequently, the carbon-14 dating method is especially useful in archaeology and has greatly helped unravel the events of the latter portion of the Pleistocene Epoch, such as where the first center for maize domestication in Mesoamerica arose, and when humans began populating North America.

Carbon 14 is constantly formed in the upper atmosphere when cosmic rays, which are high-energy particles (mostly protons), strike the atoms of upper-atmospheric gases, splitting their nuclei into protons and neutrons. When a neutron strikes the nucleus of a nitrogen atom (atomic number 7, atomic mass number 14), it may be absorbed into the nucleus and a proton emitted. Thus, the atomic number of the atom decreases by one, whereas the atomic mass number stays the same. Because the atomic number has changed, a new element, carbon 14 (atomic number 6, atomic mass number 14), is formed. The newly formed carbon 14 is rapidly assimilated into the carbon cycle and, along with carbon 12 and 13, is absorbed in a nearly constant ratio by all living organisms (▶ Figure 4.12). When an organism dies, however, carbon 14 is not replenished, and the ratio of carbon 14 to carbon 12 decreases as carbon 14 decays back to nitrogen by a single beta decay step (Figure 4.6).

Currently, the ratio of carbon 14 to carbon 12 is remarkably constant in both the atmosphere and in living organisms. There is good evidence, however, that the production of carbon 14, and thus the ratio of carbon 14 to carbon 12, has varied somewhat during the past several thousand years. This fact was determined by comparing ages established by carbon-14 dating of wood samples with ages established by counting annual tree rings in the same samples. As a result, carbon-14 ages have been corrected to reflect such variations in the past.

Tree-ring dating is another useful method for dating geologically recent events. The age of a tree can be determined by counting the growth rings in the lower part of the trunk. Each ring represents one year's growth, and the pattern of wide and narrow rings can be compared among trees to establish the exact year in which the rings were formed. The procedure of matching ring patterns from numerous trees and wood fragments in a given area is called *cross-dating*.

By correlating distinctive tree-ring sequences from living and nearby dead trees, scientists can construct a time scale that extends back approximately 14,000 years ago (▶ Figure 4.13). Wood samples whose ages are not known can be accurately dated by comparing their ring patterns to the composite ring scale.

The applicability of tree-ring dating is somewhat limited, however, because it can be used only where continuous tree records are found. It is therefore most useful in arid regions, particularly the southwestern United States, where trees live a very long time.

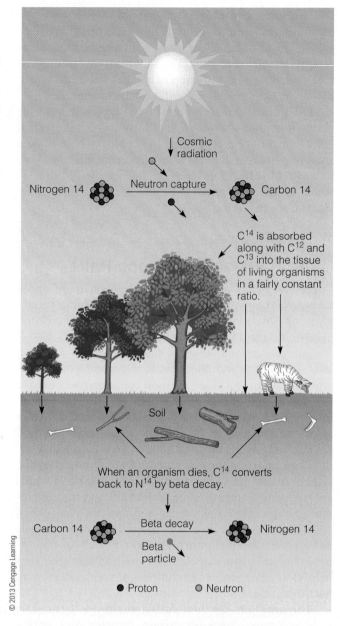

▶ Figure 4.12 **Carbon-14 Dating Method** The carbon cycle involves the formation of carbon 14 in the upper atmosphere, its dispersal and incorporation into the tissue of all living organisms, and its decay back to nitrogen 14 by beta decay.

Geologic Time and Climate Change

Given the debate concerning global warming and its possible implications, it is extremely important to be able to reconstruct past climatic regimes as accurately as possible. To model how Earth's climate system has responded to changes in the past and use that information for simulations of future climatic scenarios, geologists must have a geologic calendar that is as precise as possible. New dating techniques with greater precision are providing geologists with more accurate dates for when and how long ago past

(a) Living tree.

(b) Cross section of tree showing annular tree rings. Each tree ring represents one year's growth. The ring widths vary depending on the growing conditions during the life of the tree.

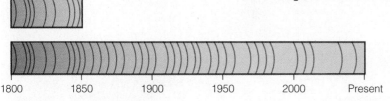

| 1800 | 1850 | 1900 | 1950 | 2000 | Present |

(c) By matching up tree-ring patterns from numerous trees in a region, a master tree-ring chronology can be constructed. Note that these tree rings are not to scale, but are diagrammatic to show how a master tree-ring chronology would work.

| 1800 | 1850 | 1900 | 1950 | 2000 | Present |

(d) By matching the tree-ring pattern from a wood beam found at an ancient site to the same pattern of the master tree-ring chronology for that region, the age of the site can be determined, at least to the time the tree died or was cut down.

climate changes occurred. The ability to accurately determine when past climate changes occurred helps geologists correlate these changes with regional and global geologic events to see whether there are any possible connections.

In addition to short-term climatic changes, it is also important to step back a bit and look at climate change from a geologic perspective (▶ Figure 4.14). We know that Earth has experienced periods of glaciation in the past— for instance, during the Proterozoic, at the end of the Ordovician Period, and most recently during the Pleistocene Epoch (see Figure 1.13). Earth has also undergone large-scale periods of soaring global temperature, such as during what is known as the *Paleocene–Eocene Thermal Maximum*, beginning around 56 million years ago and lasting more than 150,000 years (see Chapter 18). During

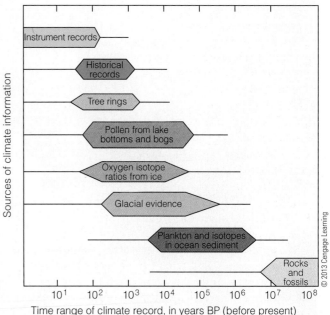

▶ **Figure 4.14 Sources of Climate Data** Some of the methods used by scientists to determine historical and ancient climates. Each method has its own useful time range and accuracy within that time range.

this time, Earth experienced drought, a rise in sea level, and changes to its biota (▸Figure 4.15).

Even though we can't physically travel back in time, geologists can reconstruct what the climate was like in the past. The distribution of plants and animals is controlled, in part, by climate. Plants are particularly sensitive to climate change, and many can live only in particular environments. The fossils of plants and animals can tell us something about the environment and climate at the time these organisms were living.

Furthermore, climate-sensitive sedimentary rocks can be used to interpret past climatic conditions. Desert dunes are typically well sorted and exhibit large-scale cross-bedding. Coals form in freshwater swamps where climatic conditions

▸ Figure 4.15 **Earth at the End of the Paleocene Epoch** At the end of the Paleocene Epoch, Earth was hot and ice free. During the Paleocene–Eocene Thermal Maximum (PETM), the average annual temperature was 25°C, compared to 20°C before (Late Paleocene) and after (Early Eocene) the PETM, and 7.2°C today. In addition, sea level was 67 m higher than it is today.

Sean Mcnaughton/National Geographic Stock

promote abundant plant growth. Evaporites such as rock salt result when evaporation exceeds precipitation, such as in desert regions or along hot, dry shorelines. Tillites (glacial sediments) result from glacial activity and indicate cold, wet environments. So, by combining all relevant geologic and paleontologic information, geologists can reconstruct what the climate was like in the past and how it has changed over time at a given locality (Figure 4.14).

Summary

▶ Time is defined by the methods used to measure it. Relative dating places geologic events in sequential order as determined from their position in the geologic record. Absolute dating provides specific dates for geologic rock units or events that are expressed in years before the present.

▶ Early Christian theologians were responsible for formulating the idea that time is linear and that Earth was very young. Based on Old Testament genealogies, Archbishop Ussher calculated Earth's age at approximately 6,000 years, thus providing the idea that Earth was very young.

▶ During the 18th and 19th centuries, attempts were made to determine Earth's age based on scientific evidence rather than revelation. Although some attempts were ingenious, they yielded a variety of ages that are now known to be much too young.

▶ James Hutton, considered by many to be the father of modern geology, thought that present-day processes operating over long periods of time could explain all of the geologic features of Earth. His observations were instrumental in establishing the principle of uniformitarianism and the fact that Earth was much older than earlier scientists thought.

▶ Uniformitarianism, as articulated by Charles Lyell, soon became the guiding principle of geology. It holds that the laws of nature have been constant through time, and that the same processes operating today have also operated in the past, although not necessarily at the same rates.

▶ Besides uniformitarianism, the principles of superposition, original horizontality, lateral continuity, and cross-cutting relationships, as well as the principles of inclusions and fossil succession, are basic for determining relative geologic ages and for interpreting Earth history.

▶ Radioactivity was discovered during the late 19th century, and soon thereafter radiometric dating techniques enabled geologists to determine absolute ages for rock units and geologic events.

▶ Absolute ages for rocks are usually obtained by determining how many half-lives of a radioactive parent element have elapsed since the sample originally crystallized. A half-life is the time it takes for one-half of the original, unstable radioactive parent element to decay to a new, more stable daughter element.

▶ The most accurate radiometric dates are obtained from long-lived radioactive isotope pairs in igneous rocks. The five common long-lived radioactive isotope pairs are uranium 238–lead 206, uranium 235–lead 207, thorium 232–lead 208, rubidium 87–strontium 87, and potassium 40–argon 40. Other useful radioactive isotope pairs include argon–argon, samarium–neodymium, and thorium–uranium.

▶ The most reliable dates are those obtained by using at least two different radioactive decay series in the same rock.

▶ Carbon-14 dating can be used only on organic matter such as wood, bones, and shells, and is effective back to approximately 70,000 years ago. Unlike the long-lived radioactive isotopic pairs, the carbon-14 dating technique determines age by the ratio of radioactive carbon 14 to stable carbon 12.

▶ To reconstruct past climate changes and link them to possible causes, geologists must have a geologic calendar that is as precise and accurate as possible. Thus, they must be able to date geologic events and the onset and duration of climate changes as precisely as possible.

Important Terms

absolute dating, p. 63

carbon-14 dating, p. 75

fission-track dating, p. 75

half-life, p. 71

principle of cross-cutting relationships, p. 70

principle of lateral continuity, p. 69

principle of original horizontality, p. 69

principle of superposition, p. 69

principle of uniformitarianism, p. 67

radioactive decay, p. 70

relative dating, p. 63

tree-ring dating, p. 76

Review Questions

1. The atomic number of an element is determined by the number of _____ in its nucleus.
   a. _____ protons; b. _____ neutrons; c. _____ electrons; d. _____ protons and neutrons; e. _____ protons and electrons.

2. If a rock is heated during metamorphism, and parent atoms migrate out of a mineral that is subsequently radiometrically dated, an inaccurate date will be obtained. This date will therefore be _____ the actual date.
   a. _____ younger than; b. _____ older than; c. _____ the same as; d. _____ impossible to determine; e. _____ none of the previous answers.

3. Who is generally considered the founder of modern geology?
   a. _____ Werner; b. _____ Lyell; c. _____ Steno; d. _____ Cuvier; e. _____ Hutton.
4. How many half-lives are required to yield a mineral with 1,250,000 atoms of thorium 232 and 18,750,000 atoms of lead 235?
   a. _____ 1; b. _____ 2; c. _____ 3; d. _____ 4; e. _____ 5.
5. In which type of radioactive decay are two protons and two neutrons emitted from the nucleus?
   a. _____ Alpha decay; b. _____ Beta decay; c. _____ Electron capture; d. _____ Fission track; e. _____ None of the previous answers.
6. Placing geologic events in sequential or chronologic order as determined by their position in the geologic record is
   a. _____ absolute dating; b. _____ correlation; c. _____ historical dating; d. _____ relative dating; e. _____ uniformitarianism.
7. What is being measured in radiometric dating is
   a. _____ the time when a radioactive isotope formed; b. _____ the time of crystallization of a mineral containing an isotope; c. _____ the amount of the parent isotope only; d. _____ when the dated mineral became part of a sedimentary rock; e. _____ when the stable daughter isotope was formed.
8. If a flake of biotite within a sedimentary rock (such as a sandstone) is radiometrically dated, the date obtained indicates when
   a. _____ the biotite crystal formed; b. _____ the sedimentary rock formed; c. _____ the parent radioactive isotope formed; d. _____ the daughter radioactive isotope(s) formed; e. _____ none of the previous answers.
9. If a radioactive element has a half-life of 32 million years, what fraction of the original amount of parent material will remain after 96 million years?
   a. _____ 1/2; b. _____ 1/8; c. _____ 1/16; d. _____ 1/4; e. _____ 1/32.
10. As carbon 14 decays back to nitrogen in radiocarbon dating, what isotopic ratio decreases?
    a. _____ Nitrogen 14 to carbon 12; b. _____ Carbon 14 to carbon 12; c. _____ Carbon 13 to carbon 12; d. _____ Nitrogen 14 to carbon 14; e. _____ Nitrogen 14 to carbon 13.
11. Can the various principles of relative dating be used to reconstruct the geologic history of Mars? Which principles might not apply to interpreting the geologic history of another planet?
12. How does metamorphism affect the potential for accurate radiometric dating using any and all techniques discussed in this chapter? How would such radiometric dates be affected by metamorphism, and why?
13. Describe the principle of uniformitarianism according to Hutton and Lyell. What is the significance of this principle?
14. A volcanic ash fall was radiometrically dated using the potassium 40 to argon 40 and rubidium 87 to strontium 87 isotope pairs. The isotope pairs yielded distinctly different ages. What possible explanation could be offered as to why these two isotope pairs yielded different ages? What would you do to rectify the discrepancy in ages?
15. What is the difference between relative dating and absolute dating?
16. Why do igneous rocks yield the most accurate radiometric dates? Why can't sedimentary rocks be dated radiometrically? What problems are encountered in dating metamorphic rocks?
17. Why were Lord Kelvin's arguments and calculations so compelling, and what was the basic flaw in his assumption? What do you think the course of geology would have been if radioactivity had not been discovered?
18. What is the major difference between the carbon-14 dating technique and the techniques used for the other radioactive isotope pairs discussed in this chapter?
19. Given the current debate over global warming and the many possible short-term consequences for humans, can you visualize how the world might look in 10,000 years or even 1 million years from now? Use what you have learned about plate tectonics and the direction and rate of movement of plates, as well as how plate movement and global warming will affect ocean currents, weather patterns, weathering rates, and other factors, to make your prediction. Do you think such short-term changes can be extrapolated to long-term trends in trying to predict what Earth will be like using a geologic time perspective?
20. This photograph was taken on Fish Creek Trail in the Sierra Nevada, California, and shows two black metavolcanic erratics on granodiorite bedrock that has been intruded by a white dike. Using your knowledge of igneous rocks and processes, glacial features, and relative dating principles, provide a geologic history of this area based on what is shown in this image.

Reed Wicander

Rocks, Fossils, and Time

Making Sense of the Geologic Record

Triassic Peak in Palo Duro Canyon State Park, near Amarillo, Texas. The rocks in this view are sedimentary, made up mostly of mud and sand that was deposited in a shallow, retreating sea and in stream channels and on floodplains. Their present surface exposure is caused by ongoing erosion, primarily by running water, but mass wasting (gravity-controlled) processes have contributed. This rock exposure is the geologic record for part of Texas, and although deposition was not continuous when the rocks formed, they nevertheless record events that took place between about 250 and 220 million years ago.

Sue Monroe

CHAPTER OBJECTIVES

At the end of this chapter, you will have learned that

▶ To analyze the geologic record, you must first determine the correct vertical sequence of rocks—that is, from oldest to youngest—even if they have been deformed.

▶ Although rocks provide our only evidence of prehistoric events, the record is incomplete at any one locality because discontinuities are common.

▶ Stratigraphy is a discipline in geology that is concerned with sedimentary rocks, most of which are layered or stratified, but many principles of stratigraphy also apply to igneous and metamorphic rocks.

▶ Several marine transgressions and regressions occurred during Earth history, at times covering much of the continents and at other times leaving the land above sea level.

▶ Fossils, the remains or traces of prehistoric organisms, are preserved in several ways, and some types of fossils are much more common than most people realize.

▶ Distinctive groups of fossils found in sedimentary rocks are useful for determining the relative ages of rocks in widely separated areas.

▶ The principles of superposition and fossil succession were used to piece together a composite geologic column that is the basis for the relative geologic time scale.

▶ Geologists have developed terminology to refer to rocks and to time.

▶ Several criteria are used to match up (correlate) similar rocks over large regions or to demonstrate that rocks in different areas are the same age.

▶ Absolute ages of sedimentary rocks are most often determined by radiometric dating of associated igneous or metamorphic rocks.

Introduction

The fact that Earth is an active planet is obvious from ongoing seismic activity, volcanism, plate movements, and processes such as running water, glaciers, wind, and waves that continually modify its surface. In short, Earth has evolved throughout its existence, and it continues to do so as a result of interactions among its systems, a fact that is abundantly clear from the **geologic record**—that is, the evidence of physical and biologic events preserved in rocks (see the chapter opening photo). In contrast, the other terrestrial planets (with the possible exception of Venus) and Earth's moon show little internal or surface activity.

As geologists decipher the geologic record, they study all of the groups of rocks, but they are especially interested in sedimentary rocks because these rocks preserve evidence of the surface processes responsible for them and, with few exceptions, they are the repository of fossils. Indeed, if it were not for fossils we would have no knowledge of now-extinct organisms such as dinosaurs and trilobites. But how is it possible to interpret events that no one witnessed? Perhaps an analogy will help.

Suppose you are walking in a forest and see a shattered, charred tree. You immediately dismiss the idea that the tree always looked as it does now, and begin to formulate hypotheses to explain your observations. Perhaps an exploding bomb can account for your observations, but in the absence of a bomb crater, or shrapnel (metallic fragments), or chemical traces from a bomb, you reject this hypothesis. Finally, you conclude that the tree was hit by lightning, a hypothesis that you can check by comparing the damaged tree to others known to have been hit by lightning. You can no doubt think of other examples of how to interpret events that you

did not witness—such as skid marks on a highway, broken glass, and a damaged power pole.

In our hypothetical example of a lightning strike, you did not witness the event, but you determined what happened because evidence of the event having occurred was present. It is irrelevant whether the event took place a few days ago or years ago; the only requirement is that the evidence still exists. Geologists use exactly the same reasoning to interpret the geologic record, although they may consider events that took place millions or billions of years ago. In short, they use the *principle of uniformitarianism* to figure out what happened during the past (see Chapters 1 and 4).

For example, we know that waves in shallow water deform sand into a series of ridges and troughs (ripple marks), and we have every reason to think that similar features in ancient rocks formed exactly the same way. However, an interesting note on deciphering Earth history comes from a publication by Philip Henry Gosse, a British naturalist, who in 1857 claimed that such features were simply created to look as if Earth had a long history. He proposed that Earth is only a few thousands of years old, and that all features in rocks, including fossil bones and shells, that looked like they formed long ago were created with that appearance.

Hardly anyone took Gosse's idea seriously, because they reasoned that if he were correct, a deceitful Creator had fabricated the geologic record, a thesis they could not accept. In addition, there was no way to check his idea because the geologic record would look exactly the same if he were correct or if it formed over a vast amount of time by processes like those operating now. Scientists opted for the uniformitarian approach, which holds that the geologic record is a faithful chronicle of past events that can be deciphered by understanding present-day processes.

Gosse's ideas on Earth history now serve only as a failed attempt to reconcile Earth history with a particular interpretation of Scripture. The uniformitarian method is much more fruitful; but keep in mind that the geologic record is complex, incomplete at any one location, and requires interpretation. In this chapter, we endeavor to explain how we make sense of this record.

Stratigraphy

Stratigraphy is a branch of geology concerned mostly with the composition, origin, age relationships, and geographic extent of layered, or stratified, rocks. It deals primarily with sedimentary rocks, because almost all of them are stratified (see the chapter opening photo), but the principles of stratigraphy also apply to layered igneous rocks as well as many metamorphic rocks.

In analyzing stratified rocks, geologists must determine both vertical and lateral relationships among individual rock layers. The vertical relationships among sedimentary rock layers are easily determined where they are well exposed, as in the walls of deep canyons. Lateral relationships are equally important, but they usually must be determined from a number of separate rock exposures, or what geologists call *outcrops*.

Vertical Stratigraphic Relationships

In vertical successions of sedimentary rocks, *bedding planes* separate individual strata from one another (▶ Figure 5.1a), or the strata grade vertically from one rock type into another. The rocks below and above a bedding plane differ in composition, texture, color, or a combination of these features,

indicating a rapid change in sedimentation or perhaps a period of nondeposition and/or erosion followed by renewed deposition. In contrast, gradually changing conditions of sedimentation account for those rocks that show a vertical gradation. Regardless of the nature of the vertical relationships among strata, the correct order in which they were deposited, that is, their relative ages, must be determined.

Superposition In Chapter 4, we discussed the *principle of superposition* proposed by Nicolas Steno during the 17th century. According to this principle, one can determine the correct relative ages of underformed strata by their position in a sequence; the oldest layer is at the bottom of the sequence with successively younger layers upward in the sequence (Figure 5.1a). If strata have been deformed by faulting, folding, or both, the task is more difficult (Figure 5.1b), but several sedimentary structures and some fossils allow geologists to resolve these kinds of problems (see Chapter 6).

The **principle of inclusions** is yet another way to figure out relative ages, because inclusions, or fragments, in a body of rock must be older than the rock itself. Obviously, the sand grains making up sandstone are older than the sandstone, but what about the relative ages of the granite and sandstone in ▶ Figure 5.2? Two interpretations are possible: the granite was intruded into the sandstone and thus is the youngest; or the granite was the source of the sand and the sandstone is youngest.

Suppose you encounter a sequence of mostly sedimentary rocks, but one layer is made up of basalt (▶ Figure 5.3). If the basalt is a buried lava flow, you can determine its relative age by the principle of superposition, but if it is a *sill*—a sheet-like intrusive body—it is younger than the layers below it and younger than the layer immediately above it. Study Figure 5.3 closely and note that the principle of inclusions as well as contact metamorphic effects help resolve this problem.

▶ Figure 5.1 **Stratified Rocks and Superposition**

(a) The individual layers, or strata, in these rocks are separated from one another by bedding planes. Superposition tells us that the oldest layer is in the lower part of the outcrop and the youngest layer is at the top.

(b) These rocks have been deformed so that you cannot determine which is oldest and youngest with the data in this image. However, sedimentary structures can be used to solve problems such as this (see Chapter 6).

From MONROE/WICANDER, The Changing Earth, 6e. Copyright © 2012 Cengage Learning

► Figure 5.2 **The Principle of Inclusions**

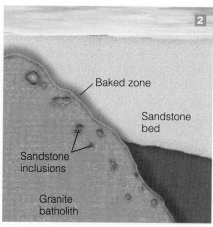

(a) The sandstone is older than the granite batholith because there are inclusions of sandstone in the granite. The sandstone also shows evidence of having been baked along its contact with the granite batholith when the granitic magma intruded the overlying sedimentary beds.

Sandstone bed

Granitic magma

Baked zone

Sandstone bed

Sandstone inclusions

Granite batholith

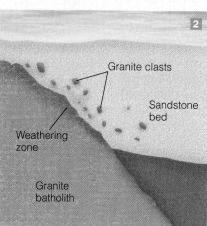

(b) The sandstone is younger than the granite batholith because it contains pieces (clasts) of granite. The granite is also weathered along the contact with the sandstone, indicating that it was the source of the granite clasts, and must therefore be older than the sandstone.

Sandstone deposition

Sandstone bed

Granite batholith

Granite clasts

Sandstone bed

Weathering zone

Granite batholith

Unconformities So far, we have discussed vertical relationships among **conformable** strata—that is, sequences of rocks in which deposition was more or less continuous. A bedding plane between strata may represent a depositional break of anywhere from minutes to tens or hundreds of years but is inconsequential in the context of geologic time. However, in many sequences of strata, surfaces known as **unconformities** are present that represent times of nondeposition and/or erosion that encompass long periods of geologic time, perhaps millions or tens of millions of years. Accordingly, the geologic record is incomplete at that particular location, just as a book with

► Figure 5.3 **How to Determine the Relative Ages of Lava Flows, Sills, and Associated Sedimentary Rocks**

© 2013 Cengage Learning

Strands and shreds of underlying sedimentary material may be in base of flow

Baking zone at bottom of flow

Lava flow

Clasts of lava may be present in overlying, younger layer

Rubble zones may be present at top and bottom of flow

Baking zones on both sides of sill

Sill

Inclusions of rock from both layers above and below may exist in sill

(a) A buried lava flow has baked underlying bed 2, and clasts of the lava were deposited along with other sediment in bed 4. The lava flow is older than beds 4, 5, and 6.

(b) The layers above and below the sill have been baked, so the sill is younger than layers 2 and 4. Can you determine its age relative to beds 5, 6, and 7?

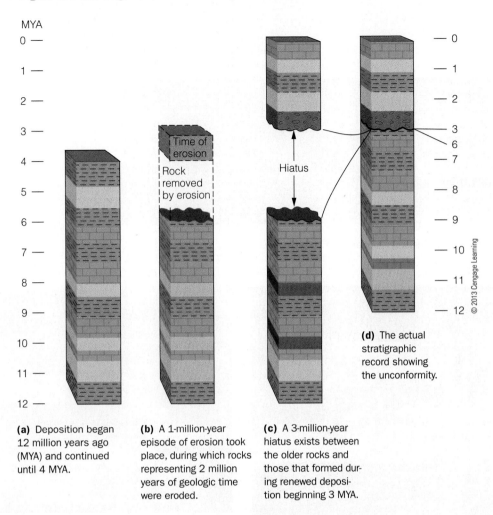

(a) Deposition began 12 million years ago (MYA) and continued until 4 MYA.

(b) A 1-million-year episode of erosion took place, during which rocks representing 2 million years of geologic time were eroded.

(c) A 3-million-year hiatus exists between the older rocks and those that formed during renewed deposition beginning 3 MYA.

(d) The actual stratigraphic record showing the unconformity.

missing pages is incomplete, and the interval of geologic time not represented by strata is a *hiatus* (►Figure 5.4).

Unconformity is a general term that encompasses three distinct types of surfaces called *disconformity*, *nonconformity*, and *angular unconformity*. Furthermore, unconformities of regional extent may change from one type to another, and they do not necessarily encompass equivalent amounts of geologic time everywhere (►Figure 5.5a). Unconformities are common, but the geologic time not recorded by rocks in one area is represented by rocks elsewhere.

A **disconformity** is an erosion surface in sedimentary rocks that separates younger rocks from older rocks, both of which are parallel to each other (Figure 5.5b). However, an erosion surface cut into plutonic rocks or metamorphic rocks that is overlain by sedimentary rocks is a **nonconformity** (Figure 5.5c). And, finally, an **angular unconformity** is present if the strata below an erosion surface are inclined at some angle to the strata above (Figure 5.5d). In this case, we can infer that the sequence of events included deposition, lithification, deformation, uplift, erosion, and, finally, renewed deposition.

Lateral Relationships—Facies

In addition to the principle of superposition, Nicholas Steno also proposed the *principle of lateral continuity*, meaning that layers of sediment or sedimentary rock extend outward in all directions until they terminate (see Figure 4.4a). Even if rock layers have been disrupted by erosion, we know that they originally extended across the intervening areas where they are no longer present (►Figure 5.6a). Rock layers may terminate abruptly where cut by faults, or they may become progressively thinner and eventually pinch out or intertongue with adjacent rocks. They may also change by *lateral gradation* as their composition and/or texture becomes increasingly different (Figure 5.6b).

Both intertonguing and lateral gradation indicate the simultaneous operation of different depositional processes in adjacent environments. For example, on the continental shelf, sand may accumulate in the high-energy nearshore environment, while at the same time mud and carbonate deposition takes place in offshore low-energy environments. Deposition in each of these laterally adjacent environments yields a **sedimentary facies**, a body of sediment with distinctive physical, chemical, and biological attributes.

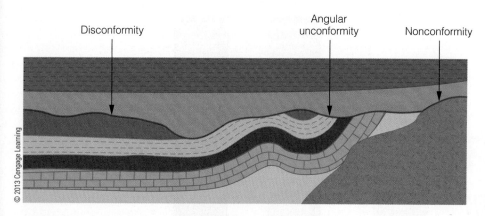

Disconformity   Angular unconformity   Nonconformity

© 2013 Cengage Learning

(a) Cross section showing the regional variation in an unconformity.

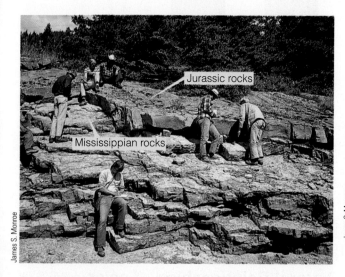

Jurassic rocks

Mississippian rocks

James S. Monroe

(b) A disconformity. The surface between the Mississippian and Jurassic rocks represents a 165-million-year break in the geologic record at this location in Montana.

Marli Bryant Miller

(d) Angular unconformity at Siccar Point, Scotland. James Hutton first realized the significance of unconformities at this site in 1788.

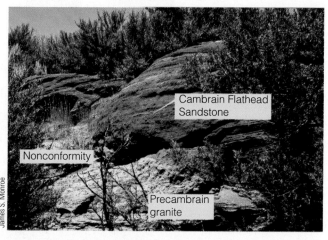

Cambrain Flathead Sandstone

Nonconformity

Precambrain granite

James S. Monroe

(c) A nonconformity between Precambrian granite and the Cambrian Flathead Formation in the Bighorn Mountains of Wyoming.

Armanz Gressly, in 1838, was the first to use the term *facies* when he carefully traced sedimentary rocks in the Jura Mountains of Switzerland and noticed lateral changes such as sandstone grading into shale. He reasoned that these changes indicated deposition in different environments that lie next to one another. Any attribute of sedimentary rocks that makes them recognizably different from laterally adjacent rocks of about the same age is sufficient to establish a sedimentary facies.

Marine Transgressions and Regressions

There are many reasons to visit Grand Canyon National Park in Arizona, not the least of which is the spectacular scenery. In addition, the Grand Canyon allows us to appreciate the magnitude of geologic time when we realize that the rocks in the canyon's walls represent more than 1 billion years of Earth history. If you look closely at the

(a) These sedimentary rocks in Monument Valley Navajo Tribal Park in Arizona originally extended outward and covered this entire area.

(b) Lateral termination of rock layers by intertonguing, pinchout, and lateral gradation.

Cambrian-age strata, you will see the Tapeats Sandstone is overlain by the Bright Angel Shale and finally by the Muav Limestone (▶Figure 5.7). Each of these rock units represents a facies, each with fossils of marine-dwelling trilobites and brachiopods. Superposition tells us the relative ages of the rocks, but how can we account for marine deposits far from the sea, and how were the facies deposited in the order observed?

Notice in ▶Figure 5.8a that sandstone, shale, and limestone facies lie adjacent to one another, and that when sea level rises, the shoreline migrates landward, as do the facies paralleling the shoreline, and offshore facies are superposed over nearshore facies. So, when sea level rises with respect to the land, a **marine transgression** takes place that accounts for the vertical sequence of facies in Figure 5.7. Notice also, in Figure 5.8a, that another aspect of a marine transgression is that each facies becomes younger

in a landward direction. In other words, the facies are *time transgressive*, meaning that as the shoreline migrated landward, the facies were not deposited simultaneously over their entire geographic extent.

Similar sequences of Cambrian-age rocks in South Dakota, Wyoming, and Montana indicate the marine transgression depicted in the Grand Canyon (Figure 5.7) was widespread. Indeed, as the sea transgressed from the west, it also transgressed from the east and covered much of North America, a topic we cover more fully in Chapter 10. But obviously the sea is no longer present in the areas just mentioned, so the Cambrian marine transgression must have ended. In fact, it did—during a **marine regression**, when sea level fell with respect to the land and the environments that paralleled the shoreline migrated seaward. In this case, nearshore facies were superposed over offshore facies, and the individual facies become younger in a seaward direction (Figure 5.8b).

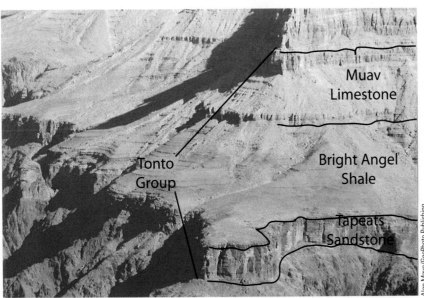

▶ Figure 5.7 **Sedimentary Rocks in the Grand Canyon** View of three formations in the Grand Canyon of Arizona. These rocks, representing three facies, were deposited during a marine transgression. Compare with the vertical sequence of rocks in Figure 5.8.

▶ Figure 5.8 **Marine Transgressions and Regressions**

(a) Three Stages of a Marine Transgression

Offshore
Low-energy | Near shore
High-energy

Limestone facies | Shale facies | Sandstone facies | Land surface

Time line

Time lines

Time lines

Limestone
Shale
Sandstone
Old land surface

(b) Three Stages of a Marine Regression

Sandstone
Shale
Limestone
Old land surface

In our discussion of marine transgressions and regressions, we considered both vertical and lateral facies relationships, the significance of which was first recognized by Johannes Walther (1860–1937). When Walther traced rock units laterally, he reasoned, as Gressly had earlier, that each sedimentary facies he encountered was deposited in laterally adjacent environments. In addition, Walther noticed that the same facies he found laterally were also present in a vertical sequence. His observations have since been formulated into **Walther's law**, which holds that the facies seen in a conformable vertical sequence will also replace one another laterally.

The application of Walther's law is well illustrated by the marine transgression and regression shown in Figure 5.8. In practice, it is usually difficult to follow rock units far enough laterally to demonstrate facies changes. It is much easier to observe vertical facies relationships and use Walther's law to work out the lateral relationships. Remember,

though, that Walther's law applies only to conformable sequences of rocks; rocks above and below an unconformity are unrelated, and Walther's law does not apply.

Extent, Rates, and Causes of Marine Transgressions and Regressions

Since the Neoproterozoic, six widespread marine transgressions and regressions have taken place in North America, each yielding unconformity-bounded rock sequences (see Figure 10.3). In fact, these rock sequences provide the stratigraphic framework for our discussions of Paleozoic, and, to a lesser degree, Mesozoic geologic history. Transgressions and regressions also took place during the Cenozoic Era, but they were not as important as during the previous eras. The paleogeographic maps in Chapters 10, 11,

© 2013 Cengage Learning

14, and 16 show the extent to which North America was covered by shallow seas or exposed during these events.

To determine how widespread these transgressions and regressions were, geologists must analyze the geologic record and thereby determine how far the sea migrated inland or how far it withdrew. The rate at which a transgression takes place is probably no more than a few centimeters per year. For instance, suppose that a shoreline moved inland 1,000 km in 20 million years. That would mean an average transgression rate of 5 cm/yr. However, neither transgressions nor regressions involve the sea moving steadily landward or seaward. Instead, there are many minor reversals in either trend, thus accounting for some of the intertonguing we see among sedimentary rock units (Figure 5.6b).

Transgressions and regressions take place in response to any tectonic movement that either elevates the land with respect to sea level or that causes subsidence (downward movement) of the land. Geologic evidence indicates that during the Cretaceous Period, sea level may have been as much as 250 m higher than it is now. As a result, large parts of the continents were covered by the oceans (see Figure 14.6). The probable cause of this inundation was comparatively rapid seafloor spreading during which heat beneath the mid-oceanic ridges caused them to expand and thus displace water onto the continents. When seafloor spreading slows, the mid-oceanic ridges subside, increasing the volume of the ocean basins, and the seas retreat from the continents.

Another cause of transgressions and regressions is the volume of water in the oceans. Widespread glaciers present during the Pennsylvanian to Late Permian periods caused several sea-level changes resulting in transgressions and regressions. And then again during the Pleistocene Epoch (2.6 million to 11,700 years ago), sea level was about 130 m lower than it is now because a huge volume of water that came from the oceans was frozen in glaciers on land. When the glaciers melted, sea level rose to its present level. If all of Earth's present-day glacial ice were to melt, sea level would rise by about 70 m.

Fossils and Fossilization

Today, it would be difficult to find anyone who doubts that bones, shells, and teeth in ancient rocks are the remains of once-living organisms, and yet this view is a rather recent one. During most of historic time in the Western world, people variously believed such objects formed in rocks by some kind of molding force, or that they were placed in rocks by the Creator to test our faith or by Satan to sow seeds of doubt. And remember that Philip Henry Gosse believed fossils were simply created aspects of rocks, not the remains of plants and animals (see the Introduction). Some perceptive observers (Leonardo da Vinci in 1508, Robert Hooke in 1665, and Nicolas Steno in 1667) recognized the true nature of these objects, but their views were largely ignored.

We now define a **fossil** as the remains or traces of prehistoric organisms preserved in rocks. They are by far most common in sedimentary rocks, but they may also be found in volcanic ash deposits and volcanic mudflows—and, rarely, in other igneous rocks or metamorphic rocks. Geologists use fossils extensively to determine environments of deposition (see Chapter 6) and as some of the evidence for evolution (see Chapter 7). Fossils are also essential for determining the relative ages of strata in widely separated areas.

You already know that you can deduce relative ages from superposition, cross-cutting relationships, and inclusions. However, suppose you encounter the situation depicted in ▸Figure 5.9. For the rocks in either column, or what geologists call *columnar sections*, the relative ages are apparent, but with the information provided you cannot determine the ages of the rocks in one column compared to those in the other one. The resolution to problems such as this involves using fossils and some physical events of short duration, such as volcanic ash falls.

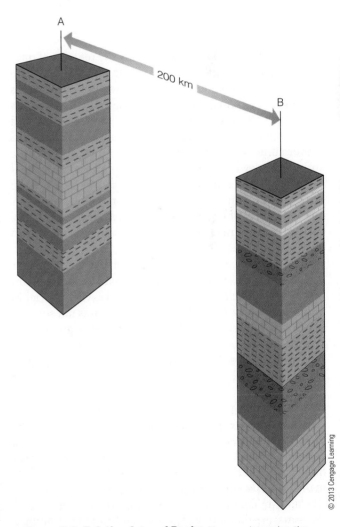

© 2013 Cengage Learning

▸ Figure 5.9 **Relative Ages of Rocks** You can determine the relative ages of rocks in either column A or B, but you cannot tell the relative ages of rocks in A compared to B. The rocks in A may be younger than those in B, the same age as some in B, or older than all of the rocks in B.

(a) Skull of the dinosaur *Allosaurus* on display at the Natural History Museum in Vienna, Austria.

(b) These fossil trilobites are on display in the Mesalands Community College's Dinosaur Museum in Tucumcari, New Mexico. Although distantly related to today's insects, spiders, crabs, and lobsters, trilobites went extinct at the end of the Paleozoic Era, 251 MYA.

How Do Fossils Form?

It is now apparent that fossils are truly the remains of once living organisms, and it is also clear that many fossils are of organisms now extinct. Our definition of *fossil* includes the term "remains," or what are called **body fossils** (► Figure 5.10), that consist mostly of shells, bones, and teeth—that is, the actual skeletal parts of organisms. Under some unusual circumstances, however, even the soft parts may be preserved by freezing or mummification. Recall that our definition of *fossil* also includes the term "traces," which are not skeletal remains but rather any indication of organic activity. These **trace fossils**, as they are called, include tracks, trails, burrows, nests, and even fossilized feces (► Figure 5.11).

As you might expect, the most favorable conditions for the preservation of body fossils are that (1) an organism has a durable skeleton and (2) that it lives where burial in sediment is likely. Given these conditions, it is not surprising that corals, clams, and brachiopods have good fossil records, whereas jellyfish with no skeleton are very poorly represented as fossils. Bats have delicate skeletons and they live where burial is not so likely, so they too have a poor fossil record.

Even given these qualifications and the fact that bacterial decay, scavenging, and metamorphism may destroy organic remains, fossils are far more common than most people realize, because so many billions of organisms have existed during so many millions of years. Indeed, at many localities you can collect dozens or hundreds of fossil clams, brachiopods, trilobites, and remains of several other marine-dwelling shelled animals. Although finding a rather complete dinosaur is unusual, pieces of fragmented bone and teeth are common in some areas. It is true that for fossilization to occur, shells, bones, or teeth must be buried rapidly in some protective medium such as sand

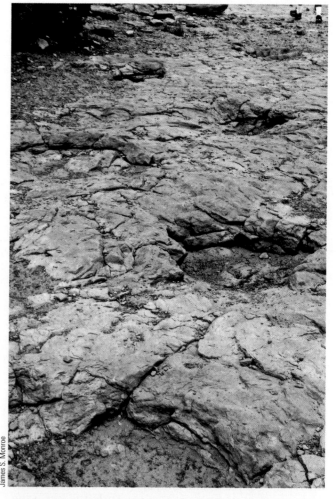

► **Figure 5.11** **Trace Fossils** These circular depressions in sandstone at Copper Ridge in Utah are trace fossils, that is, some indication of organic activity. They are tracks, which measure about 35 cm across, that were made by a large dinosaur.

▶ Figure 5.12 **Unaltered Remains** "Unaltered" means that there has been no change in composition or structure of the organism.

(a) Insects preserved in amber, which is hardened tree resin.

(b) Frozen baby mammoth found in Russia in 1989. The 44,000-year-old carcass is about 1.0 m tall.

or mud, but "rapidly" means only that remains are buried before being destroyed by physical or chemical processes, which might take many years in some environments.

Body fossils may be preserved as *unaltered remains* (▶Figure 5.12), meaning that they retain their original composition and structure, as opposed to *altered remains* (Table 5.1, ▶Figure 5.13), in which case some change has taken place. For instance, the shells of clams are made up of calcium carbonate, but during fossilization their shells may be replaced by silicon dioxide. *Molds* and *casts* represent a type of fossil preservation that does not include remains, but nevertheless preserves the physical characteristics of an organism (Table 5.1). For example, a mold forms when buried organic remains dissolve, leaving a cavity shaped like a clam or bone. If minerals or sediment fill the mold, it forms a cast—that is, a replica of the original (▶Figure 5.14).

One sometimes hears in the popular media that fossilization is a rare event requiring circumstances unlike any existing at present. Perhaps some of you have visited, or at least have heard of, the La Brea Tar Pits in Los Angeles, California, where hundreds of fossil mammals, birds, insects, and other organisms have been recovered. This remarkable concentration of fossils may seem unrelated to everyday processes, but small animals today are trapped in the sticky residue at oil seeps at the La Brea Tar Pits. Burial in volcanic ash should come as no surprise because that is exactly what happened during recent eruptions, and a similar event accounts for the dozens of Miocene-age fossil rhinoceroses, three-toed horses, saber-toothed deer, and other animals found at Ash Fall State Historic Park in Nebraska. Wildebeests by the hundreds drown in river crossings, probably much like the fate that befell horned dinosaurs in Canada millions of years ago. In short, the conditions for fossilization are readily understandable.

The record of ancient life, or simply the *fossil record*, must be analyzed and interpreted just as the geologic

TABLE 5.1 Types of Fossil Preservation

Body fossils—unaltered remains	Original composition and structure preserved
Freezing	Large Ice Age mammals frozen in sediment
Mummification	Air drying and shriveling of soft tissues
Preservation in amber	Leaves, insects, and small reptiles trapped and preserved in hardened tree resin
Preservation in tar	Bones, insects preserved in asphaltlike substance at oil seeps
Body fossils—altered remains	Change in composition and/or structure of original material
Permineralization	Addition of minerals to pores and cavities in shells and bones
Recrystallization	Change in the crystal structure—for example, aragonite in shells recrystallizes as calcite
Replacement	One chemical compound replaces another—for example, pyrite (FeS_2) replaces calcium carbonate ($CaCO_3$) of shells
Carbonization	Volatile elements lost from organic matter, leaving a carbon film; most common for leaves and insects
Trace fossils	Any indication of organic activity such as tracks, trails, burrows, droppings (coprolites), and nests
Molds and casts	Mold—a cavity with the shape of a bone or shell; cast—a mold filled by minerals or sediment

(a) Fossil trees at Petrified Forest National Park in Arizona (see Chapter 14 Perspective). Silica has replaced the organic material of the trees.

(b) Carbonized leaf on display at Florissant Fossil Beds National Monument in Colorado.

(c) This carbonized insect is from Oligocene-age rocks in Montana.

record, of which it is a part. As a repository of prehistoric organisms, the fossil record provides our only knowledge of numerous extinct organisms. Furthermore, many

fossils are interesting in their own right, and many are sought out by avid collectors and for museum displays. And, finally, the study of fossils has several practical applications that we discuss in the following section and in Chapter 6.

Fossils and Telling Time

We began the section on fossils by referring to the columnar sections in Figure 5.9 and asking how you might determine the relative ages of the strata in these geographically separate areas. Our answer was "fossils"—but exactly how do fossils resolve this problem? To fully understand the usefulness of fossils, we must examine the historical development of an important geologic principle.

The use of fossils in relative dating and geologic mapping was first demonstrated during the early 1800s in England and France. William Smith (1769–1839), an English civil engineer who was surveying and building canals in southern England, independently discovered Steno's principle of superposition (see Chapter 10 Introduction). He reasoned that in a sequence of strata, the oldest is at the bottom and the youngest is at the top, and he came to the same conclusion regarding any fossils the rocks contained. Smith made numerous observations at outcrops, mines, and quarries and discovered that the sequence of fossils, and especially groups of fossils, is consistent from area to area. In short, he discovered that the relative ages of sedimentary rocks at different locations could be determined by their fossil content (▶ Figure 5.15).

By recognizing the relationship between strata and fossils, Smith could predict the order in which fossils would appear in rocks at some locality he had not previously visited. In addition, his knowledge of rocks and fossils allowed him to predict the best route for a canal and the best areas for bridge foundations. As a result, his services were in great demand.

Smith gets much of the credit for the idea of using fossils to determine relative ages, but other geologists, such as Alexander Brongniart in France, also recognized this relationship. In any case, their observations served as the basis for what we now call the **principle of fossil succession** (also known as the *principle* or *law of faunal succession*). This important principle holds that fossil assemblages (groups of fossils) succeed one another through time in a regular and determinable order.

But why not simply match up similar rock types in Figure 5.9 and conclude they are the same age? Matching up the same kinds of rocks in the two sections and assuming that they are the same age will not work, because the same kinds of rock, such as sandstone, for example, has formed continuously throughout geologic time. Fossils have also formed continuously, but for any one kind of extinct organism, its existence marks two points in time: its time of appearance and its time of extinction. Furthermore, geologists rarely rely on a single type of fossil but rather on fossil assemblages that

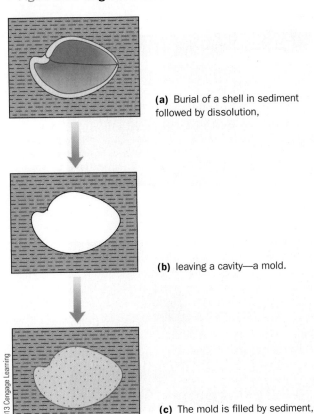

▶ Figure 5.14 **Origin of a Mold and a Cast**

(a) Burial of a shell in sediment followed by dissolution,

(b) leaving a cavity—a mold.

(c) The mold is filled by sediment, thereby forming a cast.

© 2013 Cengage Learning

have distinctive aspects compared to older or younger assemblages. Accordingly, if we match up rocks with similar fossil assemblages, then we can indeed assume they are of the same relative age.

The Relative Geologic Time Scale

Investigations of rocks by naturalists between 1830 and 1842 resulted in the recognition of rock bodies called *systems* and the construction of a composite geologic column that is the basis for the relative geologic time scale (▶Figure 5.16). A short discussion about the methods used to recognize and define the systems will help you understand how the geologic column and relative geologic time scale were established.

During the 1830s, Adam Sedgwick studied rocks in northern Wales and described the Cambrian System, and Sir Roderick Impey Murchison named the Silurian System in southern Wales. Sedgwick paid little attention to the few fossils present, and his Cambrian System could not be recognized beyond the area where it was first described. Murchison, however, carefully described fossils typical of his Silurian System, so his system could be identified elsewhere.

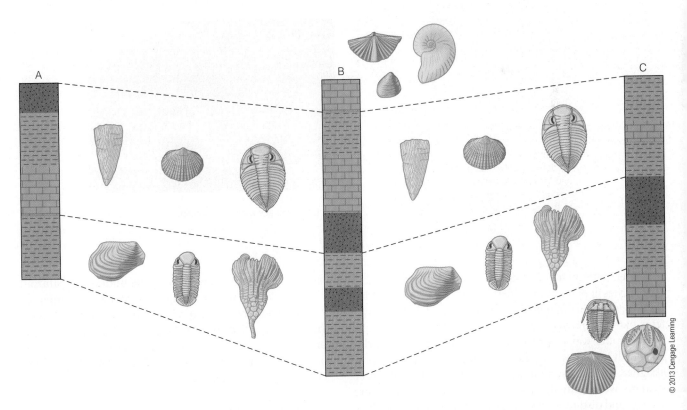

▶ Figure 5.15 **The Principle of Fossil Succession** This generalized cross section shows how geologists use the principle of fossil succession to determine the relative ages of rocks in widely separated areas. The rocks encompassed by the dashed lines contain similar fossils and are thus the same age. Note that the youngest rocks are in column B, whereas the oldest ones are in column C.

▶ **Figure 5.16** **The Geologic Column and Relative Geologic Time Scale**

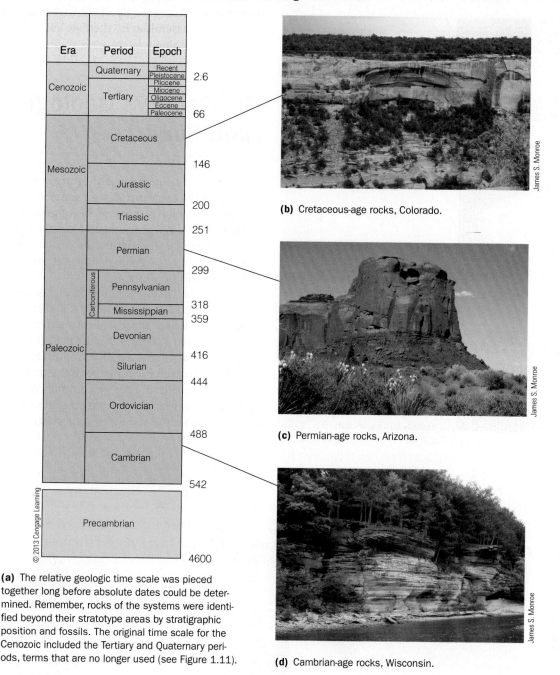

Era	Period	Epoch	
Cenozoic	Quaternary	Recent / Pleistocene	
		Pliocene	2.6
	Tertiary	Miocene / Oligocene / Eocene / Paleocene	66
Mesozoic	Cretaceous		146
	Jurassic		200
	Triassic		251
Paleozoic	Permian		299
	Pennsylvanian (Carboniferous)		318
	Mississippian (Carboniferous)		359
	Devonian		416
	Silurian		444
	Ordovician		488
	Cambrian		542
	Precambrian		4600

© 2013 Cengage Learning

(a) The relative geologic time scale was pieced together long before absolute dates could be determined. Remember, rocks of the systems were identified beyond their stratotype areas by stratigraphic position and fossils. The original time scale for the Cenozoic included the Tertiary and Quaternary periods, terms that are no longer used (see Figure 1.11).

James S. Monroe

(b) Cretaceous-age rocks, Colorado.

James S. Monroe

(c) Permian-age rocks, Arizona.

James S. Monroe

(d) Cambrian-age rocks, Wisconsin.

When both men published the results of their studies in 1835, it was apparent that Silurian rocks were younger than Cambrian strata, but it was also apparent that their two systems partially overlapped, resulting in a boundary dispute. This dispute was not resolved until 1879 when Charles Lapworth suggested that the strata in the area of overlap be assigned to a new system, the Ordovician. So finally three systems were named, their stratigraphic positions were known, and distinctive fossils of each system were described.

Before Sedgwick's and Murchison's dispute, they had named the Devonian System, which contained fossils distinctly different from those in the underlying Silurian System rocks and the overlying rocks of the Carboniferous System.* Then in 1841, Murchison visited western Russia, where he identified strata as Silurian, Devonian, and Carboniferous by their fossil content. And overlying the Carboniferous strata were fossil-bearing rocks he assigned to a Permian System (Figure 5.16a).

---

*In North America, two systems, the Mississippian and Pennsylvanian, closely correspond to the Lower and Upper Carboniferous, respectively.

TABLE 5.2 Classification of Stratigraphic Units

Units Defined by Content		Units Expressing or Related to Geologic Time	
Lithostratigraphic Units	Biostratigraphic Units	Time-Stratigraphic Units	Time Units
Supergroup	Biozone	Eonothem . . . . Eon	
Group		Erathem . . . . .Era	
Formation		System . . . . Period	
Member		Series . . . . Epoch	
Bed		Stage . . . . Age	

© 2013 Cengage Learning

We need not discuss the specifics of where and when the other systems were established, except to say that superposition and fossil content were the criteria used to define them. The important point is that geologists were piecing together a composite **geologic column**, which is in effect a **relative geologic time scale**, because the systems are arranged in their correct chronologic order (Figure 5.16a). You should be aware of another aspect of the relative geologic time scale: All rocks beneath Cambrian strata are called Precambrian. Long ago, geologists realized that these rocks contain few fossils and that they could not effectively subdivide them into systems. In Chapters 8 and 9, we will discuss the terminology for Precambrian rocks and time.

It is important to know that although scientists today accept that organic evolution accounts for the differences in organisms through time, evolution is not the basis for the geologic time scale. Indeed, the geologic column and time scale were pieced together by the 1840s by scientists who fully accepted the biblical account of creation and had little or no idea about evolution—a concept that would not be formalized until 1859. The only requirement for fossils to be useful, as we have outlined, is that they differ through time, regardless of what accounts for those differences.

Stratigraphic Terminology

Establishing systems and a relative geologic time scale brought some order to stratigraphy, but problems remained. Because many sedimentary rock bodies are time transgressive (Figure 5.8), they may belong to one system in one area and to another system elsewhere, or they may simply straddle the boundary between systems. Accordingly, geologists now use terminology for two fundamentally different kinds of units: those defined by their content and those expressing or related to geologic time (Table 5.2).

Stratigraphic Units Defined by Their Content

Rock type with no consideration of time or mode of origin is the only criterion used to define a **lithostratigraphic***

Lith and *litho* mean "stone" or "stonelike."

unit, which has unifying features that set it apart from other lithostratigraphic units (Table 5.2). The basic lithostratigraphic unit is the **formation**, which is a mappable body of rock with distinctive upper and lower boundaries (see Perspective). Formations may consist of a single rock type (the Redwall Limestone) or a variety of related rock types (the Morrison Formation), and the term *formation* may also apply to metamorphic and igneous rocks (the Lovejoy Basalt). Many formations are subdivided into *members* and *beds*, and they may be parts of more inclusive units such as *groups* and *supergroups* (▶Figure 5.17, Table 5.2).

Biostratigraphic units, in contrast, are defined solely on the basis of their fossil content with no regard to rock type or time of origin (Table 5.2), and their boundaries do not necessarily correspond with those of lithostratigraphic units. The fundamental biostratigraphic unit is the **biozone**, which is discussed more fully in the following section on correlation.

Stratigraphic Units Expressing or Related to Geologic Time

Geologists define a **time-stratigraphic unit** (also called a *chronostratigraphic* * unit*) as the rock that formed during a particular interval of geologic time (Table 5.2). The most commonly used time-stratigraphic unit is the **system**, which is based on a *stratotype* consisting of rocks in the area where the system was first described. Systems are commonly recognized beyond their stratotype areas by their fossil content. Remember our discussion on how geologists defined and recognized the systems during the 1830s and 1840s, and how these systems served as the basis for the relative geologic time scale (Figure 5.16).

Time units are simply designations for certain intervals of geologic time. The **period** is the most commonly used time unit, but two or more periods may be designated as an *era*, and two or more eras make up an *eon*. Periods may also be subdivided into shorter intervals such as *epochs* and *ages*. All of these time units have corresponding time-stratigraphic units, although the terms *erathem* and *eonothem* are not often used (Table 5.2).

These two types of units referring to time and their relationship are particularly confusing to beginning students.

Chron and *chrono* are added to words to indicate time.

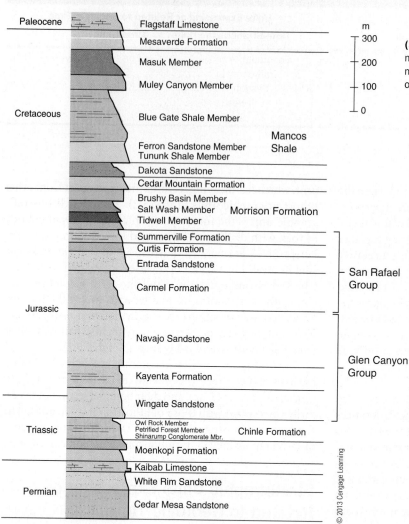

(a) Notice that some of the formations are further divided into members, and some are parts of more inclusive groups.

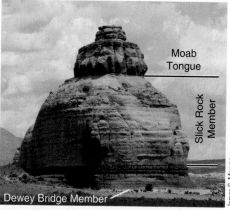

(b) More detail on the Entrada Sandstone at Church Rock near Moab, Utah. A tongue is simply a wedge-shaped member. Church Rock is an erosional remnant of the formation that originally covered this area.

James S. Monroe

© 2013 Cengage Learning

Remember, though, that time-stratigraphic units are material bodies of rock that occupy a position in a sequence of strata (for example, the Devonian System), whereas time units refer only to time (the Devonian Period). Thus, a system is a body of rock with lower, medial, and upper parts. In contrast, a time unit such as the Devonian Period is the time during which the Devonian System was deposited, and we refer to early, middle, and late subdivisions.

Consider this example and perhaps the distinction between time-stratigraphic units and time units will be clearer. Think of a three-story building that was built one floor at a time, the lowest first and so on. The building—let us call it Brooks Hall—is the Brooks Hall System, a material body that was built during a specific interval of time, the Brooks Hall Period. Lower Brooks Hall was built during the Early Brooks Hall Period and so on. And because *system* refers to a position in a sequence, we would refer to the third floor as Upper Brooks Hall but not Late Brooks Hall, and likewise we would call the first floor Lower Brooks Hall but not Early Brooks Hall.

Correlation

In geology, the term **correlation** refers to matching up geologic phenomena in two or more areas. For example, we may simply correlate the same rock units, with no regard to time, over an area in which they are no longer continuous, in which case we refer to *lithostratigraphic correlation*. In a previous section, we noted that systems were identified beyond their stratotype areas by applying the principle of fossil succession. In this case, we match up rocks of the same relative age, which is a *time-stratigraphic correlation*.

Correlation of lithostratigraphic units involves demonstrating that a formation or group was once continuous over a particular area. If outcrops are adequate, lithostratigraphic units may be traced laterally even if occasional gaps are present (▶ Figure 5.18a). Other criteria for lithostratigraphic correlation include the presence of a distinctive key bed, position in a sequence, and composition (Figure 5.18b); keep in mind that composition indicates only the

Perspective

The Goblins of Goblin Valley State Park, Utah

Southern Utah is remarkable for its spectacular scenery, vast exposures of sedimentary rocks, and its concentration of national parks and monuments. Several of these have been mentioned or featured in images in this book. Here, however, we concentrate on the stratigraphy as well as weathering and erosion that have yielded bizarre landforms in Goblin Valley State Park near the small town of Hanksville, Utah.

The Goblin Valley area is part of the much more extensive Colorado Plateau, which we discuss more fully in Chapters 14 and 16 (see Figure 16.9). Accordingly, here we will note only that it is well known for, among other features, the Grand Canyon of Arizona (see Chapter 4 Perspective). The stratigraphy of the Colorado Plateau is a study in itself, but at Goblin Valley only four sedimentary rock formations are exposed, all of Middle to Late Jurassic age (Figure 1). The oldest formation is the Entrada Sandstone, which, as its name implies, is made up mostly of sandstone, but it also contains layers of siltstone and shale, all of which were deposited on ancient tidal flats.

Figure 1 The stratigraphic succession at Goblin Valley State Park in Utah. The Entrada Sandstone as well as the Curtis and Summerville formations were deposited in marine environments, whereas the Morrison Formation was deposited on land in stream channels, floodplains, and lakes. The rock units are of Middle to Late Jurassic age.

The Entrada Sandstone is widespread in the Colorado Plateau region. Indeed, we used it in an example in our discussion of lithostratigraphic terminology (Figure 5.17b), and it is the rock unit in which the arches at Arches National Park form. At Arches, however, the formation is mostly a wind-blown dune deposit, representing another facies in the formation. Its reddish color is due to the presence of small amounts of iron oxide.

Figure 2 Some of the goblins (hoodoos) at Goblin Valley State Park in Utah in the Entrada Sandstone.

(a) These are some of the older hoodoos at Goblin Valley. The hoodoo that was once on the pillar at the left has collapsed, a fate that will take place for the other three at some future time.

(b) A close-up view of one of the more unusual hoodoos.

Although it is not obvious in Figure 1, the greenish Curtis Formation lies unconformably on the Entrada Sandstone, and in turn is overlain by the Summerville Formation, and finally by the Morrison Formation. The Curtis and Summerville formations are marine deposits, but the Morrison Formation was deposited in streams, on floodplains, and in lakes, and is famous for its fossil dinosaurs (see Figure 14.18). No dinosaurs have been found at Goblin Valley, but many have been recovered from the formation in many other areas, mostly notably at Dinosaur National Monument on the Utah–Colorado border.

In mythology, a goblin is usually portrayed as a small, grotesque creature with magical powers, as were some of the creatures in the *Harry Potter* movies. The goblins at Goblin Valley State Park are up to 3 m high and are actually *hoodoos*, a geologic name for pillars of easily eroded rock of variable width capped by more resistant rock (Figure 2). Here, they are

Figure 3 Hoodoos form as this cliff of Entrada Sandstone weathers and erodes. Some of the hoodoos are up to 3 m high. Notice that the harder sandstone layers form the tops of the hoodoos.

found in the Entrada Sandstone, although similar features are present in many other areas in other rock units. Nevertheless, the hoodoos at Goblin Valley are numerous, some of the most unusual anywhere, and they form continually.

Differential weathering and erosion along fractures account for the origin of the Goblin Valley hoodoos, but

another probable contributing factor is differences in the amount of chemical cement in the Entrada Sandstone. In any case, some parts of the Entrada Sandstone weather and erode more rapidly than others, so as these processes continue, new goblins appear and older ones eventually wear away (Figures 2 and 3).

▶ **Figure 5.18 Lithostratigraphic Correlation** This kind of correlation demonstrates the original geographic extent of a rock unit, such as a formation or group.

(a) If outcrops are adequate, as they are in the Valley of the Gods in Utah, geologists simply trace rock units laterally. It is obvious that the cliff-forming outcrop in the near distance correlates with similar rocks in the area.

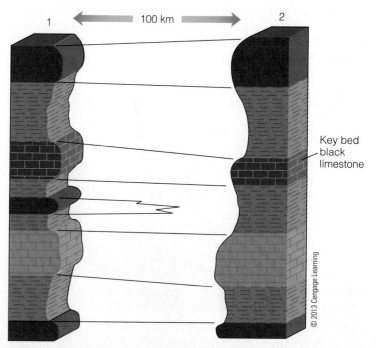

(b) Correlation by similarities in rock type, position in a sequence, and a key bed. We assume the middle sandstone in column 1 intertongues or grades laterally into the shale in column 2 because of its stratigraphic relationship with respect to the rocks above and below it.

geographic extent of a rock unit. Rock cores and well cuttings from drilling operations as well as geophysical data are useful for correlating rock units below the surface.

Many sedimentary rock units, such as widespread formations, are time transgressive, so we cannot rely on lithostratigraphic correlation to demonstrate time equivalence. For example, geologists have accurately correlated sandstone in Arizona with similar rocks in Colorado and South Dakota, but the age of the rocks varies from Early Cambrian in the west to Middle Cambrian farther east. The most effective way to demonstrate the time-stratigraphic equivalence of sedimentary rocks in different areas involves using biozones, but several other methods are useful too.

The existence of all organisms now extinct marks two points in time—time of origin and time of extinction. One type of biozone, known as the **range zone**, is defined as the geologic range (total time of existence) of a particular fossil group, such as a species or a group of related species called a *genus*. The most useful fossils are those that are easily identified, geographically widespread, and that

had a rather short geologic range. *Lingula* is a type of brachiopod that meets the first two of these criteria, but its geologic range makes it of little use, whereas the brachiopod *Atrypa* and the trilobite *Paradoxides* are **guide fossils** because they are well suited for time-stratigraphic correlation (▶Figure 5.19).

Geologists identify several types of *interval zones*, which are defined by the first and last occurrence of a species or genus, but one of the most useful is the **concurrent range zone**, which is established by plotting the overlapping ranges of two or more fossils with different ranges (▶Figure 5.20). Correlating concurrent range zones is probably the most accurate method to determine time equivalence between sedimentary rocks in widely separated areas.

Some physical events are also used to demonstrate time equivalence. Remember that rock composition in most cases is useless in this endeavor, but a few rocks, particularly lava flows and ash falls, form rapidly over their entire extent. Accordingly, correlating an ash fall, for example, is time significant (▶Figure 5.21). Furthermore, ash falls are not restricted to a specific environment, so they may extend from continental to marine environments.

The methods of time-stratigraphic correlation work well for Phanerozoic-age rocks, but Precambrian rocks have few fossils, and those present are not very useful for correlation. For these rocks, absolute ages based on radioactive dating techniques are most useful.

			Lingula	Atrypa	Paradoxides
Cenozoic	Neogene				
	Paleogene				
Mesozoic	Cretaceous				
	Jurassic				
	Triassic				
Paleozoic	Permian				
	Pennsylvanian				
	Mississippian				
	Devonian				
	Silurian				
	Ordovician				
	Cambrian				

© 2013 Cengage Learning

▶ **Figure 5.19 Comparison of the Geologic Ranges (Heavy Vertical Lines) of Three Marine Invertebrate Animals** *Lingula* is of no use in biostratigraphy because it has such a long geologic range. *Atrypa* and *Paradoxides* are good guide fossils because both are widespread, easily identified, and have short geologic ranges.

From MONROE/WICANDER, *The Changing Earth, 6e*. Copyright (c) 2012 Cengage Learning

▶ **Figure 5.20 Correlation of Two Sections Using Concurrent Range Zones** This concurrent range zone was established by the overlapping geologic ranges of fossils symbolized here by the letters A through E. The concurrent range zone is of shorter duration than any of the individual fossil geologic ranges. Correlating by concurrent range zones is probably the most accurate method of determining time equivalence.

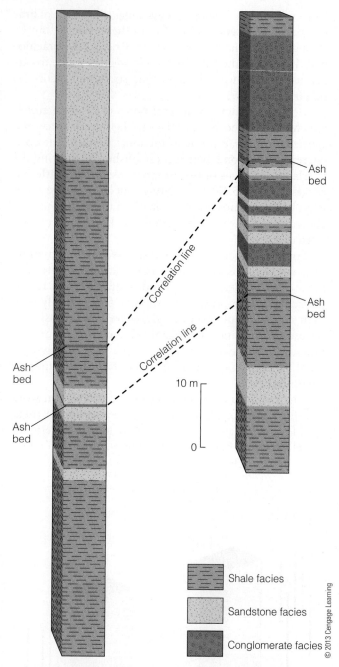

Shale facies

Sandstone facies

Conglomerate facies

© 2013 Cengage Learning

▶ Figure 5.21 **Ash Beds and Time-Stratigraphic Correlation** Volcanic ash accumulates quickly over its entire geographic extent, so correlating from outcrop to outcrop is time significant.

Absolute Dates and the Relative Geologic Time Scale

When it was first established, the relative geologic time scale depicted in Figure 5.16a had no absolute dates like the ones now shown along its right margin. Geologists have known for more than a century that rocks of Ordovician age are younger than those of the Cambrian and older than Silurian rocks, but they had no idea when the Ordovician began or ended. Furthermore, they had already accepted that Earth is very ancient, but even so there was no method for determining absolute ages for rocks. As we discussed in Chapter 4, the discovery of radioactivity provided a reliable geologic clock with which we now date many ancient events.

Beginning students often think that it should be simple to determine radiometric ages for fossil-bearing Ordovician rocks and thus resolve questions about when it took place. Unfortunately, it is not quite so simple because most minerals in sedimentary rocks cannot be dated radiometrically, and for those that can, a date on a mineral in sandstone, for instance, would give only the age of the mineral, not of the sandstone. Remember, any sand grain in sandstone must be older than the sandstone itself, so a 3-billion-year-old sand grain could be in sandstone that was deposited 20 million years ago.

The greenish mineral glauconite, however, forms in some sedimentary rocks soon after deposition, and it can be dated by the potassium–argon method (see Chapter 4). Unfortunately, glauconite easily loses its argon (the daughter element), so absolute ages for this mineral are usually too young, giving only a minimum age for the host sedimentary rock. In most cases, absolute ages for sedimentary rocks and their contained fossils must be determined indirectly by dating associated igneous and metamorphic rocks.

According to the principle of cross-cutting relationships, a dike must be younger than the rock it cuts, so an absolute age for a dike gives a minimum age for the host rock and a maximum age for any overlying rocks (▶Figure 5.22). Absolute dates obtained from granitic rocks also give a maximum age for overlying sedimentary rocks.

Lava flows and ash falls interbedded with sedimentary rocks are the most useful for determining absolute ages (Figure 5.22). Both are time equivalent, providing a maximum age for any rocks above and a minimum age for rocks below. Multiple lava flows or ash fall deposits in a stratigraphic sequence are particularly useful in this effort.

Accurate radiometric dates are now available for many lava flows, ash falls, plutons, and metamorphic rocks with associated fossil-bearing sedimentary rocks. Geologists have added these absolute ages to the geologic time scale, so we now can answer questions such as when the Ordovician Period began and ended (Figure 5.16a). In addition, we now know when a particular organism lived. For example, we know that the *Baculites reesidei* biostratigraphic zone in the Bearpaw Formation in Saskatchewan, Canada, is about 72 to 73 million years old because absolute ages have been determined for associated volcanic ash layers (▶Figure 5.23).

▶ **Figure 5.22** **Determining the Absolute Ages of Sedimentary Rocks** In this example, we know that the layer immediately above the granite is 18 to 24 million years old, whereas the rock layers between the ash fall and the lava flow are 12 to 18 million years old. Can you determine the absolute age of the layers above the ash fall?

▶ **Figure 5.23** **Rocks and Fossils of the Bearpaw Formation in Saskatchewan, Canada** The column on the left shows formations and members (lithostratigraphic units). Notice that the biozone boundaries do not coincide with lithostratigraphic boundaries. The absolute ages for the two volcanic ash layers indicate that the *Baculites reesidei* zone is about 72 to 73 million years old.

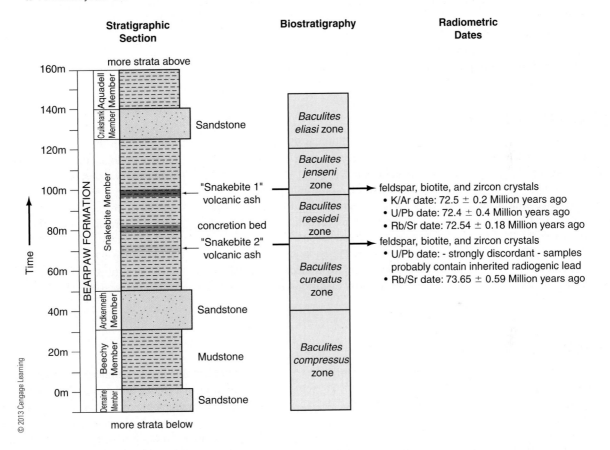

Summary

▶ The first step in deciphering the geologic history of a region involves determining the correct relative ages of the rocks present. Thus, the vertical relationships among rock layers must be ascertained even if they have been deformed.

▶ The geologic record is an accurate chronicle of ancient events, but at any single location it has discontinuities or unconformities representing times of nondeposition, erosion, or both.

▶ Simultaneous deposition in adjacent but different environments yields sedimentary facies, which are bodies of sediment, or sedimentary rock, with distinctive lithologic and biologic attributes.

▶ During a marine transgression, a vertical sequence of facies results with offshore facies superposed over nearshore facies. Just the opposite facies sequence results from a marine regression.

- According to Walther's law, the facies in a conformable vertical sequence replace one another laterally.
- Uplift and subsidence of continents, the amount of water frozen in glaciers, and the rate of seafloor spreading are responsible for marine transgressions and regressions.
- Most fossils are found in sedimentary rocks, although they might also be in volcanic ash and volcanic mudflows, but rarely in other rock types.
- Fossils of some organisms are common, but the fossil record is strongly biased toward those organisms that have durable skeletons and that lived where burial was likely.
- The work of William Smith, among others, is the basis for the principle of fossil succession, which holds that fossil assemblages succeed one another through time in a predictable order.
- Superposition and fossil succession were used to piece together a composite geologic column, which was the basis for the relative geologic time scale.

- To bring order to stratigraphic terminology, geologists recognize units based on content (lithostratigraphic and biostratigraphic units) and those related to time (time-stratigraphic and time units).
- Lithostratigraphic correlation involves demonstrating the original continuity of a rock unit over a given area even though it may not now be continuous over this area.
- Correlation of biostratigraphic zones, especially concurrent range zones, demonstrates that rocks in different areas, even though they may differ in composition, are of the same relative age.
- The best way to determine absolute ages of sedimentary rocks and their contained fossils is to obtain absolute ages for associated igneous rocks and metamorphic rocks.

Important Terms

angular unconformity, p. 85
biostratigraphic unit, p. 95
biozone, p. 95
body fossil, p. 90
concurrent range zone, p. 99
conformable, p. 84
correlation, p. 96
disconformity, p. 85
formation, p. 95
fossil, p. 89

geologic column, p. 95
geologic record, p. 82
guide fossil, p. 99
lithostratigraphic unit, p. 95
marine regression, p. 87
marine transgression, p. 87
nonconformity, p. 85
period, p. 95
principle of fossil succession, p. 92
principle of inclusions, p. 83

range zone, p. 99
relative geologic time scale, p. 95
sedimentary facies, p. 85
stratigraphy, p. 83
system, p. 95
time-stratigraphic unit, p. 95
time unit, p. 95
trace fossil, p. 90
unconformity, p. 84
Walther's law, p. 88

Review Questions

1. An erosion surface cut into plutonic or metamorphic rocks overlain by sedimentary rocks is a
   a. ____ transgression; b. ____ nonconformity; c. ____ time-stratigraphic unit; d. ____ mold; e. ____ biozone.
2. The statement that in a conformable sequence of sedimentary rocks, facies that are superposed will replace one another laterally is
   a. ____ Steno's dictum; b. ____ Darwin's theory; c. ____ Smith's hypothesis; d. ____ Neptunism; e. ____ Walther's law.
3. Which one of the following is a time unit?
   a. ____ Period; b. ____ Formation; c. ____ Member; d. ____ System; e. ____ Range.
4. The geologic column and relative geologic time scale were established by the 1840s based on
   a. ____ superposition and faunal succession; b. ____ the theory of evolution; c. ____ plate tectonic theory; d. ____ absolute dating methods; e. ____ the principle of unconformities.

5. The statement that fossil assemblages succeed one another through time in a regular and predictable order is
   a. ____ time-stratigraphic correlation; b. ____ the sedimentary facies concept; c. ____ the principle of fossil succession; d. ____ lithologic superposition; e. ____ a disconformity.
6. A group is a lithostratigraphic unit composed of at least
   a. ____ four biozones; b. ____ three periods; c. ____ two formations; d. ____ a system and a period; e. ____ three sedimentary facies.
7. Which one of the following is a body fossil?
   a. ____ Tooth; b. ____ Mold; c. ____ Track; d. ____ Spear point; e. ____ Worm burrow.
8. Which one of the following statements is correct?
   a. ____ Guide fossils must have very long geologic ranges; b. ____ An era is made up of two or more formations; c. ____ Biozone boundaries must correspond to lithostratigraphic boundaries; d. ____ Time units are defined by rock type; e. ____ Inclusions are older than the rock layer containing them.

9. Geologists use the principle of superposition to determine
   a. ____ the duration of a marine regression; b. ____ absolute ages for geologic events; c. ____ the relative ages of rocks in a vertical sequence; d. ____ how long ago a fossil organism lived; e. ____ the cause of marine transgressions.

10. A concurrent range zone is a type of
    a. ____ formation; b. ____ biostratigraphic unit; c. ____ unconformity; d. ____ body fossil; e. ____ absolute dating method.

11. Explain how the principle of uniformitarianism helps us understand how fossils were buried and preserved.

12. What are the differences between lithostratigraphic and time-stratigraphic units?

13. How were the principles of superposition and faunal succession used to piece together the relative geologic time scale?

14. Suppose that the upper ash layer in Figure 5.21 formed 1.7 million years ago and the lower one is 1.9 million years old. What was the average sedimentation rate in cm/yr for the interval between the ash beds for the column on the right? Do you think your calculated average is accurate for the actual rate of sedimentation? Explain.

15. Refer to Figure 5.9. If you correlate the limestone (blue) in the left column with the lower limestone in the right column, can you assume they are of the same relative age? Explain.

16. What are the best conditions for the preservation of body fossils?

17. Unconformities are common in the geologic record, and some are found over large areas, but none are worldwide. Why?

18. The illustration below shows an angular unconformity. What sequence of events led to the origin of this unconformity?

Angular unconformity

© 2013 Cengage Learning

19. How do time units and time-stratigraphic units differ?

20. How is it possible to get absolute ages for fossil-bearing sedimentary rocks?

6

Sedimentary Rocks
The Archives of Earth History

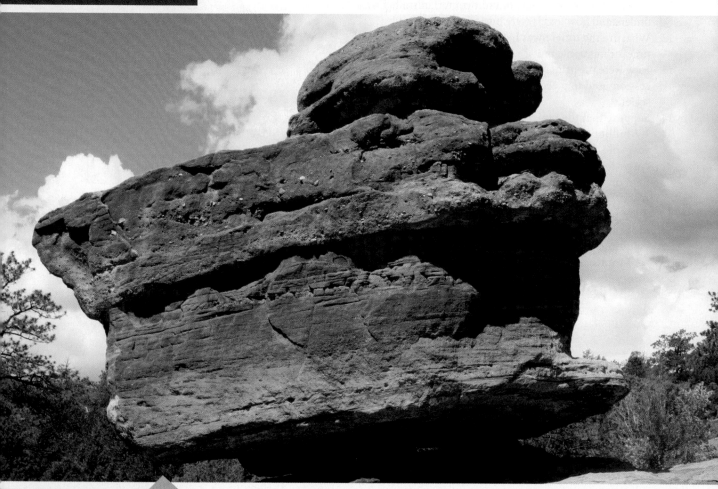

The Pennsylvanian-age Fountain Formation at Garden of the Gods near Colorado Springs, Colorado. The rocks at this exposure are mostly conglomerate and sandstone, although the sandstone contains considerable feldspar minerals and is thus more appropriately called arkose. The formation was deposited in a braided stream system that acquired its sediment load from the Ancestral Rockies, which formed during the Late Mississippian and Early Permian.

Sue Monroe

At the end of this chapter, you will have learned that

▶ Sedimentary rocks have a special place in deciphering Earth history because they preserve evidence of surface processes responsible for deposition, and they may contain fossils.

▶ Geologists recognize three broad areas of deposition—continental, transitional, and marine—each with several specific depositional environments.

▶ The distinctive attributes of sedimentary rocks result from processes operating in specific depositional environments.

▶ Textures such as sorting and rounding provide evidence about depositional processes.

▶ The three-dimensional geometry of sedimentary rock bodies and fossils are useful for determining depositional environments.

▶ Sedimentary structures such as ripple marks, cross-bedding, and mud cracks taken with other rock features allow geologists to make environmental interpretations with a high degree of confidence.

▶ The interpretations of sedimentary rocks in the chapters on geologic history are based on the considerations reviewed in this chapter.

▶ Interpretation of how and where rocks formed, especially sedimentary rocks, is the basis for determining Earth's ancient geographic features.

Introduction

The Permian- to Jurassic-age strata in ▶ Figure 6.1 were deposited in several depositional environments, including tidal flats, streams, lakes, swamps, and sand dunes, whereas the Pennsylvania-age rocks in the chapter opening photo were deposited by braided streams. In the last chapter, we noted that no one witnessed these events, and that geologists use the principle of uniformitarianism to interpret how these rocks were deposited in the first place. Our task in this chapter is to review the kinds of observations and evidence that allow us to make such inferences.

Although we concentrate on sedimentary rocks, all rocks are important in deciphering Earth history. Certainly igneous rocks help geologists identify ancient

James S. Monroe

▶ Figure 6.1 **Sedimentary Rocks and Their Depositional Environments** This rock exposure in Capitol Reef National Park in Utah is called The Cathedral. The rocks are Permian- to Jurassic-age sedimentary rocks that were deposited in sand dunes, tidal flats, lakes, swamps, and streams.

areas of volcanism, as well as convergent plate boundaries, and most of our absolute dates for sedimentary rocks come from associated igneous rocks (see Figure 5.22). Metamorphic rocks tell us something of the temperatures and pressures that prevailed in Earth's crust, especially where plates collided and mountains evolved. In short, a complete historical interpretation requires that we evaluate all evidence.

Nevertheless, sedimentary rocks are important because (1) they preserve evidence of the surface processes responsible for deposition (running water, wind, waves, glaciers, and so on), and (2) many contain fossils that are rare or absent in most other rocks. So, here we focus on areas where sediment is deposited—that is, *depositional environments*—as well as the criteria necessary to recognize deposits of specific environments. Accordingly, geologists examine the distinctive features of sedimentary rocks, always keeping in mind that processes like those operating now were responsible for them.

Sediments and sedimentary rocks are also important because many of them are resources themselves or they are the host rock for resources such as oil and natural gas. Sand and gravel are essential to the construction industry, particularly as aggregate for concrete; rock gypsum is used for wallboard; and quartz is used for abrasives and optical instruments. Most coal is burned at power plants to generate electricity, and phosphorous-rich sedimentary rocks have several uses such as for fertilizers and animal feed supplements. Indeed, Florida produces more than $2 billion worth of phosphate rock annually, making it the seventh leading state in nonfuel mineral production. Another sedimentary rock, called *banded iron formation*, is the main source of the world's iron ore and is important in the economies of Minnesota, Michigan, and Ontario, Canada (see Chapter 9).

Sedimentary Rock Properties

The first step in investigating sedimentary rocks—or any other rock type, for that matter—is observation and data gathering by visiting rock exposures (outcrops) and carefully examining textures, composition, fossils (if present), thickness, and relationships to other rocks. During these initial field studies, geologists may make some preliminary interpretations. For example, color is a useful feature of some sedimentary rocks: Red rocks likely were deposited on land (Figure 6.1), whereas greenish ones more likely were deposited in a marine environment. However, exceptions are numerous, so color must be used with caution.

After completing fieldwork, geologists study the rocks more carefully by microscopic examination, chemical analyses, fossil identification, and by making diagrams that show vertical and lateral facies relationships. A particularly important aspect of these studies is to compare the features of sedimentary rocks with those in present-day sediment accumulations that formed in known depositional environments. When all the data have been analyzed, geologists make an environmental interpretation.

Determining how sedimentary rocks were deposited and examining their fossils satisfies our curiosity about Earth and life history. For example, if you live in Kansas you might be interested to know that a widespread shallow sea populated by shelled animals known as ammonites as well as sharks and huge marine reptiles was there during the Cretaceous Period. In addition, there are economic reasons to determine environments of deposition in exploration for mineral deposits, petroleum, and natural gas.

Composition and Texture

Detrital sedimentary rocks contain more than 100 minerals, but only quartz, feldspars, and clay minerals are very common. Rock composition depends mostly on the composition of the rocks in the area from which the detrital sediment was derived (the source area), but it tells little about how deposition took place. Quartz, for example, may have been deposited in a stream channel, in desert dunes, or on a beach, so composition taken alone is not very useful for environmental determinations.

Among the chemical sedimentary rocks, limestone (composed of calcite [$CaCO_3$]) and dolostone (composed of dolomite [$CaMg(CO_3)_2$]) are the most common, and both are mostly deposited in warm, shallow seas, but a small amount forms in lakes. Evaporites such as rock salt (composed of halite [$NaCl$]) and rock gypsum (made up of gypsum [$CaSO_4 \cdot 2H_2O$]*) invariably indicate arid environments where evaporation rates are high.

Grain size in detrital sedimentary rocks tells us something of the conditions of transport and deposition. Conglomerate consists of gravel, so we can be sure that it was transported by energetic processes such as swiftly flowing streams, by waves, or by glaciers. Sand also requires high-energy transport and tends to be deposited in stream channels, desert dunes, and on beaches, but silt and clay are transported by weak currents and are deposited in low-energy environments such as lagoons, lakes, and river floodplains.

Texture refers to the size, size distribution, shape, and arrangement of clasts in detrital sedimentary rocks. Recall from Chapter 2 that gravel, sand, silt, and clay are simply size designations for detrital particles (see Table 2.2). The degree to which detrital particles have had their sharp edges and corners smoothed off by abrasion is called **rounding** (▸Figure 6.2a). Gravel tends to become rounded very quickly as particles collide with one another; and with considerable transport, sand is also rounded, but smaller particles are carried suspended in water and usually are not so well rounded.

Another textural feature is **sorting**, which refers to the size variation in a sedimentary deposit or rock. If most of the particles are of about the same size, the sediment or rock is well sorted, but if a wide range of sizes is present, the material is poorly sorted (Figure 6.2b). Wind has a limited capacity to transport and deposit sediment, so its deposits tend to be well sorted, whereas glaciers transport anything supplied to them and their deposits are poorly sorted.

Sedimentary Structures

Most sedimentary rocks have features known as **sedimentary structures** that formed during deposition or shortly thereafter, but before lithification. All are manifestations of the physical and biological processes that take place in depositional environments, so analyses of these structures is the single most important aspect of sedimentary rocks for determining how deposition took place.* The origin of sedimentary structures is well known because geologists see them forming today, and many have been produced experimentally (Table 6.1).

With very few exceptions, sedimentary rocks have a layered aspect called **stratification** or **bedding** (see the chapter opening photo and Figure 6.1); layers less than 1 cm thick are *laminations*, whereas *beds* are thicker. Laminations are most common in the mudrocks and form as silt and clay settle from suspension, but they are also found in sandstones and limestones. However, coarser-grained rocks, sandstone and conglomerate, as well as many limestones are bedded. The surfaces separating one bed from another are *bedding planes*, above and below which the rocks differ in composition, texture, color, or a combination of these features.

---

*Gypsum ($CaSO_4 \cdot 2H_2O$) is the common sulfate mineral precipitated from seawater, but when deeply buried, gypsum loses its water and is converted to anhydrite ($CaSO_4$).

*There are structures that form long after deposition, mostly by chemical processes that are of no use for determining depositional environments.

▶ Figure 6.2 **Rounding and Sorting in Detrital Sediments**

(a) Well-rounded but poorly sorted beach deposit in Washington State. "Rounded" does not mean ball-shaped or spherical—only a few of the gravel clasts are spherical.

(b) Angular, poorly sorted gravel.

Some of the deposition that takes place in the seas marginal to continents results from *turbidity currents*, which are sediment–water mixtures that move along the seafloor because they are denser than seawater. When the flow velocity in these currents diminishes, they deposit large particles followed by progressively smaller ones, thus forming **graded bedding** in which the grain size decreases upward (▶ Figure 6.3).

Deposition that takes place on the downwind, sloping side of a sand dune or a similar feature in a stream channel produces **cross-bedding**, which accumulates at an angle to the surface on which deposition occurs (Table 6.1, ▶ Figure 6.4). Transport and deposition by either wind or water produces cross-bedding, and individual cross beds are inclined downward or dip in the direction of flow. Accordingly, cross-bedding indicates the direction of flow of ancient currents.

TABLE 6.1 Summary of Sedimentary Structures

Physical Sedimentary Structures (produced by processes such as currents)	Characteristics	Origin
Laminations (or laminae)	Layers less than 1 cm thick	Form mostly as particles settle from suspension
Beds	Layers more than 1 cm thick	Form as particles settle from suspension and from moving sediment as sand in a stream channel
Graded-bedding	Individual layers with an upward decrease in grain size	Deposition by turbidity currents or during the waning stages of floods
Cross-bedding	Layers deposited at an angle to the surface on which they accumulated	Deposition on a sloping surface at the downwind side of a sand dune
Ripple marks	Small (<3 cm high) ridges and troughs on bedding planes	
Current ripple marks	Asymmetric ripple marks	Result from deposition by water or air currents flowing in one direction
Wave-formed ripple marks	Symmetric ripple marks; generally with a sharp crest and broad troughs	Formed by oscillating currents (waves)
Mud cracks	Intersecting cracks in clay-rich sediments	Drying and shrinkage of mud along a lakeshore, a floodplain, or on tidal flats
Biogenic Sedimentary Structures (produced by organisms)		
Trace fossils	Tracks, trails, tubes, and burrows	Indications of organic activity. Intense activity results in *bioturbation* involving disruption of sediment

© 2013 Cengage Learning

▸ Figure 6.3 **Turbidity Currents and Graded Bedding**

As turbidity current slows, largest particles settle followed by smaller particles

Seafloor

A graded bed

(a) Sediment–water mixtures called turbidity currents flow along the seafloor (or lake bottom) because they are more dense than sediment-free water.

(b) When a turbidity current slows, graded bedding forms as deposition of smaller and smaller particles takes place.

▸ Figure 6.4 **Cross-Bedding** Cross-bedding forms when individual beds or strata are deposited at an angle to the surface upon which they accumulate.

Wind or current direction

Deposition

Sand moves up this slope

(a) Origin of cross-bedding by deposition on the sloping surface of a desert dune. Cross-bedding also forms in dunelike structures in stream and river channels and on the continental shelf.

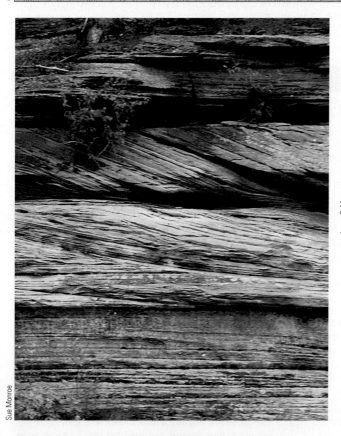

(b) Horizontal bedding and cross-bedding in sandstone at Wisconsin Dells, Wisconsin. About 10 m of strata are visible in this image.

(c) Outcrop of cross-bedding at Rock City, Kansas. These cross-beds are about 45 cm high.

Small-scale alternating ridges and troughs common on bedding planes, especially in sand, are **ripple marks** (Table 6.1). Ripple marks form either by unidirectional flow of wind or water or by the to-and-fro motion of waves. *Current ripple marks*, those formed by the flow of wind or water, have asymmetric profiles with a gentle upstream slope and a steep downstream slope (▸Figure 6.5). Because of their asymmetry, they indicate the original flow direction. *Wave-formed ripple marks* tend to have symmetrical profiles (▸Figure 6.6).

▶ Figure: 6.5 **Current Ripple Marks** Current ripple marks are small (< 3 cm high) sedimentary structures with an asymmetric profile.

(a) Current ripple marks form where water or wind flows in one direction over sand. The enlargement shows the internal cross-bedding in one ripple mark.

(b) Current ripple marks that formed in a stream channel. Which way did the current flow?

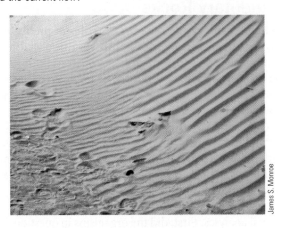

(c) Current ripple marks formed by wind.

▶ Figure 6.6 **Wave-Formed Ripple Marks** Just as with current ripple marks, these are small-scale sedimentary structures, but they tend to have symmetric profiles.

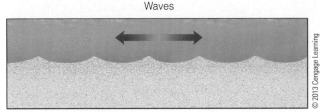

(a) Wave-formed ripples form where waves move to and fro.

(b) Wave-formed ripples in shallow seawater.

(c) These symmetric, wave-formed ripple marks are preserved in Triassic-age rocks in Capitol Reef National Park, Utah.

Clay-rich sediments tend to shrink as they dry and crack into polygonal forms called **mud cracks** (Table 6.1, ▶ Figure 6.7). If you see mud cracks in ancient rocks you can be sure that deposition took place in an environment where alternating wetting and drying took place, such as along a lakeshore, on a river floodplain, or where mud is exposed at low tide along a seashore.

The sedimentary structures discussed so far all form by physical processes, but biological processes yield **biogenic** sedimentary structures and include tracks, trails, and burrows (Table 6.1, ▶ Figure 6.8). Remember from Chapter 5 that these biologically produced features are also called *trace fossils*. Extensive burrowing by organisms, or what is known as **bioturbation**, may so thoroughly disrupt sediments that other structures are destroyed.

Sedimentary structures are important for environmental analyses, but no single structure is unique to a specific environment—current ripples can form in desert dunes, in stream channels, and in tidal channels in the sea, for

(a) Mud cracks in a present-day environment.

(b) Ancient mud cracks in Glacier National Park, Montana. Notice that the cracks have been filled in by sediment.

example. However, associations of sedimentary structures taken along with other sedimentary rock properties allow geologists to make environmental interpretations with a high degree of confidence.

We noted in Chapter 5 that to decipher the geologic history of any area geologists use the principle of superposition to determine the relative ages of sedimentary rocks, but if the rocks have been deformed, the task is more difficult. Several sedimentary structures as well as some fossils allow geologists to resolve these kinds of problems (▶ Figure 6.9).

Geometry of Sedimentary Rocks

The three-dimensional shape, or geometry, of a sedimentary rock body may be helpful in environmental analyses, but it must be used with caution because the same geometry can be found in more than one environment. Moreover, geometry can be modified by sediment compaction during lithification, and by erosion and deformation. Nevertheless, it is useful when considered in conjunction with other features.

Some of the most extensive sedimentary rocks in the geologic record resulted from marine transgressions and regressions (see Figure 5.8). These rocks cover hundreds

▶ Figure 6.8 **Bioturbation Results When Organisms Burrow Through Sediments** The vertical, dark-colored areas in this rock are sediment-filled burrows that partly disrupt the originally sedimented deposit.

or thousands of square kilometers but are perhaps only a few tens to hundreds of meters thick. That is, they are not very thick compared to their dimensions of length and width and thus, have a *blanket*, or *sheet*, *geometry*.

Some sand deposits have an *elongate*, or *shoestring, geometry*, especially those deposited in stream channels or barrier islands. Delta deposits tend to be lens-shaped when viewed in cross profile or long profile, but lobate when observed from above. Buried reefs are irregular, but many are long and narrow, although rather circular ones are also known.

Fossils—The Biologic Content of Sedimentary Rocks

We defined *fossils* as the remains or traces of prehistoric organisms preserved in rocks, and we discussed how geologists use fossils in some aspects of stratigraphy—to establish biostratigraphic units, for example (see Chapter 5). Fossils are also important constituents of some rocks, especially limestones that may be composed largely of shells of marine-dwelling animals such as brachiopods, clams, and corals (see Figures 2.15a and 2.15c), or even the droppings (pellets) of these organisms. Fossils are not present in all sedimentary rocks, but if they are, they are important for determining depositional environments.

We must consider two factors when using fossils in environmental analyses. First, did the organisms in question live where they were buried, or were their remains transported there? Fossil dinosaurs, for example, usually indicate deposition in some land environment such as a river floodplain, but if their bones are found in rocks with clams, corals, and sea lilies, we must assume a carcass was washed out to sea. Second, what kind of habitat did the organisms originally occupy? Studies of a fossil's structure and its living relatives, if any, are helpful. Clams with heavy, thick shells typically live in shallow, turbulent water, whereas those with thin shells are found in low-energy environments. Most corals live in warm, clear, shallow marine environments where symbiotic bacteria can carry out photosynthesis.

▶ Figure 6.9 **Sedimentary Structures and the Relative Ages of Deformed Strata** We can be sure that these rocks were originally deposited horizontally or nearly so, lithified, and then deformed. Our task is to figure out which layer is oldest and which is youngest. To do so, examine the sedimentary structures. The mud cracks form a "V" that opens toward the younger layers, whereas the crests of the wave-formed ripple marks point toward the next younger layer. Notice also the graded bedding. We can conclude that the youngest rock layer is the light brown sandstone in the left-center of the illustration.

Microfossils are particularly useful because many can be recovered from small rock samples. In oil-drilling operations, small rock chips called *well cuttings* are brought to the surface. These cuttings rarely contain complete fossils of large organisms, but they may have thousands of microfossils that aid in relative age determinations and environmental analyses. Even some trace fossils, which of course are not transported from their place of origin, are also characteristic of particular environments.

Depositional Environments

We defined a **depositional environment** as any area where sediment accumulates, but more specifically it entails a particular area where physical, chemical, and biological processes operate to yield a distinctive kind of deposit. Geologists recognize three broad areas of deposition—*continental*, *transitional*, and *marine*—each of which has several specific environments (▶ Figure 6.10).

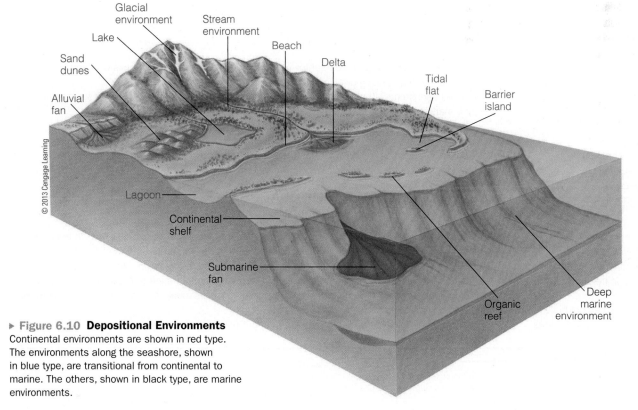

▶ Figure 6.10 **Depositional Environments**
Continental environments are shown in red type. The environments along the seashore, shown in blue type, are transitional from continental to marine. The others, shown in black type, are marine environments.

Continental Environments

Deposition on the continents—that is, on land—takes place in fluvial systems (rivers and streams), lakes, deserts, and areas covered by or adjacent to glaciers (Table 6.2). The sedimentary deposits in each of these environments show combinations of features that allow us to determine that they were in fact deposited on land and to distinguish one from the other.

The term **fluvial** refers to river and stream activity and to their deposits. Fluvial deposits may accumulate in **braided streams** that have multiple broad, shallow channels in which mostly sheets of gravel and cross-bedded sand are deposited, and mud is conspicuous by its near absence (▸Figures 6.11a and 6.11b). In contrast, **meandering streams** have a single sinuous channel and deposit mostly fine-grained sediments on floodplains but do have cross-bedded sand bodies, each with a shoestring geometry (Figures 6.11c and 6.11d). One of the most distinctive features of meandering streams is *point bar* deposits consisting of a sand body overlying an erosion surface that developed on the convex side of a meander loop (Figure 6.11d).

Desert environments are commonly inferred from an association of features found in **sand dune**, **alluvial fan**, and **playa lake** deposits (▸Figure 6.12). Alluvial fans form best along the margins of desert basins where streams and debris flows discharge from mountains onto a valley floor and form a triangular deposit of sand and gravel (Figure 6.12a). Wind-blown dunes are typically composed of well-sorted, well-rounded sand with cross-beds that are meters to tens of meters high (Figure 6.12b). And, of course, any fossils are those of land-dwelling plants and animals. The more central part of a desert basin might be the site of a temporary lake (playa lake), in which laminated mud and evaporites accumulate (Figure 6.12c).

TABLE 6.2 Summary Chart of Rocks and Features in Depositional Environments

Environment	Dominant Type of Rocks
Continental Environments	
Fluvial	
Braided stream	Mostly horizontally bedded conglomerate and cross-bedded sandstone; mudrocks not common
Meandering stream	Mostly mudrocks deposited on floodplains; subordinate but distinctive lenticular sandstones deposited in point bars
Desert	
Alluvial fan	Poorly sorted conglomerate from debris flows and sandstone- and conglomerate-filled channels
Sand dune	Well-sorted, rounded sandstone with large-scale cross-beds
Playa lake	Laminated mudstone/siltstone; evaporites such as rock salt, rock gypsum, and others
Glacial	
Outwash	Much like braided stream deposits
Moraines	Unsorted, nonstratified deposits of sand and gravel
Transitional Environments	
Delta (marine)	Mudrocks and sandstone in coarsening-upward sequences; associated rocks of marine origin; fossils of marine and land-dwelling organisms
Beach	Rounded sandstone with variable sorting, commonly with shells or shell fragments, wave-formed ripple marks, and small-scale cross-bedding
Barrier Island	
Beach	As above
Sand dunes	Much like desert dunes but with sand-sized shell fragments
Tidal flat	Mudstone and sandstone in fining-upward sequences; distinctive herringbone cross-bedding in sandstone
Marine Environments	
Continental shelf	
Inner shelf	Mostly cross-bedded sandstone with wave-formed ripples, marine fossils, and bioturbation
Outer shelf	Mostly mudrocks with subordinate sandstone; marine fossils and bioturbation
Continental slope and rise	Turbidite sequences in submarine fans with graded bedding in sandstone and mudrocks
Carbonate shelf	Limestone (dolostone). Limestone varies from coquina (made of shell fragments) to oolitic limestone to micrite (carbonate mud). Cross-beds, mud cracks, ripple marks common; marine fossils.
Deep-ocean basin	Pelagic clay and calcareous and siliceous oozes
Evaporite environments	Rock salt and rock gypsum the most common, but others, including potassium and magnesium salts, may be present

(a) Sand and gravel bars in this braided stream in Alaska divide the channel into several parts.

(c) The Sacramento River in California is a meandering stream with a single, sinuous channel. The large sand deposit is a point bar.

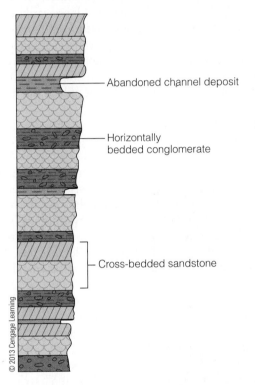

Abandoned channel deposit

Horizontally bedded conglomerate

Cross-bedded sandstone

(b) Braided stream deposits are mostly gravel and cross-bedded sand with subordinate mud.

Floodplain—mudstone with mud cracks, root casts, thin sand layers

Point bar deposit

(d) The deposits of meandering streams are mostly fine-grained floodplain deposits with subordinate sand bodies.

All sediment deposited in glacial environments is collectively called **drift**, but we must distinguish between two kinds of drift. **Till** is poorly sorted, nonstratified drift deposited directly by glacial ice, mostly in ridgelike deposits called *moraines* (▶ Figure 6.13a, b, c). A second type of drift, called **outwash**, is mostly sand and gravel deposited by braided streams issuing from melting glaciers. The association of these deposits along with scratched (striated) and polished bedrock indicates that glaciers were responsible for deposition. Another feature that may be helpful is glacial lake deposits showing alternating dark and light laminations; each dark–light couplet is a **varve** (Figure 6.13d). Varves represent yearly accumulations of sediment—the light layers form in spring and summer and consist of silt and clay, whereas the dark layers consist of clay and organic matter that settled when the lake froze over. Dropstones liberated from icebergs might also be present in glacial lake deposits (Figure 6.13d).

(a) An alluvial fan along the base of the Panamint Range in Death Valley National Park in California.

(b) This huge sand dune, called Sand Mountain, is near Fallon, Nevada.

(c) A playa lake near Fallon, Nevada. Notice deposits of rock salt that were forming when this image was taken in June 2007.

Transitional Environments

Transitional environments are those in which both marine processes and processes typical of continental environments operate (Figure 6.10, Table 6.2). For instance, deposition where a river or stream (fluvial system) enters the sea yields a body of sediment called a **delta**, but the deposit is modified by marine processes, especially waves and tides. Small deltas in lakes are also common, but the ones along seashores are much larger, far more complex, and many deltas are important economically because they form the reservoir rocks for petroleum and natural gas (see Perspective).

The simplest deltas, those in lakes, have a threefold subdivision of bottomset, foreset, and topset beds much like the one shown in ▶ Figure 6.14. As delta sedimentation takes place, bottomset beds are deposited some distance offshore, over which foreset beds come to rest. And finally, deposition also takes place in distributing channels forming the topset beds. As a result, the delta builds seaward (or lakeward), a phenomenon called **progradation**, and forms a vertical sequence of rocks that become coarser-grained from bottom to top. In addition, the bottomset beds may contain marine fossils (or fossils of lake-dwelling animals), whereas the topset beds usually have fossils of land plants or animals.

Marine deltas rarely conform precisely to this simple threefold division, because they are strongly influenced by fluvial processes, as well as by waves and tides. Thus, a stream/river-dominated delta has long distributary channels extending far seaward, as in the Mississippi River delta (▶ Figure 6.15a). Wave-dominated deltas also have distributary channels, but as their entire seaward margin progrades, it is simultaneously modified by wave action (Figure 6.15b). Tidal sand bodies that parallel the direction of tidal flow are characteristic of tide-dominated deltas (Figure 6.15c).

On broad continental margins with abundant sand, long **barrier islands** lie offshore separated from the mainland by a lagoon. These islands are common along the Gulf Coast, and much of the Atlantic Coast of the United States and many ancient deposits in those areas formed in this environment. Notice in ▶ Figure 6.16a that a barrier island has several subenvironments: Beach sands grade offshore into finer deposits of the shoreface; dune sands contain shell fragments and thus differ from dune sands of deserts; and fine-grained lagoon deposits (mostly mud) have marine fossils and bioturbation.

Tidal flats are present along many coastlines where part of the shoreline environment is periodically covered by seawater at high tide and then exposed at low tide. Many tidal flats build or prograde seaward and yield a sequence of rocks grading upward from sand to mud. One of their most distinctive features is *herringbone cross-bedding*—sets of cross-beds that dip in opposite directions (Figure 6.16b).

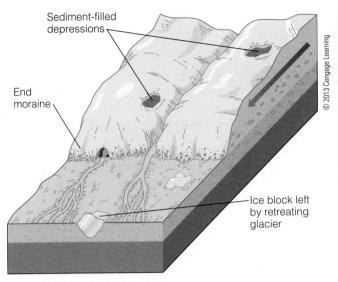

(a) Glaciers transport huge quantities of sediment that is deposited as moraines and outwash.

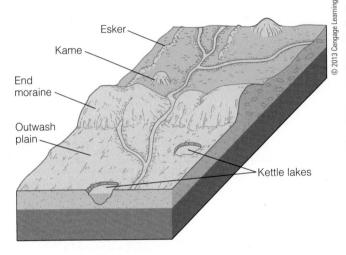

(b) The same area as shown in (a), but with the ice gone.

(c) This ridge of gravel is a small end moraine that was deposited in 1995 by the Exit Glacier near Seward, Alaska. The gravel layers in the foreground are outwash.

(d) Dropstone in glacial lake deposits showing varves in Canada.

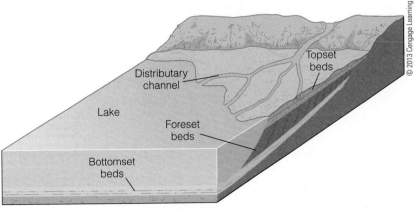

▶ Figure 6.14 **Origin of a Delta** Surface view and internal structure of the simplest type of prograding delta.

Perspective

The Mississippi River Delta—Past and Present

As we will discuss in Chapter 16, deposition has taken place along the entire Gulf Coast since the beginning of the Late Jurassic. In fact, following the breakup of Pangaea the area has had a complex history of transgressions and regressions as well as fluvial deposition, which continues on the Mississippi River delta. The lower Mississippi River basin is probably underlain by part of a rift that formed when the Gulf of Mexico opened (see Figure 14.2).

As we noted in this chapter, deltas form where a river or stream enters a standing body of water and deposits alluvium, but they may be dominated by waves, tides, or stream activity (Figure 6.15). The Mississippi River delta is a classic stream-dominated delta that lies on a shallow shelf where it is mostly protected from waves and tides. On the

delta, distributary channels prograde so far seaward that they become inefficient avenues of sediment transport, so the river abandons that channel and establishes a new one elsewhere (Figure 1). As a matter of fact, the entire coastline of Louisiana has built seaward as much as 80 km during the last 5,000 years as a result of sedimentation.

In addition to abandoning distributary channels, stream-dominated deltas periodically abandon entire deltaic lobes. For example, the Mississippi River is currently depositing sediment on a single lobe, but the entire delta is made up of several lobes that formed during the past few thousands of years (Figure 2). As a matter of fact, were it not for the efforts of the Army Corps of Engineers, the currently active lobe may have already been abandoned.

You may recall the flooding on the lower Mississippi during April and May 2011, when the Army Corps of Engineers diverted much of the water from the Mississippi into the Atchafalaya River at the Morganza Spillway. If the Mississippi were to abandon its present course, the main flow would follow the Atchafalaya River and discharge into the Gulf of Mexico about 100 km west of where it does now. Although such a diversion would be economically catastrophic for New Orleans, the river would simply continue building its delta, but in an area it occupied long ago (Figure 2).

Flood-control projects on the Mississippi River prevent much of the sediment from reaching the delta, which contributes to loss of land due to wave erosion, and, more importantly, to compaction. Delta deposits, especially mud,

Figure 1 The Mississippi River Delta

Reprinted with permission from J. M. Coleman, *Deltas: Processes of Deposition and Models for Exploration*, © 1976, p. 37, Continuing Education Company, Inc.

(a) Block diagram illustrating progradation of distributary channel sand bodies (bar-finger sands).

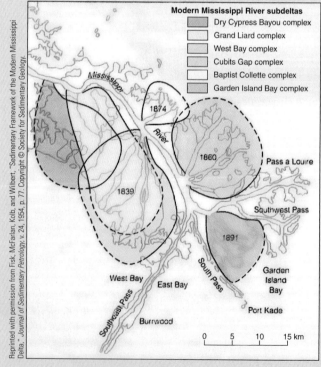

Reprinted with permission from Fisk, McFarlan, Kolb, and Wilbert, "Sedimentary Framework of the Modern Mississippi Delta," *Journal of Sedimentary Petrology,* v. 24, 1954, p. 77. Copyright © Society for Sedimentary Geology.

(b) The present lobe of the delta consists of several subdeltas that formed when distributary channels were abandoned. The area shown here corresponds to the Balize lobe in Figure 2.

Figure 2 **The Major Deltaic Lobes of the Mississippi River** Deposition of the present-day Balize lobe began about 550 years ago.

compact under their own weight, and thus the surface subsides.* Geologists estimate that during the last 4,400 years the region has subsided at the rate of about 12 cm per century. Delta subsidence is not a problem of academic interest only; a large part of New Orleans lies below sea level because of subsidence. Much of the city is situated between Lake Pontchartrain and the Mississippi River, so when Hurricane Katrina roared ashore in August 2005, nearly 80 percent of New Orleans was flooded. About 1,800 people died, and damages exceeded $100 billion, mostly in Louisiana and Mississippi.

Another recent event that has had an impact on the Mississippi River delta is the *Deepwater Horizon* oil spill that began in April 2010. You no doubt recall that an oil rig exploded and sank about 65 km off the Louisiana coast, killing 11 workers and beginning the largest oil spill in the history of the petroleum industry. The spill caused considerable damage in the Gulf and along the shoreline. Currently there are about 4,000 wells in the Gulf of Mexico and approximately 27,000 abandoned ones, many of which were not properly sealed. In fact, about 30 percent of all U.S. oil production comes from wells in the Gulf of Mexico.

---

*Pumping oil and natural gas from delta sediments also contributes to subsidence.

Marine Environments

Marine environments include the continental shelf, slope, and rise, and the deep seafloor (▶Figure 6.17, Table 6.2). Much of the detritus eroded from continents is eventually deposited in marine environments, but other types of sediments are found here, too. Much of the limestone in the geologic record, as well as many evaporites, were deposited in shallow marine environments.

Detrital Marine Environments The **continental shelf**, the gently sloping area adjacent to a continent, consists of two parts. A high-energy inner part is periodically stirred up by waves and tidal currents. Its sediment is mostly sand

shaped into large cross-bedded dunes, and bedding planes are marked by wave-formed ripple marks. Of course, marine fossils are typical, and so is bioturbation. The low-energy outer part of the shelf has mostly mud with marine fossils, but at the transition between the inner and outer shelf, layers of sand and mud intertongue.

Much of the sediment derived from continents crosses the continental shelf and is funneled into deeper water through submarine canyons. It eventually comes to rest on the **continental slope** and **continental rise** as a series of overlapping submarine fans (Figure 6.17). Once sediment passes the outer margin of the shelf, turbidity currents transport it, so sands with graded bedding are common—but so is mud that settled from seawater (Figure 6.3).

(a) The Mississippi River delta on the U.S. Gulf Coast is stream-dominated.

(c) The Ganges-Brahmaputra delta of Bangladesh is tide-dominated.

(b) The Nile delta of Egypt is wave-dominated.

Beyond the continental shelf, the seafloor is nearly covered by fine-grained deposits known as *pelagic clay* and *ooze* (▸Figure 6.18). The notable exceptions are the mid-oceanic ridges, which are where new oceanic crust forms. Sand and gravel are notable by their absence on the deep seafloor, because few mechanisms transport coarse-grained sediment far from land. However, ice rafting accounts for a band of sand and gravel adjacent to Greenland and Antarctica.

The main sources of deep-sea sediments are dust blown from continents or oceanic islands, volcanic ash, and shells of microorganisms that dwelled in the ocean's surface waters. One type of sediment called pelagic clay consists of clay covering most of the deeper parts of the seafloor. Calcareous ($CaCO_3$) and siliceous (SiO_2) oozes, in contrast, are made up of microscopic shells (Figure 6.18).

Carbonate Depositional Environments Limestone and dolostone are the only widespread carbonate rocks, and we already know that most dolostone is altered limestone. Thus, our discussion focuses on the deposition of sediment that, when lithified, becomes limestone.

In some respects limestone is similar to detrital sedimentary rocks—many limestones are made up of gravel- and sand-sized grains, and microcrystalline carbonate mud called *micrite*. However, these materials form in the environment of deposition rather than being transported there. Nevertheless, local currents and waves sort them and account for sedimentary structures such as cross-bedding and ripple marks, and micrite along shorelines commonly shows mud cracks.

▶ Figure 6.16 **Barrier Islands and Tidal Flats**

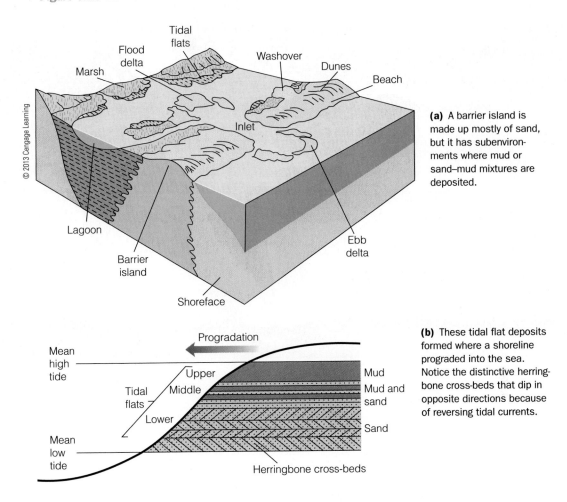

(a) A barrier island is made up mostly of sand, but it has subenvironments where mud or sand–mud mixtures are deposited.

(b) These tidal flat deposits formed where a shoreline prograded into the sea. Notice the distinctive herringbone cross-beds that dip in opposite directions because of reversing tidal currents.

▶ Figure 6.17 **Depositional Environments on the Continental Shelf, Slope, and Rise** Submarine canyons are the main avenues of sediment transport across the shelf. Turbidity currents carry sediments beyond the shelf, where they accumulate as submarine fans.

Depositional Environments **119**

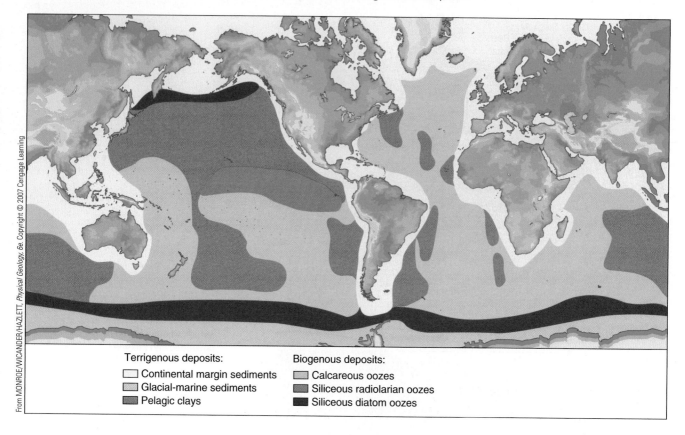

Terrigenous deposits:
☐ Continental margin sediments
☐ Glacial-marine sediments
☐ Pelagic clays

Biogenous deposits:
☐ Calcareous oozes
☐ Siliceous radiolarian oozes
☐ Siliceous diatom oozes

From MONROE/WICANDER/HAZLETT, *Physical Geology, 6e.* Copyright © 2007 Cengage Learning

Some limestone forms in lakes, but by far most of it was deposited in warm, shallow seas on carbonate shelves (▶ Figure 6.19). In any case, deposition occurs where little detrital sediment is present—and especially little mud. Carbonate barriers form in high-energy areas and may be reefs, banks of skeletal particles, or accumulations of spherical carbonate grains known as *ooids*—limestone composed mostly of ooids is called *oolitic limestone* (see Figure 2.15b). Reef rock tends to be structureless, composed of skeletons of corals, mollusks, sponges, and many other organisms, whereas carbonate banks are made up of layers with horizontal beds, cross-beds, and wave-formed ripple marks. The lagoons, however, tend to have micrite with marine fossils and bioturbation.

Evaporite Environments Evaporites consisting mostly of rock salt and rock gypsum* are found in playa lakes and saline lakes, but most of the extensive deposits were formed in the seas. Although locally abundant, evaporites are not nearly as common as sandstone, mudrocks, and limestone. Deposits more than 2 km thick lie beneath the

floor of the Mediterranean Sea, and large deposits are also known in Michigan, Ohio, and New York, as well as in several Gulf Coast states and western Canada.

Exactly how some of these deposits originated is controversial, but all agree that they formed where evaporation rates were high enough for seawater to become sufficiently concentrated for minerals to precipitate from solution. These conditions are best met in coastal environments in arid regions such as the present-day Persian Gulf. If the lagoon shown in Figure 6.19 were an arid region, inflow of normal seawater into the lagoon would be restricted, and evaporation might lead to increased salinity and evaporite deposition.

Interpreting Depositional Environments

Zion National Park in Utah has excellent exposures of several sedimentary rock formations, but the most notable one is the Jurassic-age Navajo Sandstone (▶ Figure 6.20). Geologists think that this formation originated on land as a vast blanket of sand dunes. But couldn't it just as well

---

*Remember that when gypsum ($CaSO_4 \cdot 2H_2O$) is deeply buried, it loses its water and is converted to anhydrite ($CaSO_4$).

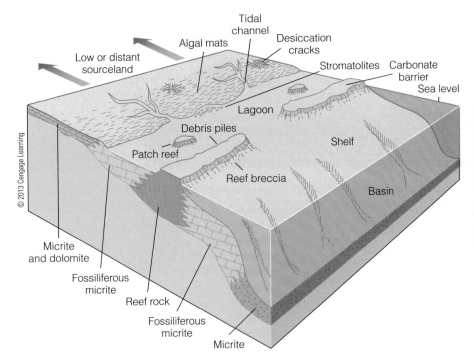

Tidal
channel
Algal mats
Desiccation
cracks
Low or distant
sourceland
Stromatolites   Carbonate
barrier
Sea level
Lagoon
Debris piles
Shelf
Patch reef
Reef breccia
Basin
Micrite
and dolomite
Fossiliferous
micrite
Reef rock
Fossiliferous
micrite
Micrite

© 2013 Cengage Learning

▶ Figure 6.19 **Carbonate Depositional Environments** Although some
limestone forms in lakes, most of it is
deposited on carbonate shelves similar
to the one in this illustration. However,
the carbonate barrier may be reefs of
corals, clams, and algae, or it may be
made up of oolitic sand or skeletal-
fragment sand. Deposition of limestone
in environments much like these is now
taking place in southern Florida and the
Persian Gulf.

Sue Monroe

▶ Figure 6.20 **The Jurassic-Age Navajo Sandstone in Zion
National Park, Utah** The evidence preserved in these rocks indicates deposition took place in a coastal dune environment.

have been deposited on the continental shelf or in river
channels? No, because the evidence indicates otherwise.

Studies show that the Navajo Sandstone is made up
mostly of well-rounded, well-sorted quartz grains measuring 0.2 to 0.5 mm in diameter. In addition, the rocks
have tracks of land-dwelling animals, including dinosaurs,
precluding the possibility of deposition on the continental
shelf. The sandstone also has cross-beds up to 30 m high
and current ripples much like those produced by wind, so
it seems very unlikely that the formation was deposited in

stream channels. All the evidence taken together justifies
the interpretation that the Navajo Sandstone is a coastal
dune deposit. In fact, the cross-beds dip generally southwest, so we can confidently say that the wind blew mostly
from the northeast.

Another good example of using the combined features
of sedimentary rocks and comparisons with present-day
deposits is Lower Silurian strata exposed in New Jersey
and Pennsylvania. The features of these rocks are summarized in ▶Figure 6.21, all of which lead to the conclusion
that they were deposited in braided streams that flowed
generally from east to west.

The preceding examples were taken from continental depositional environments, but the same reasoning is
used for deposits of transitional environments (see Perspective) and for marine environments. For example,
the Cretaceous-age Niobrara Formation is well exposed
in several areas of the Great Plains states. The formation
consists of a Fort Hays Limestone Member and an overlying Smoky Hills Chalk Member (▶Figure 6.22). The
features of this formation, including fossils of marine
invertebrates, fish, and marine reptiles such as plesiosaurs and mosasaurs, all indicate deposition in a marine
environment.

In several of the following chapters we refer to fluvial deposits, ancient deltas, carbonate shelf deposits, and
others as we examine the geologic record, especially for
North America. We cannot include the supporting evidence for all of these interpretations, but we can say that
they are based on the kinds of criteria discussed in this
chapter.

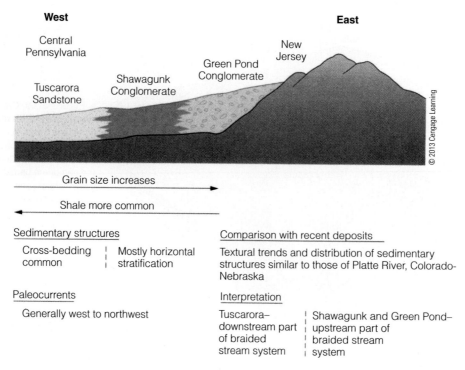

West   East

Central
Pennsylvania   New
                                                      Jersey
                            Green Pond
                            Conglomerate
           Shawagunk
Tuscarora Conglomerate
Sandstone

Grain size increases →

← Shale more common

Sedimentary structures   Comparison with recent deposits

Cross-bedding | Mostly horizontal   Textural trends and distribution of sedimentary
common | stratification   structures similar to those of Platte River, Colorado-
                | Nebraska

Paleocurrents   Interpretation

Generally west to northwest   Tuscarora– | Shawagunk and Green Pond–
                                         downstream part | upstream part of
                                         of braided | braided stream
                                         stream system | system

© 2013 Cengage Learning

▶ **Figure 6.21 Lower Silurian Strata in the Eastern United States** Simplified cross section showing the probable lateral relationships for the Green Pond Conglomerate, Shawangunk Conglomerate, and Tuscarora Sandstone, and the criteria used to interpret these rocks as braided stream deposits.

James S. Monroe

▶ **Figure 6.22 The Niobrara Formation** This formation consists of two members but only the uppermost one, the Smoky Hills Chalk Member, is exposed here at Monument Rocks National Natural Landmark, Kansas. All evidence indicates that the formation was deposited in a marine environment. The rocks here rise more than 20 m above the surrounding plain.

Paleogeography

As noted, the Navajo Sandstone is a coastal dune deposit, so we know that during the Jurassic Period a vast desert was present in what is now the southwestern United States (Figure 6.20). We also know that a marine transgression took place from the Neoproterozoic to Middle Cambrian in North America as the shoreline migrated inland from east and west (see Figure 5.7). In short, Earth's geography—that is, the distribution of its surface features—has varied through time.

Paleogeography deals with Earth's geography of the past. In the chapters on geologic history, paleogeographic maps show the distribution of continents at various times; they have not remained fixed because of plate movements. The distribution of land and sea constitutes Earth's first-order features, but paleogeography also applies on a more local scale. For example, detailed studies of rocks in several western states allow us to determine with some accuracy how the area appeared during the Late Cretaceous. According to the paleogeographic map in ▶ Figure 6.23a, a broad coastal plain sloped gently eastward from a mountainous region to the sea. The rocks also tell us that many animals lived here, including dinosaurs and mammals. Later, however, vast lakes, river floodplains, and alluvial fans covered much of this area, and the sea had withdrawn from the continent (Figure 6.23b).

▶ Figure 6.23 **Paleogeographic Maps**

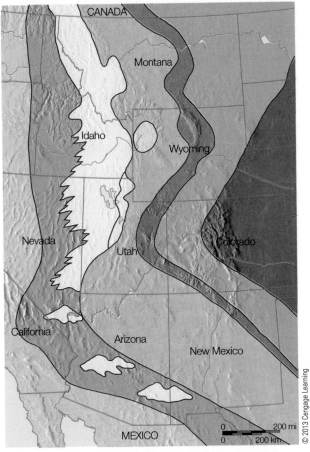

(a) Late Cretaceous paleogeography of several western states. Notice that although continental environments were present, a large part of the area was covered by the sea.

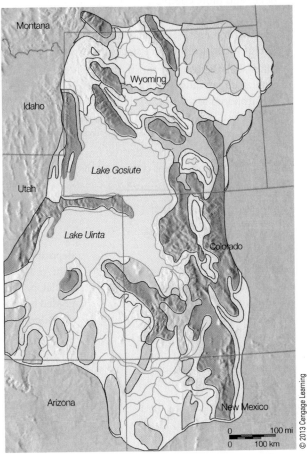

(b) Eocene paleogeography of much of the area shown in (a). All of the deposition taking place at this time was in continental environments.

	Land, high to moderate elevation; sediment source
	Land, moderate to low elevation; sediment source
	Marine sandstone, barrier island, beach or nearshore deposits
	Marine shale
	Alluvial fans
	Nonmarine rocks of coastal plain; primarily sandstone
	Coal swamps
	Marine calcareous mudstone

Areas of erosion

| | High relief |
| | Moderate to low relief |

Areas of deposition

	Alluvial fans
	Floodplains; mixed environment
	Extensive swamps; abundant thick coal beds
	Extensive lakes and rivers

Summary

▶ The physical and biologic features of sedimentary rocks provide information about the depositional processes responsible for them.

▶ Sedimentary structures and fossils are the most useful attributes of sedimentary rocks for environmental analyses, but textures and rock body geometry are also helpful.

▶ Geologists recognize three primary depositional areas—continental, transitional, and marine—each of which has several specific environments.

▶ Braided streams deposit mostly sand and gravel, whereas deposits of meandering streams are mostly mud and subordinate sand bodies with shoestring geometry.

▶ An association of alluvial fan, sand dune, and playa lake deposits is typical of desert depositional environments. Glacial deposits consist mostly of till in moraines and outwash.

▶ The simplest deltas, those in lakes, consist of a three-part sequence of rocks, grading from finest-grained rocks at the base upward to coarser-grained rocks. Marine deltas dominated by fluvial processes, waves, or tides are much larger and more complex.

▸ A barrier island system includes beach, dune, and lagoon subenvironments, each characterized by a unique association of rocks, sedimentary structures, and fossils.

▸ Inner shelf deposits are mostly sand, whereas those of the outer shelf are mostly mud; both have marine fossils and bioturbation. Much of the sediment from land crosses the shelves and is deposited on the continental slope and rise as submarine fans.

▸ Either pelagic clay or oozes derived from the shells of microscopic floating organisms cover most of the deep seafloor.

▸ Most limestone originates in shallow, warm seas where little detrital mud is present. Just like detrital rocks, carbonate rocks may possess cross-beds, ripple marks, mud cracks, and fossils that provide information about depositional processes.

▸ Evaporites form in several environments, but the most extensive ones were deposited in marine environments. In all cases, they formed in arid regions with high evaporation rates.

▸ With information from sedimentary rocks, as well as other rock types, geologists determine the past distribution of Earth's surface features.

Important Terms

alluvial fan, p. 112
barrier island, p. 114
biogenic sedimentary structure, p. 109
bioturbation, p. 109
braided stream, p. 112
continental rise, p. 117
continental shelf, p. 117
continental slope, p. 117
cross-bedding, p. 107
delta, p. 114

depositional environment, p. 111
drift, p. 113
fluvial, p. 112
graded bedding, p. 107
meandering stream, p. 112
mud crack, p. 109
outwash, p. 113
paleogeography, p. 122
playa lake, p. 112
progradation, p. 114

ripple mark, p. 108
rounding, p. 106
sand dune, p. 112
sedimentary structure, p. 106
sorting, p. 106
stratification (bedding), p. 106
tidal flat, p. 114
till, p. 113
varve, p. 113

Review Questions

1. The deposits of meandering streams are mostly
   a. ____ sheets of sand and gravel; b. ____ limestone and evaporates; c. ✓ mud with subordinate sand; d. ____ pelagic clay and ooze; e. ____ outwash.
2. Which one of the following is a sedimentary structure?
   a. ____ Tidal flat; b. ____ Drift; c. ✓ Mud crack; d. ____ Delta; e. ____ Alluvial fan.
3. Cross-beds that dip in opposite directions (herringbone cross-beds) are found in _____ environments.
   a. ✓ tidal flat; b. ____ desert dune; c. ____ glacial; d. ____ playa lake; e. ____ fluvial.
4. Where a river or stream flows into the sea or a lake it will likely deposit a(n)
   a. ____ point bar; b. ____ alluvial fan; c. ____ barrier island; d. ____ moraine; e. ✓ delta.
5. The alternating dark- and light-colored laminations that form in glacial lakes are
   a. ____ dunes; b. ____ varves; c. ✓ tills; d. ____ graded beds; e. ____ current ripples.
6. Which one of the following can you use to determine ancient current directions?
   a. ____ Cross-bedding; b. ____ Rounding; c. ____ Mud cracks; d. ____ Pelagic clay; e. ✓ Bioturbation.

7. A sand deposit that is well sorted
   a. ____ must have been deposited directly by glacial ice; b. ____ probably formed on the deep seafloor; c. ____ typically has grains with sharp corners and edges; d. ✓ has grains all of about the same size; e. ____ contains numerous fossils.
8. The process whereby organisms burrow through and thoroughly mix sediment is
   a. ____ lithification; b. ____ sorting; c. ____ bioturbation; d. ____ rounding; e. ____ cementation.
9. Many limestones are made up of sand- and gravel-size particles and calcium carbonate mud called
   a. ____ cement; b. ____ micrite; c. ____ till; d. ✓ outwash; e. ____ fossils.
10. Which one of the following would you likely find in an evaporate environment?
    a. ✓ Sandstone; b. ____ Rock gypsum; c. ____ Coal; d. ____ Gravel; e. ____ Drift.
11. How and why do sediments on the inner and outer continental shelf differ?
12. Give an example of how sedimentary structures and fossils are used to determine ancient depositional environments.

13. What kinds of deposits are typically found in areas formerly covered by glaciers?
14. Under what conditions are evaporites deposited, and what are the two most common types of evaporite rocks?
15. This image shows an outcrop at Natural Bridges National Monument in Utah. What kind of bedding is present in the layer indicated by the arrow, and can you infer anything about ancient directions?

16. Describe the sediment properties of rounding and sorting. How are they useful in environmental analyses?
17. Why are sets of cross-beds dipping in opposite directions found on tidal flats?
18. Look at Figures 6.16a and 6.16b and explain why prograding deltas yield a sequence of sediment that become coarser-grained upward, whereas in prograding tidal flats the sediments become finer-grained upward.
19. What association of sedimentary deposits would you expect to find in a desert environment?
20. Explain how graded bedding forms, and in what environment it is commonly found.

7

Evolution
The Theory and Its Supporting Evidence

Converging lines of evidence from studies in biogeography, comparative anatomy, genetics, biochemistry, classification, and paleontology support the theory of biologic evolution, that is, changes that have occurred in organisms through time. This fossil of *Archaeopteryx*, the oldest known bird from Jurassic-age rocks in Germany, provides evidence for the relationship of birds to reptiles. *Archaeopteryx* had feathers just as birds do, but in most of its other characteristics it most closely resembled a small, carnivorous dinosaur with its reptile-like teeth, claws on its wings, and a long tail.

Juan Carlos Muñoz/age fotostock/Getty Images

At the end of this chapter, you will have learned that

▶ The central point of the theory of evolution is that today's organisms descended, with modification, from ancestors that lived during the past.

▶ In 1809, Jean-Baptise de Lamarck proposed his theory of inheritance of acquired characteristics, which was the first widely accepted theory to account for evolution.

▶ In 1859, Charles Darwin and Alfred Wallace published their views on evolution and proposed natural selection as a mechanism to bring about change.

▶ Gregor Mendel carried out experiments during the 1860s and demonstrated that variations in populations are maintained during inheritance, rather than blended, as previously thought, and that organisms receive one-half of their inheritance from each parent.

▶ In the modern view of evolution, variation is accounted for mostly by sexual reproduction and mutations in sex cells. Furthermore, populations rather than individuals evolve.

▶ The fossil record provides many examples of macroevolution—that is, changes resulting in the origin of new species, genera, and so on—but these changes are simply the cumulative effect of microevolution, which involves changes within a species.

▶ Several evolutionary trends, such as size increase or changing configuration of shells, teeth, or limbs, are well known for organisms for which sufficient fossils are available.

▶ The theory of evolution is truly scientific, because we can think of experiments and observations that would support it or render it incorrect.

▶ Fossils are important as evidence for evolution, but additional evidence comes from classification, biochemistry, molecular biology, genetics, and biogeography.

Introduction

A rugged group of islands belonging to Ecuador lies in the Pacific Ocean about 1,000 km west of South America. Called the Archipelago de Colon after Christopher Columbus, the group is better known as the Galápagos Islands (▶Figure 7.1a). During Charles Robert Darwin's (Figure 7.1b) five-year voyage (1831–1836) as an unpaid naturalist aboard the research vessel HMS *Beagle*, he visited the Galápagos Islands, where he made important observations that changed his ideas about the then widely held concept called *fixity of species*. According to this idea, all present-day species were created in their present form and had changed little or not at all.

Darwin began his voyage not long after graduating from Christ's College of Cambridge University with a degree in theology, and although he was rather indifferent to religion, he fully accepted the biblical account of creation. During the voyage, though, his ideas began to change. For example, some of the fossil mammals he collected in South America were similar to present-day llamas, sloths, and armadillos, yet also differed from them. He realized that these animals had descended with modification from ancestral species, and so he began to question the idea of fixity of species.

Darwin postulated that the 13 species of finches living on the Galápagos Islands, and one on the Cocos Island, had evolved from a common ancestor species that somehow reached the islands as accidental immigrants from South America (▶Figure 7.2). Indeed, their ancestor was very likely a single species resembling the blue-back grassquit finch now living along South America's Pacific Coast. The islands' scarcity of food accounts for the ancestral species evolving different physical characteristics, especially beak shape, to survive (Figure 7.2).

Darwin became convinced that organisms descended with modification from ancestors that lived during the past, which is the central claim of the **theory of evolution**. So why should you study evolution? For one thing, evolution involving inheritable changes in organisms through time is fundamental to biology and **paleontology**, the study of life history as revealed by fossils. Furthermore, like plate tectonic theory, evolution is a unifying theory that explains an otherwise encyclopedic collection of facts. And finally, it serves as the framework for discussions of life history in the following chapters. Unfortunately, many people have a poor understanding of the theory of evolution and hold a number of misconceptions, some of which we address in this and later chapters.

Evolution: What Does It Mean?

Biologic evolution is the process whereby organisms have changed since life originated—that is, they have descended with modification from their remote ancestors. Small-scale changes, such as the changing genetic makeup of populations of organisms from generation to generation, as well as large-scale changes, including the origin of new species from a common ancestor, are examples of biologic evolution. Many things evolve, or simply change, but biologic evolution does not involve changes in nonorganic systems (rocks, mountains, and rivers), nor does it include changes that occur in the life cycle of an organism (aging, growth, and metamorphosis).

(a) The Galápagos Islands are specks of land composed of basalt in the Pacific Ocean west of Ecuador. The orange line in the map shows the route followed by Charles Darwin when he was aboard the HMS *Beagle* from 1831 to 1836.

(b) Charles Robert Darwin in 1840. During his five-year voyage aboard the HMS *Beagle,* Darwin collected fossils and made observations that would later be instrumental in his ideas about evolution.

It is also important to note that the theory of evolution does not address how life originated, only how it has changed and diversified through time. There are theories that attempt to explain life's origin, but they must be evaluated on their own merits (see Chapter 8). The idea that organisms have evolved is usually attributed solely to Charles Darwin, but it was seriously considered long before he was born, even by some ancient Greeks and philosophers and theologians during the Middle Ages (A.D. 476–1453). Nevertheless, the prevailing belief among Europeans well into the 1700s was that the works of Aristotle (384–322 B.C.) and the first two chapters of the book of Genesis contained all-important knowledge. Literally interpreted, Genesis was taken as the final word on the

origin and variety of organisms, as well as much of Earth history. To question any aspect of this interpretation was heresy, which was usually dealt with harshly.

The social and intellectual climate changed in 18th-century Europe, and the absolute authority of the Christian church in all matters declined. And ironically, the very naturalists trying to find physical evidence supporting Genesis found more and more evidence that could not be reconciled with a literal reading of Scripture. For example, observations of sedimentary rocks previously attributed to a single worldwide flood led naturalists to conclude that they were deposited during a vast amount of time, in environments like those existing now. They reasoned that this evidence truly indicated how the rocks originated, for to infer otherwise implied a deception on the part of the Creator (see Chapter 5 Introduction).

In this changing intellectual atmosphere, scientists gradually accepted the principle of uniformitarianism and Earth's great age, and the French zoologist Georges Cuvier demonstrated that many types of plants and animals had become extinct. In view of the accumulating evidence, particularly from studies of living organisms, scientists became convinced that change from one species to another actually took place. However, they lacked a theoretical framework to explain evolution.

▶ **Figure 7.2** **Darwin's Finches of the Galápagos Islands** These birds probably evolved from accidental migrants much like the blue-back grassquit finch that lives along the west coast of South America.

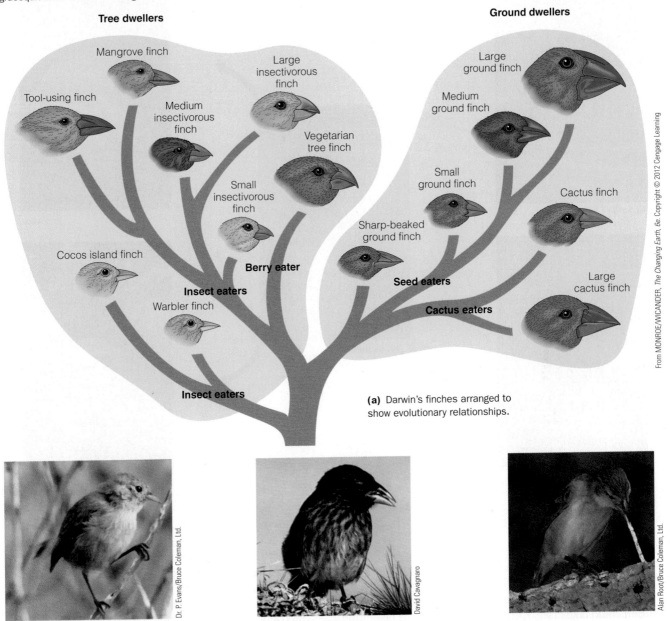

(a) Darwin's finches arranged to show evolutionary relationships.

(b) Three of the 13 species of finches on the Galápagos Islands. They are, from left to right, *Certhidea olivacea*, which has a slender beak to probe for insects in vegetation; *Geospiza scandens*, with a medium-sized beak for eating cactus flowers and fruit; and *Camarhynchus pallidus*, which uses cactus spines to probe for insects.

Jean-Baptiste de Lamarck's Ideas About Evolution

Jean-Baptiste de Lamarck (1744–1829) was not the first to propose a mechanism to account for evolution, but, in 1809, he was the first to be taken seriously. Lamarck contributed greatly to our understanding of the natural world, though unfortunately he is best remembered for his theory of **inheritance of acquired characteristics**. According to this idea, new traits arise in organisms because of their needs, and somehow these traits are passed on to their descendants. In an ancestral population of short-necked giraffes, for instance, neck stretching to browse in trees gave them the capacity to have offspring with longer necks. In short, Lamarck thought that characteristics acquired during an individual's lifetime were inheritable.

Considering the information available at the time, Lamarck's theory seemed logical and was accepted by some as a viable mechanism for evolution. Indeed, it was not totally refuted until decades later, when scientists discovered that the units of heredity known as *genes* cannot be altered by any effort by an organism during its lifetime. In fact, all attempts to demonstrate inheritance of acquired characteristics have been unsuccessful (see Perspective).

Broccoli.

Wild mustard.

▶ **Figure 7.3 Artificial Selection** Humans have practiced artificial selection for thousands of years, thereby giving rise to dozens of varieties of domestic dogs, pigeons, sheep, cereal crops, and vegetables. Wild mustard was selectively bred to yield broccoli, cauliflower, kale, and cabbage, and two other vegetables, Brussels sprouts and kohlrabi, which are not shown here.

Cauliflower.

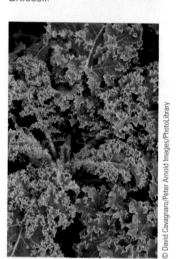
Kale.

Cabbage.

Charles Darwin and Alfred Wallace on Evolution

In 1859, Charles Robert Darwin (1809–1882) (Figure 7.1b) published *On the Origin of Species*, in which he detailed his ideas on evolution and proposed a mechanism whereby evolution could take place. Although 1859 marks the beginning of modern evolutionary thought, Darwin actually formulated his ideas more than 20 years earlier but, being aware of the furor they would cause, was reluctant to publish them.

Darwin had concluded during his 1831–1836 voyage aboard the *Beagle* that species are not immutable and fixed, but he had no idea what might bring about change. However, his observations of selection practiced by plant and animal breeders and a chance reading of Thomas Malthus's essay on population gave him the elements necessary to formulate his theory.

Plant and animal breeders practice **artificial selection** by selecting those traits they deem desirable and then breeding plants and animals with those traits, thereby bringing about a great amount of change (▶ Figure 7.3). The fantastic variety of plants and animals so produced made Darwin wonder if a process selecting among variant types

in nature could also bring about change. He came to fully appreciate the power of selection when he read in Malthus's essay that far more animals are born than reach maturity, yet the adult populations remain rather constant. Malthus reasoned that competition for resources resulted in a high infant mortality rate, thus limiting population size.

In 1858, Darwin received a letter from Alfred Russel Wallace (1823–1913), a naturalist working in southern Asia, who had also read Malthus's essay and had come to the same conclusion: a natural process was selecting only a few individuals for survival. Darwin's and Wallace's idea, called *natural selection*, was presented simultaneously in 1859 to the Linnaean Society in London.

What Is the Significance of Natural Selection?

We can summarize the salient points on **natural selection**, a mechanism that accounts for evolution, as follows:

1. Organisms in all populations possess heritable variations—size, speed, agility, visual acuity, digestive enzymes, color, and so forth.

2. Some variations are more favorable than others; that is, some variant types have a competitive edge in acquiring resources and/or avoiding predators.

3. Those with favorable variations are *more likely* to survive and pass on their favorable variations.

We should note that *sexual selection* is a special type of natural selection in which males compete for mates or females select males based on traits such as showy tail feathers, bright colors, or courtship rituals. Sexual selection accounts for many features and behaviors in living animals, but it is difficult to demonstrate in fossil animals.

In colloquial usage, natural selection is sometimes expressed as "survival of the fittest," which is misleading because it reduces to "the fittest are those that survive and are thus the fittest." But it actually involves inheritable variations leading to differential reproductive success. Having favorable variations does not guarantee survival for an individual, but in a population of perhaps thousands, those with favorable variations are more likely to survive and reproduce. In fact, today it is common to express "survival of the fittest" as "survival of the fitter," or "survival of the fit enough."

Natural selection works on the existing variation in a population, thus giving a competitive edge to some individuals (▶ Figure 7.4). So, evolution by natural selection and evolution by inheritance of acquired characteristics are both testable, but evidence supports the former, whereas attempts to verify the latter have failed (see Perspective).

Darwin was not unaware of potential problems for his newly proposed theory of natural selection. In fact, in *On the Origin of Species*, he said,

> If it could be demonstrated that any complex organ existed which could not possibly have been formed by numerous, successive, slight modifications, my theory would absolutely break down. (p. 171)

Critics have cited eyes, bird's wings, and many other features that they claim are too complex to have evolved by natural selection, because, according to them, anything less than the structure as it exists now would be useless. Eyes of vertebrate animals are a favorite example—after all, of what use is half an eye? There are no half eyes, all eyes are fully developed and functional, even eyes that are no more than light-sensitive spots, crude image makers, or more sophisticated image makers such as those of reptiles, birds, and mammals.

The "too complex to have evolved" argument fails on another score as well, because it assumes that structures have always been used exactly as they are now and anything less would be incomplete and functionless. Mature partridges can fly, although they prefer to stay on the ground. However, the wings of young partridges are small and useless for flying, so are their wings somehow incomplete and of no use? Hardly. Young partridges flap their wings vigorously when running up trees, which they cannot do if their wings are taped down. In short, this *incline running*, as it is called, is a mechanism to escape predators, so their tiny wings are actually quite useful.

One misconception about natural selection, and especially the phrase "survival of the fittest," is that among animals only the biggest, strongest, and fastest are likely to survive. Indeed, size and strength are important when male bighorn sheep compete for mates, but remember that females pass along their genes, too. Speed is certainly an advantage to some predators, but weasels and skunks are not very fast and survive quite nicely. In fact, natural selection may favor the smallest where resources are limited, as on small islands, or the most easily concealed, or those that adapt most readily to a new food source, or those having the ability to detoxify some natural or human-made substance.

▶ Figure 7.4 **Natural Selection**

(a) In preindustrial England, most peppered moths were gray and blended with the lichens on trees. Black moths were known but rare. With industrialization and pollution, the trees became soot-covered and dark, so dark moths became more common. In rural areas unaffected by pollution, the light–dark frequency did not change.

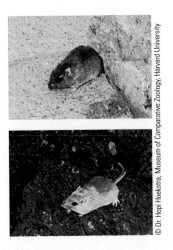

(b) These pocket mice live in the Sonora Desert, which has areas of light-brown granite and dark patches of basalt. As you might expect, mice living on the light-brown granite have light-brown coats, whereas the ones living on the basalt have dark-gray coats. The two mice here are shown on the background that contrasts with their coat color.

The Tragic Lysenko Affair

When Jean-Baptiste de Lamarck (1744–1829) proposed his *theory of inheritance of acquired characteristics,* some scientists accepted it as a viable explanation to account for evolution. In fact, it was not fully refuted for several decades, and even Charles Darwin, at least for a time, accepted some kind of Lamarckian inheritance. Nevertheless, the notion that acquired characteristics could be inherited is now no more than an interesting footnote in the history of science. There was, however, one instance in recent times when the idea enjoyed some popularity—but with tragic consequences.

One of the most notable examples of adherence to a disproved scientific theory involved Trofim Denisovich Lysenko (1898–1976), who became president of the Soviet Academy of Agricultural Sciences in 1938. He lost this position in 1953 but in the same year was appointed director of the Institute of Genetics, a post he held until 1965. Lysenko endorsed Lamarck's theory of inheritance of acquired characteristics because he thought plants and animals could be changed in desirable ways by exposing them to a new environment. Lysenko reasoned that seeds exposed to dry conditions or to cold would acquire a resistance to drought or cold weather, and these traits would be inherited by future generations.

Lysenko accepted inheritance of acquired characteristics because he thought it was compatible with Marxist-Leninist philosophy. Beginning in 1929, Soviet scientists were encouraged to develop concepts consistent with this philosophy.

As president of the Academy of Agricultural Sciences and with the endorsement of the Central Committee of the Soviet Union, Lysenko did not allow any other research concerning inheritance. Those who publicly disagreed with him lost their jobs or were sent to labor camps. In fact, beginning in 1948 it was illegal to teach or do research in genetic theories developed in the West, and even high school textbooks had no information on the role of chromosomes in heredity.

Unfortunately for the Soviet people, inheritance of acquired characteristics had been discredited more than 50 years earlier. The results of Lysenko's belief in the political correctness of this theory and its implementation were widespread crop failures and famine and the thorough dismantling of Soviet genetic research in agriculture. In fact, it took decades for Soviet genetic research programs to fully recover from this setback, after scientists finally became free to experiment with other theories of inheritance.

Lysenko's ideas on inheritance were not mandated in the Soviet Union because of their scientific merit but rather because they were deemed compatible with a belief system. The fact that Lysenko retained power and only his type of genetic research was permitted for more than 25 years is a testament to the absurdity of basing scientific theories on philosophic or political beliefs. In other words, a government mandate does not validate a scientific theory, nor does a popular vote or a decree by some ecclesiastic body. Legislating or decreeing that acquired traits are inheritable does not make it so.

Actually, the idea of evolution was accepted rather quickly, but the Darwin–Wallace theory of natural selection faced some resistance, mainly because so little was known about inheritance.

Mendel and the Birth of Genetics

Critics of natural selection were quick to point out that Darwin and Wallace could not account for the origin of variations or explain how variations were maintained in populations. These critics reasoned that should a variant trait arise, it would blend with other traits and be lost. Actually, the answers to these criticisms existed even then, but they remained in obscurity until 1900.

Mendel's Experiments

During the 1860s, Gregor Mendel, an Austrian monk, performed a series of controlled experiments with true-breeding strains of garden peas (strains that when self-fertilized always display the same trait). He concluded from these experiments that traits such as flower color are controlled by a pair of factors, or what we now call **genes**, and that these factors (genes) that control the same trait occur in alternate forms, now called **alleles**. He further realized that one allele may be dominant over another and that offspring receive one allele of each pair from each parent. For example, in ▸ Figure 7.5, *A* represents the allele for red flower color, and *a* is the allele for white flowers, so any offspring may receive the *AA, Aa,* or *aa* combinations of alleles. And because *A* is dominant over *a,* only those offspring with the *aa* combination of alleles will have white flowers.

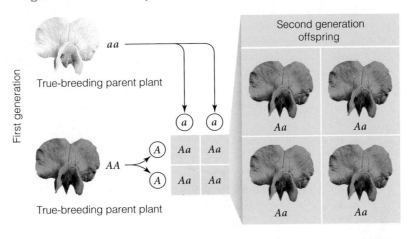

(a) If true-breeding strains, shown here as the parental generation, are cross-fertilized, they yield offspring with the *Aa* combination alleles. *A* is dominant over *a*, so all offspring in the second generation have red flowers, as does the true-breeding parental generation.

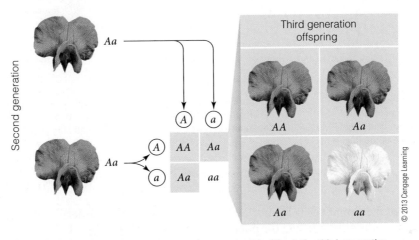

(b) If the plants from the second generation are cross-fertilized, the third generation has the allele combinations *AA*, *Aa*, and *aa*. Accordingly, this generation has a ratio of three plants with red flowers to one plant with white flowers.

We can summarize the most important aspects of Mendel's work as follows: The factors (genes) that control traits do not blend during inheritance; and even though traits may not be expressed in each generation, they are not lost (Figure 7.5). Therefore, some variation in populations is accounted for by alternate expressions of genes (alleles), because traits do not blend, as previously thought. Mendelian genetics explains much about heredity, but we now know the situation is much more complex than he realized. Our discussion focused on a single gene controlling a trait, but most traits are controlled by many genes, and some genes show incomplete dominance. Mendel was unaware of mutations (changes in the genetic material), chromosomes, and the fact that some genes control the expression of other genes. Nor had he heard of Hox genes, which were discovered during the 1980s and regulate the development of major body segments (see Chapter 12). Nevertheless, Mendel's work provided the answers Darwin and Wallace needed—but his research was published in an obscure journal and went unnoticed until 1900 when three independent researchers rediscovered them.

Genes and Chromosomes

Complex, double-stranded, helical molecules of **deoxyribonucleic acid (DNA)**, called **chromosomes**, are found in the cells of all organisms. Specific segments or regions of the DNA molecule are the hereditary units: the genes. The number of chromosomes is specific for a single species but varies among species. For instance, fruit flies have 8 chromosomes (4 pairs), humans have 46, and domestic horses have 64. Remember, chromosomes are always found in pairs carrying genes controlling the same traits.

In sexually reproducing organisms, the production of *sex cells* (pollen and ovules in plants and sperm and eggs in animals) results when cells undergo a type of cell division called **meiosis**. This process yields cells with only one chromosome of each pair, so all sex cells have only half

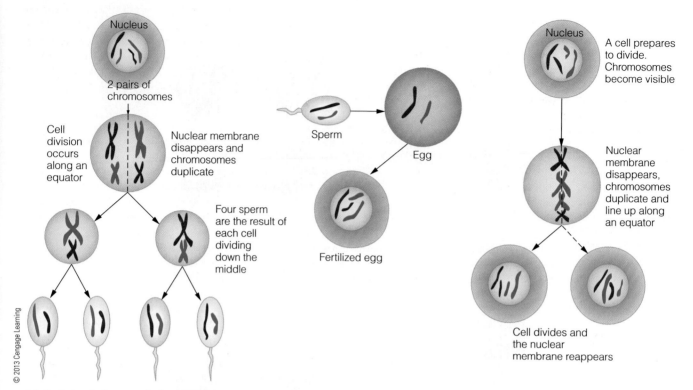

(a) During meiosis, sex cells form that contain one member of each chromosome pair. The formation of sperm is shown here. Eggs form the same way, but only one of the four final eggs is functional.

(b) The full number of chromosomes is restored when a sperm fertilizes an egg.

(c) Mitosis results in the complete duplication of a cell. In this example, a cell with four chromosomes (two pairs) produces two cells, each with four chromosomes. Mitosis takes place in all cells except sex cells. Once an egg has been fertilized, the developing embryo grows by mitosis.

© 2013 Cengage Learning

the chromosome number of the parent cell (▶ Figure 7.6a). During reproduction, a sperm fertilizes an egg (or pollen fertilizes an ovule), yielding an egg (or ovule) with the full set of chromosomes typical for that species (Figure 7.6b).

As Mendel deduced from his experiments, half of the genetic makeup of a fertilized egg comes from each parent. The fertilized egg, however, develops and grows by a cell division process called **mitosis** during which cells are simply duplicated—that is, there is no reduction in the chromosome number as in meiosis (Figure 7.6c).

The Modern View of Evolution

During the 1930s and 1940s, the ideas developed by paleontologists, geneticists, population biologists, and others were merged to form a **modern synthesis**, or neo-Darwinian view of evolution. The chromosome theory of inheritance was incorporated into evolutionary thinking, as were changes in genes (mutations) as a source of variation in populations. Lamarck's idea of inheritance of acquired characteristics was completely rejected, and the importance of natural selection was reaffirmed, although

the authors of the modern synthesis knew that other processes were also involved.

It is important to realize that in modern evolutionary theory, populations rather than individuals evolve. Individuals develop according to the genetic blueprint they inherit from their parents, but an individual may have favorable variations not present in most members of its population. In short, individuals with favorable variations in genetically varying populations are more likely to leave more descendants than others in their populations, and thus descendant populations possess these variations in greater frequency. So, an individual does not evolve during its lifetime, but evolution does take place in a succession of populations through time.

The theory of evolution is not static. New concepts now incorporated into the theory include *genetic drift* (random changes in populations due to chance), and new ideas on how speciation occurs (see the next sections). Until fairly recently, biologists and paleontologists thought that all genetic transfer was vertical—that is, from one generation to the next. They now know about the process called *lateral gene transfer* in single-celled organisms such as bacteria, in which one bacterium might incorporate genes from another bacterium but not be the offspring of that bacterium.

What Brings About Variation?

Natural selection works on variations in populations, most of which results from reshuffling of genes from generation to generation during sexual reproduction. Given that each of thousands of genes might have several alleles, and that offspring receive half of their genes from each parent, the potential for variation is enormous. However, this variation was already present, so new variations arise by **mutations**—that is, changes in the chromosomes or genes. In other words, a mutation is a change in the hereditary information. Whether a change is a *chromosomal mutation* (affecting a large segment of a chromosome) or a *point mutation* (a change in a particular gene), as long as it takes place in a sex cell, it is inheritable.

To fully understand mutations, we must explore them further. For one thing, they are random with respect to fitness, meaning they may be beneficial, neutral, or harmful. If a species is well adapted to its environment, most mutations would not be particularly useful and perhaps be harmful. Some mutations are absolutely harmful, but some that were harmful can become useful if the environment changes. For instance, some plants have developed a resistance to contaminated soils around mines. Plants of the same species from the normal environment do poorly or die in contaminated soils, whereas contaminant-resistant plants do very poorly in the normal environment. Mutations for contaminant resistance probably occurred repeatedly in the population, but they were not beneficial until contaminated soils were present.

How can a mutation be neutral? In cells, information carried on chromosomes directs the formation of proteins by selecting the appropriate amino acids and arranging them into a specific sequence. However, some mutations have no effect on the type of protein synthesized. In other words, the same protein is synthesized before and after the mutation, and thus the mutation is neutral.

What causes mutations? Some are induced by *mutagens*, agents that bring about higher mutation rates. Exposure to some chemicals, ultraviolet radiation, X-rays, and extreme temperature changes might cause mutations. But some mutations are spontaneous, taking place in the absence of any known mutagen.

Sexual reproduction and mutations account for much of the variation in populations, but another factor is *genetic drift*, a random change in the genetic makeup of a population due to chance. Genetic drift is significant in small populations but probably of little or no importance in large ones. Good examples of genetic drift include evolution in small populations that reach remote areas where only a few individuals with limited genetic diversity rebuilt a larger population.

Speciation and the Rate of Evolution

Speciation, the phenomenon of a new species arising from an ancestral species, is well documented, but the rate and ways it takes place vary.

Species Biologists use the term *species* in several ways, but here we refer to a **species** as a population of similar individuals that in nature interbreed and produce fertile offspring. Thus, a species is reproductively isolated from other species. Our definition of species does not apply to organisms such as bacteria that reproduce asexually, but it is useful for our discussions of plants, animals, fungi, and single-celled organisms called *protistans*.

Goats and sheep are easily distinguished by physical characteristics, and in nature they do not interbreed, so they are separate species. Yet, in captivity, they do produce fertile offspring. Domestic horses that have gone wild in Africa interbreed with zebras to yield a zebroid, which is sterile, and yet horses and zebras are different species. Camels and llamas do not look much alike, but they share a common ancestor and even now retain some genetic links. Indeed, an animal called a cama is a hybrid produced by artificial insemination of a female llama with sperm from a male dromedary camel (▶ Figure 7.7).

It should be obvious from these examples that reproductive barriers are not complete in all species, indicating varying degrees of change from a common ancestral species. In fact, many evolutionary changes in living populations take place within a species, a phenomenon called **microevolution**. For example, house sparrows were introduced into North America in 1852 and have evolved so that members of northern populations are larger and darker than those of southern populations, probably in response to climate. Likewise, microevolution accounts for organisms that develop resistance to insecticides and pesticides and for plants that adapt to contaminated soils.

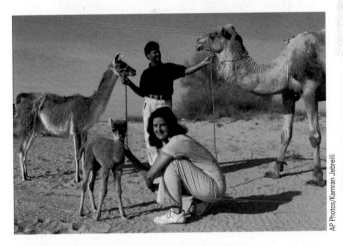

▶ Figure 7.7 **Evidence for Evolutionary Relationships** Dromedary camels and llamas do not look much alike, but they are closely related, having evolved from a common ancestor in North America. Camels became extinct in North America during the Pleistocene Epoch, but not before one ancestral group migrated to South America where it evolved into llamas and their relatives, and another group migrated to Eurasia and gave rise to the dromedary and bactrain camels. The animal in the foreground is a cama, a dromedary camel–llama hybrid. It resulted from the artificial insemination of a female llama with sperm from a male dromedary camel.

▶ Figure 7.8 **Allopatric Speciation** An example of allopatric speciation on some remote islands.

(a) A few individuals of a species on the mainland reach isolated island 1. Speciation follows genetic divergence in a new habitat.

(b) Later in time, a few individuals of the new species colonize nearby island 2. In this new habitat, speciation follows genetic divergence.

(c) Speciation may also follow colonization of islands 3 and 4. And it may follow invasion of island 1 by genetically different descendants of the ancestral species.

Akepa (*Loxous coccineus*)

Palila (*Loxioides bailleui*)

Akohekohe (*Palmeria doli*)

Iiwi (*Vestiaria coccinea*)

(d) More than 20 species of Hawaiian honeycreepers have evolved from a common ancestor, four of which are shown here, as they adapted to diverse resources on the islands.

Macroevolution, in contrast, involves changes such as the origin of new species, genera, families, orders, and classes. Many good examples of macroevolution are known from the fossil record—the origin of amphibians from fish, the evolution of mammals from mammal-like reptiles, the origin of whales from land-dwelling ancestors, and many others (see Chapters 13, 15, and 18). Although

macroevolution encompasses greater changes than microevolution, the cumulative effects of microevolution account for macroevolution. They differ only in degree of change.

Allopatric Speciation Speciation involves a change in the genetic makeup of a population, which also may bring about a change in form and structure—remember, populations rather than individuals evolve. According to the concept of **allopatric speciation**, a new species arises when a small population becomes isolated from its parent population, so that once isolated, the two populations share little if any genetic material (▶Figure 7.8).

Isolation may result from a rising mountain barrier that effectively separates a once interbreeding species, or a few individuals may somehow get to remote areas and no longer exchange genes with the parent population. Rising sea level may inundate part of a species range, thus isolating a small population from the original one. Seawater is an insurmountable barrier to amphibians, and their eggs, which will quickly perish in salt water. Scientists agree that allopatric speciation takes place, but they disagree on how rapidly a new species might evolve because of it.

Phyletic Gradualism and Punctuated Equilibrium According to the modern synthesis, the gradual accumulation of minor changes eventually brings about the origin of a new species, a phenomenon called **phyletic* gradualism**. Thus, a species arises gradually and continuously. Another view, called **punctuated equilibrium**, holds that species change little during most of their existence and then evolve rapidly, thus giving rise to a new species in as little as a few thousand years.

Advocates of punctuated equilibrium contend that few examples of gradual transitions from one species to another are found in the fossil record. Critics point out that neither Darwin nor those who formulated the modern synthesis insisted that all evolutionary change was gradual and continuous, a view shared by many biologist and paleontologists. And despite the incomplete nature of the fossil record, a number of gradual transitions from ancestral to descendant species are well known.

Regardless of how new species arise, or how rapidly they evolve, if evolution is correct, we should find examples of speciation taking place now. And, in fact, we do. This is especially true of some plant species that have arisen by *polyploidy*, which is when their chromosome count doubles. Speciation, or at least the first stages in speciation, has also taken place in populations of mosquitoes, bees, mice, salamanders, fish, and birds that are isolated or partially isolated from one another (▶Figure 7.9). Of course these are examples of microevolution, but keep in mind that macroevolution is simply the cumulative effect of microevolution.

---

*Phyletic is a term that refers to the evolutionary development of organisms.

▶ Figure 7.9 **Speciation in Songbirds** Two species of Eurasian songbirds appear to have evolved from an ancestral species. Adjacent populations of these birds, A and B, or F and G, for instance, can interbreed even though they differ slightly. However, where populations E and H overlap in Siberia, they cannot interbreed.

❶ A single ancestral species of "greenish warblers" is thought to have originated on the southern edge of the Himalayas 10,000 years ago.

❷ Over the years, the warbler established new populations on both sides of the treeless Tibetan Plateau. The birds can interbreed with adjacent populations, indicating that they are still members of the same species.

❸ But in Siberia, where birds east and west of the plateau meet, they do not interbreed, they do not recognize each other's songs, their wing coloration changes, and their genes differ. The warblers appear to have become two separate species.

Divergent, Convergent, and Parallel Evolution

Paleontologists refer to the occurrence of an ancestral species giving rise to diverse descendants adapted to various aspects of the environment as **divergent evolution**. An impressive example involves the mammals whose diversification from a common ancestor during the Late Mesozoic gave rise to such varied animals as platypuses, armadillos, rodents, bats, primates, whales, and rhinoceroses (▶ Figure 7.10).

Divergent evolution leads to descendants that differ markedly from their ancestors. But *convergent evolution* and *parallel evolution* are processes whereby similar adaptations arise in different groups. Unfortunately, they differ in degree and are not always easy to distinguish. Nevertheless,

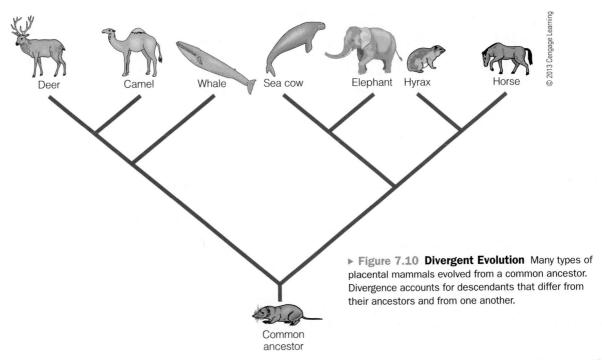

▶ Figure 7.10 **Divergent Evolution** Many types of placental mammals evolved from a common ancestor. Divergence accounts for descendants that differ from their ancestors and from one another.

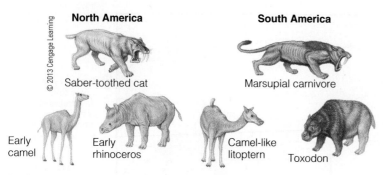

(a) Convergence in Cenozoic-age mammals from North America and South America.

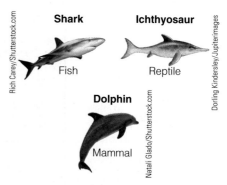

(b) Sharks, ichthyosaurs, and dolphins are all fast-swimming marine predators that resemble one another. However, they are only very distantly related.

both are common phenomena, with **convergent evolution** involving the development of similar characteristics in distantly related organisms, and **parallel evolution** involving the development of similar characteristics in closely related organisms. In both cases, similar characteristics develop independently, because the organisms in question adapt to comparable environments. Perhaps the following examples will clarify the distinction between these two concepts.

During much of the Cenozoic Era, South America was an island continent with a unique mammalian fauna that evolved in isolation. Nevertheless, several mammals in South and North America adapted in similar ways so that they superficially resembled one another (▶ Figure 7.11a)—a good example of convergent evolution.

Convergent evolution also accounts for the superficial similarities between Australian marsupial (pouched) mammals and placental mammals elsewhere. For example, the Tasmanian wolf (a marsupial) resembles placental dogs (wolves and coyotes), and hyenas, which, although doglike in appearance and habits, are most closely related to cats and mongooses. Also closely related to mongooses is the catlike fossa of Madagascar, which looks and lives much like cats elsewhere. Sharks, ichthyosaurs (extinct marine reptiles), and dolphins are morphologically similar to one another, but none are closely related (Figure 7.11b); in fact, the origin of each is separated in time by tens of millions of years.

Mosaic Evolution and Evolutionary Trends

Evolutionary changes do not involve all aspects of an organism simultaneously, because selection pressure is greater on some features than on others. As a result, a key feature we associate with a descendant group may appear before other features typical of that group. For example, the oldest known bird had feathers and the typical fused clavicles (the furcula or wishbone) of birds, but it also

retained many reptile characteristics (see Chapter 15). Accordingly, it represents the concept of **mosaic evolution**, meaning that organisms have recently evolved characteristics as well as some features of their ancestral group.

Paleontologists determine in some detail the *phylogeny*, or evolutionary history, and *evolutionary trends* for organisms if sufficient fossil material is available. For instance, one trend in ammonoids, extinct relatives of squid and octopus, was the evolution of an increasingly complex shell. Abundant fossils show that the Eocene mammals called *titanotheres* not only increased in size but also developed large nasal horns; moreover, the shape of their skull changed (▶ Figure 7.12).

Evolutionary trends are also well documented in horses (see Chapter 18) and camels. Indeed, fossil camels are common in North America, where most of their evolution took place. Trends in camel evolution include an increase in body size; longer limbs; reduction of the toes to two, which are fused together; loss of front teeth; and changes in the chewing teeth (Figure 7.12b).

Evolutionary trends are complex, they may be reversed, and not all occur at the same rate. In the evolution of horses there was a general increase in body size, but some now-extinct species showed a size decrease. We can think of these trends as a series of adaptations to a changing environment or adaptations that occur when organisms exploit new habitats. However, there are some organisms that show little obvious change for long periods. The brachiopod *Lingula* today looks much like it did during the Ordovician Period (see Figure 5.19), and the fish *Latimeria* and the ginkgo tree have also shown little change (▶ Figure 7.13).

Of course, we have no way to evaluate changes that are not obvious in bones, teeth, shells, and leaf impressions. For instance, there may have been changes in the digestive enzymes or immune system of any of the so-called **living fossils** we have mentioned, most of which are *generalized organisms*, meaning that they can live under a wide variety

34 MY
Late
37
Middle
49
Early
56

Eocene

0   1 m

© 2013 Cengage Learning

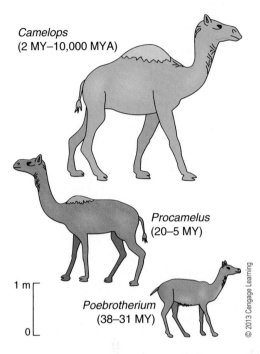

Camelops
(2 MY–10,000 MYA)

Procamelus
(20–5 MY)

1 m

Poebrotherium
(38–31 MY)

0

© 2013 Cengage Learning

(a) Titanotheres, relatives of horses and rhinoceroses, existed for about 20 million years during the Eocene Epoch. They evolved from small ancestors to giants standing 2.4 m at the shoulder. They also developed large horns, and the shape of the skull changed. Only 4 of 18 known genera are shown here.

(b) Evolutionary trends in camels are well documented by fossils. The most obvious trend is increase in size, but there were also changes in the limbs, teeth, and skull.

of environmental conditions or have persisted in an environment that has not changed significantly. Nevertheless, the remarkable stasis of these living fossils is not yet fully understood.

But isn't evolution by natural selection a random process? If so, how is it possible for a trend to continue long enough to account, just by chance, for such complex structures as eyes, wings, and hands? Actually, evolution by natural selection is a two-step process, and only the first step involves chance. First, variation must be present or arise in some population. Whether or not variations arising by mutations are favorable is indeed a matter of chance,

▶ Figure 7.13 **Two Examples of So-Called Living Fossils**

Norbert Wu/Minden Pictures

Sue Monroe

(a) *Latimeria* belongs to a group of fish known as coelacanths thought to have gone extinct at the end of the Mesozoic Era. One was caught off the east coast of Africa in 1938, and since then many more have been captured.

(b) These fan-shaped leaves from a present-day gingko tree look much like those of their ancient ancestors.

but the second step, natural selection, is not, because only individuals with favorable variations are most likely to survive and reproduce. In one sense, then, natural selection is a process of elimination, weeding out those individuals not as well adapted to a particular set of environmental circumstances.

Cladistics and Cladograms

Cladistics is a type of biological analysis in which organisms are grouped together based on derived as opposed to primitive characteristics. For instance, all land-dwelling vertebrates have bone and paired limbs, so these characteristics are primitive and of little use in establishing relationships among them. However, hair and three middle ear bones are derived characteristics, sometimes called *evolutionary novelties*, because only one subclade, the mammals, has them. If you consider only mammals, hair and middle ear bones are primitive characteristics. However, derived features, such as the shearing teeth of carnivorous mammals, serve to differentiate them from all other mammals, and feet in which the third toes bear most of the weight characterize the odd-toed hoofed mammals such as horses and rhinoceroses.

Traditionally, scientists have depicted evolutionary relationships with *phylogenetic trees*, in which the horizontal axis represents anatomic differences and the vertical axis denotes time (▶Figure 7.14a). The patterns of ancestor–descendant relationships shown are based on a variety of characteristics, although the ones used are rarely specified. In contrast, a **cladogram** shows the relationships among members of a *clade*, a group of organisms including its most recent common ancestor (Figure 7.14b).

Any number of organisms can be depicted in a cladogram, but the more shown, the more complex and difficult it is to construct. ▶Figure 7.15 shows three cladograms, each with a different interpretation of the relationships among bats, birds, and dogs. Bats and birds fly, so we might conclude that they are more closely related to one another than to dogs (Figure 7.15a). On the other hand, perhaps birds and dogs are more closely related than either is to bats (Figure 7.15b). However, if we concentrate on evolutionary novelties, such as hair and giving birth to live young, we conclude that the cladogram in Figure 7.15c shows the most probable relationships—that is, bats and dogs are more closely related to each other than they are to birds.

Scientists use cladograms to solve a variety of problems. For instance, if whales are more closely related to even-toed hoofed mammals (deer, hippos, antelope, etc.), as cladistic analyses indicate, they should possess some anatomical characteristics verifying this interpretation. And in fact some ancient whales possess the unique "double pulley" ankle typical of this group of mammals, even though the rear appendages of today's whales are only vestiges. Scientists using cladistic analysis of Pacific yews discovered a hard-to-recover compound used to treat cancer, and assumed that the same compound might be in closely related species. Indeed it was, and now the compound is widely available.

When applied to fossils, care must be taken in determining what are primitive versus derived characteristics, especially in groups with poor fossil records. Furthermore, cladistic analysis depends solely on characteristics inherited from a common ancestor, so paleontologists must be especially careful of characteristics that result from convergent evolution. Nevertheless, cladistics is a powerful tool that has more clearly elucidated the relationships among many fossil lineages and is now used extensively by paleontologists.

▶ Figure 7.14 **Phylogenetic Tree and Cladogram**

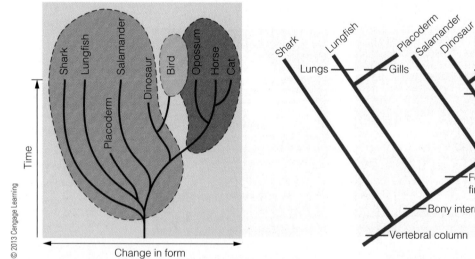

© 2013 Cengage Learning

(a) A phylogenetic tree showing the relationships among various vertebrate animals.

(b) A cladogram showing inferred relationships. Some of the characteristics used to construct this cladogram are indicated.

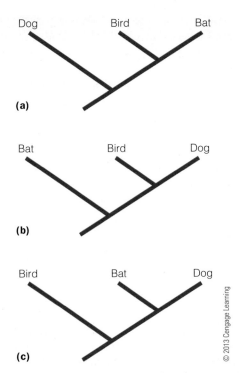

(a)

(b)

(c)

▶ **Figure 7.15 Cladograms for Dogs, Birds, and Bats** Clado-grams showing three hypotheses for the relationships among dogs, birds, and bats. Derived characteristics such as hair and giving birth to live young indicate that dogs and bats are most closely related, as shown in (c).

Extinctions

Judging from the fossil record, most organisms that ever existed are now extinct—perhaps as many as 99 percent of all species. Now, if species actually evolve as natural selection favors certain traits, shouldn't organisms be evolving toward some kind of higher order of perfection or greater complexity? Vertebrate animals are more complex, at least in overall organization, than are bacteria, but complexity does not necessarily mean they are superior in some survival sense—after all, bacteria have persisted for at least 3.0 billion years. Actually, natural selection does not yield some kind of perfect organism, but rather those adapted to a specific set of circumstances at a particular time. Thus, a clam or lizard existing now is not somehow superior to those that lived millions of years ago.

The continual extinction of species is referred to as *background extinction*, to clearly differentiate it from a **mass extinction**, during which accelerated extinction rates sharply reduce Earth's biotic diversity. Extinction is a frequent occurrence, but so is the evolution of new species that usually, but not always, quickly exploit the opportunities created by another species' extinction. When the dinosaurs and their relatives died out, the mammals began a remarkable diversification as they began occupying the niches left temporarily vacant.

Everyone has heard of the mass extinction of dinosaurs and some of their relatives at the end of the Mesozoic Era.

The greatest mass extinction, though, was at the end of the Paleozoic Era, when perhaps more than 96 percent of all marine invertebrate species and approximately 70 percent of all terrestrial vertebrate species died out.

What Kinds of Evidence Support Evolutionary Theory?

When Charles Darwin proposed his theory of evolution, he cited supporting evidence such as classification, embryology, comparative anatomy, geographic distribution, and, to a limited extent, the fossil record. He had little knowledge of the mechanism of inheritance, and both biochemistry and molecular biology were unknown during his time. Studies in these areas, coupled with a more complete and much better understood fossil record, have convinced scientists that the theory is as well supported by evidence as any other major theory.

But is the theory of evolution truly scientific? That is, can testable predictive statements be made from it? First we must be clear on what a theory is and what we mean by "predictive." Scientists propose hypotheses for natural phenomena, test them, and in some cases raise them to the status of a theory—an explanation of some natural phenomenon well supported by evidence from experiments and/or observations.

Almost everything in the sciences has some kind of theoretical underpinning—optics, heredity, the nature of matter, the present distribution of continents, diversity in the organic world, and so on. Of course, no theory is ever proven in some final sense, although it might be supported by substantial evidence; all are always open to question, revision, and occasionally to replacement by a more comprehensive theory. In his book, *Why Darwin Matters*, (pp. 1–2) Michael Shermer noted that

> A "theory" is not just someone's opinion or a wild guess made by a scientist. A theory is a well-supported and well-tested generalization that explains a set of observations. Science without theory is useless.

Prediction is commonly taken to mean to foresee some future event such as the next solar eclipse, but not all predictions are about the future. For instance, according to plate tectonic theory, oceanic crust is continually produced at spreading ridges as plates diverge. If this is true, it is logical to predict that oceanic crust should become increasingly older with distance from a spreading ridge, and that sediment thickness on the seafloor should also increase with distance from a ridge. Both of these predictive statements have been verified (see Figure 3.11).

If the theory of evolution is correct, closely related species such as wolves and coyotes should be similar not only in their anatomy but also in terms of biochemistry, genetics,

TABLE 7.1 Some Predictions From the Theory of Evolution

1. If evolution has taken place, the oldest fossil-bearing rocks should have remains of organisms very different from those existing now, and more recent rocks should have fossils more similar to today's organisms.
2. If today's organisms descended with modification from ones in the past, there should be fossils showing characteristics connecting orders, classes, and so on.
3. If evolution is true, closely related species should be similar not only in details of their anatomy, but also in their biochemistry, genetics, and embryonic development, whereas distantly related species should show fewer similarities.
4. If the theory of evolution is correct—that is, living organisms descended from a common ancestor—classification of organisms should show a nested pattern of similarities.
5. If evolution actually took place, we would expect cave-dwelling plants and animals to most closely resemble those immediately outside their respective caves rather than being most similar to those in caves elsewhere.
6. If evolution actually took place, we would expect land-dwelling organisms on oceanic islands to most closely resemble those of nearby continents rather than those on other distant islands.
7. If evolution occurred, we would expect mammals to appear in the fossil record long after the appearance of the first fish. Likewise, we would expect reptiles to appear before the first mammals or birds.
8. If we examine the fossil record of related organisms such as horses and rhinoceroses, we should find that they were quite similar when they diverged from a common ancestor but became increasingly different as their divergence continued.

© 2013 Cengage Learning

and embryonic development (number 3 in Table 7.1). Suppose that they differed in their biochemical mechanisms as well as their embryology. Obviously our prediction would fail, and we would at least have to modify our theory. If other predictions also failed—say, mammals appear in the fossil record before fishes—we would have to abandon our theory and find another explanation for our observations.

Classification—A Nested Pattern of Similarities

Carolus Linnaeus (1707–1778) proposed a classification system in which organisms are given a two-part genus and species name; the coyote, for instance, is *Canis latrans*. Table 7.2 shows Linnaeus's classification scheme, which is a hierarchy of categories that becomes more inclusive as one proceeds up the list. The coyote (*Canis latrans*) and the wolf (*Canis lupus*) share numerous characteristics, so

they are members of the same genus, whereas both share some, but fewer characteristics with the red fox (*Vulpes fulva*), and thus all three are members of the family Canidae. All canids share some characteristics with cats, bears, and weasels and are grouped together in the order Carnivora, which is one of 18 living orders of the class Mammalia, all of whom are warm-blooded, possess fur or hair, and have mammary glands.

Linnaeus recognized shared characteristics among organisms, but his intent was simply to categorize species he thought were specially created and immutable. Following the publication of Darwin's *On the Origin of Species* in 1859, however, scientists quickly realized that shared characteristics constituted a strong argument for evolution.

Linnaeus's classification is widely used, but it is arbitrary when assigning organisms to categories above the species level. Also, it does not in all cases reflect evolutionary relationships. As a consequence, scientists frequently use evolutionary relationships, or *phylogeny*, to group

TABLE 7.2 Expanded Linnaen Classification Scheme

The animal classified in this example is the coyote, *Canis latrans*

Kingdom	**Animalia**—Multicelled organisms; cells with nucleus (see Chapter 9); reproduce sexually; ingest preformed organic molecules for nutrients
Phylum	**Chordata**—Possess notochord, pharyngeal pouches, dorsal hollow nerve cord at some time during life cycle (see Chapter 13)
Subphylum	**Vertebrata**—Those chordates with a segmented vertebral column (see Chapter 13)
Class	**Mammalia**—Warm-blooded vertebrates with hair or fur and mammary glands (see Chapter 15)
Order	**Carnivora**—Mammals with teeth specialized for a diet of meat (see Chapter 18)
Family	**Canidae**—The doglike carnivores (coyotes, wolves, and foxes)
Genus	***Canis***—Made up only of closely related species—coyotes and wolves (also includes domestic dogs)
Species	***latrans***—Consists of similar individuals that in nature can interbreed and produce fertile offspring

© 2013 Cengage Learning

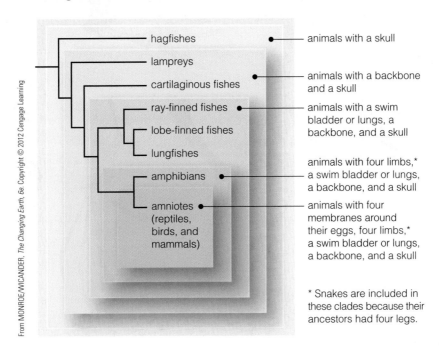

▶ **Figure 7.16 Cladograms and Shared Characteristics** This example shows the nested pattern of similarities among vertebrate animals—that is, those animals with a segmented vertebral column.

hagfishes — animals with a skull

lampreys

cartilaginous fishes — animals with a backbone and a skull

ray-finned fishes — animals with a swim bladder or lungs, a backbone, and a skull

lobe-finned fishes

lungfishes

amphibians — animals with four limbs,* a swim bladder or lungs, a backbone, and a skull

amniotes (reptiles, birds, and mammals) — animals with four membranes around their eggs, four limbs,* a swim bladder or lungs, a backbone, and a skull

* Snakes are included in these clades because their ancestors had four legs.

From MONROE/WICANDER, *The Changing Earth, 6e.* Copyright © 2012 Cengage Learning

organisms. Recall from our discussion of cladistics that a *clade* is a group of organisms, including its most recent common ancestor. Accordingly, a well-constructed cladogram shows a nested pattern of similarities (▶ Figure 7.16).

Biologic Evidence That Supports Evolution

Scientists have described at least 1.7 million species and estimate that there may be as many as several tens of millions not yet recognized. And yet, if existing organisms evolved from common ancestors, there should be similarities among all life-forms. As a matter of fact, all living things, be they bacteria, redwood trees, or whales, are composed mostly of carbon, nitrogen, hydrogen, and oxygen. Furthermore, their chromosomes consist of DNA, and all cells synthesize proteins in essentially the same way. Blood proteins are similar among all mammals but also indicate that humans are most closely related to great apes, followed in order by Old World monkeys, New World monkeys, and lower primates such as lemurs. Biochemical tests support the idea that birds descended from reptiles, a conclusion supported by evidence in the fossil record.

The forelimbs of humans, whales, dogs, and birds are superficially dissimilar (▶ Figure 7.17). Yet all are made up of the same bones; have basically the same arrangement of muscles, nerves, and blood vessels; are similarly arranged with respect to other structures; and have a similar pattern of embryonic development. These **homologous structures**, as they are called, are simply basic vertebrate

forelimbs modified for different functions. However, some similarities are unrelated to evolutionary relationships. For instance, wings of insects and birds serve the same function but are quite dissimilar in both structure and development and are thus termed **analogous structures** (▶ Figure 7.18).

Why do dogs have tiny remnants of toes, called dewclaws, on their forefeet or hind feet or, in some cases, on all four feet (▶ Figure 7.19a)? These dewclaws are **vestigial structures**—that is, remnants of structures that were fully functional in their ancestors. Ancestral dogs had five toes on each foot, each of which contacted the ground. As they evolved, though, dogs became toe walkers with only four digits on the ground, and the thumbs and big toes were either lost or reduced to their present state. Although vestigial structures need not be totally functionless, many are, or at best have a reduced function. Whales and some snakes have a pelvis but no rear limbs (Figure 7.19b), except for a rare whale with hind limbs and two Cretaceous-age fossil snakes with back legs.

No one would seriously argue that a dog's dewclaws are truly necessary, and it would be difficult to convince most people that wisdom teeth in humans are essential. After all, some people never have wisdom teeth, and in those that do, they commonly never erupt from the gums, or they come in impacted and cause significant pain. The reason? The human jaw is too short to accommodate the ancestral number of teeth. In some cases, vestigial structures are fully functional, but perform a totally different function from what they did in their ancestral condition. For instance, the incus and malleus of the mammalian

▶ Figure 7.17 **Homologous Structures** The forelimbs of mammals and a bird are compared with the forelimb of their ancestor among the reptiles. Homologous organs may serve different functions, but all are composed of the same bones, all have a similar arrangement with respect to other structures, and all undergo a similar embryonic development.

▶ Figure 7.18 **Analogous Structures** Although the fly's wings have a different structure and different embryologic development than the wings of birds and bats, they serve the same function and thus are analogous.

▶ Figure 7.19 **Vestigial Structures**

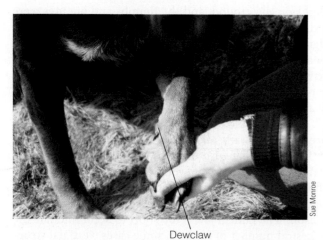

(a) Notice the dewclaw, a vestige of the thumb, on the forefoot of this dog.

(b) The Eocene-age whale *Basilosaurus* had tiny vestigial back limbs, but it did not use limbs to support its body weight. Even today, whales have a vestige of a pelvis, and on a few occasions, whales with rear limbs have been caught.

middle ear were derived from the articular and quadrate bones that formed the joint between the jaw and skull in mammal-like reptiles (see Chapter 15).

Another type of evidence for evolution is observations of microevolution in living organisms. We have already mentioned the adaptations of some plants to contaminated soils. As a matter of fact, small-scale changes take place rapidly enough that new insecticides and pesticides must be developed continually because insects and rodents develop resistance to existing ones. Development of antibiotic-resistant strains of bacteria is a continuing challenge to medicine. Whether the variation in these populations previously existed or was established by mutations is irrelevant. In either case, some variant types were more likely to survive and reproduce, bringing about a genetic change.

Biogeography

Biogeography, the geographic distribution of organisms, both past and present, provides compelling evidence for evolution. When first settled, the Hawaiian Islands had no freshwater fish, amphibians, reptiles, or mammals other than bats and monk seals. The Galápagos Islands had no native mammals, amphibians, or freshwater fish. Obviously, animals and plants introduced by humans survive on these remote islands quite nicely. In fact, oceanic islands everywhere show peculiar biotas that in almost all details differ from, but most closely resemble, the fauna and flora of the nearest continent. Furthermore, the more remote an island, the less diverse is its biota, and in a chain of islands, the fauna and flora of each island differs slightly from nearby islands.

It should come as no surprise that most organisms that colonize remote islands are the ones that are most mobile and capable of doing so. Present-day birds are sometimes spotted far out over the oceans. Many probably simply perish, but it only takes a few to reach an island and start a breeding colony. The bird called the purple gallinule was an occasional visitor to Tristan da Chuna, which lies in the Atlantic Ocean about midway between South America and Africa, but in the 1950s, a breeding colony was established on the island.

Another example involves at least 15 green iguanas that in 1995 were rafted 280 km on floating vegetation from Guadeloupe to Anguilla in the Caribbean Sea when the trees they were on were blown out to sea by hurricane winds. Given these conditions and the fact that different selective pressures are likely, they may eventually give rise to a reproductively isolated species. Of course, iguanas did not arrive at Anguilla because of their mobility, but rather by chance.

Remote islands are also nearly lacking in large trees. Most seeds and fruits are too large to be carried far by wind, and they cannot float in seawater and remain viable long enough to make a trip to an island, the notable exception being coconuts. As a result, the tree-like plants on islands are very different from those on continents, having evolved from smaller plants. Good examples are the tree-like plants in Hawaii that are related to the North American tarweeds, and the tree-sized sunflowers of the Galápagos Islands.

The islands we have discussed so far are *oceanic islands*, that is, islands that formed by volcanic activity and were never connected to a continent. *Continental islands*, in contrast, were once connected to a continent and became separated by moving plates (Madagascar) or by rising sea level (the British Isles). When these continental islands become isolated, they already had species much like those of the continent from which they became separated. However, the older the continental island, the more its biota differs from its mainland; Madagascar separated from Africa 160 million years ago, whereas the British Isles were isolated from mainland Europe by rising sea level about 300,000 years ago. In the case of oceanic islands, however, only a few hardy migrants arrived to populate an otherwise uninhabited spot of land, and they then evolved in isolation.

Fossils: What Do We Learn From Them?

Fossil marine invertebrates found far from the sea, and even high in the mountains, led early naturalists to conclude that the fossil-bearing rocks were deposited during a worldwide flood. In 1508, Leonardo da Vinci realized that the fossil distribution was not what one would expect from a rising flood, but the flood explanation persisted, and John Woodward (1665–1728) proposed a testable hypothesis. According to Woodward, fossils were sorted out as they settled from floodwaters based on their shape, size, and density, but this hypothesis was quickly rejected because observations did not support it. Fossils of various sizes, shapes, and densities are found throughout the fossil record.

The fossil record does show a sequence, but not one based on density, size, shape, agility, or habitat. Rather, the sequence shows that one-celled organisms appear before multicelled ones, plants before animals, and invertebrates before vertebrates. Among vertebrates, fish appear first, followed in succession by amphibians, reptiles, mammals, and birds.

We have noted that fossils of some organisms are much more common than many people realize, and that geologists use fossils in relative age determinations and analyses of depositional environments (see Chapters 5 and 6). Furthermore, fossils provide some of the evidence for evolution (▶ Figure 7.20). So we might ask, "Where are the fossils showing the diversification of horses, rhinoceroses, and tapirs from a common ancestor; or the origin of birds from reptiles; or the evolution of whales from a land-dwelling ancestor?" It is true that the initial diversification of many groups is poorly represented in the fossil record, but in these cases, as well as many others, fossils of just the kind we would expect are well known.

(a) Shells of marine animals known as ammonites. Ammonites were abundant during the Mesozoic Era.

(b) Bones of a duck-billed dinosaur.

Horses, rhinoceroses, and tapirs may seem an odd assortment of animals, but fossils and studies of living animals indicate that they share a common ancestor (▶ Figure 7.21). If this statement is correct, then we can predict that as we trace these animals back in the fossil record, differentiating one from the other should become increasingly difficult (Table 7.1). And, in fact, the earliest members of each are remarkably similar, differing mostly in size and details of their teeth. As their diversification proceeded, though, differences became more apparent.

Missing Links—Are They Really Missing?

The most profound misconception about the fossil record is the claim that no fossils exist bridging the gaps between ancestors and descendants. These so-called missing links, according to some outside the sciences, invalidate the theory of evolution. Indeed, there are missing links, but there are also many fossils as close to being intermediate, or transitional, between groups as we could ever hope to find.

It is true that finding "missing links" between species is difficult, because descendant species differ so little from their ancestors. When you consider genera, families, orders, and classes, though, transitional fossils are much more common. In 2006, scientists discovered the *fishapod*, an intermediate between the lobe-finned fishes and amphibians (see Chapter 13), and the transition from mammal-like reptiles to mammals is exceptionally well-documented by fossils (see Chapter 15). Living whales either have teeth or *baleen*, bristle-like material used to filter food from seawater. An Oligocene-age fossil whale had both.

Dugongs and manatees, collectively called sea cows, are thoroughly aquatic mammals with limbs modified into paddles, and yet fossil sea cows from Jamaica had four well-developed limbs and a pelvis that supported the animals on land. We have already mentioned the extinct titanotheres (Figure 7.12a), and we will discuss the fossils that show the evolution of horses, whales, and other groups of mammals in Chapter 18. In the final analysis, many "missing links" are really not missing at all.

The Evidence—A Summary

As already noted, scientists agree that the theory of evolution is as well supported by evidence as any other major theory. For example, the transitional fossils we discussed in the preceding section provide compelling evidence for evolution, but fossils, although important, are not the only evidence. Indeed, much of the evidence comes from comparative anatomy, biogeography, molecular biology, genetics, embryology, and biochemistry—that is, from converging lines of evidence.

Certainly scientists disagree on specific issues such as the rate of evolution, the significance of some fossils, and precise relationships among organisms, but they are overwhelmingly united in their support of the theory. And despite stories holding that scientists are unyieldingly committed to this theory and unwilling to investigate other explanations for the same data, just the opposite is true. If another scientific (testable) hypothesis offered a better explanation, scientists would be eager to investigate it. There is no better way in the sciences to gain respect and lasting recognition than to modify or replace an existing widely accepted theory.

▶ Figure 7.21 **Relationships Among Horses, Rhinoceroses, and Tapirs (Order Perissodactyla)** This cladogram shows the relationships among these animals, which are well established by fossils (see Chapter 18) and by studies of living animals. When they first diverged from a common ancestor, all were very similar, but as they diverged, they became increasingly different.

Tapir   Rhinoceros   Horse (zebra)

Common ancestor

Summary

▶ The central claim of the theory of evolution is that all organisms have descended with modification from ancestors that lived during the past.

▶ The first widely accepted mechanism to account for evolution—inheritance of acquired characteristics—was proposed in 1809 by Jean-Baptiste de Lamarck.

▶ Charles Darwin's observations of variation in populations and artificial selection, as well as his reading of Thomas Malthus's essay on population, helped him formulate his idea of natural selection as a mechanism for evolution.

▶ In 1859, Charles Darwin and Alfred Wallace published their ideas of natural selection, which hold that in populations of organisms, some have favorable traits that make it more likely that they will survive and reproduce and pass on these traits.

▶ Gregor Mendel's breeding experiments with garden peas provided some of the answers regarding variation and how it is maintained in populations. Mendel's work is the basis for present-day genetics.

▶ Genes, the specific segments of chromosomes, are the hereditary units in all organisms. Only the genes in sex cells are inheritable.

▶ Sexual reproduction and mutations (changes in chromosomes or genes) account for most variation in populations.

▶ Evolution by natural selection is a two-step process. First, variation must exist or arise and be maintained in interbreeding populations, and second, organisms with favorable variants must be selected for survival so that they can reproduce.

▶ Many species evolve by allopatric speciation, which involves isolation of a small population from its parent population that is then subjected to different selection pressures.

▶ When diverse species arise from a common ancestor, it is called divergent evolution. The development of similar adaptive features in different groups of organisms results from convergent evolution and parallel evolution.

▶ Microevolution involves changes within a species, whereas macroevolution encompasses all changes above the species level. Macroevolution is simply the outcome of microevolution over time.

▶ Scientists have traditionally used phylogenetic trees to depict evolutionary relationships, but now they more commonly use cladistic analyses and cladograms to show these relationships.

▶ Background extinction occurs continually, but several mass extinctions have also taken place, during which Earth's biotic diversity has decreased markedly.

▶ The theory of evolution is truly scientific because we can think of observations that would falsify it, that is, prove it wrong, yet no such observations have been made.

▶ Much of the evidence supporting evolutionary theory comes from classification, comparative anatomy, embryology, genetics, biochemistry, molecular biology, and present-day examples of microevolution.

▶ The fossil record also provides evidence for evolution in that it shows a sequence of different groups appearing through time, and some fossils show the features we would expect in the ancestors of birds, mammals, horses, whales, and so on.

Important Terms

allele, p. 132
allopatric speciation, p. 136
analogous structure, p. 143
artificial selection, p. 130
biogeography, p. 145
chromosome, p. 133
cladistics, p. 140
cladogram, p. 140
convergent evolution, p. 138
deoxyribonucleic acid (DNA), p. 133
divergent evolution, p. 137

gene, p. 132
homologous structure, p. 143
inheritance of acquired characteristics, p. 129
living fossil, p. 138
macroevolution, p. 136
mass extinction, p. 141
meiosis, p. 133
microevolution, p. 135
mitosis, p. 134
modern synthesis, p. 134

mosaic evolution, p. 138
mutation, p. 135
natural selection, p. 130
paleontology, p. 127
parallel evolution, p. 138
phyletic gradualism, p. 136
punctuated equilibrium, p. 136
species, p. 135
theory of evolution, p. 127
vestigial structure, p. 143

Review Questions

1. A good example of analogous structures are
   a. ____ eyes of dogs and cats; b. ____ wings of insects and bats; c. ____ legs of horses and deer; d. ____ scales of reptiles and amphibians; e. ____ flippers of whales and porpoises.
2. The work of _____ is the basis for genetics.
   a. ____ Alfred Wallace; b. ____ Charles Lyell; c. ____ William Smith; d. ✓ Gregor Mendel; e. ____ Trofim Lysenko.
3. The type of cell division that accounts for the origin of sperm and eggs is called _____.
   a. ✓ meiosis; b. ____ panspermia; c. ____ divergence; d. ____ vestigial splitting; e. ____ cladistics.
4. Which one of the following is involved in natural selection?
   a. ____ Growth; b. ✓ Reproduction; c. ____ Metamorphisis; d. ____ Mosaic evolution; e. ____ Convergence.
5. Large-scale evolution resulting in the origin of new species and genera is called
   a. ____ phyletic gradualism; b. ____ Wallace's law; c. ____ bionomial nomenclature; d. ____ modern synthesis; e. ✓ macroevolution.
6. A structure possessed by an animal that no longer serves any function, has a reduced function, or serves a completely different function than it did in the ancestral condition is called a(n)
   a. ____ superfluous organ; b. ____ extra appendage; c. ✓ vestigial structure; d. ____ chromosomal aberration; e. ____ living fossil.
7. The study of life history as revealed by fossils is
   a. ✓ paleontology; b. ____ herpetology; c. ____ physiology; d. ____ archaeology; e. ____ mineralogy.
8. Genes for the same trait come in alternate forms that are called
   a. ____ homologous features; b. ____ species; c. ____ mosaic chromosomes; d. ✓ alleles; e. ____ microfibers.
9. According to the concept of punctuated equilibrium,
   a. ____ organisms pass on characteristics acquired during their lifetimes; b. ____ evolution always occurs slowly but continuously; c. ____ natural selection favors the largest and strongest for survival; d. ____ animals of different species interbreed but their offspring are infertile; e. ____ species remain unchanged during most of their existence, then evolve "rapidly."
10. The fact that several marsupial mammals in Australia resemble and live much like placental mammals elsewhere is a good example of
    a. ____ microevolution; b. ____ convergent evolution; c. ____ binomial classification; d. ____ punctuated evolution; e. ____ phyletic gradualism.
11. What are homologous and analogous structures? Give some examples of each.
12. Compare phyletic gradualism and punctuated equilibrium.
13. How does divergent evolution differ from convergent evolution? Cite some examples.
14. According to scientists, present-day tapirs, horses, and rhinoceroses are more closely related to one another than they are to any other living hoofed mammal. What kinds of evidence from biology and the fossil record would support this conclusion?
15. Does natural selection really imply that only the biggest, strongest, and fastest are most likely to survive and reproduce?
16. Explain how macroevolution is simply the cumulative effect of microevolution.
17. What are vestigial structures and what is their significance to evolutionary theory?
18. How did the experiments carried out by Gregor Mendel help answer the inheritance questions that plagued Darwin and Wallace?
19. What if someone were to tell you that "the fossil record shows a sequence of organisms in older to young rocks that was determined by their density, shape, and habitat." What kinds of evidence would you cite to refute this statement?
20. Give some examples of how geographic distribution of organisms provides some evidence for the theory of evolution.

Precambrian Earth and Life History

The Hadean and the Archean Eon

8

Archean rocks in the Teton Range of Grand Teton National Park in Wyoming. The rocks visible in this image are gneiss, schist, and granite that date from 2.8 to 2.5 billion years old. The Teton Range, however, formed only about 10 million years ago. The peaks in this image are known as the Cathedral Group.
Sue Monroe

CHAPTER OBJECTIVES

At the end of this chapter, you will have learned that

▶ Precambrian time, which accounts for most geologic time, is divided into three intervals: the Hadean (an informal term), the Archean Eon, and the younger Proterozoic Eon.

▶ No rocks are known from the Hadean, but geologists can nevertheless make some reasonable inferences about events that occurred then.

▶ The Archean geologic record is difficult to interpret because many of the rocks are metamorphic, deformed, deeply buried, and contain few fossils.

▶ Each continent has at least one area of exposed Precambrian rocks called a shield and a buried extension of the shield known as a platform. A shield and its platform make up a craton, an ancient stable nucleus of a continent.

▶ The most common Archean rocks are granite-gneiss complexes with subordinate greenstone belts made up mostly of igneous rocks and some sedimentary rocks.

▶ Greenstone belts likely form in several tectonic settings, but many probably evolved in back-arc basins and in rifts within continents.

▶ During the Archean, Earth possessed more radiogenic heat and primordial heat so that plates moved more rapidly and igneous activity was much more widespread that it is now.

▶ Gases released by volcanoes were responsible for the origin of the hydrosphere and atmosphere, but the atmosphere had little free oxygen.

▶ The oldest known fossils are of single-celled bacteria and chemical traces of bacteria-like organisms. Bacteria commonly called blue-green algae produced irregular mats and moundlike structures known as stromatolites.

▶ Archean mineral resources include gold, platinum, copper, zinc, and iron.

Introduction

You know that the concept of *time* is used to specify the duration of events and the intervals between events, and you are also familiar with time from the human perspective—that is, hours, days, and years—but you probably have no frame of reference for geologic time (see Chapter 4). Indeed, geologists commonly use the phrase *deep time* to emphasize the magnitude of geologic time. Earth has existed for 4,600,000,000 years, more conveniently stated as 4.6 billion years. So, let's suppose that one second equals one year, and you want to count out Earth's history. Should you take on this task, you and your descendants would still be counting nearly 146 years from now.

In this and the next chapter, we are concerned only with that part of geologic time designated *Precambrian*, 4.6 billion to 542 million years ago. If all geologic time were represented by a 24-hour clock, the Precambrian alone would be more than 21 hours long and constitute 88 percent of all geologic time (▶ Figure 8.1). And yet we discuss this incredibly long interval in only two chapters. We devote 10 chapters to the more familiar Phanerozoic Eon, made up of the Paleozoic, Mesozoic, and Cenozoic eras, but this seemingly disproportionate treatment is justified when you consider that so much more is known about that more recent part of geologic time.

Precambrian is a widely used term referring to both time and rocks. As a time term, it includes all geologic time from Earth's origin 4.6 billion years ago to the beginning of the Phanerozoic Eon 542 million years ago. The term also refers to all rocks lying beneath those of the Cambrian System. Because of the complexities of these rocks and the scarcity of fossils, establishing formal subdivisions

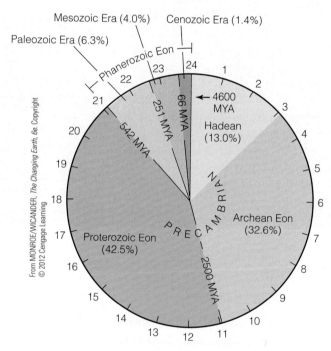

▶ Figure 8.1 **Geologic Time Represented by a 24-Hour Clock** If 24 hours represented all geologic time, the Precambrian would be more than 21 hours long, thus more than 88 percent of the total.

From MONROE/WICANDER, *The Changing Earth, 6e*. Copyright © 2012 Cengage Learning

of the Precambrian is difficult. In 1982, in an effort to standardize terminology, the North American Commission on Stratigraphic Terminology recommended the use of *Archean Eon* and *Proterozoic Eon* for most of Precambrian time, and more recently suggested the informal term

Hadean for the earliest part of the Precambrian for which there are no known rocks (▶Figure 8.2). The subdivisions in Figure 8.2 are based mostly on absolute ages rather than time-stratigraphic units, which is a departure from standard practice.

The geologic record we do have for the Precambrian, especially for the Archean (see chapter opening photo), is difficult to decipher. It was during this time that the Earth systems we discussed in Chapter 1 (see Figure 1.1) became operative, although not all at the same time or necessarily in their present form. Earth did not differentiate into a core, mantle, and crust until millions of years after it formed (see Figure 1.6), but once it did, internal heat was responsible for moving plates and for the origin and continuing evolution of the continents. Earth's early atmosphere evolved from one rich in carbon dioxide to one with free oxygen and an ozone layer, surface waters began to accumulate, and organisms appeared as much as 3.5 billion years ago. In short, Earth was very different when it formed, but during the Precambrian it began to evolve and became increasingly like it is today.

What Happened During the Hadean?

All geologic time from Earth's origin (4.6 billion years ago) to the beginning of the Archean Eon (4.0 billion years ago) is encompassed by the informal term *Hadean* (Figure 8.2). The Acasta Gneiss of Canada, at about 4.0 billion years old, is the oldest known rock on Earth, with the exception of meteorites, so we have no geologic record for the first 600 million years of Earth history. Nevertheless, geologists can make some reasonable inferences about events that took place during the Hadean. Of course, the Solar System formed and the terrestrial planets accreted from planetesimals; Earth differentiated into a core and mantle (see Figures 1.5 and 1.6); and continental crust began to form, perhaps as much as 4.4 billion years ago.

As the accreting planet grew, it swept up the debris in the vicinity, and just like the other terrestrial planets, Earth was bombarded by meteorites and comets until about 3.8 billion years ago. Unlike Mercury, Mars, and Earth's Moon, however, the evidence of this period of impacts has been obliterated by weathering, erosion, volcanism, plate movement, and mountain building. In addition to bombardment by meteorites, Earth was probably hit by a Mars-sized planetesimal 4.4 to 4.6 billion years ago, causing the ejection of a huge mass of hot material that coalesced to form the Moon.

After it first formed, Earth retained considerable heat from its origin, and much more heat was generated by radioactive decay; as a result, volcanism was ubiquitous (▶Figure 8.3). Gases emitted by volcanoes formed an atmosphere, but it was very unlike the oxygen-rich one present now, and when the planet cooled sufficiently, surface

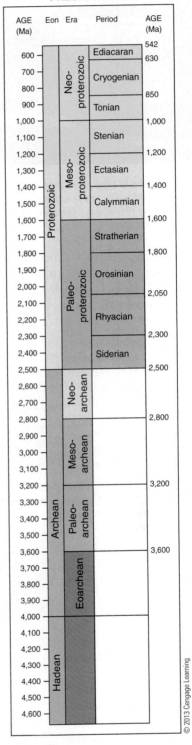

▶ **Figure 8.2 The Precambrian Geologic Time Scale** This most recent version of the geologic time scale was published by the International Commission on Stratigraphy (ICS) in 2009. See Figure 1.11 for the complete time scale. Notice the use of the prefixes *eo* (early, or dawn), *paleo* (old, or ancient), *meso* (middle), and *neo* (new, or recent). The age columns on the left and right sides of the time scale are in hundreds and thousands of millions of years (1,800 million years = 1.8 billion years, for example).

▶ **Figure 8.3 Earth As It May Have Appeared Soon After It Formed** No rocks are known from this earliest time in Earth history, but geologists can make reasonable inferences about the nature of the newly formed planet.

© Phil Degginger/Alamy

waters began to accumulate. If we could somehow go back and visit early Earth, we would see a rapidly rotating, hot, barren, waterless planet bombarded by meteorites and comets. There were no continents, cosmic radiation would have been intense, and, of course, you would see no organisms.

The age of the oldest continental crust is uncertain, but we can be sure that at least some was present by 3.8 billion years ago, and detrital sedimentary rocks in Australia have zircons ($ZrSiO_4$) 4.4 billion years old, indicating that source rocks that old must have existed. In fact, 3.8-billion-year-old rocks are known from several areas, some of which are metamorphic, so their parent rock must be even older.

The friction caused by the Moon on the oceans as well as the continents causes the rate of Earth's rotation to slow very slightly every year. When Earth formed, it may have rotated in as little as 10 hours, so there were many more days in a year; there is no evidence indicating that Earth's orbital period around the Sun has decreased. Another effect of the Earth–Moon tidal interaction is the recession of the Moon from Earth at a few centimeters per year. Thus, during the Hadean, the view of the Moon would have been spectacular!

Geologists agree that when Earth formed, it was exceedingly hot, at least hot enough that it

partially melted and differentiated into a core and mantle. Rather than the Earth being a fiery orb for more than a half billion years, as was formerly accepted, some geologists now think that it had cooled enough for surface water to accumulate by 4.4 billion years ago (see Perspective). They base this conclusion on oxygen 18 to oxygen 16 ratios in tiny inclusions in zircon crystals that indicate reactions with surface waters.

When Earth differentiated during the Hadean, the core and mantle formed, but we have made little mention so far of the crust. Remember that we have defined two types of crust, oceanic and continental, which differ in composition, density, and thickness. The first crust probably was ultramafic, but upwelling mantle currents of mafic magma disrupted the crust, subduction zones formed, and the first island arcs developed (▶ Figure 8.4a). Weathering of these island arcs yielded sediments richer in silica, and partial melting of mafic rocks yielded magma richer in silica. Collisions between island arcs formed a few continental nuclei as silica-rich materials were metamorphosed and intruded by magma (Figure 8.4b). As these larger island arcs collided, the first protocontinents formed and continued to grow by accretion along their margins (Figure 8.4c).

▶ **Figure 8.4 Origin of Granitic Continental Crust**

(a) An andesitic island arc forms by subduction of oceanic lithosphere and partial melting of basaltic oceanic crust. Partial melting of andesite yields granitic magma.

(b) The island arc in (a) collides with a previously formed island arc, thereby forming a continental core.

© 2013 Cengage Learning

(c) The process occurs again when the island arc in (b) collides with the evolving continent, thereby forming a craton, the nucleus of a continent.

The Faint Young Sun Paradox—An Unresolved Controversy

It should not be surprising that older events in written human history are more poorly understood than more recent ones, because older records are incomplete, fragmentary, and difficult to decipher. A similar situation holds for Earth history, especially for the Archean Eon for reasons already noted; alteration of the rocks, deep burial, and few fossils. Nevertheless, on some aspects of early Earth history we can be certain: Archean oceans existed; rocks that probably represent continental crust were present, and organisms appeared by at least 3.0 billion years ago, and possibly as much as 3.5 billion years ago.

However, standard models for the evolution of stars hold that Earth's early Sun was only about 70 percent to 75 percent as luminous as it is now. If this is correct, Earth's surface should have been cold enough for all water to have frozen—but the evidence for liquid water during the Archean is convincing, hence the *Faint Young Sun Paradox*. First, why has solar luminosity increased, and second, what is the evidence for and against the Faint Young Sun Paradox?

When stars such as the Sun form and evolve, their initial low luminosity increases as a function of complex changes, including nuclear reactions, taking place in their core (hydrogen fusing to form helium) and changes in chemical composition. Calculations indicate that early Earth's temperature should have been as much as 25°C lower than at present, but the evidence indicates liquid water was present and surface temperatures were not noticeably different than they are now.

The Faint Young Sun Paradox was proposed in 1972 by astronomer Carl Sagan and his colleague George Mullen. Since then, scientists have proposed several solutions to the paradox, but so far none of these proposed solutions are without critics. One of the most obvious solutions is that an early atmosphere with appreciable amounts of greenhouse gases such as carbon dioxide and methane kept Earth's surface warm even though the Sun was fainter (Figure 1). Most agree that these gases were present in the Archean atmosphere, but not all agree that they were abundant enough to account for significantly higher surface temperatures, so surface waters should have frozen.

Another suggestion is that early Earth had a lower *albedo*; that is, it reflected less solar radiation back into space than it does now and

accordingly was warm enough for water to remain liquid. We know that Archean Earth had vast oceans, but there is little evidence for extensive continents, which have a higher albedo than water. So, according to this hypothesis, oceanic waters absorbed more heat and kept Earth's surface temperature high enough to inhibit freezing (Figure 2). Also, a proposed thinner cloud cover would have allowed more solar energy to reach the surface.

Estimates of the rate at which continental growth occurs vary widely. Nevertheless, given that more residual heat and radiogenic heat were available during the Precambrian, most geologists think that continents grew more rapidly by accretion back then. In fact, some estimate that the continents had reached about 75 percent of their present volume by the beginning of the Phanerozoic Eon. And, of course, at the same time the Sun had increased its luminosity by about 30 percent. In the final analysis, geologists think that Earth's surface temperatures have remained fairly constant through most of geologic time.

Figure 1 Some scientists think that a greenhouse effect on early Earth accounts for the Faint Young Sun Paradox.

(a) Early Earth and a strong greenhouse effect.

(b) Earth today and a weaker greenhouse effect.

From *Greenhouse Puzzles* by Broecker & Peng. Copyright © 1993 by Wallace Broecker. Reprinted with permission.

Surface without snow or ice absorbs more heat

Surface with snow and ice reflects more heat

Copyright © 2012 The Map Factor. Reprinted with permission.

Figure 2 Earth's albedo varies depending on the surface materials sunlight encounters. One proposal for the Faint Young Sun Paradox is that early Earth had vast oceans that had a low albedo and thus absorbed heat, thereby offsetting the weaker energy from the Sun.

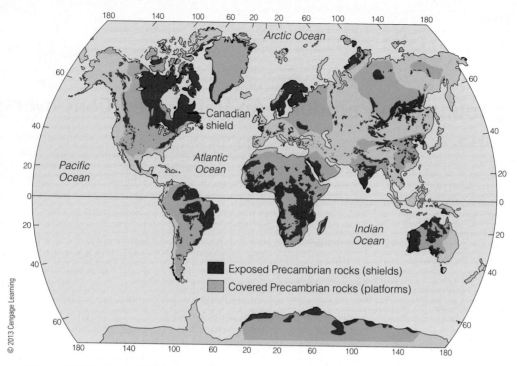

▶ **Figure 8.5** **The Distribution of Precambrian Rocks** Areas of exposed Precambrian rocks constitute the shields, whereas the platforms consist of buried Precambrian rocks. A shield and its adjoining platform are a craton. Notice that large parts of the continents existed by the end of the Precambrian—but, remember, the Precambrian makes up more than 88 percent of all geologic time.

Archean Earth History

We have already mentioned that the Precambrian is longer than we can imagine, but just the Archean alone includes 32.6 percent of all geologic time (Figure 8.1). Archean-age rocks are known from several areas, but they are difficult to interpret because (1) many are metamorphic and complexly deformed; (2) most are deeply buried beneath younger rocks; and (3) they contain few fossils, and those that are exposed are of little use in time-stratigraphic correlation.

Shields, Platforms, and Cratons

Continents are not simply parts of Earth's crust above sea level. It is true that the continents are made up of all groups of rocks, but they have an overall composition similar to granite, whereas the ocean basins are underlain by crust made up of gabbro and basalt. Furthermore, continental crust is thicker and less dense than oceanic crust. The transition from continental to oceanic crust occurs beneath the continental slope, so the margins of continents are below sea level. All continents also have a vast area of exposed ancient rocks called a **Precambrian shield**, and extending outward from shields are broad **platforms** of buried Precambrian rocks. A shield and its adjacent platform made up a **craton**, which we can think of as a continent's ancient nucleus (▶ Figure 8.5).

The cratons are the foundations of the continents, and along their margins more continental crust was added by accretion as they evolved to their present sizes and shapes. In North America, for example, the Superior, Hearne, Rae, and Slave cratons, all within the **Canadian shield**, amalgamated along deformation belts to form a larger cratonic unit during the Proterozoic Eon (see Chapter 9). Both Archean- and Proterozic-age rocks are found in cratons, many of which indicate several episodes of deformation accompanied by igneous activity, metamorphism, and mountain building. However, most of the cratons have experienced remarkably little deformation since the Precambrian.

In North America, the exposed part of the craton is the Canadian shield, which occupies most of northeastern Canada, a large part of Greenland, the Adirondack Mountains of New York, and parts of the Lake Superior region of Minnesota, Wisconsin, and Michigan (Figure 8.5). Much of the Canadian shield is an area of subdued topography, numerous lakes, exposed Archean and Proterozoic rocks thinly covered in places by Pleistocene glacial deposits. The rocks are volcanic, plutonic, and sedimentary, many of which have been altered to varying degrees by metamorphism.

Drilling and geophysical evidence indicate that Precambrian rocks, that is, rocks of the platform, underlie much of North America as well as other continents, but beyond the shields they are seen only in areas of erosion and uplift. For instance, Archean and Proterozoic rocks are present in the deeper parts of the Grand Canyon and in many ranges of the Rocky Mountains and the Appalachian Mountains (see chapter opening photo and ▶ Figure 8.6).

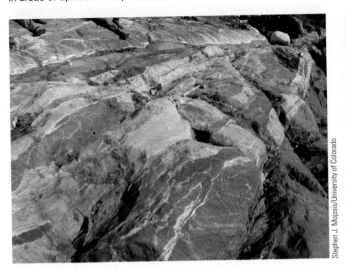

Stephen J. Mojzsis/University of Colorado

(a) Outcrop of the Acasta Gneiss in the Northwest Territories of Canada. At about 4.0 billion years old, these are the oldest known rocks on Earth except for meteorites.

Archean Rocks

Only 22 percent of Earth's exposed Precambrian crust is Archean, with the largest exposures in Africa and North America. Archean crust is made up of many kinds of rocks, but most of them are greenstone belts and granite-gneiss complexes, the latter being by far the most common. Several types of rocks are found in these **granite-gneiss complexes**, but granitic gneiss and granitic rocks predominate, both of which were probably derived from plutons that were emplaced in volcanic island arcs (Figure 8.4). Nevertheless, there are other rocks, ranging from peridotite to sedimentary rocks, all of which have been metamorphosed. Greenstone belts are subordinate, accounting for only 10 percent of Archean rocks, and yet they are important in unraveling some of the complexities of Archean tectonic events.

Greenstone Belts

A **greenstone belt** has three main rock associations; its lower and middle parts are mostly volcanic, whereas the upper rocks are mostly sedimentary (▶Figure 8.7a). Greenstone belts typically have a synclinal structure, measure anywhere from 40 to 250 km wide and 120 to 800 km long, and have been intruded by granitic magma and cut by thrust faults. Many of the igneous rocks are greenish because they contain green minerals such as chlorite, actinolite, and epidote that formed during low-grade metamorphism.

Thick accumulations of pillow lava are common in greenstone belts, indicating that much of the volcanism was subaqueous (Figure 8.7b). Pyroclastic materials, in contrast, almost certainly formed by subaerial eruptions where large volcanic centers built above sea level. The most

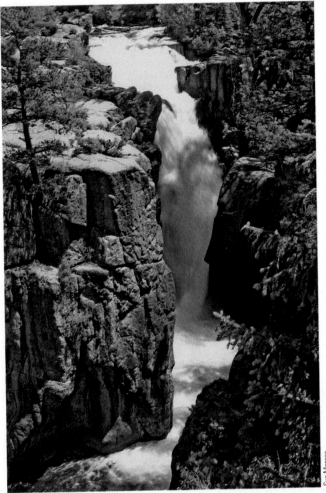

Sue Monroe

(b) Shell Creek in the Big Horn Mountains of Wyoming has cut a deep gorge into this 2.9-billion-year-old (Mesoarchean) granite.

interesting igneous rocks in greenstone belts are *komatiites* that cooled from ultramafic lava flows, which are rare in rocks younger than Archean, and none occur now.

To erupt, ultramafic magma (magma with less than 45 percent silica) requires near-surface magma temperatures of more than 1,600°C; the highest recorded surface temperature for recent lava flows is 1,350°C. During its early history, however, Earth possessed more radiogenic heat, and the mantle was as much as 300°C hotter than it is now. Given these conditions, ultramafic magma could reach the surface, but as Earth's radiogenic heat production decreased, the mantle cooled and ultramafic flows no longer occurred.

Sedimentary rocks are found throughout greenstone belts, but they predominate in the upper unit (Figure 8.7a). Many of these rocks are successions of *graywacke* (sandstone with abundant clay and rock fragments) and *argillite* (slightly metamorphosed mudrocks). Small-scale crossbedding and graded bedding indicate these rocks represent turbidity current deposition (see Figure 6.3).

(b) Pillow lava of the Ishpeming greenstone belt in Michigan.

Greenstone belt succession

- Granitic intrusives
- Upper sedimentary unit: sandstones and shales most common
- Middle volcanic unit: mainly basalt
- Lower volcanic unit: mainly peridotite and basalt
- Granite–gneiss complex

(a) Two adjacent greenstone belts. Older belts—those more than 2.8 billion years old—have an ultramafic unit overlain by a basaltic unit. In younger belts, the succession is from a basaltic lower unit to an andesite-rhyolite unit. In both cases, sedimentary rocks are found mostly in the uppermost unit.

(c) Gneiss from a granite-gneiss complex in Ontario, Canada.

Other sedimentary rocks are also present, including sandstone, conglomerate, chert, and carbonates, although none are very abundant. Iron-rich rocks known as *banded iron formations* are also found, but they are more typical of Proterozoic deposits, so we will discuss them in Chapter 9.

The oldest large, well-preserved greenstone belts are in South Africa and date from about 3.6 billion years ago. The 3.7- to 3.8-billion-year-old Isua greenstone belt of Greenland, consisting of metamorphosed lava flows, schists, quartzites, and banded iron formations (▶ Figure 8.8), has some of the oldest known rocks on Earth. And given that it contains altered sedimentary rocks, it is safe to assume that even older source rocks were present.

In North America, most greenstone belts are found in the Superior and Slave cratons of the Canadian shield (▶ Figure 8.9), but they are also found in Michigan, Minnesota, and Wyoming. Most formed between 2.7 and 2.5 billion years ago. The Abitibi greenstone belt of Ontario and Quebec, Canada, is especially well known for its extensive resources including gold, copper, and zinc.

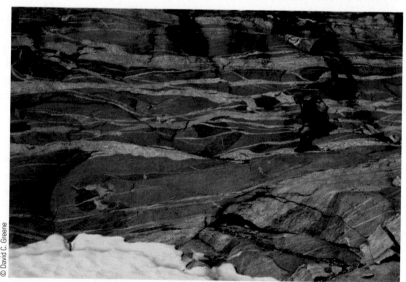

▶ Figure 8.8 **The Isua Greenstone Belt, Greenland** Felsic dikes cross-cutting the central gneiss, which is exposed between arms of the Isua Greenstone Belt. At 3.7 to 3.8 billion years old, the rocks in the Isua Greenstone Belt are some of the oldest known on Earth.

Figure 8.9 **Greenstone Belts in North America** Archean greenstone belts (shown in dark green) of the Canadian shield are mostly in the Superior and Slave cratons.

Evolution of Greenstone Belts

Most greenstone belts are found in Archean and Proterozoic terrains in multiple, parallel belts, each separated from the next by granite-gneiss complexes (Figure 8.7a, c). Most geologists would probably agree that greenstone belts originate in several tectonic settings that might include oceanic plateaus, rifted continental margins, rifts within continents, as well as in back-arc basins that subsequently close. Exactly how they form and evolve is not fully resolved, but the **back-arc basin** model—involving a basin lying between a volcanic island arc and a continent much like the present-day Sea of Japan—is appealing for the origin of some greenstone belts.

According to this model (▶ Figure 8.10), there is an early stage of extension as the back-arc basin forms, accompanied by volcanism, emplacement of plutons, and sedimentation, followed by an episode of compression when the basin closes. During this latter stage, the evolving greenstone belt rocks are deformed, metamorphosed, and intruded by granitic magma. Proponents of this model suggest that multiple episodes of the opening and closing of back-arc basins account for the parallel arrangement of greenstone belts.

The back-arc basin model for the origin of greenstone belts is accepted by many geologists, but others think that some of these belts form in intracontinental rifts above rising mantle plumes (▶ Figure 8.11). As the plume rises beneath sialic (silica- and aluminum-rich) crust, it spreads and generates tensional forces that cause rifting. The mantle plume is the source of the lower and middle volcanic units of the greenstone belt, and the upper sedimentary unit results from erosion of the volcanic rocks along the flanks of the rift. And finally, there is an episode of closure of the rift, deformation, low-grade metamorphism, and emplacement of plutons (Figure 8.11).

▶ Figure 8.10 **Origin of a Greenstone Belt in a Back-Arc Basin**

(a) Rifting on the continent side of a volcanic arc forms a back-arc basin. Partial melting of subducted oceanic lithosphere supplies andesite and diorite magmas to the island arc.

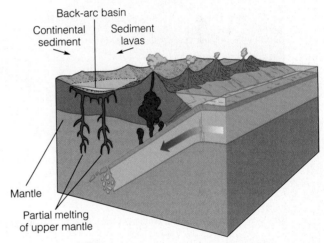

(b) Basalt lavas and sediment derived from the island arc and continent fill the back-arc basin.

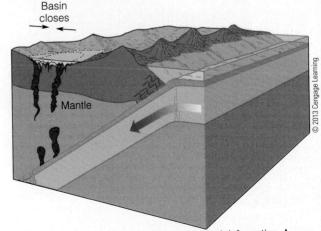

(c) Closure of the back-arc basin, compression, and deformation. A syncline-like structure forms, which is intruded by granitic magma.

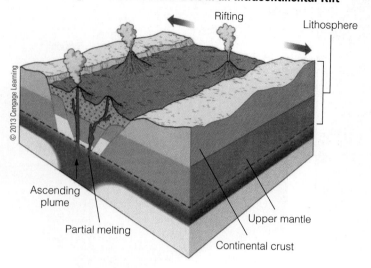

(a) An ascending mantle plume causes rifting and volcanism.

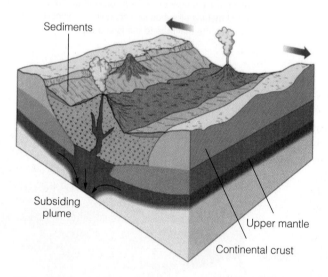

(b) As the plume subsides, erosion of the rift flanks yields sediments.

(c) Closure of the rift causes compression and deformation. Granitic magma intrudes the greenstone belt.

Archean Plate Tectonics and the Origin of Cratons

Certainly, the present plate tectonic regime of opening and closing ocean basins has been a primary agent in Earth evolution since the Paleoproterozoic. Most geologists are convinced that some kind of plate tectonic activity took place during the Archean as well, but it differed in detail from what is going on now. With more residual heat from Earth's origin and more radiogenic heat, plates must have moved faster and magma was generated more rapidly. As a result, continents no doubt grew more rapidly along their margins, a process called **continental accretion**, as plates collided with island arcs and other plates. Also, ultramafic lava flows (komatiites) were more common.

There were, however, marked differences between the Archean world and the one that followed. We have little evidence of Archean rocks deposited on broad, passive continental margins, but associations of passive continental margin sediments were widespread by Proterozoic time. Deformation belts between colliding cratons indicate that Archean plate tectonics was active, but the *ophiolites* so typical of younger convergent plate boundaries are rare.

Nevertheless, geologists have reported probable Neoarchean ophiolites from several areas. One in Russia is about 2.8 billion years old, whereas the Dongwanzi complex of China is 2.5 billion years old. Not all agree that this one in China is actually an ophiolite sequence; the rocks have been complexly deformed and intruded by magma possibly as young as Mesozoic. Another probable ophiolite in the Wind River Mountains of Wyoming is also highly deformed and difficult to interpret. Whether these rocks truly represent ophiolites may be an open question, but there is no doubt that ophiolite sequences were present by the Paleoproterozoic (see Chapter 9).

In any case, several small cratons were present during the Archean and grew by accretion along their margins. By the end of the Archean, perhaps 30 percent to 40 percent of the present volume of continental crust had formed. Remember, though, that these cratons amalgamated into a larger unit during the Proterozoic (see Chapter 9). A plate tectonic model for the Archean crustal evolution of the southern Superior craton of Canada relies on the evolution of greenstone belts, plutonism, and deformation (▶ Figure 8.12). We can take this as a model for Archean crustal evolution in general.

▶ Figure 8.12 **Origin of the Southern Superior Craton**

(a) Geologic map showing greenstone belts (dark green) and areas of granite-gneiss complexes (light green).

(b) Plate tectonic model for the evolution of the southern Superior craton. The figure is a north-south cross section, and the upper diagram shows an earlier stage of development than the lower one.

The events leading to origin of the southern Superior craton (Figure 8.12b) are part of a much more extensive orogenic episode that took place during the Mesoarchean and Neoarchean. Deformation was responsible for the origin of some of the Archean rocks in several parts of the Canadian shield as well as in Wyoming, Montana, and the Mississippi River Valley. In the northwestern part of the Canadian shield, deformation along the Snowbird tectonic zone yielded metamorphic rocks 3.2 and 2.6 billion years old that belong to the granulite metamorphic facies, which form at very high temperatures, at least 700°C, and some form at more than 1,000°C. By the time this Archean event had ended, several cratons had formed that are now found in the older parts of the Canadian shield (see Chapter 9).

The Atmosphere and Hydrosphere

In Chapter 1, we emphasized the interactions among systems, two of which, the atmosphere and hydrosphere, have had a profound impact on Earth's surface (see Figure 1.1). Shortly after Earth formed, its atmosphere and hydrosphere, although present, were quite different from the way they are now. They did, however, play an important role in the development of the biosphere.

How Did the Atmosphere Form and Evolve?

Today, Earth's atmosphere is quite unlike the noxious one we described earlier. Now it is composed of 78 percent nitrogen (N_2) and 21 percent free oxygen (O_2), meaning oxygen not combined with other elements as in carbon dioxide (CO_2) and water vapor (H_2O). It also has small but important amounts of other gases such as ozone (O_3), which, fortunately for us, is common enough in the upper atmosphere to block most of the Sun's ultraviolet radiation.

Earth's earliest atmosphere was probably composed of hydrogen and helium, the most abundant gases in the universe. If so, it would have quickly been lost into space, for two reasons. First, Earth's gravitational attraction is insufficient to retain gases with such low molecular weights. And second, before Earth differentiated, it had no core or magnetic field. Accordingly, it lacked a *magnetosphere*, the area around the planet within which the magnetic field is confined, so a strong solar wind, an outflow of ions from the Sun, would have swept away any atmospheric gases. Once Earth had differentiated and a magnetosphere was present, though, an atmosphere began accumulating as a result of **outgassing** involving the release of gases from Earth's interior during volcanism (▶Figure 8.13).

Water vapor is the most common gas emitted by volcanoes today, but they also emit carbon dioxide, sulfur dioxide, carbon monoxide, sulfur, hydrogen, chlorine, and nitrogen. No doubt Archean volcanoes emitted the same gases, and thus an atmosphere developed, but one lacking free oxygen and an ozone layer. It was, however, rich in carbon dioxide, and gases reacting in this early atmosphere probably formed ammonia (NH_3) and methane (CH_4).

This early oxygen-deficient but carbon dioxide-rich atmosphere persisted throughout the Archean. Some of the evidence for this conclusion comes from detrital deposits containing minerals such as pyrite (FeS_2) and uraninite (UO_2), both of which oxidize rapidly in the presence of free oxygen. So, the atmosphere was a chemically reducing one rather than an oxidizing one. However, oxidized iron becomes increasingly common in Proterozoic rocks, indicating that at least some free oxygen was present then (see Chapter 9).

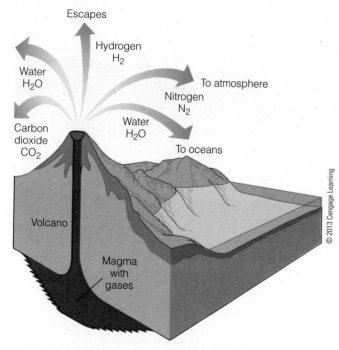

▶ **Figure 8.13 Outgassing and Earth's Early Atmosphere** Erupting volcanoes emit mostly water vapor, carbon dioxide, and several other gases but no free oxygen—that is, oxygen not combined with other elements. In addition to the gases shown here, chemical reactions in the early atmosphere probably yielded methane (CH_4) and ammonia (NH_3).

Two processes account for introducing free oxygen into the atmosphere, one or both of which began during the Hadean. The first, **photochemical dissociation**, involves ultraviolet radiation from the Sun disrupting water molecules in the upper atmosphere, thus releasing their oxygen and hydrogen (▶ Figure 8.14). This process may eventually have supplied 2 percent of the present-day oxygen level, but with this amount of free oxygen in the atmosphere, ozone forms, creating a barrier against ultraviolet radiation and the formation of more ozone. Even more important was **photosynthesis**, a metabolic process in which organisms use carbon dioxide and water to make organic molecules and then release oxygen as a waste product (Figure 8.14). Even so, probably no more than 1 percent of the free oxygen level of today was present by the end of the Archean, 2.5 billion years ago.

The Hydrosphere—Earth's Surface Waters

All water on Earth is part of the hydrosphere, but most of it—more than 97 percent—is in the oceans. Where did it come from, and how has it changed? Certainly, outgassing released water vapor from Earth's interior, and once the planet cooled sufficiently, water vapor condensed and surface waters began to accumulate. Another source of water vapor, and eventually liquid water, was meteorites and especially icy comets (Figure 8.3). It is not known whether outgassing or meteorites and comets was most important, but we do know that oceans were present by Archean time, although their volumes and geographic extent cannot be determined. Nevertheless, we can envision an Archean Earth with numerous erupting volcanoes and an early episode of intense meteorite and comet bombardment accounting for a rapid rate of surface water accumulation.

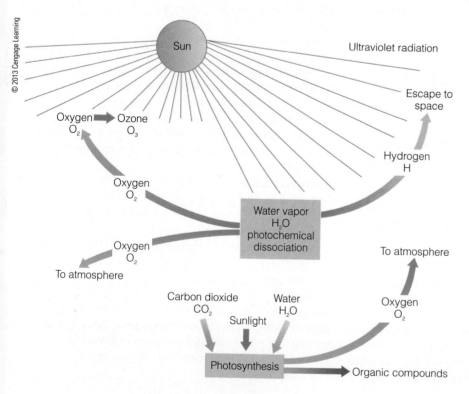

▶ **Figure 8.14 Evolution of the Atmosphere** Photochemical dissociation occurs when ultraviolet radiation disrupts water molecules that release hydrogen (H_2) and free oxygen (O_2), some of which is converted to ozone (O_3) that blocks most of the ultraviolet radiation. In the presence of sunlight, photosynthesizing organisms use carbon dioxide (CO_2) and water (H_2O) to make organic molecules and in the process release free oxygen as a waste product.

Following Earth's early episode of meteorite and comet bombardment, which ended about 3.8 billion years ago, these extraterrestrial bodies have added little to the accumulating surface waters. However, volcanoes continue to erupt and expel water vapor (much of it recycled surface water), so is the amount of water in the oceans increasing? Probably it is, but at a considerably reduced rate, because much of Earth's residual heat from its origin has dissipated and the amount of radioactive decay to generate internal heat has diminished, so volcanism is not nearly as commonplace. Accordingly, the amount of water added to the oceans now is trivial compared to their volumes.

Recall from Chapter 4 that one early attempt to determine Earth's age was to calculate how long it took for the oceans to reach their current salinity level—assuming, of course, that the oceans formed soon after Earth did, that they were freshwater to begin with, and that their salinity increased at a uniform rate. None of these assumptions is correct, so the ages determined were vastly different. We now know that the very early oceans were salty, probably about as salty as they are now. That is, very early in their history, the oceans reached chemical equilibrium and have remained in near-equilibrium conditions ever since.

Life—Its Origin and Early History

Today, the biosphere is made up of millions of species of organisms assigned to five kingdoms—animals, plants, protistans, fungi, and monera (bacteria and archaea).* Scientists have found probable fossils in rocks 3.5 billion years old, and chemical evidence in 3.8-billion-year-old rocks in Greenland convinces some that organisms were present then. Certainly by 3.0 billion years ago (Mesoarchean) organisms were present, but they represent only one of the kingdoms just listed—monera. In Chapter 7 we discussed the theory of evolution, which accounts for how organisms have changed through time, but this theory does not address how life originated in the first place. Here, we are concerned with **abiogenesis**, that is, how life formed from nonliving matter. Before proceeding, however, note that abiogenesis does not hold that a living organism such as a bacterium, or even a complex organic molecule, sprang fully developed from nonliving material. Rather than one huge step occurring, abiogenesis holds that several small steps took place, each leading to an increase in organization and complexity, much like evolutionary change occurs in living organisms.

But what is living and nonliving? The distinction is clear in most cases: dogs and trees are alive, rocks and water are not. A biologist might use several criteria to make the distinction, including growth and reaction to stimuli, but minimally, a living thing must practice some

*Archaea include microscopic organisms that resemble bacteria but differ from them genetically and biochemically.

kind of chemical activity (metabolism) to maintain itself, and it must be capable of reproduction to ensure the long-term survival of the group to which it belongs. This metabolism–reproduction criterion might seem sufficient to decide if something is living or not, and yet the distinction is not always easy to make.

Bacterial are living, but under some circumstances, they can go for long periods during which they show no signs of living and then go on living again. Are viruses living? They behave like living organisms in the appropriate host cell, but when outside a host cell, they neither metabolize nor reproduce. Some biologists think that viruses represent another way of living, but others disagree. Comparatively simple organic molecules called *microspheres* form spontaneously and grow and divide in a somewhat organism-like manner, but these processes are more like random chemical reactions, so they are not living.

So what do viruses and microspheres have to do with the origin of life? First, they show that the living versus nonliving distinction is not always clear. And second, if life originated by natural processes, it must have passed through prebiotic stages—that is, stages in which the entities would have shown signs of living organisms but were not truly living.

The Origin of Life

Investigators agree that for life to originate, an energy source must have acted upon the appropriate chemical elements from which organic molecules could synthesize. The early atmosphere, composed of carbon dioxide (CO_2), water vapor (H_2O), nitrogen (N_2), and probably ammonia (NH_3) and methane (CH_4), provided carbon, oxygen, hydrogen, and nitrogen, the primary elements that make up all organisms. Lightning and ultraviolet radiation were two possible sources of the energy necessary for these elements to combine and form rather simple organic molecules known as **monomers**, such as amino acids.

Monomers are needed as basic building blocks of more complex organic molecules, but is it plausible that they formed naturally? Experimental evidence indicates that it is. During the 1950s, Stanley Miller circulated gases approximating Earth's early atmosphere through a closed glass vessel and subjected the mixture to an electric spark to simulate lightning. Within a few days, the mixture became turbid, and an analysis showed that Miller had synthesized several amino acids common in organisms (▶Figure 8.15). More recent experiments using different gases have successfully synthesized many of the 20 amino acids in organisms.

Making monomers in a test tube is one thing, but organisms are **polymers**, such as nucleic acids (DNA and RNA) and proteins consisting of monomers linked together in a specific sequence. So how did this linking of monomers, or *polymerization*, take place? Researchers have successfully synthesized small molecules called *thermal proteins*, consisting of more than 200 linked amino acids (▶Figure 8.16). In fact, when heated, dehydrated, and concentrated, amino acids spontaneously polymerize and form thermal proteins.

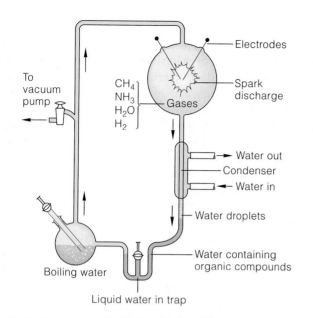

▶ **Figure 8.15 Stanley Miller's Experimental Apparatus** Miller synthesized several amino acids used by organisms. The spark discharge was to simulate lightning, and the gases approximated Earth's early atmosphere.

We call these artificially synthesized thermal proteins *protobionts*, meaning they have characteristics between those of inorganic chemical compounds and living organisms. Suppose that these protobionts came into existence as just outlined. They would have been diluted and would have ceased to exist if they had not developed some kind of outer covering. In other words, they had to be self-contained chemical systems as today's cells are. In the experiments described, proteinoids have spontaneously aggregated into microspheres that have a cell-like outer covering and grow and divide somewhat like bacteria do (Figure 8.16).

Perhaps the first steps leading to life took place as monomers formed in great abundance and polymerized, but little is known about how a reproductive mechanism came about.

Microspheres divide but do so in a nonbiologic manner. In fact, for some time, researchers were baffled because in present-day organisms either DNA or RNA is necessary for reproduction, but these nucleic acids cannot replicate without protein enzymes, yet protein enzymes cannot be made without nucleic acids. Or so it seemed until a few years ago, when researchers discovered that some small RNA molecules can in fact reproduce without the aid of protein enzymes. So, the first replicating system might have been an RNA molecule. Just how these molecules were synthesized under conditions that existed on early Earth has not been resolved.

Intractable problems such as RNA replication have a way of eventually being solved. For instance, for many years scientists thought that the ribose of RNA was "impossible" to make by prebiotic synthesis from any conceivable molecular mixture that was assumed to exist on early Earth. Finally, however, they found that it could indeed form naturally under conditions that were likely to have existed during the Archean Eon.

Researchers agree on some of the basic requirements for the origin of life, but the exact steps involved and the significance of some experimental results are debated. For example, our discussion has focused on the role of atmospheric gases in the origin of monomers and polymers, but some scientists think that life originated near hydrothermal vents on the seafloor.

Submarine Hydrothermal Vents and the Origin of Life

Seawater seeps down into the oceanic crust at or near spreading ridges, becomes heated by magma, and rises and discharges onto the seafloor as **submarine hydrothermal vents**. Some of these vents, which are now known from the Atlantic, Pacific, and Indian Oceans, are called **black smokers** because they discharge water saturated with dissolved minerals, giving them the appearance of black smoke

▶ **Figure 8.16 Experimental Production of Thermal Proteins and Microspheres**

(a) Bacteria-like thermal proteins.

(b) Thermal proteins aggregated to form microspheres.

(a) This black smoker on the East Pacific Rise is at a depth of 2,800 m. The "black smoke" is heated seawater with dissolved minerals.

(b) Several types of organisms, including these tube worms, which may be up to 3 m long, live near black smokers. Bacteria that oxidize sulfur compounds lie at the base of the food chain.

(▶ Figure 8.17a). Because Earth had more radiogenic heat during its early history, it is reasonable to infer that submarine hydrothermal vents were more common then.

Submarine hydrothermal vents are interesting because several minerals containing zinc, copper, and iron precipitate around them, and they support communities of organisms previously unknown to science (Figure 8.17b). In recent years, some investigators have proposed that the first self-replicating molecules came into existence near these vents on the seafloor. The necessary elements (C, H, O, and N), as well as sulfur (S) and phosphorous (P), are present in seawater, and heat could have provided the energy necessary for monomers to form. Those endorsing this hypothesis hold that polymerization took place on the surfaces of clay minerals, and, finally, protocells were deposited on the seafloor. In fact, amino acids have been detected in some hydrothermal vent emissions.

Not all scientists are convinced this would be a viable environment for the origin of life. They point out that polymerization would be inhibited in this hot, aqueous environment, and further claim that life here would be the result of adaptation to a hostile environment after

originating elsewhere. So, this alternate location for the origin of life remains controversial, at least for now.

Earth's Oldest Known Organisms

The first organisms were bacteria and archaea, both of which consist of **prokaryotic cells**—that is, cells that lack an internal, membrane-bounded nucleus and other structures typical of *eukaryotic cells*. (We discuss eukaryotic cells in Chapter 9.) Members of the archaea are single-celled, as are bacteria, and in fact they look much like bacteria but differ from them in the details of their genetics and chemical makeup. In addition, archaea can live in extremely hot, acidic, and saline environments where even bacteria cannot survive. In any case, prior to the 1950s, scientists assumed that the fossils common in Cambrian-age rocks had a long, earlier history, but the fossil record offered little to support this idea. In fact, it appeared that Precambrian rocks were so devoid of fossils that this interval of time was formerly called the *Azoic*, meaning "without life."

During the early 1900s, Charles Walcott described layered moundlike structures from the Paleoproterozoic-age

▶ Figure 8.18 **Stromatolites**

(a) Present-day stromatolites exposed at low tide at Shark Bay, Australia.

(b) Different types of stromatolites include (1) irregular mats, (2) isolated columns, and (3) columns linked by mats.

Gunflint Chert of Ontario, Canada. He proposed that these structures, now called **stromatolites**, represented reefs constructed by algae (▶ Figure 8.18), but not until 1954 did paleontologists demonstrate that stromatolites are the product of organic activity. Present-day stromatolites form and grow as sediment grains are trapped on sticky mats of photosynthesizing cyanobacteria (blue-green algae), and they now are restricted to environments where snails cannot live.

Currently, the oldest known undisputed stromatolites are found in 3.0-billion-year-old rocks in South Africa, but probable ones are also known from the 3.3- to 3.5-billion-year-old Warrawoona Group in Australia. And, as noted previously, chemical evidence in rocks 3.8 billion years old in Greenland indicates life was present then. These oldest known cyanobacteria were photosynthesizing organisms, but photosynthesis is a complex metabolic process that must have been preceded by a simpler type of metabolism.

No fossils are known of these earliest organisms, but they must have resembled tiny bacteria that needed no free oxygen, so they were **anaerobic**. In addition, they must have been totally dependent on an external source of nutrients—that is, they were **heterotrophic**, as opposed to **autotrophic** organisms, which make their own nutrients, as in photosynthesis. And, finally, they were all prokaryotic cells. The nutrient source for these earliest organisms was most likely *adenosine triphosphate* (ATP), which was used to drive the energy-requiring reactions in cells. ATP can easily be synthesized from simple gases and phosphate, so it was no doubt available in the early Earth environment. These organisms must have acquired their ATP from their surroundings, but this situation could not have persisted for long as more and more cells competed for the same resources. The first organisms to develop a more sophisticated metabolism probably used *fermentation* to meet their energy needs. Fermentation is an anaerobic process in which molecules such as sugars are split, releasing carbon dioxide, alcohol, and energy. In fact, most living prokaryotic cells ferment.

Other than the origin of life itself, the most significant biologic event of the Archean was the development of the autotrophic process of photosynthesis as much as

3.5 billion years ago. The cells were still prokaryotic and anaerobic, but as autotrophs, they were no longer dependent on preformed organic molecules as a source of nutrients. The Archean microfossils in ▶ Figure 8.19 are anaerobic, autotrophic prokaryotes.

Archean Mineral Resources

Of the several types of mineral deposits found in Archean rocks, gold is the one most commonly associated with them, although it is also found in Proterozoic and Phanerozoic rocks. This soft yellow metal has been the cause of feuds and wars and was one incentive for European exploration of the Americas. It is prized for jewelry, but it is or has also been used as a monetary standard, and it is still used in glass making, electrical circuitry, and the chemical industry. Since 1886, Archean and Proterozoic rocks in South Africa have yielded about half the world's gold. Gold mines are also found in Archean rocks of the Superior craton in Canada.

Archean sulfide deposits of zinc, copper, and nickel are known in several areas, such as Australia, Zimbabwe, and the Abitibi greenstone belt in Ontario, Canada. At least some of these deposits probably formed as mineral accumulations adjacent to hydrothermal vents on the seafloor, much as they do now around black smokers (Figure 8.17a).

About one-fourth of Earth's chrome reserves are in Archean rocks, especially in Zimbabwe. These ore deposits are found in the volcanic units of greenstone belts, where they probably formed when mineral crystals settled and became concentrated in the lower parts of plutons such as mafic and ultramafic sills. Chrome is needed in the steel industry, but the United States has very few commercial chrome deposits so must import most of what it uses.

One chrome deposit in the United States is in the Stillwater Complex in Montana. Low-grade ores were mined there during both World Wars as well as during the Korean War, but they were simply stockpiled and never refined for

▶ **Figure 8.19 Probable Archean Microfossils** These fossils and their restorations from the Warrawoona Group of Western Australia resemble bacteria. They were found in layers of chert between lava flows that have been dated; the chert layer is 3.465 billion years old, plus or minus 5 million years.

their chrome content. These same rocks are also a source of platinum, one of the precious metals, which in the automotive industry is used in catalytic converters. It is also used in the chemical industry and for cancer chemotherapy. Most of it mined today comes from the Bushveld Complex in South Africa.

Most of Earth's banded iron formations, the primary source of iron ore, were deposited during the Proterozoic, but about 6 percent of them are of Archean age. Some Archean-age banded iron formations are mined, but they

are neither as thick nor as extensive as those of the Proterozoic (see Chapter 9).

The term *pegmatite* refers to very coarsely crystalline plutonic rock, most of which approximates the composition of granite and has little commercial value. But some Archean pegmatites, such as those in the Herb Lake District in Manitoba, Canada, and the Rhodesian Province of Africa, contain valuable minerals. In addition to minerals of gem quality, Archean pegmatites also contain minerals mined for their lithium, beryllium, rubidium, and cesium.

Summary

Table 8.1 provides a summary of the Archean geologic and biologic events discussed in the text.

▶ All geologic time from Earth's origin to the beginning of the Phanerozoic Eon is included in the Precambrian. Precambrian also refers to rocks lying stratigraphically below Cambrian-age rocks.

▶ Geologists use a threefold division of the Precambrian, dividing it into the Hadean, the Archean, and Proterozoic.

▶ Rocks from the early part of the Eoarchean indicate that crust existed then, but very little of it has been preserved.

▶ All continents have an ancient craton made up of an exposed shield and a buried platform. In North America, the Canadian shield is made up of smaller units delineated by their ages and structural trends.

▶ Archean greenstone belts are linear, syncline-like bodies of rock found within much more extensive granite-gneiss complexes.

▶ Ideal greenstone belts consist of two lower units of mostly igneous rocks and an upper sedimentary unit. They probably formed in several settings, including back-arc basins and intracontinental rifts.

▶ Many geologists are convinced that Archean plate tectonics took place, but plates probably moved faster, and igneous activity was more common then because Earth had more radiogenic heat.

▶ Outgassing was responsible for the early atmosphere and hydrosphere, but comets also contributed to the latter. The atmosphere that formed lacked free oxygen but contained abundant carbon dioxide and water vapor.

- Models for the origin of life require an oxygen-deficient atmosphere, the necessary elements for organic molecules, and energy to promote the synthesis of organic molecules.
- The first naturally formed organic molecules were probably monomers, such as amino acids, that linked together to form more complex polymers, including nucleic acids and proteins.
- RNA molecules may have been the first self-replicating molecules. However, the method whereby a reproductive system formed is not yet known.
- The only known Archean fossils are of single-celled, prokaryotic bacteria such as blue-green algae, but chemical compounds in some Archean rocks may indicate the presence of archaea.
- Stromatolites that formed by the activities of photosynthesizing bacteria are found in rocks as much as 3.5 billion years old.
- Archean mineral resources include gold, chrome, zinc, copper, and nickel.

Important Terms

abiogenesis, p. 161
anaerobic, p. 164
autotrophic, p. 164
back-arc basin, p. 157
black smoker, p. 162
Canadian shield, p. 154
continental accretion, p. 158

craton, p. 154
granite-gneiss complex, p. 155
greenstone belt, p. 155
heterotrophic, p. 164
monomer, p. 161
outgassing, p. 159
photochemical dissociation, p. 160

photosynthesis, p. 160
platform, p. 154
polymer, p. 161
Precambrian shield, p. 154
prokaryotic cell, p. 163
stromatolite, p. 164
submarine hydrothermal vent, p. 162

Review Questions

1. Outgassing is the process whereby _____ formed.
   a. ____ greenstone belts; b. ____ the atmosphere; c. ____ monomers and polymers; d. ____ continental crust; e. ____ stromatolites.
2. An area in North America consisting of the Canadian shield and its adjacent platform is a
   a. ____ syncline; b. ____ prokaryote; c. ____ basin; d. ____ terrane; e. ____ craton.
3. Any organism that depends on preformed organic molecules for its nutrients is
   a. ____ polymeric; b. ____ abiogenic; c. ____ heterotropic; d. ____ prokaryotic; e. ____ photochemical.
4. Granite-gneiss complexes are
   a. ____ the most common Archean-age rocks; b. ____ found at oceanic spreading ridges; c. ____ noted for their fossil plants and animals; d. ____ green because they contain the minerals chlorite, epidote, and actinolite; e. ____ most likely turbidity current deposits.
5. Many scientists think that the first self-replicating system was a(n)
   a. ____ RNA molecule; b. ____ ATP cell; c. ____ thermal protein; d. ____ stromatolite; e. ____ black smoker.
6. Which one of the following geologic time designations from oldest to youngest is correct?
   a. ____ Proterozoic-Archean-Hadean; b. ____ Archean-Hadean-Proterozoic; c. ____ Hadean-Archean-Proterozoic; d. ____ Phanerozoic-Precambrian-Komatiitic; e. ____ Archean-Proterozoic-Hadean.
7. The origin of life from nonliving matter is known as
   a. ____ photosynthesis; b. ____ abiogenesis; c. ____ polymerization; d. ____ biotic accretion; e. ____ outgassing.
8. Black smokers are
   a. ____ the sites where greenstone belts formed; b. ____ complexes of igneous and sedimentary rocks; c. ____

hydrothermal vents on the seafloor; d. ____ found only in Archean-age rocks; e. ____ heterotropic prokaryotes.
9. One process that led to free oxygen in the atmosphere was
   a. ____ stromatolitic conversion; b. ____ cratonization; c. ____ back-arc basin deposition; d. ____ probiogenesis; e. ____ photochemical dissociation.
10. All known Archean organisms belong to the kingdom
    a. ____ Eukaryota; b. ____ Fungi; c. ____ Protozoa; d. ____ Bacteria and archea; e. ____ Protoctista.
11. As a high school science teacher, how would you explain to your students that an initially hot Earth is cooling, and at some time in the far distant future its internally driven processes such as volcanism, seismic activity, and plate movements will slow down and eventually stop?
12. Explain how two processes account for adding free oxygen to Earth's early atmosphere. Which process was most important?
13. Why do some scientists think that life originated near submarine hydrothermal vents?
14. What are monomers and polymers, and how might they have formed naturally?
15. Describe the process of continental accretion and explain how it takes place.
16. What are greenstone belts, and how might they have formed?
17. How can you account for the fact that ultramafic lava flows occurred during the Archean but only rarely thereafter and none form now?
18. Explain what it means to say that the earliest organisms were heterotrophic, anaerobic prokaryotes.
19. What evidence indicates that some continental crust existed as early as 4.0 billion years ago?
20. Why do you think it is so important for geologists to study Archean-age rocks?

TABLE 8.1 **Summary of Archean Geologic and Biologic Events Discussed in the Text**

© 2013 Cengage Learning

9

Precambrian Earth and Life History

The Proterozoic Eon

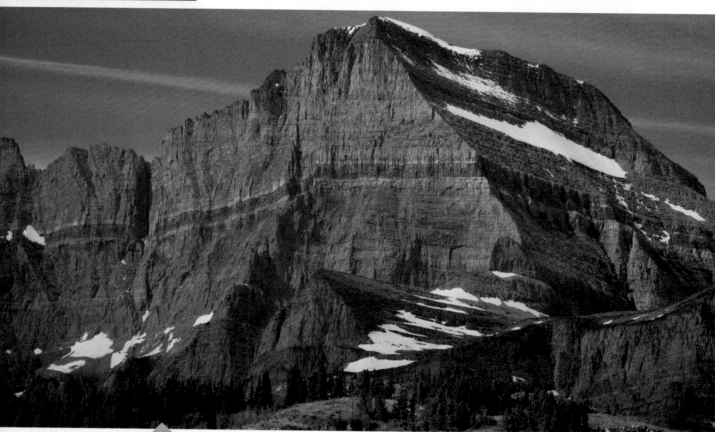

View of the Purcell Sill (dark layer) intruded into the Proterozoic-age Siyeh Limestone in Glacier National Park, Montana. The 750-million-year-old Purcell Sill is composed of diorite and is 40 to 90 m thick. Geologists know this is a sill rather than a buried lava flow because contact metamorphism above and below the intrusion has altered the limestone to marble, visible as white layers in this image. In addition to the Siyeh Limestone, several other Proterozoic-age sedimentary rocks are exposed in the park, which have been altered only slightly by metamorphism.

Dr. Marli Miller/Visuals Unlimited, Inc.

At the end of this chapter, you will have learned that

▶ A large landmass called Laurentia, made up mostly of Greenland and North America, formed by the amalgamation of Archean cratons along deformation belts during the Paleoproterozoic.

▶ Following its initial stage of amalgamation, Laurentia grew by accretion along its southern and eastern margins.

▶ The Mesoproterozoic of Laurentia was a time of widespread igneous activity, orogenies, and rifting.

▶ Widespread Proterozoic assemblages of sandstone, shale, and carbonate rocks look much like the rocks deposited now on passive continental margins.

▶ Plate tectonics, essentially like that occurring now, was operating during the Proterozoic and one or possibly two supercontinents formed.

▶ The presence of banded iron formations and continental red beds indicate that at least some free oxygen was present in the atmosphere.

▶ Extensive glaciation took place during the Paleoproterozoic and the Neoproterozoic.

▶ The first eukaryotic cells, that is, cells with a nucleus and other internal structures, evolved from prokaryotic cells by 1.2 billion years ago.

▶ Impressions of multicelled animals are found on all continents except Antarctica.

▶ Banded iron formations, as sources of iron ore, are important Proterozoic resources, as are deposits of gold, copper, platinum, and nickel.

Introduction

In Chapter 8, we emphasized the magnitude of geologic time and noted that the Precambrian encompasses more than 88 percent of all geologic time. In this chapter, we review Earth and life history for the Proterozoic Eon, which, like the Hadean and Archean, is another inconceivably long interval, in this case lasting from 2.5 billion years ago to 542 million years ago (see Figures 8.1 and 8.2). Remember that Archean-age rocks are difficult to study because they are mostly metamorphic and deformed, deeply buried, and contain few fossils. Some Proterozoic rocks are also metamorphic and likewise difficult to interpret, but there are vast exposures of Proterozoic rocks that show little or no effects of metamorphism (see the chapter opening photo). Unfortunately, rocks of Proterozoic age do not contain many fossils, although some of those present are important.

Just as with Archean terminology, the Proterozoic terms are not based on time-stratigraphic units with stratotypes (see Chapter 5). The Archean-Proterozoic boundary at 2.5 billion years ago is somewhat arbitrarily placed but it marks the approximate time of changes in the style of crustal evolution. However, we must emphasize "approximate," because Archean-style crustal evolution was not completed at the same time in all areas. Also recall that Archean greenstone belts contain ultramafic lava flows (komatiites) which are rare in Proterozoic rocks and none form now.

But what do we mean by a change in style of crustal evolution? First, the Archean was a time of origin of granite-gneiss complexes and greenstone belts that were shaped into cratons. These same rock associations continued to form during the Proterozoic, but they did so at a slower rate. Furthermore, as we have noted, most Archean rocks have been metamorphosed, whereas many Proterozoic rocks are unaltered or nearly so (▶Figure 9.1), and in many areas Archean and Proterozoic rocks are separated by an unconformity. The Proterozoic was also the time when widespread associations of sedimentary rocks typical of passive continental margins became common, and the style of plate tectonics was essentially the same as it is now.

During the Proterozoic, important events occurred in the evolution of the atmosphere and biosphere, as well as the formation of some essential natural resources, such as the large deposits of banded iron formations. The first continental red beds were deposited during the Proterozoic, which along with the banded iron formations indicate that the atmosphere contained at least some free oxygen. Also, two episodes of widespread glaciation took place.

Oxygen-dependent organisms also made their appearance during the Proterozoic, as did cells with an internal, membrane-bounded nucleus, found in all plants and animals. Another remarkable development was the evolution of the first multicelled organisms and animals. Indeed, some Proterozoic animals were likely the ancestors of those that appeared in great abundance at the beginning of the Phanerozoic Eon (see Chapter 12). The Proterozoic fossil record is still poor compared to that of the Phanerozoic, but it is much better than the one for the Archean Eon.

Before we begin our formal discussion of Proterozoic Earth and life history, recall that the subdivisions of this eon have the prefixes *paleo* (old, or ancient), *meso* (middle), and *neo* (new, or recent) (see Figure 8.2). Thus, we discuss events that took place during the Paleoproterozoic, Mesoproterozoic, and so on.

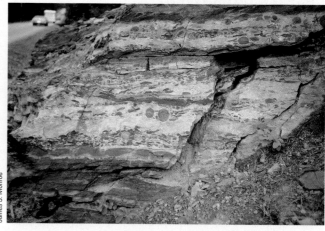

(a) The Vishnu Schist (black) in the Grand Canyon of Arizona was intruded by the Zoraster Granite about 1.7 billion years ago. The Vishnu Schist was originally lava flows and sedimentary rocks.

(b) This 1.0-billion-year-old outcrop of sandstone and mudstone of the Spokane Formation in Glacier National Park in Montana has been only slightly altered by metamorphism.

Proterozoic History of Laurentia

Geologists use the term *basement rock* for the crystalline igneous and metamorphic rocks lying beneath sedimentary rocks. Basement rock can be of any age, but much of it is Archean- and Proterozoic-age rock, like the rocks at Black Canyon of the Gunnison National Park in Colorado where Proterozoic metamorphic rocks, granite, and pegmatites are well exposed (see Perspective).

The Proterozoic was an important time in the evolution of a large landmass called **Laurentia**, which is made up of North America, Greenland, parts of northwest Scotland, and perhaps some of the Baltic shield of Scandinavia. The cratons that formed during the Archean collided with one another and formed the nucleus of Laurentia; it was along the eastern and southern boundaries of this landmass that more crust was accreted during the Proterozoic.

Laurentia During the Paleoproterozoic

As you might suspect, the evolution of Laurentia was long, complex, and is still not fully understood. Nevertheless, geologists have made remarkable progress in deciphering this episode in Earth history, which involves the amalgamation of cratons, the accretion of volcanic arcs and oceanic terranes, and extensive plutonism, metamorphism, and volcanism.

The time between 2.0 and 1.8 billion years ago (bya) was especially important, because this is when collisions among Archean cratons formed several **orogens**, which are linear to arcuate deformation belts in which many of the rocks have been metamorphosed and intruded by magma that cools to form plutons, especially batholiths.

As a result, these ancient cratons were sutured along these orogens, thereby forming a much larger landmass (▶ Figure 9.2a), which now makes up much of Greenland, central Canada, and the north-central United States.

Excellent examples of these craton-forming processes are recorded in rocks of the Thelon orogen (1.92–1.96 bya) in northwestern Canada where the Slave and Rae cratons collided, and the Trans-Hudson orogen (1.82–1.84 bya), in the United States and Canada where the Superior, Hearne, and Wyoming cratons were sutured (Figure 9.2a). In addition, the Penokean orogen formed along the southern margin of Laurentia over tens of millions of years, although the most intense episode of this event was about 1.85 bya.

Sedimentary rocks in the Wopmay orogen of northwestern Canada record the opening and closing of an ocean basin, or what geologists call a Wilson cycle. A complete **Wilson cycle**, named after the Canadian geologist J. Tuzo Wilson, includes rifting and the opening of an ocean basin with passive continental margins on both sides (▶ Figure 9.3). As a result of rifting, an expansive ocean basin forms, but eventually it begins to close, and subduction zones and volcanic island arcs form on both sides of the ocean basin. Finally, a continent-continent collision takes place, resulting in deformation of the passive margin deposits as well as rocks of the oceanic crust (Figure 9.3f).

Some of the rocks in the Wopmay orogen consist of a suite of sedimentary rocks called a **sandstone-carbonate-shale assemblage** that forms on passive continental margins. Remember that passive margin deposits are rare or absent in Archean rocks, but they become common during the Proterozoic and thereafter. This assemblage of rocks is also well represented in the Penokean orogen of the Great Lakes region of the United States and Canada (▶ Figure 9.4).

Following the initial episodes of amalgamation of cratons, additions to Laurentia occurred along its

▶ **Figure 9.2** **Proterozoic Evolution of Laurentia** These three illustrations show the overall trends in the Proterozoic evolution of Laurentia, but they do not show many of the details of this long, complex episode in Earth history.

(a) During the Paleoproterozoic, Archean cratons were sutured along deformation belts called orogens.

(b) Laurentia grew along its southeastern margin by accretion of the Yavapai and Mazatzal provinces.

1.3–1.0 billion
1.55–1.35 billion } Mesoproterozoic
1.69–1.65 billion
1.76–1.72 billion
1.9–1.8 billion } Paleoproterozoic
2.0–1.8 billion
>2.5 billion } Archean

From MONROE/WICANDER, *The Changing Earth, 6e.*
Copyright © 2012 Cengage Learning

(c) The last episodes in the Proterozoic accretion of Laurentia involved the origin of the Granite-Rhyolite province and the Grenville-Llano provinces.

Some geologists think that the Wopmay orogen in Canada respresents a complete Wilson cycle, which is shown here diagrammatically.

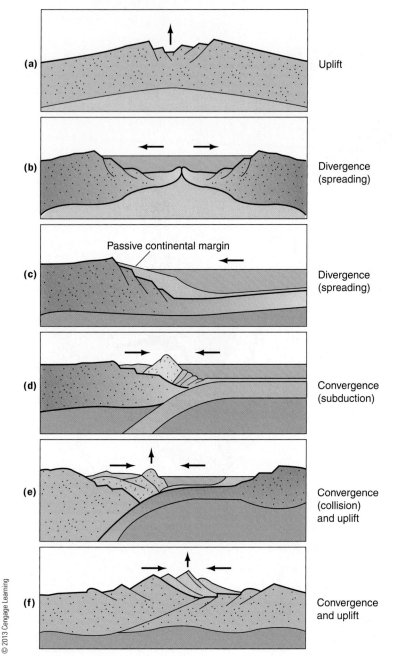

(a)   Uplift

(b)   Divergence (spreading)

(c)   Passive continental margin   Divergence (spreading)

(d)   Convergence (subduction)

(e)   Convergence (collision) and uplift

(f)   Convergence and uplift

© 2013 Cengage Learning

(a–f) In fact, some of the sedimentary rocks in the Wopmay orogen were probably deposited on a passive continental margin as shown in (c).

southeastern margin as it collided with volcanic island arcs and oceanic terranes. As a consequence, the Yavapai and Mazatzal orogens were added to the evolving continent between about 1.65 and 1.76 bya (Figure 9.2b). Of course, these were major orogenies, so the rocks have been deformed, altered by metamorphism, intruded by granitic batholiths, and incorporated into Laurentia.

The Paleoproterozoic events and rocks just reviewed are important in the evolution of Laurentia. However, this was also the time when most of Earth's *banded iron formations* (BIFs), and the first *continental red beds*—sandstone and shale with oxidized iron—were deposited. We will have more to say about BIFs and red beds in the section headed "The Evolving Atmosphere."

Mesoproterozoic Accretion and Igneous Activity

Following a lull of several millions of years, tectonism and continental accretion resumed along Laurentia's southeastern margin. In addition, the interval from about 1.35 to 1.55 billion years ago was a time of extensive igneous activity, some of it unrelated to orogenic activity that accounted for the addition of the Granite-Rhyolite province (Figure 9.2c). Some of the igneous activity resulted in plutons being emplaced in already existing continental crust. The resulting igneous rocks are exposed in eastern Canada, extend across Greenland, and are also found in the Baltic shield of Scandinavia, but they are buried beneath younger rocks in most areas.

The origins of these granitic and anorthosite* plutons, calderas and their fill, and vast sheets of rhyolite and ashflows are the subject of debate. According to one hypothesis, large-scale upwelling of magma beneath a Proterozoic supercontinent was responsible for these rocks. According to this hypothesis, the mantle temperature beneath a Proterozoic supercontinent would have been considerably higher than beneath later supercontinents because radiogenic heat production within Earth has decreased. Accordingly, nonorogenic igneous activity would have occurred following the amalgamation of the first supercontinent.

Mesoproterozoic Orogeny and Rifting

A Mesoproterozoic orogenic event in Laurentia, the 1.3-to 1.0-billion-year-old **Grenville orogeny**, took place in the eastern part of the evolving continent (Figure 9.2c). Grenville rocks are well exposed in the present-day northern Appalachian Mountains, as well as in eastern Canada (▶Figure 9.5a), Greenland, and Scandinavia. The Llano province in Texas is probably a westward extension of the Grenville orogenic belt (Figure 9.5b). Many geologists think the Grenville orogen resulted from closure of an ocean basin, the final stage in a Wilson cycle. Indeed, the Grenville

---

*Anorthosite is a plutonic rock composed almost entirely of plagioclase feldspars.

► Figure 9.4 **Paleoproterozoic Sedimentary Rocks of the Great Lakes Region**

(b) Outcrop of the Mesnard Quartzite. The crests of the ripples point toward the observer.

(a) Geographic distribution of Paleoproterozoic-aged rocks in the Great Lakes region.

(c) Outcrop of the Kona Dolomite in northern Michigan, showing bulbous structures known as stromatolites.

orogeny may have been the final episode in the assembly of the supercontinent Rodinia, which persisted into the Neoproterozoic.

Whatever the cause of the Grenville orogeny, it was the final stage in the Proterozoic continental accretion of Laurentia (Figure 9.2c). By then, about 75 percent of present-day North America existed. The remaining 25 percent accreted along its margins, particularly its eastern and western margins, during the Phanerozoic Eon.

Beginning about 1.1 billion years ago, tensional forces opened the **Midcontinent Rift**, a long, narrow trough bounded by faults that outline two branches; one branch

Black Canyon of the Gunnison National Park, Colorado

Certainly the Grand Canyon is notable for its depth and width; Hells Canyon on the Idaho–Oregon border is not as scenic but is even deeper; and Yosemite Valley in California is magnificent by any measure. However, for depth and narrowness there is little to compare with the Black Canyon of the Gunnison in Colorado (Figure 1). At one location, it is 845 m deep but as little as 395 m from rim to rim, so it is a classic gorge—that is, a narrow, deep canyon with vertical or nearly vertical walls. Indeed, at its base the canyon narrows to as little as 12 m. The Black Canyon of the Gunnison River is 77 km long, but only 39 km of it lies within the park boundaries. Another remarkable fact about the canyon is that the Gunnison River has a gradient of 18 m/km within the park, which is quite steep when compared with other rivers in North America.

There are many reasons to visit Black Canyon of the Gunnison National Park, including the wildlife, flora, scenic vistas, and, of course, the geology. If you are interested in Proterozoic-age rocks, this is certainly the place to see them. In fact, the canyon walls are made up mostly of the Black Canyon Gneiss and some schist that formed about 1.8 billion years ago (Paleoproterozoic). Heat and pressure have so thoroughly altered the rocks from their original condition that they are difficult to interpret, but

geologists think that they may have been part of an island arc that was sutured to the Wyoming craton 1.7 billion years ago.

Gneiss is the most obvious rock in the canyon walls, but about 1.4 billion years ago these metamorphic rocks were intruded by large masses of granite. Watery magma from the cooling granitic magma invaded cracks and crevices in the gneiss and formed pegmatites consisting of especially large minerals. Most pegmatites here and elsewhere have an overall composition similar to granite (quartz and potassium and plagioclase feldspars), but their mineral crystals are at least 1 cm across and commonly much larger. The best place to see the Black Canyon pegmatites is at the Painted Wall, which at 700 m is Colorado's highest cliff (Figure 2).

Paleozoic-age rocks were deposited in what is now the park, but they have long since been eroded. The same is nearly true for the Mesozoic rocks, although small outcrops of the Jurassic-age Entrada Sandstone, deposited as sand dunes, are present. The Cenozoic Era was a time of extensive volcanism in this part of Colorado, during which lava flows, ashfalls, and volcanic mudflows covered the area. Then, beginning about 2 million years ago, the entire area was uplifted and the Gunnison River began cutting down into the bedrock, forming the present-day gorge.

Valleys are common landforms, and with few exceptions, they form and evolve in response to erosion by running water, although other processes, especially mass wasting, contribute. Their sizes and shapes vary from small, steep-sided gullies to broad valleys with gently sloping walls, but not many are deep and narrow enough to be called gorges.

The reason that the Black Canyon of the Gunnison has these features is that the river has eroded downward into very hard rock, and once entrenched in this rock, it continued to erode downward at perhaps 2.5 cm per century.

Another important Paleoproterozoic event was a huge meteorite impact that occurred 1.85 billion years ago in southern Ontario, Canada, and formed what is now called the Sudbury Basin. The blast and heat from this impact must have been phenomenal—debris was scattered 800 km from the impact site, and thousands of tons of rock were vaporized. Nothing in the vicinity of the impact could have survived, but in one way it was a fortunate incident for us because the Sudbury Basin is one of the richest mining areas in the world.

In addition, some Paleoproterozoic-age rocks and associated features indicate that widespread glaciers were present at that time. We will cover this topic, as well as the evidence for more recent Neoproterozoic glaciers, later in this chapter.

Figure 1 Black Canyon of the Gunnison National Park in Colorado. This canyon is truly a gorge, that is, a narrow, deep canyon with steep to near-vertical walls. It is about 845 m deep but measures only 395 m from rim to rim.

Figure 2 View of the Painted Wall, which, at 700 m, is Colorado's highest cliff. The light layers in the Black Canyon Gneiss are 1.4-billion-year-old pegmatites.

(b) At this outcrop near Alexandria, New York, the metamorphic rocks of the Grenville orogen in the lower part of the image are unconformably overlain by the Upper Cambrian Potsdam Formation.

(a) Grenville gneiss unconformably overlain by the Paleozoic-age Nepean Sandstone in Quebec, Canada. The gneiss is 1 billion years old, whereas the sandstone is about 500 million years old.

extended southeast as far as Kansas, and one extended southeasterly through Michigan and into Ohio (▶ Figure 9.6). The rift cuts through Archean and Proterozoic rocks and terminates against the Grenville orogen in the east. Although not all geologists agree, many think that the Midcontinent Rift is a failed rift where Laurentia began splitting apart. Had this rifting continued, Laurentia would have split into two separate landmasses, but the rifting ceased after about 20 million years.

Most of the Midcontinent Rift is buried beneath younger rocks except in the Lake Superior region, where igneous and sedimentary rocks are well exposed (Figures 9.6b, 9.6c, and 9.6d). The central part of the rift contains numerous overlapping basalt lava flows, forming a volcanic pile several kilometers thick. Along the rift's margins, conglomerate was deposited in large alluvial fans that grade into sandstone and shale with increasing distance from the sediment source.

Meso- and Neoproterozoic Sedimentation

The Grenville orogeny (Figure 9.2c) was the final episode of large-scale deformation in Laurentia until the Ordovician Period (see Chapter 10). Nevertheless, other important geologic events were taking place, such as sediment deposition in what is now the eastern United States and Canada, in the Death Valley region of California and Nevada, and in three huge basins in the west (▶ Figure 9.7a).

Meso- and Neoproterozoic-age sedimentary rocks are exceptionally well exposed in the northern Rocky Mountains of Montana, Idaho, and Alberta, Canada. As a result, their colors, deformation features, and erosion by

Pleistocene and recent glaciers have yielded some fantastic scenery. Like the Paleoproterozoic rocks in the Great Lakes region, they are mostly sandstones, shales, and stromatolite-bearing carbonates.

Proterozoic rocks of the Grand Canyon Supergroup lie unconformably on Archean rocks and in turn are overlain unconformably by Phanerozoic-age rocks (Figure 9.7c). The rocks, consisting mostly of sandstone, shale, and dolostone, were deposited in shallow-water marine and fluvial environments between 740 million years ago and 1.2 billion years ago. The presence of stromatolites and carbonaceous impressions of algae in some of these rocks indicate probable marine deposition.

To summarize this entire section on the Proterozoic history of Laurentia, the events described did take place in the order discussed. But, as you might expect for such a long episode in Earth evolution, the actual history is much more complex than indicated here. We did not mention the multiple episodes of deformation, plutonism, metamorphism, and the suturing of small crustal blocks to Laurentia, nor did we talk about the vast amounts of erosion that occurred when mountains formed and were subsequently worn away. In short, we compressed nearly 2.0 billion years of Earth history into a few pages of text.

Proterozoic Supercontinents

Continents are not simply areas of land above sea level. Indeed, continents consist of granitic crust that is thicker and less dense than basalt/gabbro oceanic crust. And whereas large parts of continents are now above sea level, they were inundated by several marine transgressions during the Phanerozoic Eon (see Figure 10.3). A **supercontinent** consists of at least two continents merged into one, and in the context of Earth history, a supercontinent consists of all, or most, of Earth's landmasses except oceanic

▶ Figure 9.6 The Midcontinent Rift

(a) The rocks filling the rift are exposed around Lake Superior in the United States and Canada, but they are deeply buried elsewhere.

Legend:
- Detrital sediments
- Igneous rocks

© 2013 Cengage Learning

dvande/Shutterstock.com

(b) Image of the Portage Lake Volcanics that partially filled the more central part of the Midcontinent Rift. This image is from the Porcupine Mountains in Michigan.

Mona Sirbescu

(c) The Copper Harbor Conglomerate was deposited along the margins of the rift. With increasing distance from the source area, the conglomerate graded into sandstone and finally shale.

Photo by Dan Smith

(d) The Nonesuch Shale along the shore of Lake Superior in Michigan.

islands. You know of the supercontinent Pangaea that existed at the end of the Paleozoic Era, but you probably have not heard of older supercontinents.

Before we discuss supercontinents, it is important to know that the present style of plate tectonics of opening and then closing ocean basins was operative by the Paleoproterozoic. In fact, *ophiolites* (see Figure 3.17), the presence of which indicates plate convergence, is known from Neoarchean and Paleoproterozoic rocks from Russia and probably China, respectively. They compare closely with younger, well-documented ophiolites, such as the deformed, but complete, Jormua ophiolite in Finland (▶Figure 9.8).

A supercontinent called *Nuna* may have existed as much as 1.8 billion years ago during the Paleoproterozoic,

but the first one that geologists recognize with some certainty is **Rodinia** (▶Figure 9.9), which assembled 1.3 to 1.0 billion years ago and began fragmenting 750 million years ago. Judging by the large-scale deformation during the *Pan-African orogeny* that took place in the Southern Hemisphere continents, Rodinia's separate pieces reassembled about 650 million years ago and formed the supercontinent **Pannotia**. During the latest Neoproterozoic, about 550 million years ago, Pannotia fragmented, giving rise to the continental configuration that existed at the onset of the Phanerozoic Eon (see Chapter 10).

Ancient Glaciers and Their Deposits

Very few times of widespread glaciation have occurred during Earth history. The most recent one, during the Pleistocene Epoch (2.6 million to 11,700 years ago), is the best known, but we also have evidence for Paleozoic glaciers (see Chapters 10 and 11) and two major episodes of Proterozoic glaciation. But how can we be sure there were

► Figure 9.7 **Meso- and Neoproterozoic Sedimentary Rocks in the West**

(a) Proterozoic basins in the west.

(b) Red mudrock in the Belt basin in Montana.

(c) Angular unconformity between Paleozoic rocks in the Grand Canyon and the underlying sedimentary rocks of the Grand Canyon Supergroup.

► Figure 9.8 **The Jormua Mafic-Ultramafic Complex in Finland** Although these 1.96-billion-year-old rocks are highly deformed, they are one of the oldest known complete ophiolites.

Lithology	Estimated thickness m	Interpretation
Turbiditic graywacke and mudrocks	> 500	Deep-sea sediments
Tuff, schist, chert, carbonate rock	< 200	
Pillow breccia pillow basalt massive lava	> 300	Submarine eruptions
Locally no host rock, 100% mafic dikes, abundant interdikes gabbro and/or serpentinite	< 1,000	Sheet dikes grading down into gabbro
Gabbro	> 100	
Serpentinite	> 1,000	Meta-morphosed peridotite

Oceanic crust — Upper mantle

(a) Stratigraphic column showing the sequence of units in the ophiolite.

(b) Metamorphosed basaltic pillow lava.

Grenville orogenic belts

pre-Grenville cratons

|||| Rifts where Rodinia started to fragment 750 million years ago

© 2013 Cengage Learning

Glaciers of the Neoproterozoic

Tillites and other glacial features dating from between 900 and 600 million years ago are found on all continents except Antarctica. Glaciation was not continuous during this entire time but was episodic, with four major glacial episodes so far recognized. Figure 9.10c shows the approximate distribution of these glaciers, but we must emphasize "approximate," because the actual geographic extent of these glaciers is unknown. In addition, the glaciers covering the areas in Figure 9.10c were not all present at the same time. Despite these uncertainties, this Neoproterozoic glaciation was the most extensive in Earth history. In fact, glaciers seem to have been present even in near-equatorial areas.

Because Neoproterozoic glacial deposits are so widespread, some geologists think that glaciers covered all land and the seas were frozen—a *snowball Earth*, as it has come to be known. The snowball Earth hypothesis is controversial, but proponents claim that the onset of this glacial episode may have been triggered by the near-equatorial location of all continents, and as a result, accelerated weathering would absorb huge quantities of carbon dioxide from the atmosphere. With little CO_2 in the atmosphere, glaciers would form and reflect solar radiation back into space, leading to more glacial ice forming.

So if there actually was a snowball Earth, why wouldn't it stay frozen? Of course, volcanoes would continue to erupt, spewing volcanic gases, which include the greenhouse gases carbon dioxide and methane, which would warm the atmosphere and end the glacial episode. In fact, proponents of this hypothesis note that several such snowball Earths may have occurred until the continents moved into higher latitudes. One criticism of the hypothesis is that if all land was ice covered and the sea froze, how would life survive? Several suggestions have been made to account for this: life persisted at hydrothermal vents on the seafloor; even photosynthesis can take place beneath thin glacial ice; perhaps life persisted in subglacial lakes as it does now in Antarctica; and there may have been pools of liquid water near active volcanoes.

The Evolving Atmosphere

Geologists agree that the Archean atmosphere contained little or no free oxygen (see Chapter 8), so the atmosphere was not strongly oxidizing as it is now. Photochemical dissociation and photosynthesis were adding free oxygen to the atmosphere (see Figure 8.14), but the amount present at the beginning of the Proterozoic was probably no more than 1 percent of that present now. In fact, it might not have exceeded 10 percent of present levels even at the end of the

Proterozoic glaciers? After all, their most common deposit, called *tillite*, is simply a type of conglomerate that may look much like conglomerate that originated by other processes.

Paleoproterozoic Glaciers

Tillite or deposits that look like tillite are known from at least 300 Precambrian localities, and some of these are undoubtedly not glacial deposits. But the extensive geographic distribution of others and their associated features, such as striated and polished bedrock, are distinctive (▶ Figure 9.10a). Based on this kind of evidence, geologists are now convinced that widespread glaciation took place during the Paleoproterozoic.

Tillites of about the same age in Michigan, Wyoming, and Quebec indicate that North America may have had an ice sheet centered southwest of Hudson Bay (Figure 9.10b). Tillites of about this age are also found in Australia and South Africa, but dating is not precise enough to determine if there was a single widespread glacial episode or a number of glacial events at different times in different areas. One tillite in the Bruce Formation in Ontario, Canada, is about 2.7 billion years old, thus making it Neoarchean in age.

▶ Figure 9.10 **Paleo- and Neoproterozoic Glaciers**

(a) Evidence for Proterozoic glaciations in Norway. This deposit of glacial till lies upon a striated bedrock surface of sandstone.

(b) Paleoproterozoic glacial deposits in the United States and Canada indicate that Laurentia had an extensive ice sheet.

☐ Paleoproterozoic

☐ Minimum areas of continental glaciation

© 2013 Cengage Learning

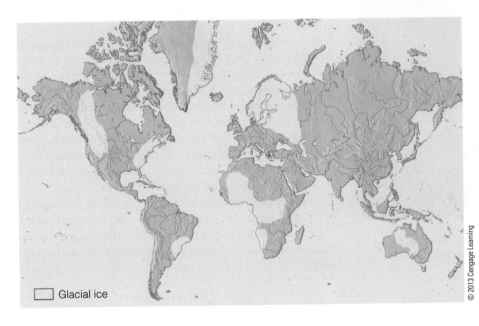

☐ Glacial ice

© 2013 Cengage Learning

(c) Neoproterozoic glacial centers shown with the continents in their present positions. The extent of the glaciers is approximate.

Proterozoic. Remember from our previous discussions that cyanobacteria (blue-green algae) were present during the Archean, but the structures they formed, called *stromatolites*, did not become common until about 2.3 billion years ago—that is, during the Paleoproterozoic. These photosynthesizing organisms and, to a lesser degree, photochemical dissociation both added free oxygen to the evolving atmosphere.

In Chapter 8 we cited some of the evidence indicating that Earth's early atmosphere had little or no free oxygen, but abundant carbon dioxide. Here we contend that as more oxygen became available, the amount of carbon dioxide decreased. So what evidence indicates that the atmosphere became an oxidizing one, and where is the carbon dioxide now? Of course, a small amount of CO_2 is

present in today's atmosphere; it is one of the greenhouse gases partly responsible for global warming. Much of it, however, is now tied up in minerals and rocks, especially the carbonate rocks limestone and dolostone, and in the biosphere. As for evidence that the Proterozoic atmosphere was evolving from a chemically reducing one to an oxidizing one, we must discuss two types of Proterozoic sedimentary rocks that we already alluded to briefly.

Banded Iron Formations (BIFs)

Most of the world's iron ores come from **banded iron formations (BIFs)** consisting of alternating millimeter- to centimeter-thick layers of iron-rich minerals and chert

(►Figure 9.11a). Some BIFs are found in Archean-age rocks, but these are small deposits mostly in greenstone belts and appear to have formed near submarine volcanoes. Most BIFs, fully 92 percent, were deposited in shallow-water shelf environments during the interval from 2.5 to 2.0 billion years ago—that is, during the earlier part of the Paleoproterozoic. These deposits are much more extensive than those of the Archean, and they have important implications for the evolving atmosphere.

The iron in the Proterozoic BIFs is iron oxide in the minerals hematite (Fe_2O_3) and magnetite (Fe_3O_4), and, of course, the chert layers are mostly silicon dioxide (SiO_2). Iron is a highly reactive element, and in an oxidizing atmosphere it combines with oxygen to form rustlike oxides that do not readily dissolve in water. If oxygen is absent, though, iron is easily taken into solution and can accumulate in large quantities in the oceans, which it probably did during the Archean. The Archean atmosphere was deficient in free oxygen, so it is doubtful that seawater had very much dissolved oxygen.

However, as free oxygen accumulated in the oceans as waste from photosynthesizing organisms, it triggered the precipitation of iron oxides and silica and thus the origin of BIFs.

One model that accounts for the details of BIF precipitation involves a Paleoproterozoic ocean with an upper oxygenated layer above a large volume of oxygen-deficient (anoxic) water that contained dissolved iron and silica. Some of the dissolved iron probably came from the weathering of rocks on land, but a likely source for much of it was submarine hydrothermal vents similar to those on the sea floors today (see Figure 8.17a). Transfer of water from the anoxic zone to the surface, or upwelling, brought iron- and silica-rich waters onto the developing shallow marine shelves, where iron and silica combined with oxygen and widespread precipitation of BIFs took place (Figure 9.11b). Precipitation of these rocks continued until the iron in seawater was largely depleted, and the atmosphere now contained some free oxygen, so iron was no longer taken into solution in large quantities.

Continental Red Beds

Obviously, the term **continental red beds** refers to red rocks on the continents, but more specifically it means red sandstone or shale colored by iron oxides, especially hematite (Fe_2O_3) (Figures 9.7b and 9.7c). These deposits first appear in the geologic record about 1.8 billion years ago. They increase in abundance throughout the rest of the Proterozoic, and are quite common in rocks thereafter.

The onset of red bed deposition coincides with the introduction of free oxygen into the Proterozoic atmosphere. But the atmosphere at that time may have had only 1 percent or perhaps 2 percent of present levels. Is this sufficient to account for oxidized iron in sediment? Probably not, but we must also consider other attributes of this atmosphere.

No ozone (O_3) layer existed in the upper atmosphere before free oxygen (O_2) was present. But as photosynthesizing organisms released free oxygen into the atmosphere, ultraviolet radiation converted some of it to elemental oxygen (O) and ozone (O_3), both of which oxidize minerals more effectively

► **Figure 9.11** **Paleoproterozoic Banded Iron Formation (BIF)**

Courtesy of Megan Babich/NOAA

(a) At this outcrop on Jasper Knob in Ishpeming, Michigan, the Negaunee Iron Formation rocks are made up of alternating layers of red chert and silvery iron minerals.

(b) Depositional model for the origin of banded iron formations.

Ultraviolet radiation

Oxygen produced by photochemical dissociation

Oxidation and photo-oxidation of Fe^{++}

Oxygen produced by planktonic photosynthetic cyanobacteria

Ferric hydroxide and silica precipitate

Iron formation accumulates

Ferric hydroxide reduced, dissolve

Upwelling

Silica precipitate accumulates

10^3 km

© 2013 Cengage Learning

▶ **Figure 9.12** **Proterozoic Fossil Bacteria and Stromatolites**

Reed Wicander

Byron Richter and Mary McCart

(a) These spherical and filamentous microfossils from the Gunflint Chert of Ontario, Canada, resemble bacteria living today. The filaments measure about 1/1,000th of a millimeter across.

(b) This rock has been eroded but it clearly shows two stromatolites that have grown together (see Figure 8.18b).

than does O_2. Once an ozone layer became established, most ultraviolet radiation failed to penetrate to the surface, and O_2 became the primary agent for oxidizing minerals.

Proterozoic Life

The fossil record of the Archean is sparse, consisting of bacteria and archaea, although there were undoubtedly many types of these organisms. Likewise, the Paleoproterozoic fossil record is characterized by these same organisms (▶ Figure 9.12a), although stromatolites, structures produced by cyanobacteria, became common (Figures 9.4c and 9.12b). Actually, the lack of biotic diversity is not too surprising, because prokaryotic cells reproduce asexually. As a result, they do not share their genetic material* as sexually reproducing organisms do, so evolution was a

*Prokaryotic cells reproduce by binary fission and give rise to two genetically identical cells. Under some conditions, they engage in conjugation, during which some genetic material is transferred from cell to cell.

comparatively slow process. Organisms that reproduced sexually probably evolved by the Mesoproterozoic, and the tempo of evolution increased markedly, although from our perspective it was still exceedingly slow.

Eukaryotic Cells Evolve

Eukaryotic cells are much larger than prokaryotic cells, they have an internal membrane-bounded nucleus that contains the chromosomes, and they have other internal structures not found in prokaryotes (▶ Figure 9.13). Furthermore, many eukaryotes are multicelled and aerobic, so they could not have existed until some free oxygen was present in the atmosphere. And lastly, most of them reproduce sexually.

The distinction between prokaryotic and eukaryotic cells is one of the most fundamental in the entire biotic realm. Biologists recognize five kingdoms of organisms— monera (archaea and bacteria), protists, fungi, plants, and animals (Table 9.1), but a system of classification that has become increasingly popular recognizes three broad groups or *domains* of all living things (▶ Figure 9.14). Two

TABLE 9.1 **The Five-Kingdom Classification of Organisms**		

Notice that only Archaea and Bacteria have prokaryotic cells.

Kingdom	Characteristics	Examples
Monera		
Archaea	Prokaryotic cells; single-celled, differ from bacteria in genetics and chemistry	Methanogens, halophiles, thermophiles
Bacteria	Prokaryotic cells; single-celled, cell walls different from archaea and eukaryotic cells	Cyanobacteria (also called blue-green algae), mycoplasmas
Protista	Eukaryotic cells; single-celled; greater internal complexity than bacteria	Various types of algae, diatoms, protozoans
Fungi	Eukaryotic cells; multicelled; major decomposers and nutrient recyclers	Fungi, yeasts, molds, mushrooms
Plantae	Eukaryotic cells; multicelled; obtain nutrients by photosynthesis	Trees, grasses, roses, rushes, palms, broccoli, poison ivy
Animalia	Eukaryotic cells; multicelled; obtain nutrients by ingestion of preformed organic molecules	Worms, clams, corals, sponges, jellyfish, fishes, amphibians, reptiles, birds, mammals

▶ **Figure 9.13 Prokaryotic and Eukaryotic Cells** Eukaryotic cells have a cell nucleus containing the genetic material and organelles such as mitochondria and plastids, as well as chloroplasts in plant cells. In contrast, prokaryotic cells are smaller and not nearly as complex as eukaryotic cells.

Prokaryotic cell

Prokaryotic cells

Bacteria

Archaea

DNA

Bacterial flagellum

Pili

Capsule

Cell wall

Ribosomes

Cytoplasm

Plasma membrane

Eukaryotic cell

Eukaryotic cells

Protista
Protozoans, algae

Fungi
Spores

Plantae
Plant tissue cells

Animalia
Animal tissue cells

Nuclear envelope
Nucleoplasm + DNA — Nucleus
Nucleolus

Microtubules

Microfilaments

Vesicle

Lysosome

Rough endoplasmic reticulum

Smooth endoplasmic reticulum

Plasma membrane

Mitochondrion

Pair of centrioles

Golgi body

Vesicle

© 2013 Cengage Learning

▶ **Figure 9.14 The Three-Domain Classification System** The inferred relationships among these organisms are based on analyses of a type of ribosomal RNA. Notice that the eukarya is more closely related to the archaea than it is to bacteria.

EUKARYA

Protozoans **Plants** **Animals**

Green Non-Sulfur Bacteria

ARCHAEA

Fungi

Trichomonads

Microsporidia

BACTERIA

Methanosarcina Halophiles

Methanobacterium

Methanococcus

T. Ceier

Purple Bacteria

Cyanobacteria

Thermoproteus

Pyrodictium

Gram Positive Bacteria

Flavobacteria

Thermotogales

Last Common Ancestor

© 2013 Cengage Learning

► Figure 9.15 The Oldest Known Eukaryote and Megafossil

(a) At 1.2 billion years old, *Bangiomorpha* is the oldest known eukaryotic organism. It is only a few millimeters long.

(b) *Grypania,* at 2.1 billion years old, is the oldest known megafossil. It was probably a bacterium or some kind of algae. It is about 1 cm across.

of the domains consist of organisms with prokaryotic cells, and the other group has members with eukaryotic cells.

Mesoproterozoic rocks 1.2 billion years old in Canada contain fossils of the oldest known eukaryotes. These tiny organisms called *Bangiomorpha* were multicelled, probably reproduced sexually, and look remarkably similar to living red algae (►Figure 9.15a). Other Proterozoic rocks have yielded even older fossils, but their affinities are unknown. For instance, the 2.1-billion-year-old Negaunee Iron Formation in Michigan has fossils known as *Grypania,* the oldest known megafossil, but it was very likely a single-celled bacterium or some kind of algae (Figure 9.15b).

Although *Bangiomorpha* is the oldest accepted eukaryote, cells larger than 60 microns appear in abundance at least 1.4 billion years ago. Many of them show an increase in organizational complexity, and an internal membrane-bounded nucleus is present in some. In addition, hollow fossils known as *acritarchs* that were probably cysts of

planktonic algae become common during the Meso- and Neoproterozoic (►Figures 9.16a and 9.16b).

Other intriguing evidence of Neoproterozoic life comes from the Grand Canyon of Arizona, where vase-shaped microfossils tentatively identified as cysts of algae have been found (Figure 9.16c).

Endosymbiosis and the Origin of Eukaryotic Cells

According to a widely accepted theory, eukaryotic cells formed from prokaryotic cells that entered into a symbiotic relationship. Symbiosis, involving a prolonged association of two or more dissimilar organisms, is common today. In many cases, both symbionts benefit from the association, as in lichens, which were once thought to be plants but actually are symbiotic associations between fungi and algae.

In a symbiotic relationship, each symbiont is usually capable of metabolism and reproduction, but the degree of dependence in some relationships is such that one or both symbionts cannot live independently. This may have been the case with Proterozoic symbiotic prokaryotes that became increasingly interdependent until the unit could exist only as a whole. In this relationship, though, one symbiont lived within the other, which is a special type of symbiosis called **endosymbiosis** (►Figure 9.17). Endosymbiosis was proposed as early as 1905, but it was the work of Lynn Margulis and her 1970 publication that convinced scientists that it was a viable theory accounting for the origin of eukaryotic cells.

Supporting evidence for the endosymbiosis theory comes from studies of living eukaryotic cells that contain internal structures, or organelles, such as mitochondria and plastids. These organelles have their own genetic material, and they synthesize proteins just as prokaryotic cells do, so they probably represent free-living bacteria that entered into a symbiotic relationship, eventually giving rise to eukaryotic cells. That such a relationship among cells is viable was demonstrated in 1966, when a microbiologist studying amoeba found that his samples had been infected by a bacterium that invaded and killed most of the amoeba. After several months, though, the amoeba and bacterium coexisted in an endosymbiotic relationship; the amoeba could no longer live without the bacterium that produces an essential protein.

► Figure 9.16 Proterozoic Fossils These are probably from eukaryotic organisms and measure only 40 to 50 microns across.

(a) The acritarch *Tappania piana* is from Mesoproterozoic rocks in China.

(b) This acritarch, known as *Octoedryxium truncatum,* was found in Neoproterozoic rocks in Sweden.

(c) This vase-shaped microfossil from Neoproterozoic rocks in the Grand Canyon in Arizona is a cyst from some kind of algae.

From RUSSELL/WOLFE/HERTZ/STARR, *Biology, 1e.* Copyright © 2008 Cengage Learning

The Dawn of Multicelled Organisms

We certainly live in a world dominated by microscopic organisms, but we are most familiar with multicelled plants and animals. However, **multicelled organisms** are not simply composed of many cells, but also have cells specialized to perform specific functions such as reproduction and respiration. We know that multicelled organisms were present by the Neoproterozoic, but the fossil record does not tell us how this transition took place. Nevertheless, studies of present-day organisms give some clues about what probably happened. In fact, there are some living organisms that, while multicelled, possess as few as four identical cells, all of which are capable of living on their own (▶ Figure 9.18a).

Suppose that a single-celled organism divided and formed a group of cells that did not disperse, but remained together as a colony. If so, the cells in some colonies might have become somewhat specialized, as are the cells of some colonial organisms today. Further specialization might have led to simple multicelled organisms such as sponges, consisting of cells that carry out functions such as reproduction, respiration, and food gathering. Carbonaceous impressions of Proterozoic multicelled algae are known from many areas (Figure 9.18b).

But is there any particular advantage to being multicelled? After all, for about 1.5 billion years all organisms were single-celled and life thrived. In fact, single-celled organisms are quite good at what they do, but what they do is limited. For example, they can't become very large, because as size increases, proportionately less of a cell is exposed to the external environment in relation to its volume. So as volume increases, the proportion of surface area decreases and transferring materials from the exterior

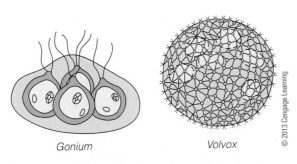

Gonium   *Volvox*

© 2013 Cengage Learning

(a) *Gonium* consists of as few as four cells. All cells are alike and can reproduce a new colony. *Volvox* has some cells specialized for specific functions, so it has crossed the threshold that separates single-celled from multicelled organisms.

Courtesy of Robert Hordyski

(b) This carbonaceous impression in the Little Belt Mountains of Montana may be from multicelled algae.

to the interior becomes less efficient. Also, multicelled organisms live longer, because cells can be replaced and more offspring can be produced. And finally, cells have increased functional efficiency when they are specialized into organs with specific functions.

Neoproterozoic Animals

Criteria such as method of reproduction and type of metabolism allow us to easily distinguish between animals and plants. Or so it would seem—but some present-day organisms blur this distinction, and the same is true for some Proterozoic fossils. Nevertheless, the first fairly controversy-free fossils of animals come from the Ediacaran fauna of Australia and similar faunas of about the same age elsewhere.

The Ediacaran Fauna
In 1947, an Australian geologist, R. C. Sprigg, discovered impressions of soft-bodied animals in the Pound Quartzite in the Ediacara Hills of South Australia. Additional discoveries by others turned up what appeared to be impressions of algae and several animals, many bearing no resemblance to any animals existing now (►Figure 9.19). Before these discoveries, geologists were perplexed by the apparent absence of fossil-bearing rocks predating the Phanerozoic. They had assumed the fossils so common in Cambrian rocks must have had a long previous history but had little evidence to support this conclusion. This discovery and subsequent ones have not answered all questions about pre-Phanerozoic animals, but they have certainly increased our knowledge about this chapter in the history of life.

Some investigators think that three present-day phyla are represented in the Ediacaran fauna: jellyfish and sea pens (phylum Cnidaria), segmented worms (phylum Annelida), and primitive members of the phylum Arthropoda (the phylum with insects, spiders, crabs, and others). One Ediacaran fossil, *Spriggina*, has been cited as a possible ancestor of trilobites (Figure 9.19b), and another may be a primitive member of the phylum Echinodermata. However, some scientists think these Ediacara animals represent an early evolutionary group quite distinct from the ancestry of today's invertebrate animals.

The **Ediacaran fauna**, the collective name for fossil associations similar to those in the Ediacara Hills, is now known from all continents except Antarctica. For example, excellent fossils from this 545- to 600-million-year-old fauna have been found in Namibia, Africa, and in Newfoundland, Canada (►Figure 9.20). Thus, Ediacaran animals were widespread, but because they lacked durable skeletons, their fossils are not very common.

Other Proterozoic Animal Fossils Although scarce, a few animal fossils older than those of the Ediacaran fauna are known. A jellyfish-like impression is present in rocks 2,000 m below the Pound Quartzite, and in many areas burrows, presumably made by worms, are found in rocks at least 700 million years old. Some possible fossil worms from China may be 700 to 900 million years old (►Figure 9.21a).

All known Proterozoic animals were soft-bodied, but there is some evidence that the earliest stages in the evolution of skeletons were under way. Even some Ediacaran animals may have had a chitinous carapace, and others appear to have spots or small areas of calcium carbonate. Small branching tubes preserved in 590- to 600-million-year-old rocks in China may be early relatives of corals (Figure 9.21b). The odd creature known as *Kimberella* from the Neoproterozoic of Russia had a tough outer covering similar to that of some present-day marine invertebrates (Figure 9.21c). Exactly what *Kimberella* was is uncertain; some think it was a mollusk.

Minute scraps of shell-like material and small tooth-like denticles and spicules, presumably from sponges, indicate that several animals with skeletons, or at least partial

(a) The affinities of *Tribachidium* remain uncertain. It may be either a primitive echinoderm or a cnidarian.

(b) *Spriggina* was originally thought be a segmented worm (annelid), but now it appears more closely related to arthropods, possibly even an ancestor of trilobites.

(c) Shield-shaped *Parvanconrina* is perhaps related to the arthropods.

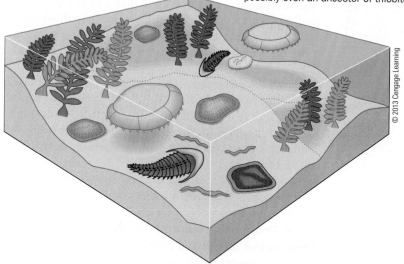

(d) Reconstruction of the Ediacaran environment.

▶ Figure 9.20 **Ediacaran-type Fossils from the Mistaken Point Formation of Newfoundland**

(a) These fossils have not been given scientific names. The most obvious ones are informally called *spindles*, whereas the one toward the lower center is a *feather duster*.

(b) This fossil, known as *Bradgatia*, has also been found in England. These fossils are about 575 million years old.

(a) Possible worm fossils from 700- to 900-million-year-old rocks in China.

(b) These small branching tubes measure only 0.1 to 0.3 mm across. The animals that made the tubes may have been early relatives of corals.

(c) Many think that *Kimberella*, of Neoproterozoic age from Russia, was some kind of mollusk.

skeletons, existed by latest Neoprotoerozoic time. Nevertheless, more durable skeletons of silica, calcium carbonate, and chitin (a complex organic substance) did not appear in abundance until the beginning of the Phanerozoic Eon 542 million years ago (see Chapter 12).

Proterozoic Mineral Resources

In an earlier section, we mentioned that most of the world's iron ore comes from Paleoproterozoic banded iron formations (Figure 9.11) and that Canada and the United States have large deposits of these rocks in the Lake Superior region and in eastern Canada. Both rank among the ten leading nations in iron ore production.

In the Sudbury mining district in Ontario, Canada, nickel and platinum are extracted from Proterozoic rocks. Nickel is essential for the production of nickel alloys such as stainless steel and Monel metal (nickel plus copper), which are valued for their strength and resistance to corrosion and heat. The United States must import more than 50 percent of all nickel used, mostly from the Sudbury mining district.

Some platinum for jewelry, surgical instruments, and chemical and electrical equipment is also exported to the United States from Canada, but the major exporter is South Africa. The Bushveld Complex of South Africa is a layered complex of igneous rocks from which both platinum and chromite, the only ore of chromium, are mined. Much chromium used in the United States is imported from South Africa; it is used mostly in manufacturing stainless steel.

Economically recoverable oil and gas have been discovered in Proterozoic rocks in China and Siberia, arousing some interest in the Midcontinent Rift as a potential source of hydrocarbons (Figure 9.6). So far, land has been leased for exploration, and numerous geophysical studies have been done. However, even though some rocks within the rift are known to contain petroleum, no producing oil or gas wells are operating.

A number of Proterozoic pegmatites are important economically. The Dunton pegmatite in Maine, whose age is generally considered Neoproterozoic, has yielded magnifcent gem-quality specimens of tourmaline and other minerals. Other pegmatites are mined for gemstones as well as for tin; for industrial minerals, such as feldspars, micas, and quartz; and for minerals containing such elements as cesium, rubidium, lithium, and beryllium.

Geologists have identified more than 20,000 pegmatites in the country rocks adjacent to the Harney Peak Granite in the Black Hills of South Dakota. These pegmatites formed about 1.7 billion years ago when the granite was emplaced as a complex of dikes and sills. A few have been mined for gemstones, tin, lithium, and micas, and some of the world's largest known mineral crystals were discovered in these pegmatites.

Summary

▶ Table 9.2 provides a summary of the Proterozoic geologic and biologic events discussed in the text.

▶ The crust-forming processes that yielded Archean granite-gneiss complexes and greenstone belts continued into the Proterozoic but at a considerably reduced rate.

▶ Paleoproterozoic collisions between Archean cratons formed larger cratons that served as nuclei around which crust accreted. One large landmass so formed was Laurentia, consisting mostly of North America and Greenland.

▶ Paleoproterozoic amalgamation of cratons, followed by Mesoproterozoic igneous activity, the Grenville orogeny, and the Midcontinent Rift, were important events in the evolution of Laurentia.

▶ Ophiolite sequences marking convergent plate boundaries are first well documented from the Neoarchean and Paleoproterozoic, indicating that a plate tectonic style similar to that operating now had become established.

▶ Sandstone-carbonate-shale assemblages deposited on passive continental margins were very common by Proterozoic time.

▶ The supercontinent Rodinia assembled between 1.3 and 1.0 billion years ago, fragmented, and then reassembled to form Pannotia about 650 million years ago, which began fragmenting about 550 million years ago.

▶ Glaciers were widespread during both the Paleoproterozoic and the Neoproterozoic.

▶ Photosynthesis continued to release free oxygen into the atmosphere, which became increasingly rich in oxygen through the Proterozoic.

▶ Fully 92 percent of Earth's iron ore deposits in the form of banded iron formations (BIFs) were deposited between 2.5 and 2.0 billion years ago.

▶ Widespread continental red beds dating from 1.8 billion years ago indicate that Earth's atmosphere had enough free oxygen for oxidation of iron compounds.

▶ Most of the known Proterozoic organisms are single-celled prokaryotes (bacteria). When eukaryotic cells first appeared is uncertain, but they were probably present by 1.2 billion years ago. Endosymbiosis is a widely accepted theory for their origin.

▶ The oldest known multicelled organisms are probably algae, some of which might date back to the Paleoproterozoic.

▶ Well-documented multicelled animals are found in several Neoproterozoic localities. Animals were widespread at this time, but because all lacked durable skeletons their fossils are not common.

▶ Most of the world's iron ore production is from Proterozoic banded iron formations. Other important resources include nickel and platinum.

Important Terms

banded iron formation (BIF), p. 179
continental red beds, p. 180
Ediacaran fauna, p. 185
endosymbiosis, p. 183
eukaryotic cell, p. 181

Grenville orogeny, p. 172
Laurentia, p. 170
Midcontinent Rift, p. 173
multicelled organism, p. 184
orogen, p. 170
Pannotia, p. 176

Rodinia, p. 176
sandstone-carbonate-shale assemblage, p. 170
supercontinent, p. 175
Wilson cycle, p. 170

Review Questions

1. The presence of Proterozoic-age continental red beds indicates that
   a. ____ prokaryotic cells had evolved; b. ____ the atmosphere contained some free oxygen; c. ____ ultramafic lava flows no longer formed; d. ____ greenstone belts formed in back-arc basins; e. ____ passive continental margins were common.

2. Laurentia was a Proterozoic landmass made up mostly of
   a. ____ Baltica and Asia; b. ____ South America and Pannotia; c. ____ Rodinia and Romania; d. ____ North America and Greenland; e. ____ Africa and Eurasia.

3. Cells with a membrane-bounded nucleus and internal structures called organelles are _____ cells.
   a. ____ komatiitic; b. ____ endosymbiotic; c. ____ eukaryotic; d. ____ aphanitic; e. ____ stromatolitic.

4. The final orogenic event to affect Laurentia during the Proterozoic was the _____ orogeny.
   a. ____ Trans-Hudson; b. ____ Wilson; c. ____ Penokean; d. ____ Wopmay; e. ____ Grenville.

5. The oldest known animal fossils are found in the _____ fauna of Australia.
   a. ____ Alberta; b. ____ Ediacaran; c. ____ Grenvillian; d. ____ Wilsonian; e. ____ Sudbury.

6. The widely accepted theory for the origin of eukaryotic cells holds that these cells formed by
   a. ____ endosymbiosis; b. ____ parthenogenesis; c. ____ binary fission; d. ____ autotrophism; e. ____ heterotrophism.

7. Mats and columnar masses of rock resulting from the activities of cyanobacteria (blue-green algae) are
   a. ____ orogens; b. ____ stromatolites; c. ____ symbionts; d. ____ trilobites; e. ____ prokaryotes.

8. The middle part of the Midcontinent Rift is filled with a. ____ layers of limestone; b. ____ overlapping basalt lava flows; c. ____ greenstones and granite-gneiss complexes; d. ____ ophiolites; e. ____ banded iron formations.

9. Banded iron formations (BIFs) are made up of thin, iron-rich layers alternating with layers of a. ____ limestone; b. ____ chert; c. ____ gneiss; d. ____ greenstone; e. ____ basalt.

10. The oldest supercontinent that is known with some certainty is a. ____ Amazonia; b. ____ Laurentia; c. ____ Scandinavia; d. ____ Rodinia; e. ____ Ediacara.

11. Neoproterozoic sedimentary rocks in the northern Rocky Mountains are 4,000 m thick and were deposited between 1.45 billion and 850 million years ago. What was the average rate of sediment accumulation in centimeters per year? Why is this figure unlikely to represent the actual rate of sedimentation?

12. What is the Midcontinent Rift, when did it form, and what kinds of rocks are found in it?

13. Explain how endosymbiosis is thought to have been responsible for the origin of eukaryotic cells.

14. What is a Wilson cycle? Is there any evidence of such cycles in Precambrian rocks?

15. Outline the events leading to the Proterozoic development of Laurentia.

16. When did the first animals appear in the fossil record, and why are their fossils not very common?

17. What evidence indicates that two episodes of glaciation took place during the Proterozoic?

18. What is the significance of stromatolites, and how do they form?

19. How do banded iron formations and continental red beds provide evidence for changes in Earth's Proterozoic atmosphere?

20. How did the style of crustal evolution for the Archean and Proterozoic differ?

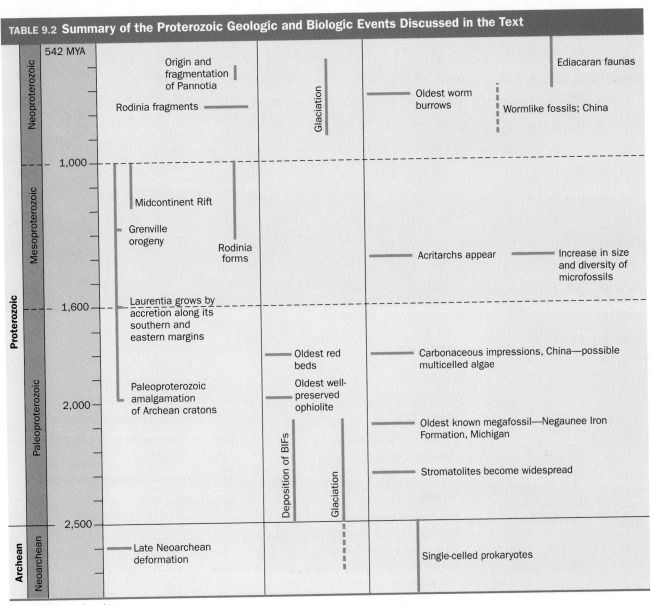

TABLE 9.2 Summary of the Proterozoic Geologic and Biologic Events Discussed in the Text

© 2013 Cengage Learning

10

Early Paleozoic Earth History

The need to cheaply transport coal from where it was mined to where it was needed resulted in widespread canal building in England during the late 1700s and early 1800s. William Smith, who started his career mapping various coal mines and later produced the first geologic map of England, was instrumental in helping to find the most efficient canal routes to bring coal to market. Canals like the Grand Junction Canal, shown here in this 1819 woodcut, were critical not only for transporting coal from the mines to market, but also for the movement of people and goods.

Hulton Archive/Stringer/Getty Images

CHAPTER OBJECTIVES

At the end of this chapter, you will have learned that

▶ Six major continents were present at the beginning of the Paleozoic Era, and plate movements during the Early Paleozoic resulted in the first of several continental collisions leading to the formation of the supercontinent Pangaea at the end of the Paleozoic.

▶ The Paleozoic history of North America can be subdivided into six cratonic sequences, which represent major craton-wide transgressive–regressive cycles.

▶ During the Sauk Sequence, warm, shallow seas covered most of North America, leaving only a portion of the Canadian shield and a few large islands, known as the Transcontinental Arch, above sea level.

▶ Like the Sauk Sequence, the Tippecanoe Sequence began with a major transgression resulting in widespread sandstones, followed by extensive carbonate and evaporite deposition.

▶ During the Tippecanoe Sequence, an oceanic–continental convergent plate boundary formed along the eastern margin of North America (known as the Appalachian mobile belt), resulting in the Taconic orogeny, the first of several orogenies to affect this area.

▶ Lower Paleozoic rocks contain a variety of important mineral resources.

Introduction

August 1, 1815, is an important date in the history of geology. On that date William Smith, a canal builder, published the world's first true geologic map. Measuring more than eight feet high and six feet wide, Smith's hand-painted geologic map of England represented more than 20 years of detailed study of England's rocks and fossils.

England is a country rich in geologic history. Five of the six Paleozoic geologic systems (Cambrian, Ordovician, Silurian, Devonian, and Carboniferous) were described and named for rocks exposed in England (see Chapter 5). The Carboniferous coal beds of England helped fuel the Industrial Revolution, and the need to transport coal cheaply from where it was mined to where it was used set off a flurry of canal building during the late 1700s and early 1800s (see chapter opening photo). During this time, William Smith, who was mapping various coal mines, first began to notice how rocks and fossils repeated themselves in a predictable manner. During the ensuing years, Smith surveyed the English countryside for the most efficient canal routes to bring the coal to market. Much of his success was based on his ability to predict what rocks the canal diggers would encounter. Realizing that his observations allowed him to unravel the geologic history of an area and correlate rocks from one region to another, William Smith set out to make the first geologic map of an entire country!

The story of how William Smith came to publish the world's first geologic map is a fascinating tale of determination and perseverance. However, instead of finding fame and success after that accomplishment, Smith found himself, slightly less than four years later, in debtors' prison, and—upon his release after spending more than two months in prison—homeless. If such a story can have a happy ending, William Smith at least lived long enough to finally be recognized and honored for his seminal contributions to the then-fledgling science of geology.

Just as William Smith applied basic geologic principles in deciphering the geology of England, we use these same principles in this and the next chapter to interpret the geology of the Paleozoic Era. The Paleozoic history of most continents involves major mountain-building activity along their margins and numerous shallow-water marine transgressions and regressions over their interiors. These transgressions and regressions were caused by global changes in sea level that most probably were related to seafloor spreading and glaciation.

In Chapters 10 and 11, we initially provide an overview of the geologic history of the world during the Paleozoic Era in order to place into context the geologic events taking place in North America during this time. We then focus our attention on the geologic history of North America—not in a period-by-period chronology, but in terms of the major transgressions and regressions taking place on the continent, as well as the mountain-building activity occurring during this time. Such an approach allows us to place the North American geologic events within a global context.

Continental Architecture: Cratons and Mobile Belts

During the Precambrian, continental accretion and orogenic activity led to the formation of sizable continents. Recall that movement of these continents during the Mesoproterozoic led to the first known supercontinent that geologists call Rodinia (see Figure 9.9). Following fragmentation of Rodinia, sometime around 750 million years ago, a second Pangaea-like supercontinent, referred to as Pannotia, began forming approximately

650 million years ago. This supercontinent started breaking apart during the latest Proterozoic (550 million years ago), and by the beginning of the Paleozoic Era, six major continents were present. Each continent can be divided into two major components: a craton and one or more mobile belts.

Cratons are the relatively stable and immobile parts of continents and form the foundation on which Phanerozoic sediments were deposited (▶ Figure 10.1). Cratons typically consist of two parts: a shield and a platform. *Shields* are the exposed portion of the crystalline basement rocks of a continent and are composed of Precambrian metamorphic and igneous rocks that reveal a history of extensive orogenic activity during the Precambrian (see Figure 8.5). During the Phanerozoic, however, shields were extremely stable and formed the foundations of the continents.

Extending outward from the shields are buried Precambrian rocks that constitute a *platform,* another part of the craton (see Figure 8.5). Overlying the platform are flat-lying or gently dipping Phanerozoic detrital and chemical sedimentary rocks that were deposited in widespread shallow seas that transgressed and regressed over the craton.

These seas, called **epeiric seas**, were a common feature of most Paleozoic cratonic histories. Changes in sea level caused primarily by continental glaciation, as well as by plate movement, were responsible for the advance and retreat of these epeiric seas.

Whereas most Paleozoic platform rocks are still essentially flat-lying, in some places they were gently folded into regional arches, domes, and basins (Figure 10.1). In many cases, some of these structures stood out as low islands during the Paleozoic Era and supplied sediments to the surrounding epeiric seas.

Mobile belts are elongated areas of mountain-building activity. They are located along the margins of continents where sediments are deposited in the relatively shallow waters of the continental shelf and the deeper waters at the base of the continental slope. During plate convergence along these margins, the sediments are deformed and intruded by magma, creating mountain ranges.

Four mobile belts formed around the margin of the North American craton during the Paleozoic: the Franklin, **Cordilleran**, **Ouachita**, and **Appalachian mobile belts** (Figure 10.1). Each was the site of mountain building in response to compressional forces along a convergent plate boundary and formed mountain ranges such as the Appalachians and Ouachitas.

Paleozoic Paleogeography

One result of plate tectonics is that Earth's geography is constantly changing. The present-day configuration of the continents and ocean basins is merely a snapshot in time. As the plates move about, the location of continents and ocean basins constantly changes. One of the goals of historical geology is to provide paleogeographic reconstructions of the world during the geologic past. By synthesizing all of the pertinent paleoclimatic, paleomagnetic, paleontologic, sedimentologic, stratigraphic, and tectonic data available, geologists construct paleogeographic maps. Such maps are simply interpretations of the geography of an area for a particular time in the geologic past and usually show the distribution of land and sea, possible climatic regimes, and such geographic features as mountain ranges, swamps, and glaciers.

© 2013 Cengage Learning

▶ Figure 10.1 **Major Cratonic Structures and Mobile Belts** The major cratonic structures and mobile belts of North America that formed during the Paleozoic Era.

The paleogeographic history of the Paleozoic Era, for example, is not as precisely known as that of the Mesozoic and Cenozoic eras, in part because the magnetic anomaly patterns preserved in the oceanic crust were destroyed when much of the Paleozoic oceanic crust was subducted during the formation of Pangaea. Paleozoic paleogeographic reconstructions are, therefore, based primarily on structural relationships, climate-sensitive sediments such as red beds, evaporites, and coals, as well as the distribution of plants and animals.

At the beginning of the Paleozoic, six major continents were present. Besides these large landmasses, geologists have also identified numerous microcontinents such as *Avalonia* (composed of parts of present-day Belgium, northern France, England, Wales, Ireland, the Maritime Provinces and Newfoundland of Canada, as well as parts of the New England area of the United States), and various island arcs associated with microplates. We are primarily concerned, however, with the history of the six major continents and their relationships to each other. The six major Paleozoic continents are **Baltica** (Russia west of the Ural Mountains and the major part of northern Europe), **China** (a complex area consisting of at least three Paleozoic continents that were not widely separated and are here considered to include China, Indochina, and the Malay Peninsula), **Gondwana** (Africa, Antarctica, Australia, Florida, India, Madagascar, and parts of the Middle East and southern Europe), **Kazakhstania** (a triangular continent centered on Kazakhstan but considered by some to be an extension of the Paleozoic Siberian continent), **Laurentia** (most of present North America, Greenland, northwestern Ireland, and Scotland), and **Siberia** (Russia east of the Ural Mountains and Asia north of Kazakhstan and south of Mongolia). The paleogeographic reconstructions that follow (▶ Figure 10.2) are based on the methods used to determine and interpret the location, geographic features, and environmental conditions on the paleocontinents.

Early Paleozoic Global History

In contrast to today's global geography, the Cambrian world consisted of six major continents dispersed around the globe at low tropical latitudes (Figure 10.2a). Water circulated freely among ocean basins, and the polar regions were mostly ice free. By the Late Cambrian, epeiric seas had covered areas of Laurentia, Baltica, Siberia, Kazakhstania, and China, whereas highlands were present in northeastern Gondwana, eastern Siberia, and central Kazakhstania.

During the Ordovician and Silurian periods, plate movement played a major role in the changing global geography (Figures 10.2b and 10.2c). Gondwana moved southward during the Ordovician and began to cross the South Pole, as indicated by Upper Ordovician tillites found today in the Sahara Desert. During the Early Ordovician,

▶ **Figure 10.2 Paleozoic Paleogeography** Paleogeography of the world during the (a) Late Cambrian Period, (b) Late Ordovician Period, and (c) Middle Silurian Period.

Shallow sea   Deep ocean   Lowlands   Mountains

© 2013 Cengage Learning

(a) Late Cambrian Period.

(b) Late Ordovician Period.

(c) Middle Silurian Period.

the microcontinent Avalonia separated from Gondwana and began moving northeastward, where it would finally collide with Baltica during the Late Ordovician–Early Silurian. In contrast to the passive continental margin Laurentia exhibited during the Cambrian, an active convergent plate boundary formed along its eastern margin during the Ordovician, as indicated by the Late Ordovician Taconic orogeny that occurred in New England.

During the Silurian, Baltica, along with the newly attached Avalonia, moved northwestward relative to Laurentia and collided with it to form the larger continent of Laurasia. This collision, which closed the northern Iapetus Ocean, is marked by the Caledonian orogeny. After this orogeny, the southern part of the Iapetus Ocean still remained open between Laurentia and Avalonia-Baltica (Figure 10.2c). Siberia and Kazakhstania moved from a southern equatorial position during the Cambrian to north temperate latitudes by the end of the Silurian Period.

With this plate tectonic paleogeographic overview in mind, we now focus our attention on North America (Laurentia) and its role in the Early Paleozoic geologic history of the world.

Early Paleozoic Evolution of North America

It is convenient to divide the geologic history of the North American craton into two parts, the first dealing with the relatively stable continental interior over which epeiric seas transgressed and regressed, and the other with the mobile belts where mountain building occurred.

The sedimentary rock record of the North American craton can be divided into six cratonic sequences. A **cratonic sequence** is a large-scale (greater than supergroup) lithostratigraphic unit representing a major transgressive–regressive cycle bounded by craton-wide unconformities (▶Figure 10.3). The transgressive phase, which is usually covered by younger sediments, commonly is well

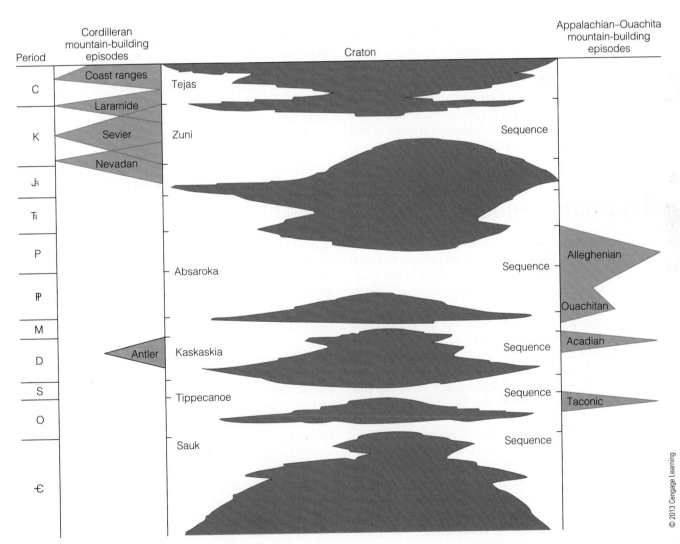

▶ Figure 10.3 **Cratonic Sequences of North America** A cratonic sequence is a large-scale lithostratigraphic unit representing a major transgressive–regressive cycle and bounded by craton-wide unconformities. The white areas represent sequences of rocks separated by large-scale unconformities (brown areas). The major Cordilleran orogenies are shown on the left side, and the major Appalachian orogenies are on the right.

preserved, whereas the regressive phase of each sequence is marked by an unconformity. Where rocks of the appropriate age are preserved, each of the six unconformities are seen to extend across the various sedimentary basins of the North American craton and into the mobile belts along the cratonic margin.

Geologists have also recognized major unconformity-bounded sequences in cratonic areas outside North America. Such global transgressive and regressive cycles of sea level changes are thought to result from major tectonic and glacial events.

The realization that rock units can be divided into cratonic sequences, and that these sequences can be further subdivided into smaller units and correlated, provides the foundation for an important concept in geology that allows high-resolution analysis of time and facies relationships within sedimentary rocks. **Sequence stratigraphy** is the study of rock relationships within a time-stratigraphic framework of related facies bounded by erosional or non-depositional surfaces. The basic unit of sequence stratigraphy is the *sequence*, which is a succession of rocks bounded by unconformities and their equivalent conformable strata. Sequence boundaries result from a relative drop in sea level. Sequence stratigraphy is an important tool in geology because it allows geologists to subdivide sedimentary rocks into related units that are bounded by time-stratigraphically significant boundaries. Geologists currently use sequence stratigraphy, particularly in the petroleum industry, for high-resolution correlation and mapping, as well as interpreting and predicting depositional environments.

The Sauk Sequence

Rocks of the **Sauk Sequence** (Neoproterozoic–Early Ordovician) record the first major transgression onto the North American craton (Figure 10.3). During the Neoproterozoic and Early Cambrian, deposition of marine sediments was limited to the passive shelf areas of the Appalachian and Cordilleran borders of the craton. The craton itself was above sea level and experiencing extensive weathering and erosion. Because North America was located in a tropical climate at this time and there was very limited, if any, terrestrial vegetation, weathering and erosion of the exposed Precambrian basement rocks must have proceeded rapidly.

During the Middle Cambrian, the transgressive phase of the Sauk began with epeiric seas encroaching over the craton (see Perspective). By the Late Cambrian, these epeiric seas had covered most of North America, leaving only a portion of the Canadian shield and a few large islands above sea level (▶ Figure 10.4). These islands, collectively named the **Transcontinental Arch**, extended from New Mexico to Minnesota and the Lake Superior region.

The sediments deposited both on the craton and along the shelf area of the craton margin show abundant evidence of shallow-water deposition. The only difference between the shelf and craton deposits is that the shelf deposits are thicker. In both areas, the sands are generally clean and well sorted, and commonly contain ripple marks and small-scale cross-bedding. Many of the carbonates are bioclastic (composed of fragments of organic remains), contain stromatolites, or have oolitic (small, spherical calcium carbonate grains) textures. Such sedimentary structures and textures indicate shallow-water deposition.

The Cambrian of the Grand Canyon Region: A Transgressive Facies Model

Recall from Chapter 5 that sediments become increasingly finer the farther away from land they are. Therefore, in a stable environment where sea level remains the same, coarse detrital sediments are typically deposited in the nearshore environment, and finer-grained sediments are deposited in the offshore environment. Carbonates form farthest from land, in the area beyond the reach of detrital sediments. During a transgression, these facies (sediments that represent a particular environment) migrate in a landward direction (see Figure 5.8a).

The Cambrian rocks of the Grand Canyon region (see Chapter 4 opening photo and Figure 5.7) provide an excellent model of the sedimentation patterns of a transgressing sea. The Grand Canyon region occupied the passive shelf and western margin of the craton during the time of the Sauk Sequence. During the Neoproterozoic and Early Cambrian, most of the craton was above sea level and deposition of marine sediments was mainly restricted to the margins of the craton (continental shelves and slopes).

In the Grand Canyon region, the Tapeats Sandstone represents the basal transgressive shoreline deposits that accumulated as marine waters transgressed across the shelf and just onto the western margin of the craton during the Early Cambrian (▶ Figure 10.5). These sediments are clean and well-sorted sands of the type one would find on a beach today. As the transgression continued into the Middle Cambrian, muds of the Bright Angel Shale were deposited over the Tapeats Sandstone. By the Late Cambrian, the Sauk Sea had transgressed so far onto the craton that, in the Grand Canyon region, carbonates of the Muav Limestone were being deposited over the Bright Angel Shale.

This vertical succession of sandstone (Tapeats), shale (Bright Angel), and limestone (Muav) forms a typical transgressive sequence and represents a progressive migration of offshore facies toward the craton through time (Figure 10.5; see Figure 5.7).

Cambrian rocks of the Grand Canyon region also illustrate that many formations are time transgressive; that is,

Islands
Land
Epeiric sea
Deep ocean

Cordilleran mobile belt

Muddy bottom

Carbonate bottom

Sandy bottom

Canadian Shield

Sandy bottom

Carbonate bottom

Ouachita - Appalachian mobile belt

Muddy bottom

Transcontinental Arch

Paleoequator

▶ **Figure 10.4 Cambrian Paleogeography of North America** Note the position of the Cambrian paleoequator. During this time, North America straddled the equator as indicated in Figure 10.2a.

their age is not the same in every place they are found. Mapping and correlations based on faunal evidence indicate that deposition of the Muav Limestone had already started on the shelf before deposition of the Tapeats Sandstone was completed on the craton. Faunal analysis of the Bright Angel Shale indicates that its age is Early Cambrian in California and Middle Cambrian in the Grand Canyon region, thus illustrating the time-transgressive nature of formations and facies.

This same facies relationship also occurred elsewhere on the craton as the seas encroached from the Appalachian and Ouachita mobile belts onto the craton interior (▶ Figure 10.6). Carbonate deposition dominated on the craton as the Sauk transgression continued during the Early Ordovician, and the advancing Sauk Sea soon covered the islands of the Transcontinental Arch. By the end of the Sauk Sequence, much of the craton was submerged beneath a warm, equatorial epeiric sea (Figure 10.2a).

The Grand Canyon—A Geologist's Paradise

"The Grand Canyon is the one great sight which every American should see," declared President Theodore Roosevelt. "We must do nothing to mar its grandeur." And so, in 1908, he named the Grand Canyon a national monument to protect it from exploitation. In 1919, Grand Canyon National Monument was upgraded to a national park, primarily because its scenery and the geology exposed in the canyon are unparalleled.

Located in the Colorado Plateau in northwestern Arizona, Grand Canyon National Park encompasses 1,218,375 acres and contains several major ecosystems (Figure 1). Most of the more than 5,000 km² of the park is maintained as wilderness, with many trails affording visitors both day and overnight backcountry hiking opportunities.

Formed by the erosive power of the Colorado River, the Grand Canyon winds more than 500 km through northwestern Arizona, averages a depth of 1,220 m, and is 1,830 m deep at its deepest point (Figure 2). The Colorado River is responsible for carving the Grand Canyon and other canyons during the past 23 million years or so in response to the general uplift of the Colorado Plateau, primarily during the Neogene Epoch (see Chapter 16).

Figure 1 Grand Canyon National Park, Arizona, consists of three distinct sections: the South Rim, the North Rim, and the Inner Canyon. Each section has a different climate and vegetation, and offers the visitor different experiences.

© 2013 Cengage Learning

Reed Wicander

Figure 2 The Grand Canyon is world famous for its grandeur and beauty, and it attracts approximately 5 million visitors a year. Mohave Point, on West Rim Drive, is a popular spot to view the South Rim of the Grand Canyon and the surrounding plateau area.

Figure 3 In 1869, Major John Wesley Powell led the first geologic expedition down the Colorado River through the Grand Canyon in Arizona.

Library of Congress, Miscellaneous Items in High Demand, #LC-USZ62-20230

Major John Wesley Powell, a Civil War veteran, led the first geologic expedition down the Colorado River through the Grand Canyon in 1869 (Figure 3; see Chapter 4 Introduction). Following this first exploration, he led a second expedition in 1871, which included a photographer, a surveyor, and three topographers. This expedition made detailed topographic and geologic maps of the Grand Canyon area, as well as the first photographic record of the region.

The rocks of the Grand Canyon record more than 1 billion years of Earth history (Figure 4). The Vishnu Schist represents a major mountain-building episode that occurred during the Precambrian. Following erosion of this mountain range, sediments were deposited in a variety of marine, coastal, and terrestrial settings during the Paleozoic Era and were lithified into the rock formations that are now exposed in the canyon walls (Figure 2).

Although best known for its geology, Grand Canyon National Park is home to a wide variety of mammals, reptiles, amphibians, birds, fish, and more than 1,500 known plant species (Figure 5). Its great biological diversity can be attributed to the fact that within its boundaries are five of the seven life zones and three of the four desert types in North America.

Figure 4 The rocks of the Grand Canyon preserve more than 1 billion years of Earth history, including periods of mountain building and erosion, as well as periods of transgressions and regressions of shallow seas over the area.

Figure 5 White aspen and pinyon pine, shown here in the DeMotte campground, 28 km north of the North Rim, are the typical trees found in the Grand Canyon.

▶ Figure 10.5 **Cambrian Rocks of the Grand Canyon**

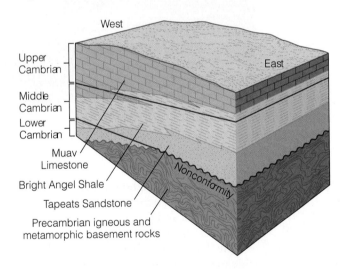

(a) Block diagram of Cambrian strata exposed in the Grand Canyon, illustrating the transgressive nature of the three formations.

(b) Block diagram of Cambrian rocks exposed along the Bright Angel Trail, Grand Canyon, Arizona.

The Sauk Sequence **199**

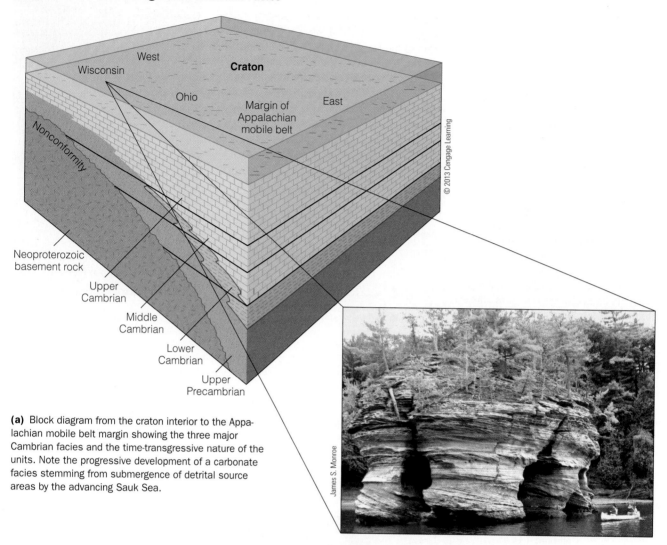

(a) Block diagram from the craton interior to the Appalachian mobile belt margin showing the three major Cambrian facies and the time-transgressive nature of the units. Note the progressive development of a carbonate facies stemming from submergence of detrital source areas by the advancing Sauk Sea.

(b) Outcrop of cross-bedded Upper Cambrian sandstone in the Dells area of Wisconsin.

The Tippecanoe Sequence

As the Sauk Sea regressed from the craton during the Early Ordovician, a landscape of low relief emerged. The exposed rocks were predominantly limestones and dolostones deposited earlier as part of the Sauk transgression. Because North America was still located in a tropical environment when the seas regressed, these carbonates experienced extensive erosion at that time (▶Figure 10.7). The resulting craton-wide unconformity thus marks the boundary between the Sauk and Tippecanoe sequences.

Just as in the Sauk Sequence, deposition of the **Tippecanoe Sequence** (Middle Ordovician–Early Devonian) began with a major transgression onto the craton. This transgressing sea deposited clean and well-sorted quartz sands over most of the craton. The best known of the Tippecanoe basal sandstones is the St. Peter Sandstone, an almost pure quartz sandstone used in manufacturing

glass. It occurs throughout much of the midcontinent and resulted from numerous cycles of weathering and erosion of Proterozoic and Cambrian sandstones deposited during the Sauk transgression (▶Figure 10.8).

The Tippecanoe basal sandstones were followed by widespread carbonate deposition (Figure 10.7). The limestones were generally the result of deposition by calcium carbonate–secreting organisms such as corals, brachiopods, stromatoporoids, and bryozoans. Besides the limestones, there were also many dolostones. Most of the dolostones formed as a result of magnesium replacing calcium in calcite, thus converting limestones into dolostones.

In the eastern portion of the craton, the carbonates grade laterally into shales. These shales mark the farthest extent of detrital sediments derived from weathering and erosion of the Taconic Highlands, which resulted from a tectonic event taking place in the Appalachian mobile belt, and which we will discuss later.

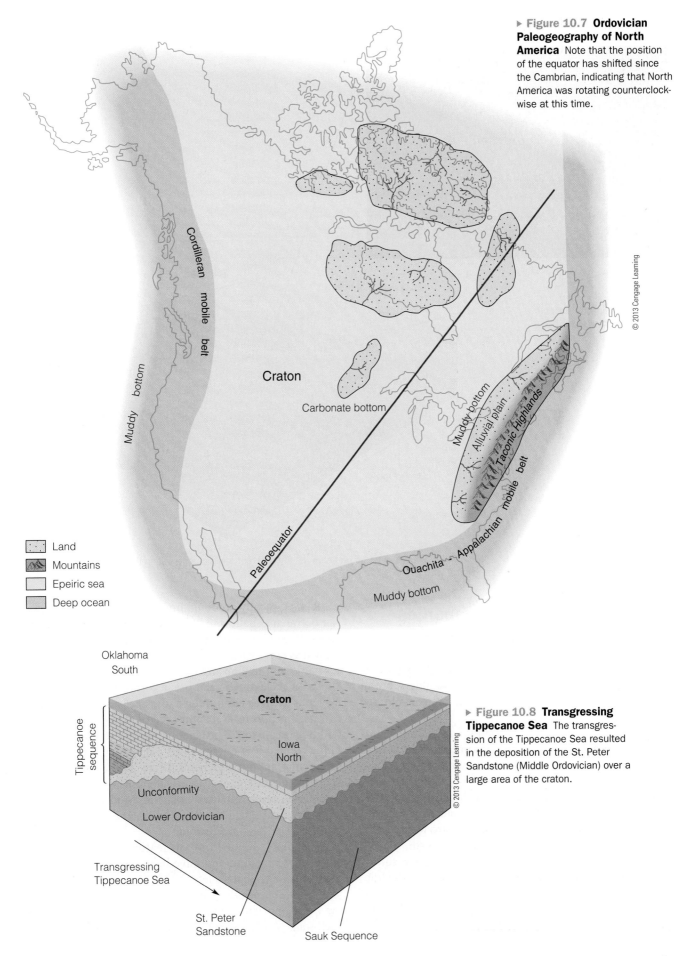

▶ Figure 10.7 **Ordovician Paleogeography of North America** Note that the position of the equator has shifted since the Cambrian, indicating that North America was rotating counterclockwise at this time.

© 2013 Cengage Learning

Legend:
- Land
- Mountains
- Epeiric sea
- Deep ocean

Map labels: Cordilleran mobile belt, Muddy bottom, Craton, Carbonate bottom, Paleoequator, Muddy bottom, Alluvial plain, Taconic Highlands, Appalachian mobile belt, Ouachita, Muddy bottom

▶ Figure 10.8 **Transgressing Tippecanoe Sea** The transgression of the Tippecanoe Sea resulted in the deposition of the St. Peter Sandstone (Middle Ordovician) over a large area of the craton.

© 2013 Cengage Learning

Block diagram labels: Oklahoma South, Craton, Iowa North, Tippecanoe sequence, Unconformity, Lower Ordovician, Transgressing Tippecanoe Sea, St. Peter Sandstone, Sauk Sequence

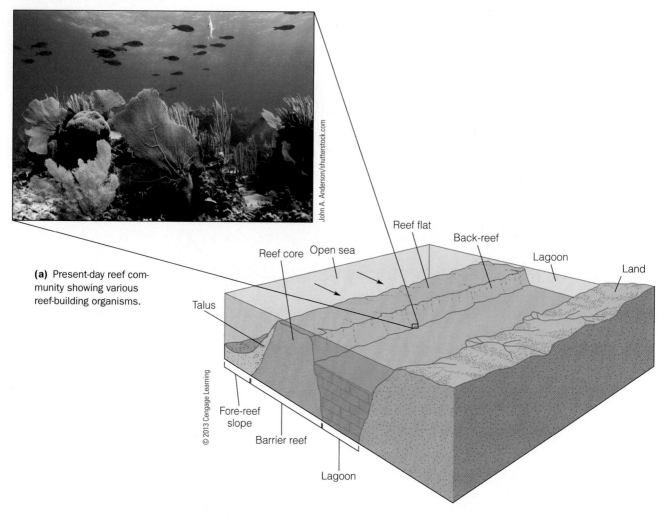

► Figure 10.9 **Organic Reefs**

(a) Present-day reef community showing various reef-building organisms.

John A. Anderson/shutterstock.com

© 2013 Cengage Learning

(b) Block diagram of a reef showing the different environments within the reef complex.

Tippecanoe Reefs and Evaporites

Organic reefs are limestone structures constructed by living organisms, some of which contribute skeletal materials to the reef framework (►Figure 10.9). Today, corals and calcareous algae are the most prominent reef builders, but in the geologic past other organisms played a major role (see Chapter 12).

Regardless of the organisms dominating reef communities, reefs appear to have occupied the same ecologic niche in the geologic past that they do today. Because of the ecologic requirements of reef-building organisms, present-day reefs are confined to a narrow latitudinal belt between approximately 30 degrees north and south of the equator. Corals, the major reef-building organisms today, require warm, clear, shallow water of normal salinity for optimal growth.

The size and shape of a reef are mostly the result of interactions among the reef-building organisms, the bottom topography, wind and wave action, and subsidence of the seafloor. Reefs also alter the area around them by forming barriers to water circulation or wave action.

Reefs typically are long, linear masses forming a barrier between a shallow platform on one side and a comparatively deep marine basin on the other side. Such reefs are known as *barrier reefs* and they create and maintain a steep seaward front that absorbs incoming wave energy (Figure 10.9). As skeletal material breaks off from the reef front, it accumulates as talus along a fore-reef slope. The barrier reef itself is porous and composed of many different reef-building organisms. The lagoon area on the landward side of the reef is a low-energy, quiet-water zone where fragile, sediment-trapping organisms thrive. The lagoon area can also become the site of evaporite deposits when circulation to the open sea is restricted or cut off. Modern examples of barrier reefs are the Florida Keys, Bahama Islands, and the Great Barrier Reef of Australia.

Reefs have been common features in low latitudes since the Cambrian and have been built by a variety of organisms. The first skeletal builders of reeflike structures were *archaeocyathids*. These conical organisms lived during the Cambrian and had double, perforated, calcareous shell walls. Archaeocyathids built small mounds that have

Land

Mountains

Barrier reefs

Evaporites

Epeiric sea

Deep ocean

▶ Figure 10.10 **Silurian Paleogeography of North America** Note the development of reefs in the Michigan, Ohio, and Indiana–Illinois–Kentucky areas.

been found on all continents except South America (see Figure 12.6).

Beginning in the Middle Ordovician, reefs made up of stromatoporoids and corals became common in the low latitudes, and similar reefs remained so throughout the rest of the Phanerozoic Eon. The burst of reef building seen in the Late Ordovician through Devonian probably occurred in response to evolutionary changes triggered by the appearance of extensive carbonate seafloors and platforms beyond the influence of detrital sediments.

The Middle Silurian rocks (Tippecanoe Sequence) of the present-day Great Lakes region are world famous for their reef and evaporite deposits (▶ Figure 10.10). The best-known structure in the region, the Michigan Basin, is a broad, circular basin surrounded by large barrier reefs. No doubt these reefs contributed to increasingly restricted circulation and the precipitation of Upper Silurian evaporites within the basin (▶ Figure 10.11).

Within the rapidly subsiding interior of the basin, other types of reefs are found. *Pinnacle reefs* are tall, spindly structures up to 100 m high. They reflect the rapid upward growth needed to maintain themselves near sea level during subsidence of the basin (Figure 10.11a). Besides the pinnacle reefs, bedded carbonates and thick sequences of rock salt and rock anhydrite are also found in the Michigan Basin (Figure 10.11).

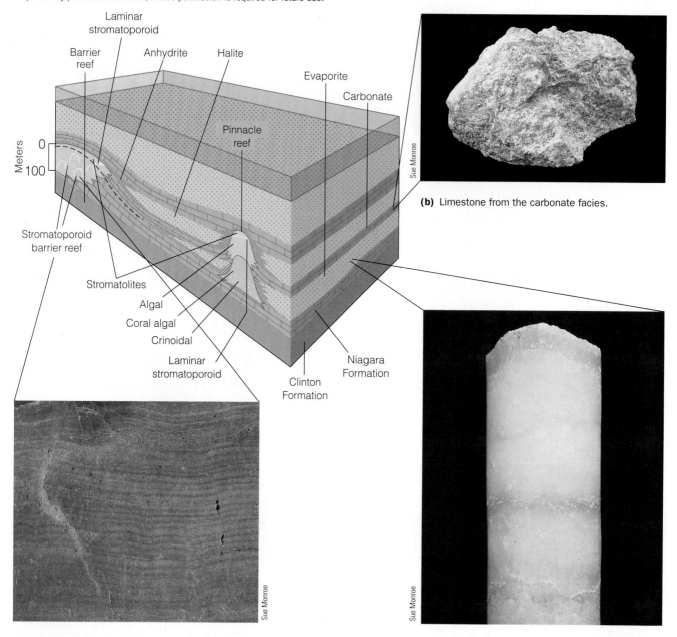

▶ Figure 10.11 **The Michigan Basin**

(a) Generalized block diagram of the northern Michigan Basin during the Silurian Period. From K. J. Mesolella, J. D. Robinson, L. M. McCormick and A. R. Ormiston, "Cyclic Deposition of Silurian Carbonates and Evaporates in Michigan Basin," AAPG Bulletin, Vol. 58, No. 1 (Fig. 6, p. 40). Copyright © 1958 AAPG. Reprinted by permission of AAPG, whose permission is required for future use.

(b) Limestone from the carbonate facies.

(c) Cross section of a stromatoporoid colony from the stromatoporoid barrier reef facies.

(d) Core of rock salt from the evaporite facies.

As the Tippecanoe Sea gradually regressed from the craton during the Late Silurian, precipitation of evaporite minerals occurred in the Appalachian, Ohio, and Michigan basins (Figure 10.1). In the Michigan Basin alone, approximately 1,500 m of sediments were deposited, nearly half of which are halite and anhydrite. How did such thick sequences of evaporites accumulate? One possibility is that when sea level dropped, the tops of the barrier reefs were as high as, or above sea level, thus preventing the influx of new seawater into the basin. Evaporation of the basinal seawater would result in the formation of brine, and as the brine became increasingly concentrated, the precipitation

of salts would occur. A second possibility is that the reefs grew upward so close to sea level that they formed a sill, or barrier, that eliminated interior circulation and allowed for the evaporation of the seawater that produced a dense brine, eventually resulting in evaporite deposits (▶ Figure 10.12).

With North America still near the equator during the Silurian Period (Figure 10.2c), temperatures were probably high. As circulation to the Michigan Basin was restricted or ceased altogether, seawater within the basin evaporated and began forming brine. Because the brine was heavy, it concentrated near the bottom, with minerals precipitating on the basin floor and forming evaporite deposits. When

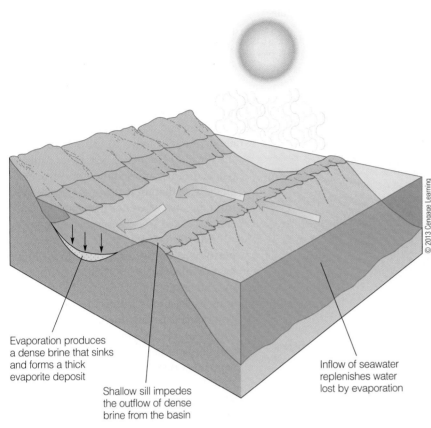

Evaporation produces
a dense brine that sinks
and forms a thick
evaporite deposit

Shallow sill impedes
the outflow of dense
brine from the basin

Inflow of seawater
replenishes water
lost by evaporation

© 2013 Cengage Learning

▶ Figure 10.12 **Evaporite Sedimentation** Silled basin model for evaporite sedimentation by direct precipitation from seawater. Vertical scale is greatly exaggerated.

growth followed by evaporite deposition, to alternation of reef growth and evaporite deposition. Although the Michigan Basin has been studied extensively for years, no model yet proposed completely explains the genesis and relationship of its various reef, carbonate, and evaporite facies.

The End of the Tippecanoe Sequence

By the Early Devonian, the regressing Tippecanoe Sea had retreated to the craton margin, exposing an extensive lowland topography. During this regression, marine deposition was initially restricted to a few interconnected cratonic basins and, finally, by the end of the Tippecanoe Sequence, to only the mobile belts surrounding the craton.

As the Tippecanoe Sea regressed during the Early Devonian, the craton experienced mild deformation, forming many domes, arches, and basins (Figure 10.1). These structures were mostly eroded during the time the craton was exposed, and deposits from the ensuing and encroaching Kaskaskia Sea eventually covered them.

seawater flowed back into the Michigan Basin, either over the sill or through channels cut in the barrier reefs, this replenishment added new seawater, allowing the process of brine formation and precipitation of evaporites to repeat itself.

The order and type of salts precipitating from seawater depend on their solubility, the original concentration of seawater, and local conditions of the basin. In general, salts precipitate in a sequential order, beginning with the least soluble and ending with the most soluble. Therefore, calcium carbonate usually precipitates first, followed by gypsum,* and lastly halite. Many lateral shifts and interfingering of the limestone, anhydrite, and halite facies may occur, however, because of variations in the amount of seawater entering the basin and changing geologic conditions.

Thus, the periodic evaporation of seawater as just discussed could account for the observed vertical and lateral distribution of evaporites in the Michigan Basin. Associated with those evaporites, however, are pinnacle reefs, and the organisms constructing those reefs could not have lived in such a highly saline environment (Figure 10.11). How, then, can such contradictory features be explained? Numerous models have been proposed, ranging from cessation of reef

---

*Recall from Chapter 6 that gypsum ($CaSO_4 \cdot 2H_2O$) is the common sulfate precipitated from seawater, but when deeply buried, gypsum loses its water and is converted to anhydrite ($CaSO_4$).

The Appalachian Mobile Belt and the Taconic Orogeny

Having examined the Sauk and Tippecanoe geologic history of the craton, we now turn our attention to the Appalachian mobile belt, where the first Phanerozoic orogeny began during the Middle Ordovician. The mountain building occurring during the Paleozoic Era had a profound influence on the climate and sedimentary history of the craton. In addition, it was part of the global tectonic regime that sutured the continents together, forming Pangaea by the end of the Paleozoic.

Throughout Sauk time (Neoproterozoic–Early Ordovician), the Appalachian region was a broad, passive, continental margin. Sedimentation was closely balanced by subsidence as extensive carbonate deposits succeeded thick, shallow marine sands. During this time, movement along a divergent plate boundary was widening the **Iapetus Ocean** (▶Figure 10.13a). Beginning with the subduction of the Iapetus plate beneath Laurentia (an oceanic–continental convergent plate boundary), the Appalachian mobile belt was born (Figure 10.13b). The resulting **Taconic orogeny**—named after the present-day Taconic Mountains of eastern New York, central Massachusetts, and Vermont—was the first of several orogenies to affect the Appalachian region.

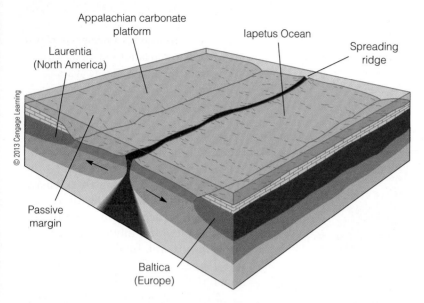

(a) During the Neoproterozoic to the Early Ordovician, the Iapetus Ocean was opening along a divergent plate boundary. Both the east coast of Laurentia and the west coast of Baltica were passive continental margins with large carbonate platforms.

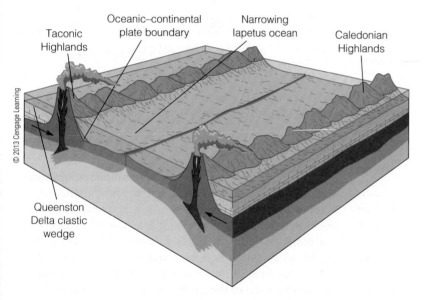

(b) Beginning in the Middle Ordovician, the passive margins of Laurentia and Baltica changed to active oceanic–continental plate boundaries, resulting in orogenic activity (Figure 10.2b).

The Appalachian mobile belt can be divided into two depositional environments. The first is the extensive, shallow-water carbonate platform that formed the broad eastern continental shelf and stretched from Newfoundland to Alabama (Figure 10.13a). It formed during the transgression of the Sauk Sea onto the craton when carbonates were deposited in a vast, shallow sea. Stromatolites, mud cracks, and other sedimentary structures and fossils are evidence of the shallow water depth on the platform.

Carbonate deposition ceased along the east coast during the Middle Ordovician and was replaced by deepwater deposits characterized by thinly bedded black shales, graded beds, coarse sandstones, graywackes, and associated volcanic material. This suite of sediments marks the onset of mountain building, in this case, the Taconic orogeny. The subduction of the Iapetus plate beneath Laurentia resulted in volcanism and downwarping of the carbonate platform (Figure 10.13b). Throughout the Appalachian mobile belt, facies patterns, paleocurrents, and sedimentary structures all indicate that these deposits were derived from the east, where the Taconic Highlands and associated volcanoes were rising.

Additional structural, stratigraphic, petrologic, and sedimentologic evidence has provided much information on the timing and origin of this orogeny. For example, at many locations within the Taconic belt, pronounced angular unconformities occur where steeply dipping Lower Ordovician rocks are overlain by gently dipping or horizontal Silurian and younger rocks.

Other evidence includes volcanic activity in the form of deep-sea lava flows, volcanic ash layers, and intrusive bodies in the area from present-day Georgia to Newfoundland. These igneous rocks show a clustering of radiometric ages corresponding to the Middle to Late Ordovician. In addition, regional metamorphism coincides with the radiometric dates.

The final piece of evidence for the Taconic orogeny is the development of a large **clastic wedge**, an extensive accumulation of mostly detrital sediments deposited adjacent to an uplifted area. These deposits are thickest and coarsest nearest the highland area and become thinner and finer grained away from the source area, eventually grading into the carbonate cratonic facies (▶Figure 10.14). The clastic wedge resulting from the erosion of the Taconic Highlands is referred to as the **Queenston Delta**. Careful mapping and correlation of its deposits indicate that more than 600,000 km³ of rock were eroded from the Taconic Highlands. Based on this figure, geologists estimate that the Taconic Highlands were at least 4,000 m high.

The Taconic orogeny marked the first pulse of mountain building in the Appalachian mobile belt and was a response to the subduction taking place beneath the east coast of Laurentia. As the Iapetus Ocean narrowed and closed, another orogeny occurred in Europe during the Silurian. The *Caledonian orogeny* was essentially a mirror image of the Taconic and the Acadian orogenies (see Chapter 11), and was part of the global mountain-building episode that occurred during the Paleozoic Era (Figure 10.2c). Even

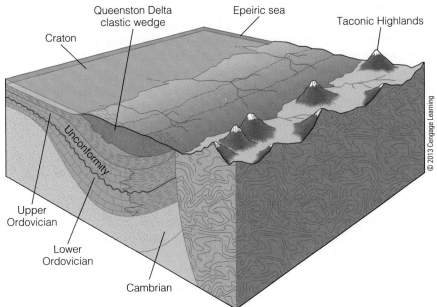

Craton
Queenston Delta clastic wedge
Epeiric sea
Taconic Highlands
Unconformity
Upper Ordovician
Lower Ordovician
Cambrian

© 2013 Cengage Learning

▶ Figure 10.14 **Reconstruction of the Taconic Highlands and Queenston Delta Clastic Wedge** The Queenston Delta clastic wedge, resulting from the erosion of the Taconic Highlands, consists of thick, coarse-grained detrital sediments nearest the highlands and thins laterally into more finely grained sediments in the epeiric seas covering the craton.

though the Caledonian orogeny occurred during the Tippecanoe Sequence, we discuss it in the next chapter because it was intimately related to the Devonian Acadian orogeny.

Early Paleozoic Mineral Resources

Early Paleozoic-age rocks contain a variety of important mineral resources, including sand and gravel for construction, building stone, and limestone used in the manufacture of cement. Important sources of industrial, or silica, sand are the Upper Cambrian Jordan Sandstone of Minnesota and Wisconsin, the Lower Silurian Tuscarora Sandstone in Pennsylvania and Virginia, and the Middle Ordovician St. Peter Sandstone. The latter, the basal sandstone of the Tippecanoe Sequence (Figure 10.8), occurs in several states, but the best-known area of production is in La Salle County, Illinois.

Silica sand has several uses, including the manufacture of glass, refractory bricks for blast furnaces, and molds for casting iron, aluminum, and copper alloys. Some silica

sands, called *hydraulic fracturing sands*, are pumped into wells to fracture oil- or gas-bearing rocks and provide permeable passageways for the oil or gas to migrate to the well.

Thick deposits of Silurian evaporites, mostly rock salt (NaCl) and rock gypsum ($CaSO_4 \cdot 2H_2O$) altered to rock anhydrite ($CaSO_4$), underlie parts of Michigan, Ohio, New York, and adjacent areas in Ontario, Canada. These rocks are important sources of various salts. In addition, barrier and pinnacle reefs in carbonate rocks associated with these evaporites are the reservoirs for oil and gas in Michigan and Ohio.

The host rocks for deposits of lead and zinc in southeast Missouri are Cambrian dolostones, although some Ordovician rocks contain these metals as well. These deposits have been mined since 1720 but have been largely depleted. Now most lead and zinc mined in Missouri comes from Mississippian-age sedimentary rocks.

The Silurian Clinton Formation crops out from Alabama north to New York, and equivalent rocks are found in Newfoundland. This formation has been mined for iron in many places. In the United States, the richest ores and most extensive mining occurred near Birmingham, Alabama, but only a small amount of ore is currently produced in that area.

Summary

Table 10.1 summarizes the Early Paleozoic geologic history of the North American craton and mobile belts, as well as global events, sea level changes, and major evolutionary events.

▶ Most continents consist of two major components: a relatively stable craton over which epeiric seas transgressed and regressed, surrounded by mobile belts in which mountain building took place.

▶ Six major continents and numerous microcontinents and island arcs existed at the beginning of the Paleozoic Era; all of these were dispersed around the globe at low latitudes during the Cambrian.

▶ During the Ordovician and Silurian, plate movements resulted in a changing global geography. Gondwana moved southward and began to cross the South Pole as indicated by Upper Ordovician tillite deposits; the microcontinent Avalonia separated from Gondwana during the Early Ordovician and collided with Baltica during the Late Ordovician–Early Silurian; Baltica, along with the newly attached Avalonia, moved northwestward relative to Laurentia and collided with it to form Laurasia during the Silurian.

▶ Geologists divide the geologic history of North America into cratonic sequences that formed as a result of craton-wide transgressions and regressions.

- The Sauk Sequence began with a major marine transgression onto the craton. At its maximum, the Sauk Sea covered the craton except for parts of the Canadian shield and the Transcontinental Arch, a series of large, northeast–southwest trending islands.
- The Tippecanoe Sequence began with deposition of an extensive sand unit over the exposed and eroded Sauk landscape, and was followed by extensive carbonate deposition. In addition, large barrier reefs surrounded many cratonic basins, resulting in evaporite deposition within these basins.
- The eastern edge of North America was a stable carbonate platform during the Sauk Sequence. During the Tippecanoe Sequence, an oceanic–continental convergent plate boundary formed, resulting in the Taconic orogeny, the first of three major orogenies to affect the Appalachian mobile belt.
- The newly formed Taconic Highlands shed sediments into the western epeiric sea, producing a clastic wedge that geologists call the Queenston Delta.
- Early Paleozoic-age rocks contain a variety of mineral resources, including building stone, limestone for cement, silica sand, hydrocarbons, evaporites, and iron ore.

Important Terms

Appalachian mobile belt, p. 192
Baltica, p. 193
China, p. 193
clastic wedge, p. 206
Cordilleran mobile belt, p. 192
craton, p. 192
cratonic sequence, p. 195
epeiric sea, p. 192

Gondwana, p. 193
Iapetus Ocean, p. 205
Kazakhstania, p. 193
Laurentia, p. 193
mobile belt, p. 192
organic reef, p. 202
Ouachita mobile belt, p. 192
Queenston Delta, p. 206

Sauk Sequence, p. 196
sequence stratigraphy, p. 196
Siberia, p. 193
Taconic orogeny, p. 205
Tippecanoe Sequence, p. 200
Transcontinental Arch, p. 196

Review Questions

1. The Taconic orogeny resulted from what type of plate boundary activity?
   a. ____ Continental–continental convergent; b. ____ Transform; c. ____ Oceanic–oceanic convergent; d. ____ Divergent; e. ____ Oceanic–continental convergent.
2. The relatively stable and immobile parts of continents, which form the foundation on which Phanerozoic sediments were deposited, make up the
   a. ____ shield; b. ____ platform; c. ____ mobile belt; d. ____ craton; e. ____ none of the previous answers.
3. A major transgressive–regressive cycle bounded by craton-wide unconformities is a(n)
   a. ____ biostratigraphic unit; b. ____ cratonic sequence; c. ____ orogeny; d. ____ shallow sea; e. ____ cyclothem.
4. Evaporites and reef carbonates were the predominant cratonic rocks during which sequence?
   a. ____ Kaskaskia; b. ____ Zuni; c. ____ Sauk; d. ____ Absaroka; e. ____ Tippecanoe.
5. Which was the first major transgressive sequence onto the North American craton?
   a. ____ Absaroka; b. ____ Sauk; c. ____ Zuni; d. ____ Kaskaskia; e. ____ Tippecanoe.
6. The ocean separating Laurentia from Baltica is called the
   a. ____ Panthalassa; b. ____ Tethys; c. ____ Iapetus; d. ____ Atlantis; e. ____ Perunica.
7. An elongated area marking the site of mountain building is a
   a. ____ cyclothem; b. ____ mobile belt; c. ____ platform; d. ____ shield; e. ____ craton.
8. Which mobile belt is located along the western side of North America?
   a. ____ Franklin; b. ____ Cordilleran; c. ____ Ouachita; d. ____ Appalachian; e. ____ Answers a and b.
9. Besides the Canadian shield, the only area above sea level during deposition of the Sauk Sequence was the
   a. ____ cratonic margin; b. ____ Transcontinental Arch; c. ____ Queenston Delta; d. ____ Appalachian mobile belt; e. ____ Taconic Highlands.
10. The eastern margin of Laurentia changed from a passive plate margin to an active plate margin during which sequence?
    a. ____ Zuni; b. ____ Tippecanoe; c. ____ Sauk; d. ____ Kaskaskia; e. ____ Absaroka.
11. Which of the following was an Early Paleozoic microcontinent?
    a. ____ Avalonia; b. ____ Baltica; c. ____ China; d. ____ Laurentia; e. ____ Gondwana.
12. The Queenston Delta clastic wedge resulted from the erosion of which highlands?
    a. ____ Caledonian; b. ____ Acadian; c. ____ Sevier; d. ____ Taconic; e. ____ Transcontinental Arch.
13. How many major continents were there at the beginning of the Cambrian?
    a. ____ Three; b. ____ Four; c. ____ Five; d. ____ Six; e. ____ Seven.
14. What are some methods geologists can use to determine the locations of continents during the Paleozoic Era?
15. Discuss why cratonic sequences are a convenient way to study the geologic history of the Paleozoic Era.
16. Discuss how the Cambrian rocks of the Grand Canyon illustrate the sedimentation patterns of a transgressive sea.
17. Discuss how sequence stratigraphy can be used to make global correlations, and explain why it is so useful in reconstructing past events.
18. What is the economic benefit to the automobile industry in having Paleozoic silica sand deposits near Toledo, Ohio?

19. What evidence in the geologic record indicates that the Taconic orogeny occurred?
20. According to estimates made from mapping and correlation, the Queenston Delta contains more than 600,000 km³ of rock eroded from the Taconic Highlands. Based on this figure, geologists estimate the Taconic Highlands were at least 4,000 m high. They also estimate that the Catskill Delta (see Chapter 11) contains three times as much sediment as the Queenston Delta. From what you know about the geographic distribution of the Taconic Highlands and the Acadian Highlands (see Chapter 11), can you estimate how high the Acadian Highlands might have been?

TABLE 10.1 Summary of Early Paleozoic Geologic and Evolutionary Events

Age (millions of years)	Geologic Period	Sequence	Relative Changes in Sea Level (Rising / Falling)		Cordilleran Mobile Belt	Craton	Ouachita Mobile Belt	Appalachian Mobile Belt	Major Events Outside North America	Major Evolutionary Events
416	Silurian	Tippecanoe				Extensive barrier reefs and evaporites common		Acadian orogeny	Caledonian orogeny	First jawed fish evolve
										Early land plants—seedless vascular plants
444	Ordovician		Present sea level			Queenston Delta clastic wedge		Taconic orogeny	Continental glaciation in Southern Hemisphere	Extinction of many marine invertebrates near end of Ordovician
						Transgression of Tippecanoe Sea				Plants move to land?
						Regression exposing large areas to erosion				Major adaptive radiation of all invertebrate groups
488	Cambrian	Sauk				Canadian shield and Transcontinental Arch only areas above sea level				Many trilobites become extinct near end of Cambrian
										Earliest vertebrates—jawless fish evolve
542						Transgression of Sauk Sea				

© 2013 Cengage Learning

11

Late Paleozoic Earth History

An aerial view of an open-pit coal mine northeast of Somerset, Pennsylvania. The coal being mined is Pennsylvanian in age and formed as a result of tremendous accumulations of plant material deposited in a swampy environment as part of a cyclothem, or a repetitive sequence of marine and nonmarine cycles of deposition. Notice the dragline (digging machine) on top of the coal deposit for scale.

© H. Mark Weidman Photography/Alamy

At the end of this chapter, you will have learned that

► Movement of the six major continents during the Paleozoic Era resulted in the formation of the supercontinent Pangaea at the end of the Paleozoic.

► In addition to the large-scale plate interactions during the Paleozoic, microplate and terrane activity also played an important role in forming Pangaea.

► Most of the Kaskaskia Sequence is dominated by carbonates and associated evaporites.

► Transgressions and regressions over the low-lying craton during the Absaroka Sequence resulted in cyclothems and the formation of coals.

► During the Late Paleozoic Era, mountain-building activity took place in the Appalachian, Ouachita, and Cordilleran mobile belts.

► The Caledonian, Acadian, and Hercynian (Variscan)–Alleghenian orogenies were all part of the global tectonic activity resulting from the assembly of Pangaea.

► Late Paleozoic-age rocks contain a variety of mineral resources, including petroleum, coal, evaporites, and various metallic deposits.

Introduction

Coal. The name conjures up various images. For some, it is what helped fuel the Industrial Revolution in England during the 18th and 19th centuries. For others, it brings to mind the many hazards associated with underground coal mining, and the denudation of the landscape caused by open-pit surface mining. And, more recently, coal has been closely linked with global warming as a result of its use in industry and power plants.

Coal is a biochemical sedimentary rock composed of the partially altered, compressed remains of land plants (see Figure 2.16b). It forms in anaerobic swamps and bogs, where the decomposition of the plant matter is interrupted and the accumulating vegetation is altered into peat. If the peat is later covered and buried by sediments, it is then converted into coal.

Much of the world's coal formed during the Late Carboniferous (Pennsylvanian Period), when widespread swampy conditions provided the ideal conditions for coal formation. It is the coal mined from these Carboniferous deposits that has fueled the economies of many of the world's nations. But, coal did not become an important comodity until it was needed in quantity to fuel the steam engines and furnaces of factories during the Industrial Revolution. Recall that William Smith, who published the world's first geologic map, got his start mapping various coal mines, after which he turned his attention to finding the most efficient canal routes for bringing coal to market. In fact, coal production in England increased from 2.7 million tons in 1700 to 50 million tons by 1850 as a direct result of the Industrial Revolution.

Coal is primarily used today to generate electricity. In the United States, coal-fired power plants provide approximately 50 percent of the electricity generated. Other uses of coal are for iron and steel production; liquefaction, in which it is converted into liquid fuels such as diesel or gasoline; and home heating, as well as any type of industrial application requiring large amounts of heat, such as in cement manufacturing.

Coal is generally mined in two ways, and the method used depends on the geology of the coal deposit. In underground mining, shafts are drilled into the *coal seam* (a layer or bed of coal), and the seam is then followed until as much of the coal that can be extracted is removed. Underground mining currently accounts for about 60 percent of the world's coal production.

When the coal is close to the surface, and of widespread extent, then open-pit, or surface mining is used. This method involves removing the surface rocks and exposing the coal seam, where it can be drilled or blasted and loaded onto trucks or conveyers (see chapter opening photo). One type of surface mining, called mountaintop-removal mining, literally involves the removal of the top of a mountain and filling in the adjacent valleys with the overburden. This method of mining is highly controversial because of the damage done to the surrounding environment as a result of the denudation of the surface.

Currently, world production of coal amounts to more than 6,185 million tons of hard coal, that is, bituminous grade or higher. The top producer of coal and one of its largest consumers is China, which, according to 2011 statistics, mined 3,162 million tons of hard coal. The United States is next, having mined 932 million tons, and India, Australia, and South Africa follow in turn. As impressive as these numbers are, geologists estimate that, based on current usage, the United States has an approximately 300-year supply of coal.

Despite its abundance and contribution to the world's energy supply, coal has disadvantages. The extraction and burning of coal releases many heavy metals into the environment that are harmful to human health. Furthermore, the burning of coal, especially in power plants, releases tremendous quantities of carbon dioxide, which cause climate change and global warming.

Late Paleozoic Paleogeography

The Late Paleozoic was a time marked by continental collisions, mountain building, fluctuating sea levels, and varied climates. Coals, evaporites, and tillites testify to the variety of climatic conditions experienced by the different continents during the Late Paleozoic. Major glacial and interglacial episodes took place over much of Gondwana as it continued moving over the South Pole during the Late Mississippian to Early Permian.

The growth and retreat of continental glaciers during this time profoundly affected the world's biota as well as contributing to global sea-level changes. Collisions between continents not only led to the formation of the supercontinent Pangaea by the end of the Permian, but also resulted in mountain building that strongly influenced oceanic and atmospheric circulation patterns. By the end of the Paleozoic, widespread arid and semiarid conditions governed much of Pangaea.

The Devonian Period

Recall from Chapter 10 that Baltica, which had earlier united with Avalonia, collided along a convergent plate boundary during the Silurian to form the larger continent of **Laurasia**. This collision, which closed the northern Iapetus Ocean, is marked by the Caledonian orogeny.

During the Devonian, as the southern Iapetus Ocean narrowed between Laurasia and Gondwana, mountain building continued along the eastern margin of Laurasia as a result of the Acadian orogeny (▶ Figure 11.1a). Erosion of the ensuing highlands spread vast amounts of reddish fluvial sediments over large areas of northern Europe (Old Red Sandstone) and eastern North America (the Catskill Delta).

Other Devonian tectonic events, probably related to the collision of Laurentia and Baltica, include the Cordilleran Antler orogeny, the Ellesmere orogeny along the northern margin of Laurentia (which may reflect the collision of Laurentia with Siberia), and the change from a passive continental margin to an active convergent plate boundary in the Uralian mobile belt of eastern Baltica. The distribution of reefs, evaporites, and red beds, as well as the existence of similar floras throughout the world, suggest a rather uniform global climate for the Devonian Period.

The Carboniferous Period

During the Carboniferous Period, southern Gondwana moved over the South Pole, resulting in extensive continental glaciation (Figure 11.1b and ▶ Figure 11.2a). The advance and retreat of these glaciers produced global changes in sea level that affected sedimentation patterns on the cratons. As Gondwana moved northward, it began colliding with Laurasia during the Early Carboniferous and continued suturing with it during the rest of the Carboniferous (Figures 11.1b and 11.2a). Because Gondwana rotated clockwise relative to Laurasia, deformation generally progressed in a northeast-to-southwest direction along the Hercynian, Appalachian, and Ouachita mobile belts of the two continents. The final phase of collision between Gondwana and Laurasia is indicated by the Ouachita Mountains of Oklahoma and Arkansas, formed by thrusting during the Late Carboniferous and Early Permian.

Elsewhere, Siberia collided with Kazakhstania and moved toward the Uralian margin of Laurasia (Baltica), colliding with it during the Early Permian. By the end of the Carboniferous, the various continental landmasses were fairly close together as Pangaea began taking shape.

The Carboniferous coal basins of eastern North America, western Europe, and the Donets Basin of the Ukraine all lay in the equatorial zone, where rainfall was high and temperatures were consistently warm. The absence of strong seasonal growth rings in fossil plants from these coal basins indicates such a climate. The fossil plants found in the coals of Siberia, however, show well-developed growth rings, signifying seasonal growth with abundant rainfall and distinct seasons such as in the temperate zones (latitudes 40 degrees to 60 degrees north).

Glacial conditions and the movement of large continental ice sheets in the high southern latitudes are indicated by widespread tillites and glacial striations in southern Gondwana (see Figure 3.5). These ice sheets spread toward the equator and, at their maximum growth, extended well into the middle temperate latitudes.

The Permian Period

The assembly of Pangaea was essentially completed during the Permian as a result of the many continental collisions that began during the Carboniferous (Figure 11.2b). Although geologists generally agree on the configuration and location of the western half of the supercontinent, there is no consensus on the number or configuration of the various terranes and continental blocks that composed the eastern half of Pangaea. Regardless of the exact configuration of the eastern portion, geologists know that the supercontinent was surrounded by various subduction zones and moved steadily northward during the Permian. Furthermore, an enormous single ocean, **Panthalassa**, surrounded Pangaea and spanned Earth from pole to pole (Figure 11.2b). Waters of this ocean probably circulated more freely than at present, resulting in more equable water temperatures.

The formation of a single large landmass had climatic consequences for the terrestrial environment as well. Terrestrial Permian sediments indicate that arid and

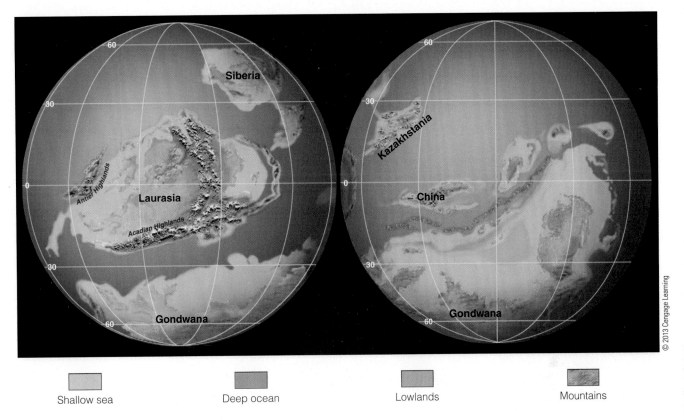

© 2013 Cengage Learning

Shallow sea	Deep ocean	Lowlands	Mountains

(a) Late Devonian Period.

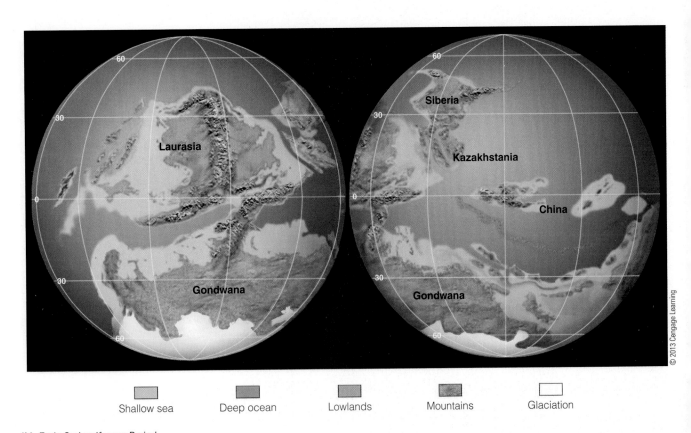

© 2013 Cengage Learning

Shallow sea	Deep ocean	Lowlands	Mountains	Glaciation

(b) Early Carboniferous Period.

(a) Late Carboniferous Period.

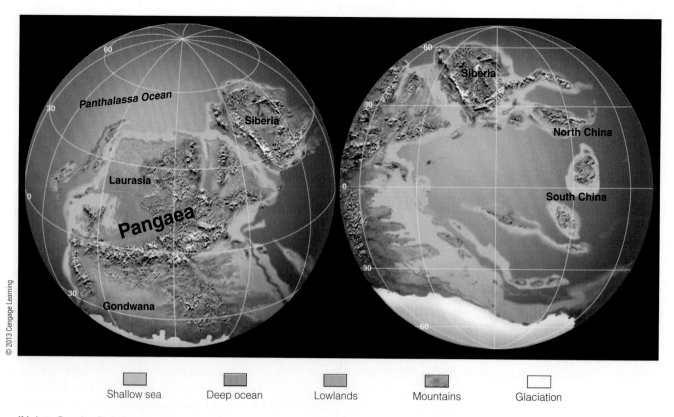

(b) Late Permian Period.

semiarid conditions were widespread over Pangaea. The mountain ranges produced by the Hercynian (Variscan)–Alleghenian and Ouachita orogenies were high enough to create rain shadows that blocked the moist, subtropical, easterly winds—much as the southern Andes Mountains do in western South America today. This produced very dry conditions in North America and Europe, as evident from the extensive Permian red beds and evaporites found in western North America, central Europe, and parts of Russia. Permian coals, indicating abundant rainfall, were mostly limited to the northern temperate belts (latitude 40 degrees to 60 degrees north), whereas the last remnants of the Carboniferous ice sheets continued to recede.

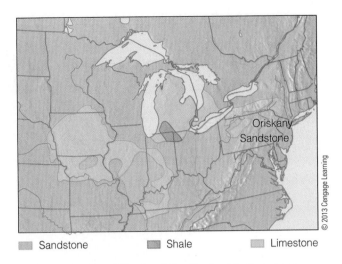

Sandstone   Shale   Limestone

▶ Figure 11.3 **Basal Rocks of the Kaskaskia Sequence**
Extent of the basal rocks of the Kaskaskia Sequence in the eastern and north–central United States.

Late Paleozoic Evolution of North America

The Late Paleozoic cratonic history of North America included periods of extensive shallow-marine carbonate deposition and large coal-forming swamps as well as dry, evaporite-forming, terrestrial conditions. Cratonic events largely resulted from sea-level changes caused by Gondwanan glaciation and tectonic events related to the assemblage of Pangaea.

Mountain building that began with the Ordovician Taconic orogeny continued with the Caledonian, Acadian, Alleghenian, and Ouachita orogenies. These orogenies were part of the global tectonic process that resulted in the formation of Pangaea by the end of the Paleozoic Era.

The Kaskaskia Sequence

The boundary between the Tippecanoe Sequence and the overlying **Kaskaskia Sequence** (Middle Devonian–Late Mississippian) is marked by a major unconformity. As the Kaskaskia Sea transgressed over the low-relief landscape of the craton, most basal bed deposits consisted of clean, well-sorted quartz sandstones. A good example is the Oriskany Sandstone of New York and Pennsylvania and its lateral equivalents (▶Figure 11.3). The Oriskany Sandstone, like the basal Tippecanoe St. Peter Sandstone, is an important glass sand as well as a good gas-reservoir rock.

The source areas for the basal Kaskaskia sandstones were primarily the eroding highlands of the Appalachian mobile belt area (▶Figure 11.4), exhumed Cambrian and Ordovician sandstones cropping out along the flanks of the Ozark Dome, and exposures of the Canadian shield in the Wisconsin area. The lack of similar sands in the Silurian carbonate beds below the Tippecanoe–Kaskaskia unconformity indicates that the source areas of the basal

Kaskaskia detrital rocks were submerged when sediments of the Tippecanoe Sequence were deposited. Stratigraphic studies indicate that these source areas were uplifted and the Tippecanoe carbonates removed by erosion before the Kaskaskia transgression.

Kaskaskian basal rocks elsewhere on the craton consist of carbonates that are frequently difficult to differentiate from the underlying Tippecanoe carbonates unless they are fossiliferous.

Except for widespread Upper Devonian and Lower Mississippian black shales, the majority of Kaskaskian rocks are carbonates, including reefs, and associated evaporite deposits. In many other parts of the world, such as southern England, Belgium, central Europe, Australia, and Russia, the Middle and early Late Devonian epochs were times of major reef building.

Reef Development in Western Canada

The Middle and Late Devonian reefs of western Canada contain large reserves of petroleum and have been widely studied from outcrops and in the subsurface (▶Figure 11.5). These reefs began forming as the Kaskaskia Sea transgressed southward into western Canada. By the end of the Middle Devonian, they had coalesced into a large barrier reef system that restricted the flow of oceanic water into the back-reef platform, creating conditions for evaporite precipitation (Figure 11.5). In the back-reef area, up to 300 m of evaporites precipitated in much the same way as in the Michigan Basin during the Silurian (see Figure 10.11). More than half the world's potash, which is used in fertilizers, comes from these Devonian evaporites. By the middle of the Late Devonian, reef growth had stopped in the western Canada region, although nonreef carbonate deposition continued.

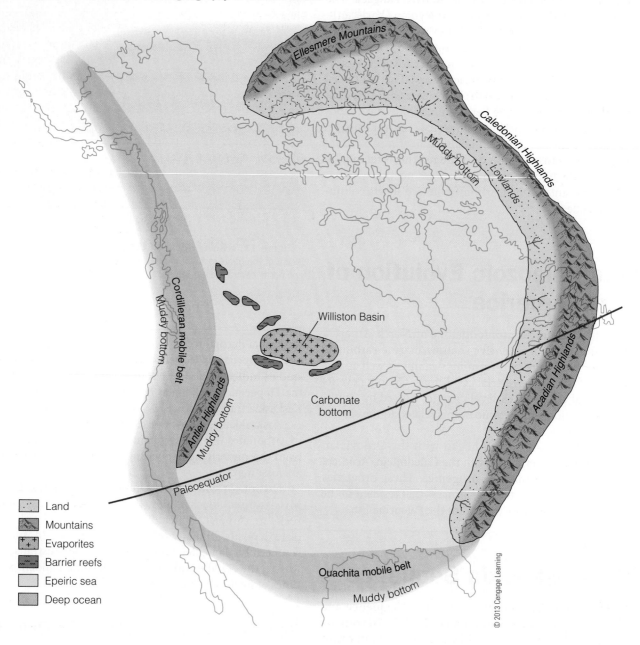

Ellesmere Mountains

Caledonian Highlands

Muddy bottom

Lowlands

Cordilleran mobile belt

Muddy bottom

Williston Basin

Acadian Highlands

Antler Highlands

Muddy bottom

Carbonate bottom

Paleoequator

- Land
- Mountains
- Evaporites
- Barrier reefs
- Epeiric sea
- Deep ocean

Ouachita mobile belt

Muddy bottom

© 2013 Cengage Learning

Black Shales

In North America, many areas of carbonate–evaporite deposition gave way to a greater proportion of shales and coarser detrital rocks beginning in the Middle Devonian and continuing into the Late Devonian. This change to detrital deposition resulted from the formation of new source areas brought about by the mountain-building activity associated with the Acadian orogeny in North America (Figure 11.4).

As the Devonian Period ended, a conspicuous change in sedimentation took place over the North American craton with the appearance of widespread black shales (▶ Figure 11.6a). In the eastern United States, these black shales

are commonly called Chattanooga Shale, but they are known by a variety of local names elsewhere (for example, New Albany Shale and Antrim Shale). Although these black shales are best developed from the cratonic margins along the Appalachian mobile belt to the Mississippi Valley, correlative units can also be found in many western states and in western Canada (Figure 11.6a).

The Upper Devonian–Lower Mississippian black shales of North America are typically noncalcareous, thinly bedded, and less than 10 m thick (Figure 11.6b). Fossils are usually rare, but some Upper Devonian black shales do contain rich *conodont* (microscopic animals) faunas (see Figure 12.13). Because most black shales lack body fossils, they are difficult to date and correlate. However, in places where conodonts, *acritarchs* (microscopic algae, see Figure 12.9), or plant spores are found, these fossils indicate that the lower beds are Late Devonian and the upper beds are Early Mississippian in age.

Although the origin of these extensive black shales is still being debated, the essential features required to produce them include undisturbed anaerobic bottom water, a reduced supply of coarser detrital sediment, and high organic productivity in the overlying oxygenated waters. High productivity in the surface waters leads to a shower of organic material, which decomposes on the undisturbed seafloor and depletes the dissolved oxygen at the sediment–water interface.

The wide extent of such apparently shallow-water black shales in North America remains puzzling. Nonetheless, these shales are rich in uranium and are an important potential source rock for oil and gas in the Appalachian region.

The Late Kaskaskia—A Return to Extensive Carbonate Deposition

Following deposition of the widespread Upper Devonian–Lower Mississippian black shales, carbonate sedimentation on the craton dominated the remainder of the Mississippian Period (▶Figure 11.7). During this time, a variety of carbonate sediments were deposited in the epeiric sea, as indicated by the extensive deposits of crinoidal limestones (rich in crinoid fragments), oolitic limestones, and various other limestones and dolostones. These Mississippian carbonates display cross-bedding, ripple marks, and well-sorted fossil fragments, all of which indicate a shallow-water environment. Analogous features can be observed on the present-day Bahama Banks. In addition, numerous

▶ **Figure 11.5 Devonian Reef Complex of Western Canada** Reconstruction of the extensive Devonian Reef complex of western Canada. These extensive reefs controlled the regional facies of the Devonian epeiric seas.

© 2013 Cengage Learning

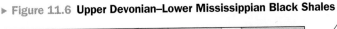
▶ **Figure 11.6 Upper Devonian–Lower Mississippian Black Shales**

© 2013 Cengage Learning

(a) The extent of the Upper Devonian to Lower Mississippian Chattanooga Shale and its equivalent units (such as the Antrim Shale and New Albany Shale) in North America.

Reed Wicander

(b) Upper Devonian New Albany Shale, Button Mold Knob Quarry, Kentucky.

▶ Figure 11.7 **Mississippian Paleogeography of North America**

Ellesmere Mountains

Caledonian Highlands

Uplands

Lowlands

Muddy bottom

Cordilleran mobile belt

Muddy bottom

Williston Basin

Acadian Highlands

Antler Highlands

Muddy bottom

Carbonate bottom

Paleoequator

© 2013 Cengage Learning

Ouachita mobile belt

Muddy bottom

:::: Land

Mountains

Evaporites

Barrier reefs

Epeiric sea

Deep ocean

small organic reefs occurred throughout the craton during the Mississippian. However, these were all much smaller than the large barrier reef complexes that dominated the earlier Paleozoic seas.

During the Late Mississippian regression of the Kaskaskia Sea from the craton, vast quantities of detrital sediments replaced carbonate deposition. The resulting sandstones, particularly in the Illinois Basin, have been studied in great detail because they are excellent petroleum reservoirs. Before the end of the Mississippian, the epeiric sea had retreated to the craton margin, once again exposing the craton to widespread weathering and erosion, resulting in a craton-wide unconformity at the end of the Kaskaskia Sequence.

The Absaroka Sequence

The **Absaroka Sequence** includes rocks deposited during the Pennsylvanian through Early Jurassic. In this chapter, however, we are concerned only with the Paleozoic rocks of the Absaroka Sequence. The extensive unconformity separating the Kaskaskia and Absaroka sequences essentially divides the strata into the North American Mississippian and Pennsylvanian systems. These two systems are closely equivalent to the European Lower and Upper Carboniferous systems, respectively. The rocks of the Absaroka Sequence not only differ from those of the Kaskaskia Sequence, but also result from different tectonic regimes.

▶ **Figure 11.8 Pennsylvanian Paleogeography of North America**

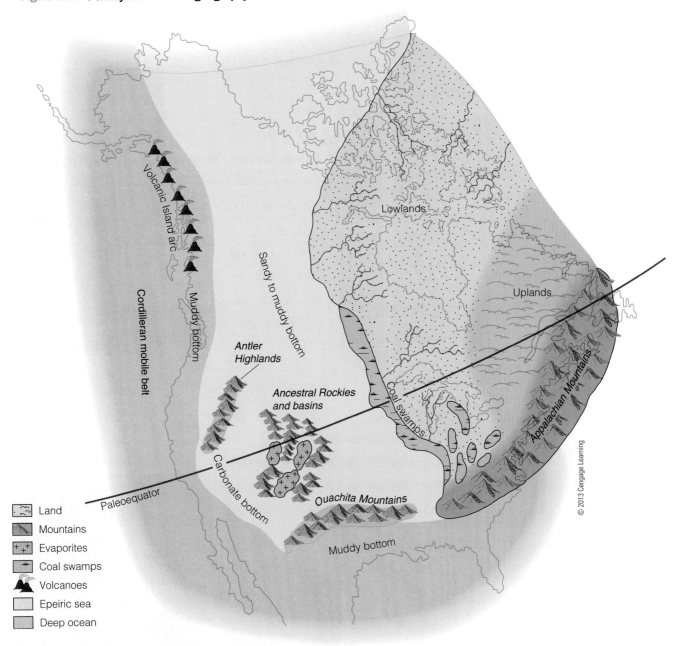

Land
Mountains
Evaporites
Coal swamps
Volcanoes
Epeiric sea
Deep ocean

© 2013 Cengage Learning

The lowermost sediments of the Absaroka Sequence are confined to the margins of the craton. These deposits are generally thickest in the east and southeast, near the emerging highlands of the Appalachian and Ouachita mobile belts, and thin westward onto the craton. The rocks also reveal lateral changes from nonmarine detrital rocks and coals in the east, through transitional marine–nonmarine beds, to largely marine detrital rocks and limestones farther west (▶Figure 11.8).

What Are Cyclothems and Why Are They Important?

One characteristic feature of Pennsylvanian rocks is their repetitive pattern of alternating marine and nonmarine strata known as **cyclothems**. They result from repeated alternations of deposition in marine and nonmarine environments, usually in areas of low relief. Although seemingly simple, cyclothems reflect a delicate interplay between nonmarine deltaic and shallow-marine interdeltaic and shelf environments.

For illustration, look at a typical coal-bearing cyclothem from the Illinois Basin (▶Figure 11.9a). Such a cyclothem contains nonmarine units, capped by a coal unit and overlain by marine units. Figure 11.9a shows the depositional environments that produced the cyclothem. The initial units represent deltaic and fluvial deposits. Above them is an underclay that frequently contains root casts from the plants and trees that comprise the overlying coal. The coal bed results from accumulations of plant material and is overlain by marine units of alternating limestones and

▶ Figure 11.9 **Cyclothems**

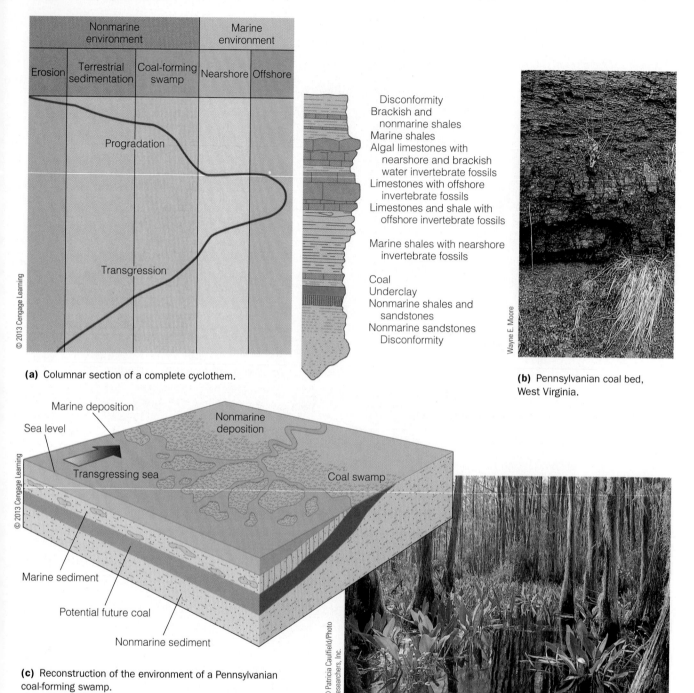

(a) Columnar section of a complete cyclothem.

Disconformity
Brackish and
  nonmarine shales
Marine shales
Algal limestones with
  nearshore and brackish
  water invertebrate fossils
Limestones with offshore
  invertebrate fossils
Limestones and shale with
  offshore invertebrate fossils

Marine shales with nearshore
  invertebrate fossils

Coal
Underclay
Nonmarine shales and
  sandstones
Nonmarine sandstones
  Disconformity

(b) Pennsylvanian coal bed,
West Virginia.

(c) Reconstruction of the environment of a Pennsylvanian coal-forming swamp.

(d) The Okefenokee Swamp, Georgia, is a modern example of a coal-forming environment, similar to those that occurred during the Pennsylvanian Period.

shales, usually with an abundant marine invertebrate fauna. The marine interval ends with an erosion surface. A new cyclothem begins with a nonmarine deltaic sandstone. All the beds illustrated in the idealized cyclothems are not always preserved because of abrupt changes from marine to nonmarine conditions or removal of some units by erosion.

Cyclothems represent transgressive and regressive sequences with an erosional surface separating one cyclothem from another. Thus, an idealized cyclothem passes upward from fluvial-deltaic deposits, through coals, to detrital shallow-water marine sediments, and finally to limestones typical of an open marine environment.

Such places as the Mississippi River delta, the Florida Everglades, the Okefenokee Swamp, Georgia, and the Dutch lowlands represent modern coal-forming environments similar to those existing during the Pennsylvanian Period (Figure 11.9d). By studying these modern analogs, geologists can make reasonable deductions about conditions existing in the geologic past.

The Pennsylvanian coal swamps must have been widespread lowland areas with little topographic relief neighboring the sea (Figure 11.9c). In such cases, a very slight rise in sea level would have flooded these large lowland areas, whereas slight drops would have exposed vast areas, resulting in alternating marine and nonmarine environments. The same result could also have been caused by a combination of rising sea level and *progradation* (the seaward extension of a delta by the accumulation of sediment) of a large delta, such as occurs today in Louisiana.

Such repetitious sedimentation over a widespread area requires an explanation. In most cases, local cyclothems of limited extent can be explained by rapid but slight changes in sea level in a swamp-delta complex of low relief near the sea such as that resulting from progradation or localized crustal movement.

Explaining widespread cyclothems is more difficult. The hypothesis currently favored by many geologists is a rise and fall of sea level related to advances and retreats of Gondwanan continental glaciers. When the Gondwanan ice sheets advanced, sea level dropped; when they melted, sea level rose. Late Paleozoic cyclothem activity on all of the cratons closely corresponds to Gondwanan glacial–interglacial cycles.

Cratonic Uplift—The Ancestral Rockies

Recall that cratons are stable areas, and what deformation they do experience is usually mild. The Pennsylvanian Period, however, was a time of unusually severe cratonic deformation, resulting in uplifts of sufficient magnitude to expose Precambrian basement rocks. In addition to newly formed highlands and basins, many previously formed arches and domes, such as the Cincinnati Arch, Nashville Dome, and Ozark Dome, were also reactivated (see Figure 10.1).

During the Late Absaroka (Pennsylvanian), the area of greatest deformation was in the southwestern part of the North American craton, where a series of fault-bounded uplifted blocks formed the **Ancestral Rockies** (Figure 11.10a). These mountain ranges had diverse geologic histories and were not all elevated at the same time. Uplift of these mountains, some of which were elevated more than 2 km along near-vertical faults, resulted in erosion of overlying Paleozoic sediments and exposure of the Precambrian igneous and metamorphic basement rocks (Figure 11.10b). As the mountains eroded, tremendous quantities of coarse, red arkosic sand and conglomerate were deposited in the surrounding basins (see Chapter 6 opening photo). These sediments are preserved in many areas, including the rocks of the Garden of the Gods near Colorado Springs (Figure 11.10c) and at the Red Rocks Amphitheatre near Morrison, Colorado.

Intracratonic mountain ranges are unusual, and their cause has long been debated. It is currently thought that the collision of Gondwana with Laurasia along the Ouachita mobile belt (Figure 11.2a) generated great stresses in the southwestern region of the North American craton. These crustal stresses were relieved by faulting. Movement along these faults produced uplifted cratonic blocks and downwarped adjacent basins, forming a series of related ranges and basins.

The Middle Absaroka—More Evaporite Deposits and Reefs

While the various intracratonic basins were filling with sediment during the Late Pennsylvanian, the epeiric sea slowly began retreating from the craton. During the Early Permian, the Absaroka Sea occupied a narrow region from Nebraska, south through West Texas (Figure 11.11). By the Middle Permian, it had retreated to the West Texas and southern New Mexico area. The thick evaporite deposits in Kansas and Oklahoma show the restricted nature of the Absaroka Sea during the Early and Middle Permian and its southwestward retreat from the central craton.

During the Middle and Late Permian, the Absaroka Sea was restricted to West Texas and southern New Mexico, forming an interrelated complex of lagoonal, reef, and open-shelf environments (Figure 11.12). Three basins separated by two submerged platforms developed in this area during the Permian. Massive reefs grew around the basin margins (Figure 11.13), and limestones, evaporites, and red beds were deposited in the lagoonal areas behind the reefs. As the barrier reefs grew and the passageways between the basins became more restricted, Late Permian evaporites gradually filled the individual basins.

Spectacular deposits representing the geologic history of this region can be seen today in the Guadalupe Mountains of Texas and New Mexico where the Capitan Limestone forms the caprock of these mountains (Figure 11.14). These reefs have been extensively studied because tremendous oil production comes from this region.

By the end of the Permian Period, the Absaroka Sea had retreated from the craton, exposing continental red beds that had been deposited over most of the southwestern and eastern region (Figure 11.2b).

History of the Late Paleozoic Mobile Belts

Having examined the Kaskaskian and Absarokian history of the craton, we now turn our attention to the orogenic activity in the mobile belts. The mountain building occurring during this time had a profound influence on the climate and sedimentary history of the craton. In addition, it was part of the global tectonic regime that sutured the continents together, forming Pangaea by the end of the Paleozoic Era.

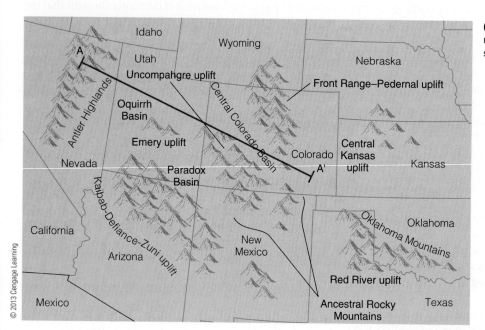

(a) Location of the principal Pennsylvanian highland areas and basins of the southwestern part of the craton.

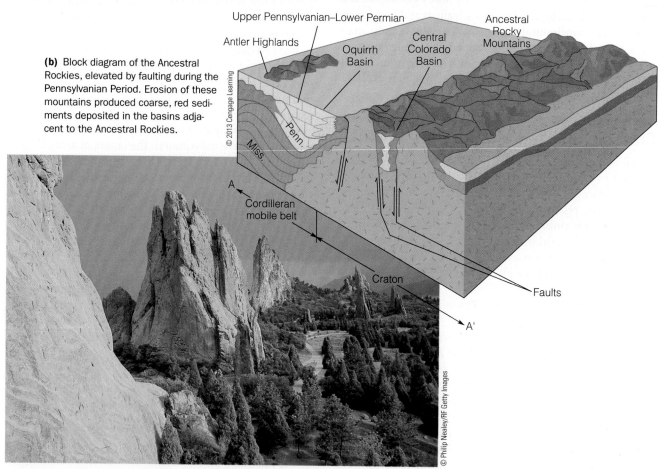

(b) Block diagram of the Ancestral Rockies, elevated by faulting during the Pennsylvanian Period. Erosion of these mountains produced coarse, red sediments deposited in the basins adjacent to the Ancestral Rockies.

(c) Garden of the Gods, Storm Sky View from Near Hidden Inn, Colorado Springs, Colorado.

▶ Figure 11.11 **Permian Paleogeography of North America**

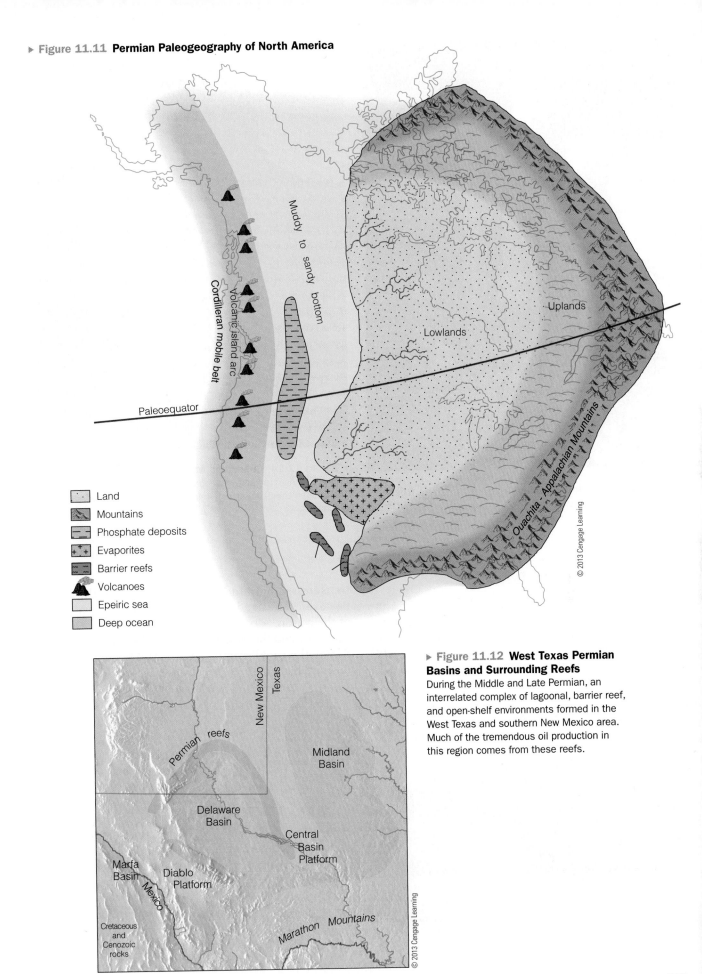

Cordilleran mobile belt

Volcanic island arc

Muddy to sandy bottom

Paleoequator

Uplands

Lowlands

Ouachita-Appalachian Mountains

Land

Mountains

Phosphate deposits

Evaporites

Barrier reefs

Volcanoes

Epeiric sea

Deep ocean

© 2013 Cengage Learning

New Mexico

Texas

Permian reefs

Midland Basin

Delaware Basin

Central Basin Platform

Marfa Basin

Diablo Platform

Mexico

Cretaceous and Cenozoic rocks

Marathon Mountains

© 2013 Cengage Learning

▶ Figure 11.12 **West Texas Permian Basins and Surrounding Reefs**
During the Middle and Late Permian, an interrelated complex of lagoonal, barrier reef, and open-shelf environments formed in the West Texas and southern New Mexico area. Much of the tremendous oil production in this region comes from these reefs.

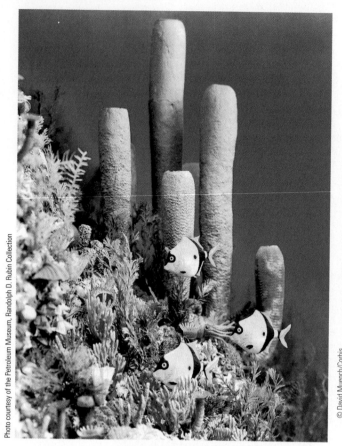

▶ Figure 11.13 **Middle Permian Capitan Limestone Reef Environment** A reconstruction of the Middle Permian Capitan Limestone reef environment. Shown are brachiopods, corals, bryozoans, and large glass sponges.

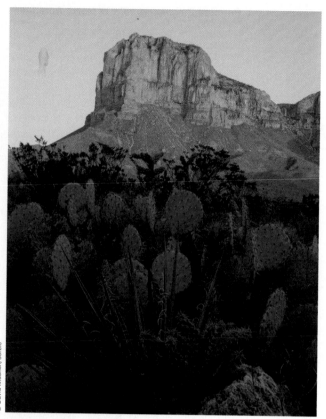

▶ Figure 11.14 **Guadalupe Mountains, Texas** The prominent Capitan Limestone forms the caprock of the Guadalupe Mountains. The Capitan Limestone is rich in fossil corals and associated reef organisms.

Cordilleran Mobile Belt

During the Neoproterozoic and Early Paleozoic, the Cordilleran area was a passive continental margin along which extensive continental shelf sediments were deposited (see Figures 9.7 and 10.5). Thick deposits of marine sediments graded laterally into thin cratonic units as the Sauk Sea transgressed onto the craton. Beginning in the Middle Paleozoic, an island arc formed off the western margin of the craton. A collision between this eastward-moving island arc and the western border of the craton took place during the Late Devonian and Early Mississippian, producing a highland area (▶Figure 11.15).

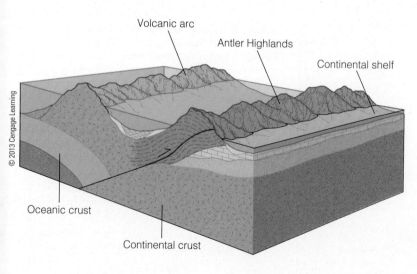

Volcanic arc

Antler Highlands

Continental shelf

Oceanic crust

Continental crust

▶ Figure 11.15 **Antler Orogeny** Reconstruction of the Cordilleran mobile belt during the Early Mississippian in which deep-water continental slope deposits were thrust eastward over shallow-water continental shelf carbonates, forming the Antler Highlands.

This orogenic event, the **Antler orogeny**, was caused by subduction, resulting in deep-water deposits and oceanic crustal rocks being thrust eastward over shallow-water continental shelf sediments, thus closing the narrow ocean basin that separated the island arc from the craton (Figure 11.15). Erosion of the resulting Antler Highlands produced large quantities of sediment that were deposited to the east in the epeiric sea covering the craton, and to the west in the adjacent deep sea. The Antler orogeny was the first in a series of orogenic events to affect the Cordilleran mobile belt. During the Mesozoic and Cenozoic, this area was the site of major tectonic activity caused by oceanic–continental convergence and accretion of various terranes.

Ouachita Mobile Belt

The Ouachita mobile belt extends for approximately 2,100 km from the subsurface of Mississippi to the Marathon region of Texas. Approximately 80 percent of the former mobile belt is buried beneath a Mesozoic and Cenozoic sedimentary cover (see Chapters 14 and 16). The two major exposed areas in this region are the Ouachita Mountains of Oklahoma and Arkansas and the Marathon Mountains of Texas.

During the Neoproterozoic to Early Mississippian, shallow-water detrital and carbonate sediments were deposited on a broad continental shelf, and in the deeper-water portion of the adjoining mobile belt, bedded cherts and shales were accumulating (▶ Figure 11.16a). Beginning in the Mississippian Period, the sedimentation rate increased dramatically as the region changed from a passive continental margin to an active convergent plate boundary, marking the beginning of the **Ouachita orogeny** (Figure 11.16b).

Thrusting of sediments continued throughout the Pennsylvanian and Early Permian, driven by the compressive forces generated along the zone of subduction as Gondwana collided with Laurasia (Figure 11.16c). The collision of Gondwana and Laurasia is marked by the formation of a large mountain range, most of which eroded during the Mesozoic Era. Only the rejuvenated Ouachita and Marathon Mountains remain of this once lofty mountain range.

The Ouachita deformation was part of the general worldwide tectonic activity that occurred when Gondwana united with Laurasia. The Hercynian, Appalachian, and Ouachita mobile belts were continuous and marked the southern boundary of Laurasia (Figure 11.2). The tectonic activity that uplifted the Ouachita mobile belt was very complex and involved not only the collision of Laurasia and Gondwana, but also several microplates and terranes between the continents that eventually became part of Central America. The compressive forces impinging on the Ouachita mobile belt also affected the craton by broadly uplifting the southwestern part of North America.

Figure 11.16 Ouachita Mobile Belt Plate tectonic model for deformation of the Ouachita mobile belt.

© 2013 Cengage Learning

(a) Depositional environment prior to the beginning of orogenic activity.

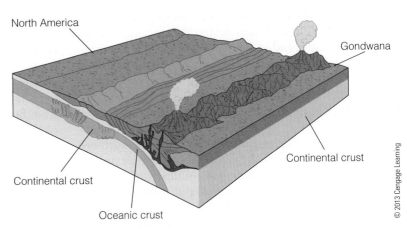

© 2013 Cengage Learning

(b) Incipient continental collision between North America and Gondwana began during the Mississippian Period.

© 2013 Cengage Learning

(c) Continental collision continued during the Pennsylvanian and Permian periods.

Appalachian Mobile Belt

Caledonian Orogeny The Caledonian mobile belt stretches along the western border of Baltica and includes the present-day countries of Scotland, Ireland, and Norway (see Figure 10.2c). During the Middle Ordovician, subduction along the boundary between the Iapetus plate and Baltica (Europe) began, forming a mirror image of the convergent plate boundary off the east coast of Laurentia (North America).

The culmination of the **Caledonian orogeny** (mentioned earlier) occurred during the Late Silurian and Early Devonian with the formation of a mountain range along the western margin of Baltica (see Figure 10.2c). Red-colored sediments deposited along the front of the Caledonian Highlands formed a large clastic wedge known as the *Old Red Sandstone* (see Perspective).

Acadian Orogeny The third Paleozoic orogeny to affect Laurentia and Baltica began during the Late Silurian and concluded at the end of the Devonian Period. The **Acadian orogeny** affected the Appalachian mobile belt from Newfoundland to Pennsylvania as sedimentary rocks were folded and thrust against the craton.

As with the preceding Taconic and Caledonian orogenies, the Acadian orogeny occurred along an oceanic–continental convergent plate boundary. As the northern Iapetus Ocean continued to close during the Devonian, the plate carrying Baltica finally collided with Laurentia, forming a continental–continental convergent plate boundary along the zone of collision (Figure 11.1a).

As the increased metamorphic and igneous activity indicates, the Acadian orogeny was more intense and lasted longer than the Taconic orogeny. Radiometric dates from the metamorphic and igneous rocks associated with the Acadian orogeny cluster between 360 and 410 million years ago. Just as with the Taconic orogeny, deep-water sediments were folded and thrust northwestward during the Acadian orogeny, producing angular unconformities separating Upper Silurian from Mississippian rocks.

Weathering and erosion of the Acadian Highlands produced the **Catskill Delta**, a thick clastic wedge named for the Catskill Mountains in upstate New York, where it is well exposed. Composed of red, coarse conglomerates, sandstones, and shales, the Catskill Delta contains nearly three times as much sediment as the Queenston Delta (see Figure 10.14).

The Devonian rocks of New York are among the best studied on the continent. A cross section of the Devonian strata clearly reflects an eastern source (Acadian Highlands) for the Catskill facies (▶Figure 11.17). These detrital rocks can be traced from eastern Pennsylvania, where the coarse-grained deposits are approximately 3 km thick, to Ohio, where the deltaic facies are only about 100 m thick and consist of cratonic shales and carbonates.

The red beds of the Catskill Delta derive their color from hematite (Fe_2O_3) in the sediments. Plant fossils and oxidation of the hematite indicate the beds were deposited in a continental environment. Toward the west, the red beds grade laterally into gray sandstones and shales containing fossil tree trunks, which indicate a swamp or marsh environment.

The Old Red Sandstone The red beds of the Catskill Delta have a European counterpart in the Devonian **Old Red Sandstone** of the British Isles (Figure 11.17). This formation was a Devonian clastic wedge that grew eastward from the Caledonian Highlands onto the Baltica craton and, like its North American Catskill counterpart, contains numerous fossils of freshwater fish, early amphibians, and land plants (see Perspective).

By the end of the Devonian Period, Baltica and Laurentia were sutured together, forming Laurasia

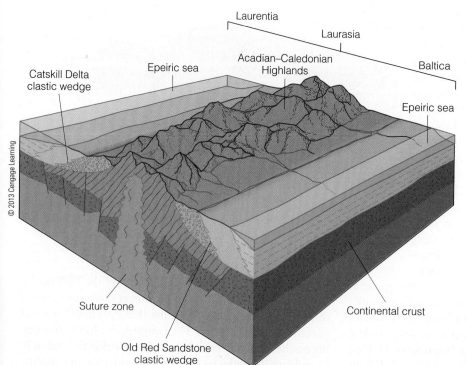

© 2013 Cengage Learning

Laurentia

Laurasia

Acadian–Caledonian Highlands

Baltica

Epeiric sea

Epeiric sea

Catskill Delta clastic wedge

Suture zone

Old Red Sandstone clastic wedge

Continental crust

▶ **Figure 11.17 Formation of Laurasia**
Block diagram showing the area of collision between Laurentia and Baltica. Note the bilateral symmetry of the Catskill Delta clastic wedge and the Old Red Sandstone and their relationship to the Acadian and Caledonian Highlands.

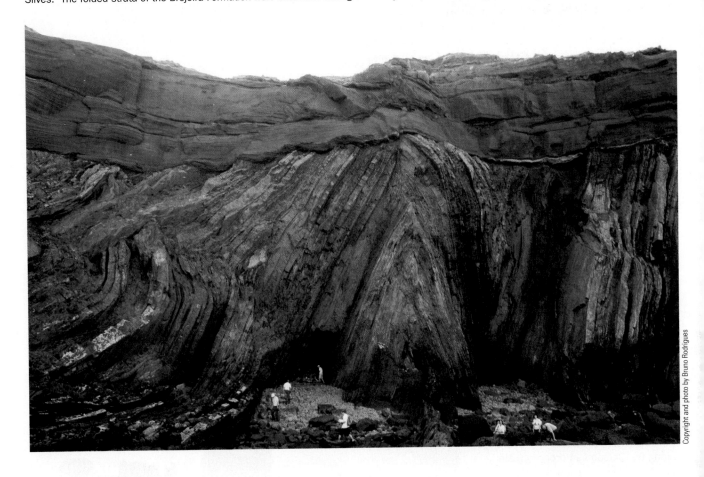

▶ **Figure 11.18 The Variscan Orogeny** The Variscan unconformity exposed at Telheiro Beach in southwestern Portugal shows the highly folded rocks of the Pennsylvanian Brejeira Formation, unconformably overlain by the red-colored, terrestrial beds of the Upper Triassic "Grés de Silves." The folded strata of the Brejeira Formation were deformed during the Hercynian (Variscan)–Alleghenian orogeny.

(Figure 11.17). The red beds of the Catskill Delta can be traced north, through Canada and Greenland, to the Old Red Sandstone of the British Isles, and into northern Europe. These beds were all deposited in similar environments along the flanks of developing mountain chains formed at convergent plate boundaries.

The Taconic, Caledonian, and Acadian orogenies were all part of the same major orogenic event related to the closing of the Iapetus Ocean (see Figures 10.13b and 11.17). This event began with paired oceanic–continental convergent plate boundaries during the Taconic and Caledonian orogenies and culminated along a continental–continental convergent plate boundary during the Acadian orogeny as Laurentia and Baltica became sutured. After this, the Hercynian (Variscan)–Alleghenian orogeny began, followed by orogenic activity in the Ouachita mobile belt.

Hercynian (Variscan)–Alleghenian Orogeny The Hercynian mobile belt of Europe, and the Appalachian and Ouachita mobile belts of North America mark the zone along which Europe (part of Laurasia) collided with Gondwana (Figure 11.1). While Gondwana and southern Laurasia collided during the Pennsylvanian and Permian in the area of the Ouachita mobile belt, eastern Laurasia (Europe and southeastern North America) joined together with Gondwana (Africa) as part of the **Hercynian (Variscan)–Alleghenian orogeny** (Figure 11.2). The terms *Hercynian* and *Variscan* are frequently used interchangeably, and originally referred to the direction of fold belts in Europe.

Initial contact between eastern Laurasia and Gondwana began during the Mississippian Period along the Hercynian mobile belt. The greatest deformation occurred during the Pennsylvanian and Permian periods and is referred to as the *Hercynian* or *Variscan orogeny* (▶ Figure 11.18). Conversely, and contemporaneously, the central and southern parts of the Appalachian mobile belt (from New York to Alabama) were folded and thrust toward the craton as eastern Laurasia and Gondwana were sutured. This tectonic event in North America is referred to as the *Alleghenian orogeny*.

These three Late Paleozoic orogenies (Hercynian [Variscan], Alleghenian, and Ouachita) represent the final joining of Laurasia and Gondwana into the supercontinent Pangaea during the Permian.

The Devonian Old Red Sandstone

The Devonian System was named by Roderick Murchison and Adam Sedgwick in 1839 for rock exposures in Devon County, England (Figure 1). The Old Red Sandstone, which was deposited during the Devonian Period, and crops out extensively throughout the British Isles, is predominantly of terrestrial origin.

The Old Red Sandstone consists of upwards of 11,000 m of clastic sediments deposited in various structural basins throughout Scandinavia, the British Isles, Greenland, and northeastern Canada. The sediments are poorly sorted and variable, and they consist of conglomerates, coarse-grained sandstones, cross-bedded sandstones, siltstones, shales, and thin limestones, deposited in several continental environments. The predominant red color comes from the presence of oxidized iron minerals, but colors also range from gray and green, through orange and purple.

The collision of Baltica and Laurentia during the Caledonian orogeny (Late Silurian and Early Devonian), produced a mountain range along the western margin of Baltica (see Figures 10.2c and 11.1a) that is referred to as the Caledonian Mountain chain or Caledonian Highlands. The weathering and erosion of these highlands resulted in the formation of a large clastic wedge, whose sediments were named the Old Red Sandstone (Figure 11.17).

It was in 1788 at Siccar Point, Scotland, that James Hutton first recognized the significance of unconformities in interpreting Earth history. At this location he observed the Devonian Old Red Sandstone overlying the steeply dipping Silurian graywackes (see Figure 4.3).

Not only is The Old Red Sandstone important in British and global stratigraphy, but it is also important for the many fossils of jawless armored fish (ostracoderms), jawed armored fish (placoderms), early bony fish, as well as arthropods and primitive seedless vascular plants found throughout its extent (see Chapter 13).

Lastly, the Old Red Sandstone has been widely used as a building stone.

kolo5/Shutterstock.com

Figure 1 Red cliffs composed of the Old Red Sandstone crop out along the shoreline in Devon County, England.

David Hughes/Shutterstock.com

Figure 2 Remains of Goodrich Castle in Herefordshire, England, dated to the period between 1160 and 1270. Goodrich Castle is one of the many castles built from rocks quarried from the Old Red Sandstone, a Devonian-age formation. Identical red sandstone is found in the Catskill Mountains of New York. It too has been used as a building stone for many structures in the New York area.

For example, many of the castles, cathedrals, and abbeys in Scotland, Wales, and England were built using Old Red Sandstone. Some examples include Muchalls Castle, Aberdeenshire, Scotland, St. Magnus Cathedral, Orkney, Scotland, and Goodrich Castle, Herefordshire, England (Figure 2). Interestingly, the New York Life Insurance Building (also known as the Quebec Bank Building) in Montreal, Canada, was built in 1887–1889 using the Old Red Sandstone, imported from Dumfriesshire, Scotland.

What Role Did Microplates and Terranes Play in the Formation of Pangaea?

We have presented the geologic history of the mobile belts bordering the Paleozoic continents in terms of subduction along convergent plate boundaries. It is becoming increasingly clear, however, that accretion along the continental margins is more complicated than the somewhat simple, large-scale plate interactions we have described. Geologists now recognize that numerous terranes or microplates existed during the Paleozoic and were involved in the orogenic events that occurred during that time.

In this chapter and the previous one, we have been concerned only with the six major Paleozoic continents. However, terranes and microplates of varying size were present during the Paleozoic and participated in the formation of Pangaea. For example, as we mentioned in the previous chapter, the microcontinent of *Avalonia* consisted of some coastal parts of New England, southern New Brunswick, much of Nova Scotia, the Avalon Peninsula of eastern Newfoundland, southeastern Ireland, Wales, England, and parts of Belgium and northern France. This microcontinent separated from Gondwana in the Early Ordovician and existed as a small, separate continent until it collided with Baltica during the Late Ordovician–Early Silurian and then with Laurentia (as part of Baltica) during the Silurian (see Figures 10.2b and 10.2c).

Other terranes and microplates include *Iberia-Armorica* (a portion of southern France, Sardinia, and most of the Iberian peninsula), *Perunica* (Bohemia), numerous Alpine fragments (especially in Austria), as well as many other bits and pieces of island arcs and suture zones. Not only did these terranes and microplates separate and move away from the larger continental landmasses during the Paleozoic, but they usually developed their own unique faunal and floral assemblages.

Thus, although the basic history of the formation of Pangaea during the Paleozoic remains essentially the same, geologists now realize that microplates and terranes also played an important role in the formation of Pangaea. Furthermore, they help explain some previously anomalous geologic and paleontologic situations.

Late Paleozoic Mineral Resources

Late Paleozoic-age rocks contain a variety of important mineral resources including energy resources and metallic and nonmetallic mineral deposits. Petroleum and natural gas are recovered in commercial quantities from rocks ranging in age from the Devonian through Permian. For example, Devonian rocks in the Michigan Basin, Illinois Basin, and the Williston Basin of Montana, North Dakota, South Dakota, and adjacent parts of Alberta, Canada, have yielded considerable amounts of hydrocarbons. Middle Devonian black shales, such as the Marcellus Shale in the Appalachian Basin, are being actively explored because of their potential to yield enormous quantities of natural gas. Permian reefs and other strata in the western United States, particularly Texas, have also been prolific producers.

Although Permian coal beds are known from several areas, including Asia, Africa, and Australia, much of the coal in North America and Europe comes from Pennsylvanian (Upper Carboniferous) deposits. Large areas in the Appalachian region and the midwestern United States are underlain by vast coal deposits (▶Figure 11.19). These coal deposits formed from the lush vegetation that flourished in Pennsylvanian coal-forming swamps (Figure 11.9).

Much of this coal is *bituminous coal*, which contains about 80 percent carbon. It is a dense, black coal that has been so thoroughly altered that plant remains can be seen only rarely. Bituminous coal is used to make *coke*, a hard, gray substance made up of the fused ash of bituminous coal. Coke is used to fire blast furnaces for steel production.

Some Pennsylvanian coal from North America is *anthracite*, a metamorphic type of coal containing up to 98 percent carbon. Most anthracite is in the Appalachian region (Figure 11.19). It is especially desirable because it burns with a smokeless flame and yields more heat per unit volume than other types of coal. Unfortunately, it is the least common type—much of the coal used in the United States is bituminous.

Several Late Paleozoic-age evaporite deposits are important nonmetallic mineral resources. The Zechstein evaporites of Europe extend from Great Britain across the North Sea and into Denmark, the Netherlands, Germany, eastern Poland, and Lithuania. Besides the evaporites themselves, Zechstein deposits form the caprock for the large reservoirs of the gas fields of the Netherlands and part of the North Sea region.

Other important evaporite mineral resources include those of the Permian Delaware Basin of West Texas and New Mexico and Devonian evaporites in the Elk Point Basin of Canada. In Michigan, gypsum is mined and used in the construction of sheetrock. Late Paleozoic-age limestones from many areas in North America are used in manufacturing cement. Limestone is also mined and used in blast furnaces for steel production.

Most silica sand mined in the United States comes from east of the Mississippi River, and much of this comes from Late Paleozoic-age rocks. For example, the Devonian Ridgeley Formation is mined in West Virginia, Maryland, and Pennsylvania, and the Devonian Sylvania Sandstone is mined near Toledo, Ohio. Recall from Chapter 10 that silica sand is used in manufacturing glass; for refractory bricks in blast furnaces; for molds for casting aluminum, iron, and copper alloys; and for a variety of other uses.

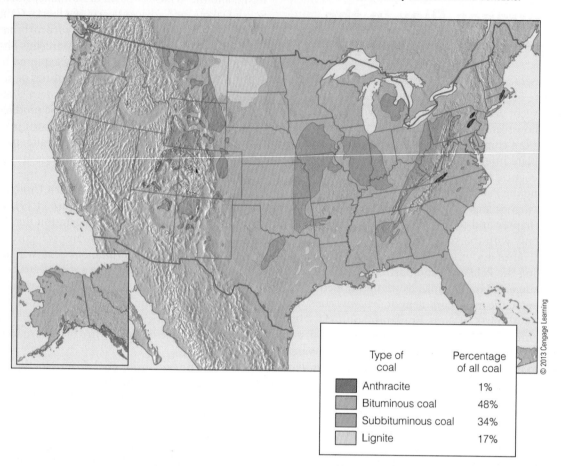

▶ Figure 11.19 **Distribution of Coal Deposits in the United States** The age of the coals in the midwestern states and the Appalachian region are mostly Pennsylvanian, whereas those in the West are mostly Cretaceous and Cenozoic.

Type of coal	Percentage of all coal
Anthracite	1%
Bituminous coal	48%
Subbituminous coal	34%
Lignite	17%

© 2013 Cengage Learning

Metallic mineral resources, including tin, copper, gold, and silver, are also known from Late Paleozoic-age rocks, especially those deformed during mountain building. Although the precise origin of the Missouri lead and zinc deposits remains unresolved, much of the ores of these metals come from Mississippian-age rocks. In fact, mines in Missouri account for a substantial amount of all domestic production of lead ores.

Summary

Table 11.1 summarizes the Late Paleozoic geologic history of the North American craton and mobile belts as well as global events, sea-level changes, and major evolutionary events.

▶ During the Late Paleozoic, Baltica and Laurentia collided, forming Laurasia. Siberia and Kazakhstania collided and were then sutured to Laurasia. Gondwana continued moving over the South Pole and experienced several glacial–interglacial periods, resulting in global sea-level changes and transgressions and regressions along the low-lying craton margins.

▶ Laurasia and Gondwana underwent a series of collisions beginning in the Carboniferous, and the formation of Pangaea was completed during the Permian. Surrounding the supercontinent was the global ocean, Panthalassa.

▶ The Late Paleozoic history of the North American craton can be deciphered from the rocks of the Kaskaskia and Absaroka sequences.

▶ The basal beds of the Kaskaskia Sequence that were deposited on the exposed Tippecanoe surface consisted either of sandstones derived from the eroding Taconic Highlands or of carbonate rocks.

▶ Most of the Kaskaskia Sequence is dominated by carbonates and associated evaporites. The Devonian Period was a time of major reef building in western Canada, southern England, Belgium, Australia, and Russia.

▶ Widespread black shales were deposited over large areas of the North American craton during the Late Devonian and Early Mississippian.

▶ The Mississippian Period was dominated, for the most part, by carbonate deposition.

▶ Transgressions and regressions over the low-lying North American craton, probably caused by advancing and retreating Gondwanan ice sheets, resulted in cyclothems and the formation of coals during the Pennsylvanian Period.

- Cratonic mountain building, specifically the Ancestral Rockies, occurred during the Pennsylvanian Period and resulted in thick nonmarine detrital sediments and evaporites being deposited in the intervening basins.
- By the Early Permian, the Absaroka Sea occupied a narrow zone of the south–central craton. Here, several large reefs and associated evaporites developed. By the end of the Permian Period, this epeiric sea had retreated from the craton.
- The Cordilleran mobile belt was the site of the Antler orogeny, a minor Devonian orogeny, during which deep-water sediments were thrust eastward over shallow-water sediments.
- During the Pennsylvanian and Early Permian, mountain building occurred in the Ouachita mobile belt. In addition to the mountain range the Ouachita orogeny produced, this tectonic activity was partly responsible for the cratonic uplift in the southwest that resulted in the Ancestral Rockies.
- The Caledonian (Baltica) and Acadian (Laurentia) orogenies occurred from the Late Silurian through Late Devonian. Both began as oceanic–continental convergent plate boundaries, culminating along a continental–continental convergent plate boundary when Baltica and Laurentia collided to form Laurasia.
- The Caledonian, Acadian, and Hercynian (Variscan)–Alleghenian orogenies were all part of the global tectonic activity that resulted in the assembly of Pangaea.
- During the Paleozoic Era, numerous microplates and terranes—such as Avalonia, Iberia–Armorica, and Perunica—existed and played an important role in the formation of Pangaea.
- Late Paleozoic-age rocks contain a variety of mineral resources, including petroleum, coal, evaporites, silica sand, lead, zinc, and other metallic deposits.

Important Terms

Absaroka Sequence, p. 218
Acadian orogeny, p. 226
Ancestral Rockies, p. 221
Antler orogeny, p. 225
Caledonian orogeny, p. 226

Catskill Delta, p. 226
cyclothem, p. 219
Hercynian (Variscan)–Alleghenian orogeny, p. 227
Kaskaskia Sequence, p. 215

Laurasia, p. 212
Old Red Sandstone, p. 226
Ouachita orogeny, p. 225
Panthalassa, p. 212

Review Questions

1. The European Old Red Sandstone is the equivalent of the North American
   a. _____ Queenston Delta; b. _____ Capitan Limestone; c. _____ Phosphoria Formation; d. _____Oriskany Sandstone; e. _____ Catskill Delta.
2. The enormous single ocean surrounding Pangaea and spanning Earth from pole to pole at the end of the Permian is called
   a. _____ Panthalassa; b. _____ Tethys; c. _____ Iapetus; d. _____ Atlantis; e. _____ Avalonia.
3. Which was the first Paleozoic orogeny to occur in the Cordilleran mobile belt?
   a. _____ Acadian; b. _____ Alleghenian; c. _____ Antler; d. _____ Caledonian; e. _____ Ellesmere.
4. Which of the following resulted from intracratonic deformation?
   a. _____ Antler Highlands; b. _____ Ancestral Rockies; c. _____ Acadian Highlands; d. _____ Caledonian Highlands; e. _____ Taconic Highlands.
5. Which mobile belt was located along southern Europe, and marked the zone in which Europe (as part of Laurasia) collided with Gondwana during the Late Paleozoic?
   a. _____ Cordilleran; b. _____ Ouachita; c. _____ Appalachian; d. _____ Hercynian; e. _____ Alleghenian.
6. The economically valuable deposit in a cyclothem is
   a. _____ gravel; b. _____ metallic ore; c. _____ coal; d. _____ carbonates; e. _____ evaporites.
7. During which period did extensive continental glaciation of the Gondwana continent occur?
   a. _____ Cambrian; b. _____ Silurian; c. _____ Devonian; d. _____ Carboniferous; e. _____ Permian.
8. The Late Devonian–Early Mississippian is known for what type of widespread deposits?
   a. _____ Carbonates; b. _____ Black shales; c. _____ Evaporites; d. _____ Volcanics; e. _____ Coals.
9. The Catskill Delta clastic wedge resulted from weathering and erosion of the _____ highlands.
   a. _____ Taconic; b. _____ Nevadan; c. _____ Transcontinental Arch; d. _____ Acadian; e. _____ Sevier.
10. Which orogeny was *not* involved in the closing of the Iapetus Ocean?
    a. _____ Alleghenian; b. _____ Acadian; c. _____ Taconic; d. _____ Caledonian; e. _____ Antler.
11. Discuss how plate movement during the Paleozoic Era affected worldwide weather patterns.
12. What were the major differences between the Appalachian, Ouachita, and Cordilleran mobile belts during the Paleozoic Era?
13. In your travels you notice that many buildings in the eastern United States, as well as numerous castles in the United Kingdom, seem to be constructed of the same coarse-grained red sandstones and conglomerates. How would you account for such a coincidence? Or is it really a coincidence? Explain.

14. How does the origin of evaporite deposits of the Kaskaskia Sequence compare with the origin of evaporites of the Tippecanoe Sequence?

15. Describe the geologic history of the Iapetus Ocean during the Paleozoic Era.

16. What was the relationship between the Ouachita orogeny and the cratonic uplifts on the craton during the Pennsylvanian Period?

17. How are the Caledonian, Acadian, Ouachita, and Hercynian (Variscan)–Alleghenian orogenies related to modern concepts of plate tectonics?

18. Based on the discussion of Milankovitch cycles and their role in causing glacial–interglacial cycles (see Chapter 17), could these cycles be partly responsible for the transgressive–regressive cycles that resulted in cyclothems during the Pennsylvanian Period?

19. How did the formation of Pangaea and Panthalassa affect the world's climate at the end of the Paleozoic Era?

20. Paleogeographic maps of what the world looked like during the Paleozoic Era are found in almost every Earth history book and in numerous scientific journals. What criteria are used to determine the location of ancient continents and ocean basins, and why are there minor differences in the Paleozoic paleogeography among the different books and articles?

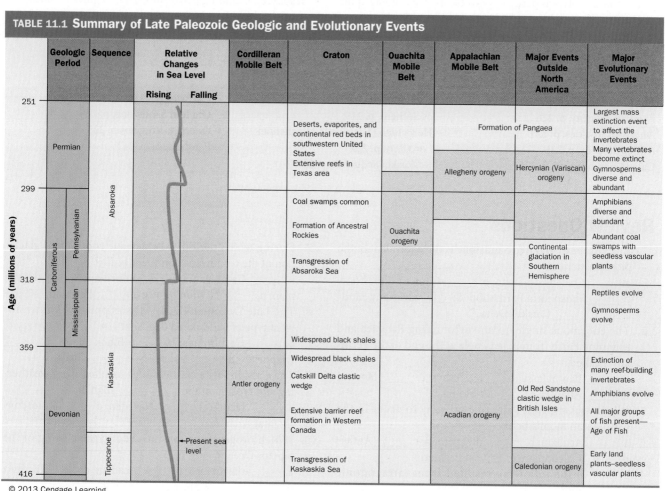

TABLE 11.1 Summary of Late Paleozoic Geologic and Evolutionary Events

© 2013 Cengage Learning

Paleozoic Life History
Invertebrates

12

Diorama of the environment and biota of the Phyllopod bed of the Middle Cambrian Burgess Shale, British Columbia, Canada. In the background is a vertical wall of a submarine escarpment with algae growing on it. The large, cylindrical, ribbed organisms on the muddy bottom in the foreground are sponges.

Smithsonian Institution, Transparency No. 86-13471A

CHAPTER OBJECTIVES

At the end of this chapter, you will have learned that

▶ Animals with skeletons appeared rather abruptly at the beginning of the Paleozoic Era and experienced a short period of rapid evolutionary diversification.

▶ The present marine ecosystem is a complex organization of organisms that interrelate and interact not only with each other, but also with the physical environment.

▶ The Cambrian Period was a time of many evolutionary innovations during which almost all of the major invertebrate phyla evolved.

▶ The Ordovician Period witnessed striking changes in the marine community, resulting in a dramatic increase in diversity of the shelly fauna, followed by a mass extinction at the end of the Ordovician.

▶ The Silurian and Devonian periods were a time of rediversification and recovery for many of the invertebrate phyla as well as a time of major reef building.

▶ Following the Late Devonian extinctions, the marine community again experienced renewed adaptive radiation and diversification during the Carboniferous and Permian periods.

▶ Mass extinctions occur when anomalously high numbers of species go extinct in a short period of time. The greatest recorded mass extinction in Earth's history occurred at the end of the Permian Period.

Introduction

On August 30 and 31, 1909, near the end of the summer field season, Charles D. Walcott, geologist and head of the Smithsonian Institution, was searching for fossils along a trail on Burgess Ridge between Mount Field and Mount Wapta, near Field, British Columbia, Canada. On the west slope of this ridge, he discovered the first soft-bodied fossils from the Burgess Shale, a discovery of immense importance in deciphering the early history of life. During the following week, Walcott and his collecting party split open numerous blocks of shale, many of which yielded carbonized impressions of a number of soft-bodied organisms beautifully preserved on bedding planes. Walcott returned to the site the following summer and located the shale stratum that was the source of his fossil-bearing rocks in the steep slope above the trail. He quarried the site and shipped back thousands of fossil specimens to the United States National Museum of Natural History, where he later catalogued and studied them.

The importance of Walcott's discovery is not that it was another collection of well-preserved Cambrian fossils, but rather that it allowed geologists a rare glimpse into a world previously almost unknown: that of the soft-bodied animals that lived during the Middle Cambrian. The beautifully preserved fossils from the Burgess Shale present a much more complete picture of a Middle Cambrian community than do deposits containing only fossils of the hard parts of organisms. In fact, 60 percent of the total fossil assemblage of more than 100 genera is composed of soft-bodied animals, a percentage comparable to present-day marine communities.

What conditions led to the remarkable preservation of the Burgess Shale fauna? The depositional site of the Burgess Shale lay at the base of a steep submarine escarpment.

The animals whose exquisitely preserved fossil remains are found in the Burgess Shale lived in and on mud banks that formed along the top of this escarpment. Periodically, this unstable area would slump and slide down the escarpment as a turbidity current. At the base, the mud and the animals carried with it were deposited in a deep-water anaerobic environment devoid of life. In such an environment, bacterial degradation did not destroy the buried animals, and thus they were compressed by the weight of the overlying sediments and eventually preserved as carbonaceous films.

In the following two chapters, we examine the history of Paleozoic life as a system in which its parts consist of a series of interconnected biologic and geologic events. The underlying processes of evolution and plate tectonics are the forces that drove this system. The opening and closing of ocean basins, transgressions and regressions of epeiric seas, the formation of mountain ranges, and the changing positions of the continents profoundly affected the evolution of the marine and terrestrial communities.

In this chapter, we will look at the tremendous biologic change that began with the appearance of skeletonized animals near the Precambrian–Cambrian boundary. Following this event, marine invertebrates began a period of adaptive radiation and evolution during which the Paleozoic marine invertebrate community greatly diversified.

The end of the Paleozoic Era witnessed the greatest mass extinction in Earth's history. The marine invertebrate community was nearly completely decimated, and many amphibians and reptiles, whose evolutionary history, along with plants, will be covered in the next chapter, also became extinct.

What Was the Cambrian Explosion?

At the beginning of the Paleozoic Era, animals with skeletons appeared rather abruptly in the fossil record. In fact, their appearance is described as an explosive development of new types of animals and is referred to as the "Cambrian explosion" by most scientists. This sudden appearance of new animals in the fossil record is rapid, however, only in the context of geologic time, having taken place over millions of years during the Early Cambrian Period. Moreover, it is not a recent discovery. Early geologists observed that the remains of skeletonized animals appeared rather abruptly in the fossil record. In fact, Charles Darwin addressed this problem in *On the Origin of Species by Means of Natural Selection* (1859), and observed that, without a convincing explanation, such an event was difficult to reconcile with his newly expounded evolutionary theory.

The sudden appearance of shelled animals during the Early Cambrian contrasts sharply with the biota living during the preceding Proterozoic Eon. Up until the evolution of the Ediacaran fauna, Earth was populated primarily by single-celled organisms. Recall from Chapter 9 that the Ediacaran fauna, which is found on all continents except Antarctica, consists primarily of multicelled soft-bodied organisms (see Figures 9.19 and 9.20). Microscopic calcareous tubes, presumably housing wormlike suspension-feeding organisms, have also been found at some localities (see Figure 9.21b). In addition, trails and burrows, which represent the activities of worms and other sluglike animals, are also found associated with Ediacaran faunas throughout the world. The trails and burrows are similar to those made by present-day soft-bodied organisms.

Until recently, it appeared that there was a fairly long period of time between the extinction of the Ediacaran fauna and the first Cambrian fossils. That gap has been considerably narrowed in recent years with the discovery of new Proterozoic fossiliferous localities. Now, Proterozoic fossil assemblages continue right to the base of the Cambrian.

Nonetheless, the cause of the sudden appearance of so many different animal phyla during the Early Cambrian is still hotly debated. Newly developed molecular techniques that allow evolutionary biologists to compare molecular sequences of the same gene from different species are being applied to the phylogeny, or evolutionary history, of many organisms. In addition, new fossil sites and detailed stratigraphic studies are shedding light on the early history and ancestry of the various invertebrate phyla.

The Cambrian explosion probably had its roots firmly planted in the Proterozoic. However, the mechanism or mechanisms that triggered this event are still being investigated. Although some would argue for a single causal event, it is more likely that the Cambrian explosion was a combination of factors, both biologic and geologic. For example, geologic evidence indicates that Earth was glaciated one or more times during the Proterozoic, followed by global warming during the Cambrian. These global environmental changes may have stimulated evolution and contributed to the Cambrian explosion.

Others would argue that a change in the chemistry of the oceans favored the evolution of mineralized skeletons. In this scenario, an increase in the concentration of calcium from the Neoproterozoic through the Early Cambrian, in part due to volcanism along mid-oceanic ridges, allowed for the precipitation of calcium carbonate and calcium phosphate, compounds that comprise the shells of most invertebrates.

Another hypothesis is that the rapid evolution of a skeletonized fauna was a response to the evolution of predators. A shell or mineralized covering would provide protection against predation by the various predators evolving during this time. The conflict between predators, whose goal is to consume prey, and prey, whose goal is to avoid becoming a meal for a predator, is one of the strongest factors in natural selection.

A further line of research related to the Cambrian explosion involves what are known as *homeobox* or, simply, Hox, genes (see Chapter 7). These are sequences of regulatory genes that control the development of individual regions of the body, and thus the basic body plan in organisms. Studies indicate that the basic body plans for all animals were apparently established by the end of the Cambrian explosion, and only minor modifications have occurred since then. Whatever the ultimate cause of the Cambrian explosion, the appearance of a skeletonized fauna and the rapid diversification of that fauna during the Early Cambrian were major events in life history.

The Emergence of a Shelly Fauna

The earliest organisms with hard parts are small, calcareous tubes found associated with Ediacaran faunas (Neoproterozoic) from several locations throughout the world. These were followed by other microscopic skeletonized fossils from the Early Cambrian (►Figure 12.1) and are collectively referred to as the *small shelly fauna* or *small shelly fossils*. This small shelly fauna was in turn followed by the appearance of large skeletonized animals during the Cambrian explosion. Along with the question of why animals appeared so suddenly in the fossil record is the equally intriguing question of why they initially acquired skeletons and what selective advantage skeletons provided. A variety of explanations about why marine organisms evolved skeletons have been proposed, but none are completely satisfactory or universally accepted.

The formation of an exoskeleton, or shell, confers many advantages on an organism: (1) It provides protection against ultraviolet radiation, allowing animals to move into shallower waters; (2) it helps prevent drying out in an

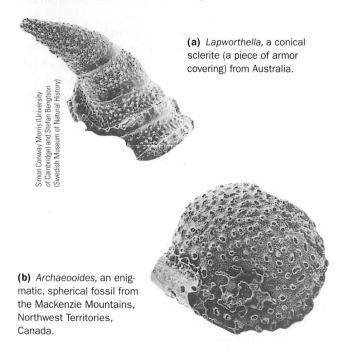

Figure 12.1 Lower Cambrian Shelly Fossils Two small (several millimeters in size) Lower Cambrian shelly fossils.

(a) *Lapworthella,* a conical sclerite (a piece of armor covering) from Australia.

Simon Conway Morris (University of Cambridge) and Stefan Bengtson (Swedish Museum of Natural History)

(b) *Archaeooides,* an enigmatic, spherical fossil from the Mackenzie Mountains, Northwest Territories, Canada.

intertidal environment; (3) a supporting skeleton, whether an exo- or endoskeleton, allows animals to increase their size and provides attachment sites for muscles; and (4) it provides protection against predators. Evidence of actual fossils of predators and specimens of damaged prey, as well as antipredatory adaptations in some animals, indicates that the impact of predation during the Cambrian was great, leading some scientists to hypothesize that the rapid evolution of a shelly invertebrate fauna was a response to the rise of predators (▶ Figure 12.2). With predators playing

▶ **Figure 12.2 Cambrian Predator** *Anomalocaris,* a predator from the Early and Middle Cambrian Period, is shown feeding on *Opabinia,* one of many extinct invertebrates found in the Burgess Shale. *Anomalocaris* was approximately 45 cm long and used its gripping appendages to bring its prey to its circular mouth structure.

De Agostini/Getty Images

an important role in the Cambrian marine ecosystem, any mechanism or feature that protected an animal would certainly be advantageous and confer an adaptive advantage to the organism.

Scientists currently have no clear answer as to why marine organisms initially evolved mineralized skeletons (small shelly fauna) during the Proterozoic and Early Cambrian, followed by a variety of large shelly invertebrates during the Cambrian explosion and shortly thereafter. They undoubtedly evolved because of a variety of biologic and environmental factors. Whatever the reason, the acquisition of a mineralized skeleton was a major evolutionary innovation that allowed invertebrates to increase in size and successfully occupy a wide variety of marine habitats.

The Present-Day Marine Ecosystem

Before we can begin to understand ancient ecosystems, we must first analyze the present-day marine ecosystem and look at where organisms live, how they get around, and how they feed (▶ Figure 12.3). Organisms that live in the water column above the seafloor are called *pelagic.* They can be divided into two main groups: the floaters, or **plankton**, and the swimmers, or **nekton**.

Plankton are mostly passive and go where currents carry them. Plant plankton such as diatoms, dinoflagellates, and various algae are called *phytoplankton* and are mostly microscopic. Animal plankton are called *zooplankton* and are also mostly microscopic. Examples of zooplankton include foraminifera, radiolarians, and jellyfish. The nekton are swimmers and are mainly vertebrates such as fish; the invertebrate nekton include cephalopods.

Organisms that live on or in the seafloor make up the **benthos**. They are characterized as *epifauna* (animals) or *epiflora* (plants)—those that live on the seafloor—or as *infauna*—animals that live in and move through the sediments. The benthos are further divided into those organisms that stay in one place, called *sessile,* and those that move around on or in the seafloor, called *mobile.*

The feeding strategies of organisms are also important in terms of their relationships with other organisms in the marine ecosystem. There are basically four feeding groups: **suspension-feeding** animals remove or consume microscopic plants and animals as well as dissolved nutrients from the water; **herbivores** are plant eaters; **carnivore-scavengers** are meat eaters; and **sediment-deposit feeders** ingest sediment and extract the nutrients from it.

We can define an organism's place in the marine ecosystem by where it lives and how it eats. For example, an articulate brachiopod is a benthonic, epifaunal, suspension feeder, whereas a cephalopod is a nektonic carnivore.

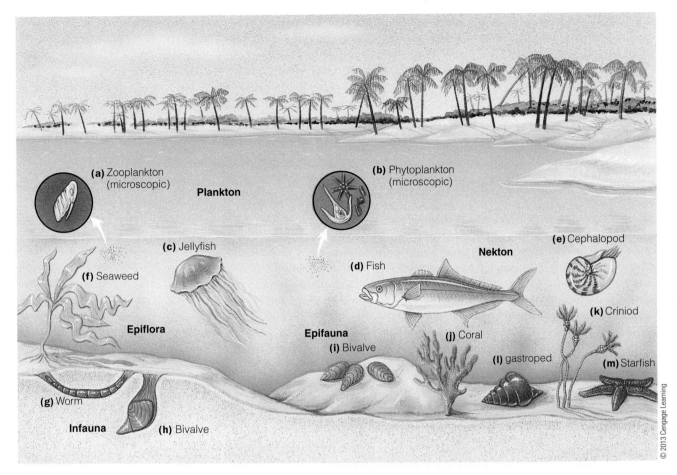

▶ **Figure 12.3 Marine Ecosystem** Where and how animals and plants live in the marine ecosystem. Plankton: (a) through (c). Nekton: (d) and (e). Benthos: (f) through (m). Sessile epiflora: (f). Sessile epifauna: (i), (j), and (k). Mobile epifauna: (l) and (m). Infauna: (g) and (h). Suspension feeders: (i), (j), and (k). Herbivores: (l). Carnivore-scavengers: (m). Sediment-deposit feeders: (g).

An ecosystem includes several *trophic levels,* which are tiers of food production and consumption within a feeding hierarchy. The feeding hierarchy, and hence energy flow, in an ecosystem comprise a food web of complex interrelationships among the producers, consumers, and decomposers (▶Figure 12.4). The **primary producers,** or *autotrophs,* are those organisms that manufacture their own food. Virtually all marine primary producers are phytoplankton. Feeding on the primary producers are the *primary consumers,* which are mostly suspension feeders. *Secondary consumers* feed on the primary consumers and thus are predators, whereas *tertiary consumers,* which are also predators, feed on the secondary consumers. Besides the producers and consumers, there are also transformers and decomposers. These are bacteria that break down the dead organisms that have not been consumed into organic compounds, which are then recycled.

When we look at the marine realm today, we see a complex organization of organisms interrelated by trophic interactions and affected by changes in the physical environment. When one part of the system changes, the whole structure changes, sometimes almost insignificantly, other times catastrophically.

Paleozoic Invertebrate Marine Life

Having already considered the origin, differentiation, and evolution of the Precambrian–Cambrian marine biota, and having examined the nature and structure of living marine communities for context, we will now explore the changes that occurred in the marine invertebrate community during the Paleozoic Era. Rather than focusing on the history of each invertebrate phylum (Table 12.1), we will survey the evolution of the marine invertebrate community through time, concentrating on the major features and changes that took place.

As we study the evolution of the Paleozoic marine ecosystem, keep in mind how geologic and evolutionary changes have had a significant impact on its composition and structure. For example, the major marine transgressions over the world's cratons opened up vast areas of shallow seas that could be inhabited and exploited. Furthermore, the movement of continents not only affected oceanic circulation patterns, but also resulted in environmental changes and shifting biotic provinces.

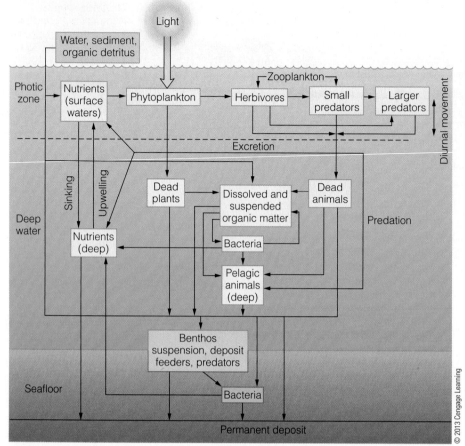

▶ Figure 12.4 **Marine Food Web**
Marine food web showing the relationship between the producers (phytoplankton), consumers (herbivores, small and large predators, pelagic animals, and benthos), and decomposers (bacteria).

© 2013 Cengage Learning

TABLE 12.1 The Major Invertebrate Groups and Their Stratigraphic Ranges

Phylum Protozoa	Cambrian—Recent	**Phylum Mollusca**	Cambrian—Recent
Class Sarcodina	Cambrian—Recent	Class Monoplacophora	Cambrian—Recent
Order Foraminifera	Cambrian—Recent	Class Gastropoda	Cambrian—Recent
Order Radiolaria	Cambrian—Recent	Class Bivalvia	Cambrian—Recent
Phylum Porifera	Cambrian—Recent	Class Cephalopoda	Cambrian—Recent
Class Demospongea	Cambrian—Recent	**Phylum Annelida**	Precambrian—Recent
Order Stromatoporida	Cambrian—Oligocene	**Phylum Arthropoda**	Cambrian—Recent
Phylum Archaeocyatha	Cambrian	Class Trilobita	Cambrian—Permian
Phylum Cnidaria	Cambrian—Recent	Class Crustacea	Cambrian—Recent
Class Anthozoa	Ordovician—Recent	Class Insecta	Silurian—Recent
Order Tabulata	Ordovician—Permian	**Phylum Enchinodermata**	Cambrian—Recent
Order Rugosa	Ordovician—Permian	Class Blastoidea	Ordovician—Permian
Order Scleractinia	Triassic—Recent	Class Crinoidea	Cambrian—Recent
Phylum Bryozoa	Ordovician—Recent	Class Echinoidea	Ordovician—Recent
Phylum Brachiopoda	Cambrian—Recent	Class Asteroidea	Ordovician—Recent
Class Inarticulata	Cambrian—Recent	**Phylum Hemichordata**	Cambrian—Recent
Class Articulata	Cambrian—Recent	Class Graptolithina	Cambrian—Mississippian

© 2013 Cengage Learning

Cambrian Marine Community

The Cambrian Period was a time during which many new body plans evolved and animals moved into new ecologic niches. As might be expected, the Cambrian witnessed a higher percentage of such experiments than any other period of geologic history.

Although almost all of the major invertebrate phyla evolved during the Cambrian Period (Table 12.1), many were represented by only a few species. Whereas trace fossils are common and echinoderms diverse, trilobites, inarticulate brachiopods, and archaeocyathids comprised the majority of Cambrian skeletonized life (▶ Figure 12.5).

▶ Figure 12.5
**Cambrian Marine
Community** Reconstruc-
tion of a Cambrian marine
community showing float-
ing jellyfish, swimming
arthropods, benthonic
sponges, and scavenging
trilobites.

Trilobites were by far the most conspicuous element of the Cambrian marine invertebrate community and made up approximately half of the total fauna. Most trilobites were benthonic, mobile, sediment-deposit feeders that crawled or swam along the seafloor (see Perspective).

Cambrian *brachiopods* were mostly primitive types known as *inarticulates* that secreted a chitinophosphate shell that was composed of the organic compound chitin combined with calcium phosphate. Inarticulate brachiopods also lacked a tooth-and-socket arrangement along the hinge line where the two shells pivot. The other group of brachiopods, the *articulates*, secreted a calcium carbonate shell and had a tooth-and-socket arrangement along their hinge line. The articulate brachiopods were also present during the Cambrian but did not become abundant until the Ordovician period. Both groups of brachiopods were benthonic, sessile, suspension feeders.

The third major group of Cambrian organisms was the *archaeocyathids* (▶ Figure 12.6). These organisms were benthonic, sessile, presumably suspension feeders that constructed reeflike structures beginning in the Early Cambrian, were abundant during the Early and Middle Cambrian, and went extinct at the end of the Cambrian.

The remainder of the Cambrian invertebrate fauna consisted of representatives of most of the other major invertebrate phyla, including many organisms that were short-lived evolutionary experiments. As might be expected during times of adaptive radiation and evolutionary experimentation, many of the invertebrates that evolved during the Cambrian soon became extinct.

The Burgess Shale Biota

No discussion of Cambrian life would be complete without mentioning one of the best examples of a preserved soft-bodied fauna and flora: the Burgess Shale biota. As the Sauk Sea transgressed from the Cordilleran shelf onto the western edge of the craton, Early Cambrian–age sands were covered by Middle Cambrian–age black muds that allowed a diverse soft-bodied benthonic community to be preserved. As we discussed in the Introduction, these fossils were discovered in 1909 by Charles Walcott near Field, British Columbia, Canada. They represent one of the most significant fossil finds of the 20th century because they consist of carbonized impressions of soft-bodied animals and plants that are rarely preserved in the fossil record.

▶ Figure 12.6 **Archaeocyathids** Models of archaeocyathids in life position and lying on the seafloor after dying and falling over. Archaeocyathids were vase-shaped, benthonic, sessile, presumably suspension feeders that constructed reeflike structures during the Cambrian. Visible in these models is the porate nature of the walls, and the holdfasts that kept the organism anchored to the substrate.

Trilobites—Paleozoic Arthropods

Trilobites, an extinct class of arthropods, are probably the favorite and most sought after of invertebrate fossils. They lived from the Early Cambrian until the end of the Permian and were most diverse during the Late Cambrian. More than 15,000 species of trilobites have been described, and they are currently grouped into nine orders.

Trilobites had a worldwide distribution throughout the Paleozoic, and they lived in all marine environments from shallow, nearshore waters to deep oceanic settings. They occupied a wide variety of habitats. Most were bottom-dwellers, crawling around the seafloor and scavenging organic detritus or feeding on microorganisms or algae. Others were free-swimming predators or filter-feeders living throughout the water column.

The trilobite body is divided into three parts (Figure 1): a cephalon (head), containing the eyes, mouth, and sensory organs; the thorax (body), composed of individual segments; and the pygidium (tail). Interestingly, the name *trilobite* does not refer to its three main body parts, but means *three-lobed,* which corresponds to the three longitudinal lobes of the thorax—the axial lobe, and the two flanking pleural lobes.

The appendages of trilobites are rarely preserved. However, from specimens in which the soft-part anatomy is preserved as an impression, paleontologists know that beneath each thoracic segment was a two-part appendage consisting of a gill-bearing outer branch used for respiration and an inner branch, or walking leg, composed of articulating limb segments (Figure 2).

Trilobites range in size from a few millimeters in length, as found in many of the agnostid trilobites (Figure 3), to more than 70 cm, as recorded by the world's largest trilobite

Figure 1 *Cedaria minor,* from the Cambrian Weeks Formation, Utah, illustrates the major body parts of a trilobite.

Labels: Axial lobe, Cephalon, Thorax, Pygidium, Pleural lobes

(Figure 4). Most trilobites, however, range between 3 and 10 cm long.

Trilobites were either blind, as in the agnostids (Figure 3), or possessed

Figure 2 Model of dorsal (a) and ventral (b) anatomy of *Triarthrus eatoni,* a Late Ordovician trilobite from the Frankfort Shale, New York.

(a) Dorsal view.

(b) Ventral view.

Figure 3 *Hypagnostus parvifrons,* a small (5–8 mm long) agnostid trilobite from the Cambrian Marjum Formation, Utah.

holochroal or schizochroal paired eyes. *Holochroal eyes* are characterized by generally hexagonal-shaped, biconvex lenses that are packed closely together and are all covered by a single corneal layer (Figure 5). This type of eye is similar to a modern compound eye and results in mosaic vision. *Schizochroal eyes*, in contrast, are composed of relatively large, individual lenses, each of which is covered by its own cornea and separated from adjacent lenses by a thick dividing wall (Figure 6). Schizochroal lenses are arranged in rows and columns, and they most likely produced a visual field consisting of shadows.

Although trilobites had a hard dorsal exoskeleton, useful, in part, as a protection against predators, some trilobites also had the ability to enroll, presumably to protect their antennae, limbs, and soft ventral appendages. By doing so, trilobites could view their surroundings while protecting their soft ventral anatomy (Figure 7).

As mentioned earlier, trilobites were a very successful group. They first appeared in the Early Cambrian, rapidly diversified, reached their maximum diversity in the Late Cambrian, and then suffered major reductions in diversity near the end of the Cambrian, and

again at the end of the Ordovician, and once again near the end of the Devonian. As yet, no consensus exists on what caused the trilobite extinctions, but a combination of factors was likely involved, possibly including a reduction of shelf space, increased competition, and a rise in predators. Some scientists have also suggested that a cooling of the seas may have played a role, particularly for the extinctions at the end of the Ordovician Period. The trilobites, like most of their Permian marine invertebrate contemporaries, finally fell victim to the Permian mass extinction event.

Figure 4 *Isotellus rex,* the world's largest trilobite (more than 70 cm long), from the Ordovician Churchill River Group, Manitoba, Canada.

Figure 5 Holochroal eyes from the trilobite *Scutellum campaniferum.*

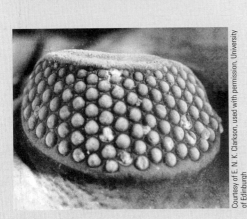

Figure 6 Schizochroal eyes from the trilobite *Eophacops trapeziceps.*

Figure 7 An enrolled specimen of the Ordovician trilobite *Pibmera fisheri* from Putilowa, Poland.

Figure 12.7 Fossils from the Burgess Shale Some of the fossil animals preserved in the Burgess Shale.

(b) *Hallucigenia*, a velvet worm.

(a) *Wiwaxia,* a scaly armored, sluglike creature whose affinities remain controversial.

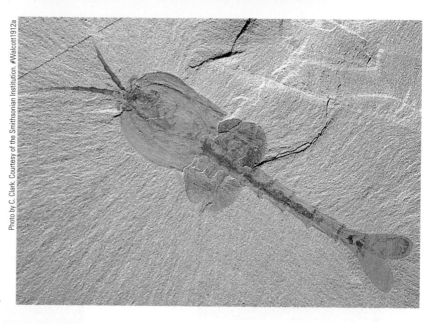

(c) *Waptia,* an arthropod.

Consequently, they thus provide us with a valuable glimpse of rarely preserved organisms, as well as the soft-part anatomy of many extinct groups (▶ Figures 12.7 and 12.2).

In recent years, the reconstruction, classification, and interpretation of many of the Burgess Shale fossils have undergone a major change that has led to new theories and explanations of the Cambrian explosion of life. Recall that during the Neoproterozoic, multicelled organisms evolved, and shortly thereafter animals with hard parts made their first appearance. These were followed by an explosion of invertebrate groups during the Cambrian, many of which are now extinct. These Cambrian organisms represent the root-stock and basic body plans from which all present-day invertebrates evolved.

The question that paleontologists are still debating is how many phyla arose during the Cambrian, and at the center of that debate are the Burgess Shale fossils. For years, most paleontologists placed the bulk of the Burgess Shale organisms into existing phyla, with only a few assigned to phyla that are now extinct. Thus, the phyla of the Cambrian world were viewed as being essentially the same in number as the phyla of the present-day world but with fewer species in each phylum. According to this view, the history of life has been simply a gradual increase in the diversity of species within each phylum through time. The number of basic body plans has therefore remained more or less constant since the initial radiation of multicelled organisms.

This view, however, has been challenged by other paleontologists who think the initial explosion of varied life-forms in the Cambrian was promptly followed by a short period of experimentation and then extinction of many phyla. According to this opposing view, the richness and diversity of modern life-forms are the result of repeated variations of the basic body plans that survived the Cambrian extinctions. In other words, life was much more diverse in terms of phyla during the Cambrian than it is today. The reason why many members of the Burgess Shale biota look so strange to us is that no living organisms possess their basic body plans, and therefore many of them have been reassigned into new phyla.

Discoveries of additional Cambrian fossils at localities such as Sirius Passet, Greenland, and Yunnan, China, have resulted in reassignment of some Burgess Shale specimens back into extant phyla. If these reassignments to known phyla prove to be correct, then no massive extinction event followed the Cambrian explosion, and life has gradually increased in diversity through time. Currently, there is no clear answer to this debate, and the outcome will probably be decided as more fossil discoveries are made.

Ordovician Marine Community

A major transgression that began during the Middle Ordovician (Tippecanoe Sequence) resulted in widespread inundation of the North American craton (see Figure 10.7). This vast, uniformly warm epeiric sea opened many new marine habitats that were soon filled by a variety of organisms.

Not only did sedimentation patterns change dramatically from the Cambrian to the Ordovician, but the fauna underwent equally striking changes. Whereas trilobites, primitive brachiopods, and archaeocyathids dominated the Cambrian invertebrate community, the Ordovician was characterized by the continued diversification of brachiopods and by the adaptive radiation of many other invertebrate phyla (such as bryozoans, and corals), with a consequent dramatic increase in the diversity of the total shelly fauna (▶Figure 12.8). In fact, so spectacular was this increase in marine invertebrate diversity, it is called "The Great Ordovician Biodiversification Event." Concomitant with the increased diversity in the Ordovician marine invertebrate fauna was an equally striking rise in diversity and abundance of the *acritarchs* (organic-walled phytoplankton of unknown affinity), which were the major phytoplankton group of the Paleozoic Era and the primary food source of the suspension feeders (▶Figure 12.9).

During the Cambrian, archaeocyathids were the main builders of reeflike structures, but bryozoans, stromatoporoids, and tabulate and rugose corals (▶Figure 12.10) assumed that role beginning in the Middle Ordovician. Many of these reefs were small patch reefs similar in size to those of the Cambrian but of a different composition, whereas others were quite large. As with present-day reefs, Ordovician reefs showed a high diversity of organisms and were dominated by suspension feeders.

▶ Figure 12.8 **Middle Ordovician Marine Community** Reconstruction of a Middle Ordovician seafloor fauna. Cephalopods, crinoids, colonial corals, bryozoans, trilobites, and brachiopods are shown.

(a) *Aremoricanium* sp. Ordovician.

(b) *Navifusa* sp. Devonian.

(a) *Halysites*, an Ordovician–Silurian colonial tabulate coral from Michigan.

(c) *Multiplicisphaeridium* sp. Ordovician.

(d) *Carminella* sp. Silurian.

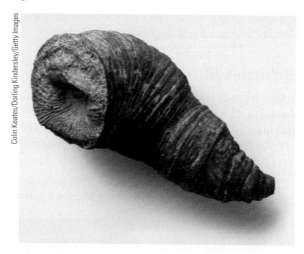

(b) *Heliophylum*, a Devonian solitary rugose coral. Solitary rugose corals are popularly called horn corals.

In addition to an increase in diversity among the various reef-building organisms, three Ordovician fossil groups have proved particularly useful for biostratigraphic correlation—the articulate brachiopods, graptolites, and conodonts. The *articulate brachiopods*, present since the Cambrian, began a period of major diversification in the shallow-water marine environment during the Ordovician (▶ Figure 12.11). They became a conspicuous element of the invertebrate fauna during the Ordovician and in suc-ceeding Paleozoic periods.

The vast majority of *graptolites* were planktonic and thus carried about by ocean currents. Furthermore, most species existed for less than a million years, thus making graptolites excellent guide fossils (see Figure 5.19), par-ticularly for the Ordovician and Silurian periods, during which they were especially abundant (▶ Figure 12.12).

Conodonts are a group of well-known, small, tooth-like fossils composed of the mineral apatite (calcium phosphate), the same mineral that composes bone (▶ Fig-ure 12.13). Their wide geographic distribution and short stratigraphic range of individual elements, or species, make conodonts excellent fossils for biostratigraphic zona-tion and correlation.

Although conodonts have been known for more than 150 years, it was not until the discovery of a conodont ani-mal in the 1980s that their biological affinity was revealed. Conodonts are now considered to be the elements that make up a bilaterally symmetrical feeding apparatus located in the head area of an eel-like swimming organism. This creature, based on numerous carbonized impressions, is presently classified as a member of a group of primitive jawless ani-mals assigned to the phylum Chordata (see Chapter 13).

▶ Figure 12.13 **Conodonts** Conodonts are microscopic, toothlike fossils. Shown are 20 different Devonian conodonts from Rockford, Iowa.

▶ Figure 12.11 **Brachiopods** Brachiopods are benthonic, sessile, suspension feeders. Shown here are various Ordovician articulate brachiopods from Ohio. The largest brachiopod specimens on this limestone slab are *Rafinesquina alternata*.

▶ Figure 12.12 **Graptolites** Graptolites are predominantly planktonic, suspension-feeding organisms, and are excellent guide fossils. These tuning-fork–shaped graptolites are *Didymograptus murchisoni* from Wales, United Kingdom.

The end of the Ordovician was a time of mass extinctions in the marine realm. More than 100 families of marine invertebrates became extinct, and in North America alone, about half of the brachiopods and bryozoans died out. What caused such an event? Many geologists think these extinctions were the result of extensive glaciation at the end of the Ordovician Period as Gondwana moved over the south pole area (see Chapter 10). This widespread glaciation resulted in decreased sea level, and a cooling of the surface waters, thus creating stressful conditions for the near-surface phytoplankton community (acritarchs) and the shallow-water marine faunas, especially the suspension-feeding organisms (see Chapter 10).

Silurian and Devonian Marine Communities

The mass extinction at the end of the Ordovician was followed by rediversification and recovery of many of the decimated groups. Brachiopods, bryozoans, gastropods, bivalves, corals, crinoids, and graptolites were just some of the groups that rediversified beginning in the Silurian.

As we discussed in Chapters 10 and 11, the Silurian and Devonian were times of major reef building. Whereas most of the Silurian radiations of invertebrates represented the repopulation of niches, organic reef builders diversified in new ways, building massive reefs larger than any produced during the Cambrian or Ordovician. This repopulation was probably caused, in part, by renewed transgressions over the cratons. Although a major drop in sea level occurred at the end of the Silurian, the Middle Paleozoic sea level was generally high (see Table 10.1).

The Silurian and Devonian reefs were dominated by tabulate and colonial rugose corals and stromatoporoids (▶ Figure 12.14). Although the fauna of these Silurian and Devonian reefs was somewhat different from that of earlier reefs and reeflike structures, the general composition and structure are the same as in present-day reefs (see Figure 10.9).

The Silurian and Devonian periods were also the time when *eurypterids* (arthropods with scorpion-like bodies and impressive pincers) were abundant. Unlike many other marine invertebrates, eurypterids expanded into brackish and freshwater habitats (▶ Figure 12.15). *Ammonoids,* a subclass of cephalopods, evolved from nautiloids during the Early Devonian and rapidly diversified. With their distinctive suture patterns, short stratigraphic ranges, and widespread distribution, ammonoids are excellent guide fossils for the Devonian through Cretaceous periods (▶ Figure 12.16).

Near the end of the Devonian, another mass extinction occurred that resulted in a worldwide near-total collapse

► Figure 12.14 **Middle Devonian Marine Reef Community** Reconstruction of a Middle Devonian reef from the Great Lakes area of North America. Shown are corals, bryozoans, cephalopods, trilobites, crinoids, and brachiopods.

► Figure 12.15 **Silurian Brackish Water Community** Restoration of a Silurian brackish water scene near Buffalo, New York. Shown are algae, eurypterids, gastropods, worms, and shrimp.

of the massive marine reef communities. On land, however, the seedless vascular plants were seemingly unaffected (see Chapter 13). Thus, extinctions at this time were most extensive among marine life, particularly in the reef and pelagic communities.

The demise of the Middle Paleozoic reef communities highlights the geographic aspects of the Late Devonian mass extinction. The tropical groups were most severely affected; in contrast, the higher-latitude communities were seemingly little affected. Apparently, an episode of global cooling was largely responsible for the extinctions near the

► Figure 12.16 **Ammonoid Cephalopod** A Late Devonian-age ammonoid cephalopod from Erfoud, Morocco. The distinctive suture pattern, short stratigraphic range, and wide geographic distribution make ammonoids excellent guide fossils.

The American Museum of Natural History, #K10257

▶ Figure 12.17 **Late Mississippian Marine Community** Reconstruction of marine life during the Mississippian, based on an Upper Mississippian fossil site at Crawfordsville, Indiana. Invertebrate animals shown include crinoids, blastoids, lacy bryozoans, brachiopods, and small corals.

end of the Devonian. During such a cooling, the disappearance of tropical conditions would have had a severe effect on reef and other warm-water organisms. Cool-water species, in contrast, could have simply migrated toward the equator. Although cooling temperatures certainly played an important role in the Late Devonian extinctions, the closing of the Iapetus Ocean and the orogenic events of this time (see Figure 11.1a) undoubtedly also played a role by reducing the area of shallow shelf environments where many marine invertebrates lived.

Carboniferous and Permian Marine Communities

The Carboniferous invertebrate marine community responded to the Late Devonian extinctions in much the same way that the Silurian invertebrate marine community responded to the Late Ordovician extinctions—that is, by renewed adaptive radiation and rediversification. The brachiopods and ammonoids quickly recovered and again assumed important ecologic roles. Other groups, such as the lacy bryozoans and crinoids, reached their greatest diversity during the Carboniferous. With the decline of the stromatoporoids and the tabulate and rugose corals, large organic reefs such as those existing earlier in the Paleozoic virtually disappeared and were replaced by small patch reefs. These reefs were dominated by crinoids, blastoids, lacy bryozoans, brachiopods, and calcareous algae and flourished during the Late Paleozoic (▶Figure 12.17). In addition, bryozoans and crinoids contributed large amounts of skeletal debris to the formation of the vast bedded limestones that constitute the majority of Mississippian sedimentary rocks.

The Permian invertebrate marine faunas resembled those of the Carboniferous, but were not as widely distributed because of the restricted size of the shallow seas on the cratons and the reduced shelf space along the continental margins (see Figure 11.11). The spiny and odd-shaped *productids* dominated the brachiopod assemblage and constituted an important part of the reef complexes that formed in the Texas region during the Permian (▶Figure 12.18). The *fusulinids* (spindle-shaped foraminifera), which first evolved during the Late Mississippian and greatly diversified during the Pennsylvanian (▶Figure 12.19), experienced a further diversification during the Permian. Because of their abundance, diversity, and worldwide occurrence, fusulinids are important guide fossils for Pennsylvanian and Permian strata. Bryozoans, sponges, and various types of calcareous algae also were common elements of the Permian invertebrate fauna.

Mass Extinctions

Throughout geologic history, various plant and animal species have become extinct. In fact, extinction is a common feature of the fossil record, and the rate of extinction through time has fluctuated only slightly. Just as new species evolve, others become extinct. There have, however, been brief intervals in the geologic past during which mass extinctions have eliminated large numbers of species. Extinctions of this magnitude could only occur due to radical changes in the environment on a regional or global scale.

When we look at the different mass extinctions that have occurred during the geologic past, several common

► Figure 12.18 **Permian Patch-Reef Marine Community** Reconstruction of a Permian patch-reef community from the Glass Mountains of West Texas. Shown are algae, productid brachiopods, cephalopods, sponges, and corals.

► Figure 12.19 **Fusulinids** Fusulinids are spindle-shaped, microscopic, benthonic foraminifera that are excellent guide fossils for the Pennsylvanian and Permian periods. Shown here are three natural cross sections of fusulinids from the Lower Permian Owens Valley Group, Inyo County, California. The two elongated specimens are *Parafusulina* sp., and the circular specimen in the lower right area (view is a cross section perpendicular to the long axis of the specimen) is an unidentified fusulinid.

themes stand out. The first is that mass extinctions typically have affected life both in the sea and on land. Second, tropical organisms, particularly in the marine realm, apparently are more affected than organisms from the temperate and high-latitude regions. Third, some groups within different animal phyla experienced extinctions, but the phyla themselves persisted.

When we examine the mass extinctions for the past 650 million years, we see that the first major extinction involved only the acritarchs. Several extinction events occurred during the Cambrian, and these affected only marine invertebrates, particularly trilobites. Three other marine mass extinctions took place during the Paleozoic Era: one at the end of the Ordovician, involving many invertebrates; one near the end of the Devonian, affecting the major barrier reef-building organisms, as well as the primitive armored fish; and the most severe at the end of the Permian, when about 96 percent of all marine invertebrate species and more than 65 percent of all land animals became extinct.

The Mesozoic Era also experienced several mass extinctions, the most devastating occurring at the end of the Cretaceous, when almost all large animals, including dinosaurs, flying reptiles, and seagoing animals such as plesiosaurs and ichthyosaurs, became extinct. Many scientists think the terminal Cretaceous mass extinction was caused by a meteorite impact (see Chapter 15).

Several mass extinctions also occurred during the Cenozoic Era (see Chapter 18). The most severe was near the end of the Eocene Epoch and is correlated with global cooling and climatic change. The most recent extinction occurred near the end of the Pleistocene Epoch and affected mostly the large mammalian fauna.

Although many scientists think of the marine mass extinctions as sudden events from a geologic perspective, they were rather gradual from a human viewpoint, occurring over hundreds of thousands and even millions of years. Furthermore, many geologists think that climatic changes, rather than a single catastrophic event, were primarily responsible for the extinctions, particularly in the marine realm. Evidence of glacial episodes or other signs of climatic change, such as global warming, have been correlated with the various extinctions recorded in the geologic record.

The Permian Mass Extinction

The greatest recorded mass extinction to affect Earth occurred at the end of the Permian Period (▶Figure 12.20). By the time the Permian ended, roughly 50 percent of all marine invertebrate families and about 96 percent of all marine invertebrate species became extinct. Fusulinids, rugose and tabulate corals, and several bryozoan and brachiopod orders, as well as trilobites and blastoids, did not survive the end of the Permian. All of these groups had been very successful during the Paleozoic Era. In addition, approximately 70 percent of all terrestrial vertebrate species, as well as nearly 33 percent of insects on land, also became extinct.

What caused such a crisis for both marine and land-dwelling organisms? Various hypotheses have been proposed, but no completely satisfactory answer has yet been found. Because the extinction event extended over many millions of years at the end of the Permian, a meteorite impact such as occurred at the end of the Cretaceous Period (see Chapter 15) can be reasonably discounted.

A reduction in the habitable marine shelf area caused by the formation of Pangaea and a widespread marine regression resulting from glaciation can also be rejected as the primary cause of the Permian extinctions. By the end of the Permian, most collisions of the continents had already taken place, such that the reduction in the shelf area had already occurred before the mass extinctions began in earnest. Furthermore, the widespread glaciation that took place during the Carboniferous was now waning in the Permian.

Many scientists think that an episode of deep-sea anoxia and increased oceanic carbon dioxide levels resulted in a highly stratified ocean during the Late Permian. In other words, there was very little, if any, circulation of oxygen-rich surface waters into the deep ocean. Also, during this time, stagnant waters covered the shelf regions, thus affecting the shallow marine fauna.

In addition, there is evidence of increased global warming during the Late Permian. This factor would contribute to a stratified global ocean because warming of the high latitudes would significantly reduce or even eliminate the downwelling of cold, dense, oxygenated waters from the polar areas into the deep oceans at lower latitudes that occurs today. Increased global warming would also result in stagnant, stratified oceans rather than a well-mixed, oxygenated oceanic system.

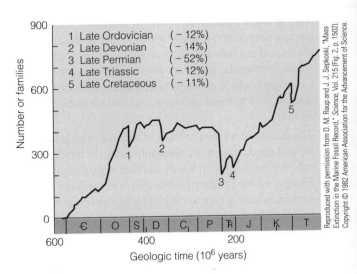

▶ Figure 12.20 **Phanerozoic Marine Diversity** Phanerozoic diversity for marine invertebrate and vertebrate families. Note the three episodes of Paleozoic mass extinctions, with the greatest occurring at the end of the Permian Period.

During the Late Permian, widespread volcanic and continental fissure eruptions such as the Siberian Traps, where lava flows covered more than 2 million km², were also taking place. These eruptions released not only large amounts of additional carbon dioxide into the atmosphere, but also high amounts of fluorine and chlorine, which could have damaged the ozone layer and helped to increase climatic instability and ecologic collapse.

Several recent studies have examined the role that the Siberian Traps flood basalts may have played in causing widespread ozone depletion, thus allowing an increase in UV-B radiation at the end of the Permian. These studies have included using a two-dimensional atmospheric chemistry-transport model to assess the impact of the Siberian Traps eruption in altering the end-Permian stratospheric ozone layer as well as an analysis of the relationship of morphologic abnormalities in gymnosperm (flowerless seed plant) pollen grains to deteriorating conditions around the Permian–Triassic boundary.

By the end of the Permian, a near collapse of both the marine and terrestrial ecosystem had occurred. Although the ultimate cause of such devastation is still being debated and investigated, it is safe to say that it was probably a combination of interconnected and related geologic and biologic events.

Summary

Table 12.2 summarizes the major evolutionary and geologic events of the Paleozoic Era and shows their relationships to each other.

▶ Multicelled organisms presumably had a long Precambrian history during which they lacked hard parts.

▶ Invertebrates with hard parts "suddenly" appeared during the Early Cambrian in what is called the "Cambrian explosion." Skeletons provided such advantages as protection against predators and support for muscles, enabling organisms to grow large and increase locomotor efficiency. Hard parts probably evolved as a result of several geologic and biologic factors rather than a single cause.

▶ Marine organisms are classified as plankton if they are floaters, nekton if they swim, and benthos if they live on or in the seafloor.

- Marine organisms are divided into four basic feeding groups: suspension feeders, which consume microscopic plants and animals as well as dissolved nutrients from water; herbivores, which are plant eaters; carnivore-scavengers, which are meat eaters; and sediment-deposit feeders, which ingest sediment and extract nutrients from it.

- The marine ecosystem consists of various trophic levels of food production and consumption. At the base are primary producers, on which all other organisms depend. Feeding on the primary producers are the primary consumers, which in turn are fed on by higher levels of consumers. The decomposers are bacteria that break down the complex organic compounds of dead organisms and recycle them within the ecosystem.

- The Cambrian invertebrate community was dominated by three major groups: the trilobites, inarticulate brachiopods, and archaeocyathids. Little specialization existed among the invertebrates, and most phyla were represented by only a few species.

- The Middle Cambrian Burgess Shale contains one of the finest examples of a well-preserved, soft-bodied biota in the world and provides us with an important glimpse of rarely preserved organisms, plus the soft-part anatomy of many extinct groups.

- The Ordovician marine invertebrate community marked the beginning of the dominance of the shelly fauna and the start of large-scale reef building. The end of the Ordovician Period was a time of major extinction of many invertebrate phyla.

- The Silurian and Devonian periods were times of diverse faunas dominated by reef-building animals, whereas the Carboniferous and Permian periods saw a great decline in invertebrate diversity.

- Mass extinctions are times when anomalously high numbers of organisms go extinct in a short period of time. Such events have occurred several times during the past 650 million years.

- A major extinction occurred at the end of the Paleozoic Era, affecting the invertebrates as well as the vertebrates. Its cause is still the subject of debate.

Important Terms

benthos, p. 236
carnivore-scavenger, p. 236
herbivore, p. 236

nekton, p. 236
plankton, p. 236
primary producer, p. 237

sediment-deposit feeder, p. 236
suspension feeder, p. 236

Review Questions

1. The age of the Burgess Shale is
   a. _____ Cambrian; b. _____ Ordovician; c. _____ Silurian; d. _____ Devonian; e. _____ Mississippian.

2. Which group of benthonic, mobile, sediment-deposit feeders made up approximately half of the total invertebrate marine fauna during the Cambrian?
   a. _____ Graptolites; b. _____ Brachiopods; c. _____ Trilobites; d. _____ Cephalopods; e. _____ Fusulinids.

3. Organisms living on or in the seafloor are
   a. _____ pelagic; b. _____ epifaunal; c. _____ infaunal; d. _____ planktic; e. _____ benthonic.

4. The _____ and _____ were times of major reef building.
   a. _____ Cambrian, Ordovician; b. _____ Ordovician, Silurian; c. _____ Silurian, Devonian; d. _____ Devonian, Mississippian; e. _____ Mississippian, Pennsylvanian.

5. Organisms that manufacture their own food are
   a. _____ autotrophs; b. _____ herbivores; c. _____ benthos; d. _____ epifaunal; e. _____ none of the previous answers.

6. The Burgess Shale fauna is significant because it contains the
   a. _____ first shelled animals; b. _____ carbonized impressions of many extinct soft-bodied animals; c. _____ fossils of rare marine plants; d. _____ earliest known benthonic community; e. _____ conodont animal.

7. An exoskeleton is advantageous because it
   a. _____ prevents drying out in an intertidal environment; b. _____ provides protection against ultraviolet radiation; c. _____ provides protection against predators; d. _____ provides attachment sites for development of strong muscles; e. _____ all of the previous answers.

8. What type of invertebrates dominated the Ordovician invertebrate community?
   a. _____ Epifloral, planktonic, primary producers; b. _____ Infaunal, nektonic carnivores; c. _____ Infaunal, benthonic, sessile, suspension feeders; d. _____ Epifaunal, benthonic, mobile, suspension feeders; e. _____ Epifaunal, benthonic, sessile, suspension feeders.

9. Mass extinctions occurred at, or near the end, of which three periods?
   a. _____ Cambrian, Ordovician, Permian; b. _____ Cambrian, Silurian, Devonian; c. _____ Ordovician, Devonian, Permian; d. _____ Silurian, Devonian, Permian; e. _____ Cambrian, Devonian, Permian.

10. Which group was the major phytoplankton group during the Paleozoic Era, and thus the primary food source of the suspension feeders?
    a. _____ Dinoflagellats; b. _____ Coccolithophorids; c. _____ Graptolites; d. _____ Diatoms; e. _____ Acritarchs.

11. The earliest reeflike structures were constructed by
    a. _____ bryozoans; b. _____ mollusks; c. _____ archaeocyathids; d. _____ sponges; e. _____ corals.

12. The greatest recorded mass extinction in Earth history took place at the end of which period?
    a. _____ Cambrian; b. _____ Ordovician; c. _____ Devonian; d. _____ Permian; e. _____ Cretaceous.

13. Discuss how the incompleteness of the fossil record may play a role in what is known as the Cambrian explosion.

14. Draw a marine food web for the Paleozoic that shows the relationships among the producers, consumers, and decomposers.

15. If the Cambrian explosion of life was partly the result of filling unoccupied niches, why don't we see such rapid evolution following mass extinctions such as those that occurred at the end of the Permian and Cretaceous periods?

16. Discuss some of the possible causes for the Permian mass extinction.

17. Discuss how changing geologic conditions affected the evolution of invertebrate life during the Paleozoic Era.

18. What are the major differences between the Cambrian marine community and the Ordovician marine community?

19. Discuss the significance of the appearance of the first shelled animals and possible causes for the acquisition of a mineralized exoskeleton.

20. One concern of environmentalists is that environmental degradation is leading to vast reductions in global biodiversity. As a paleontologist, you are aware mass extinctions have taken place throughout Earth history. What facts and information can you provide from your geologic perspective that will help focus the debate as to whether or not Earth's biota is being adversely affected by such human activities as industrialization, and what the possible outcome(s) might be if global biodiversity is severely reduced?

TABLE 12.2 Major Evolutionary and Geologic Events of the Paleozoic Era

Age (millions of years)	Geologic Period		Invertebrates	Vertebrates	Plants	Major Geologic Events
251	Permian		Largest mass extinction event to affect the invertebrates	Acanthodians, placoderms, and pelycosaurs become extinct Therapsids and pelycosaurs the most abundant reptiles	Gymnosperms diverse and abundant	Formation of Pangaea Alleghenian orogeny Hercynian (Variscan) orogeny
299	Carboniferous	Pennsylvanian	Fusulinids diversify	Amphibians abundant and diverse	Coal swamps with flora of seedless vascular plants and gymnosperms	Coal-forming swamps common Formation of Ancestral Rockies Continental glaciation in Gondwana
318		Mississippian	Crinoids, lacy bryozoans, and blastoids become abundant Renewed adaptive radiation following extinctions of many reef-builders	Reptiles evolve	Gymnosperms appear (may have evolved during Late Devonian)	Ouachita orogeny Widespread deposition of black shale
359	Devonian		Extinctions of many reef-building invertebrates near end of Devonian Reef building continues Eurypterids abundant	Amphibians evolve All major groups of fish present—Age of Fish	First seeds evolve Seedless vascular plants diversify	Widespread deposition of black shale Acadian orogeny Antler orogeny
416	Silurian		Major reef building Diversity of invertebrates remains high	Ostracoderms common Acanthodians, the first jawed fish, fish evolve	Early land plants— seedless vascular plants	Caledonian orogeny Extensive barrier reefs and evaporites
444	Ordovician		Extinctions of a variety of marine invertebrates near end of Ordovician Major adaptive radiation of all invertebrate groups Suspension feeders dominant	Ostracoderms diversify	Plants move to land?	Continental glaciation in Gondwana Taconic orogeny
488	Cambrian		Many trilobites become extinct near end of Cambrian Trilobites, brachiopods, and archaeocyathids are most abundant	Earliest vertebrates— jawless fish called ostracoderms		First Phanerozoic transgression (Sauk) onto North American craton
542						

© 2013 Cengage Learning

13

Paleozoic Life History
Vertebrates and Plants

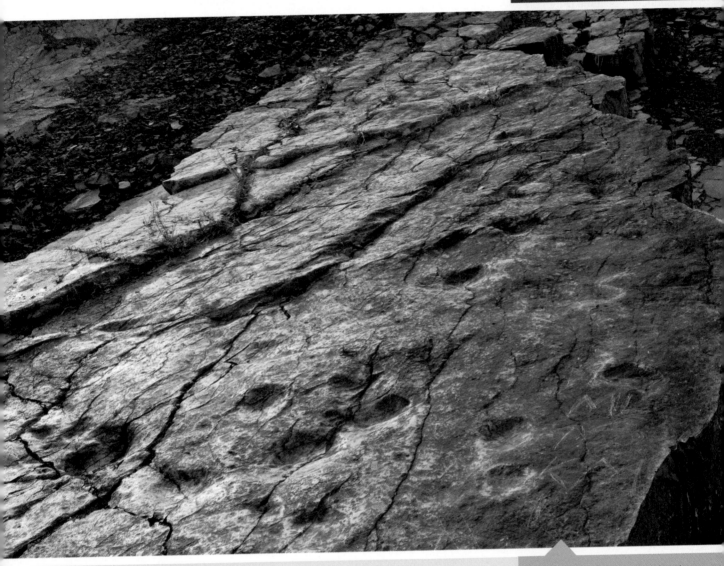

Tetrapod trackway exposed at the Zachelmie Quarry in the Holy Cross Mountains of Poland. The tracks, which measure up to 26 cm wide, are exposed on the bedding surface of 397-million-year-old (earliest Middle Devonian) shallow, marine rocks. These tracks represent the oldest known evidence of four-legged animals.

Courtesy of Ken Higgs, Department of Geology, University College Cork, Ireland

At the end of this chapter, you will have learned that

▶ Vertebrates first evolved during the Cambrian Period, and fish diversified rapidly during the Paleozoic Era.

▶ Amphibians first appear in the fossil record during the Late Devonian, having made the transition from water to land, and became extremely abundant during the Pennsylvanian Period when coal-forming swamps were widespread.

▶ The evolution of the amniote egg allowed reptiles to colonize all parts of the land beginning in the Late Mississippian.

▶ The pelycosaurs, or finback reptiles, were the dominant reptiles during the Permian and were the ancestors to the therapsids, or mammal-like reptiles.

▶ The earliest land plants are known from the Ordovician Period, whereas the oldest known vascular land plants first appear in the Middle Silurian.

▶ Seedless vascular plants, such as ferns, were very abundant during the Pennsylvanian Period.

▶ With the onset of arid conditions during the Permian Period, the gymnosperms became the dominant element of the world's flora.

Introduction

The recent discovery of fossilized tetrapod footprints and trackways from 397-million-year-old (earliest Middle Devonian) shallow, marine rocks in Poland has forced paleontologists into a reappraisal of the origin and early evolution of tetrapods (from the Greek *tetra,* meaning "four," and *podos,* meaning "foot"). The footprints, made by several individuals, measure up to 26 cm wide and are indicative of animals ranging in size from 0.5 to 2.5 m in length (see the chapter opening photo).

The significance of this discovery is that the transition from lobe-finned fish to tetrapod might have taken place even earlier than previously thought, and furthermore, that the earliest tetrapods moved around in the mud of a coral-reef lagoon, and not in streams, lakes, or swamps, as previously thought. A discovery such as this, with its implications regarding the lobe-finned fish to tetrapod transition, once again illustrates how science works, and how theories have to be modified as new information becomes available. The transition from water to land, and the role limbs played in this adaptation to a new environment, is further discussed in this chapter.

In Chapter 12, we examined the Paleozoic history of invertebrates, beginning with the acquisition of hard parts and concluding with the massive Permian extinctions. In this chapter, we examine the Paleozoic evolutionary history of vertebrates and plants. One of the striking parallels between plants and animals is that in making the transition from water to land, they had to solve the same basic problems. For both groups, the method of reproduction proved the major barrier to expansion into the various terrestrial environments. With the evolution of the seed in plants and the amniote egg in animals, this limitation was removed, and both groups expanded into all terrestrial habitats.

Vertebrate Evolution

A **chordate** (phylum Chordata) is an animal that has, at least during part of its life cycle, a notochord, a dorsal hollow nerve cord, and pharyngeal slits (▶ Figure 13.1). **Vertebrates**, which are animals with backbones, are simply a subphylum of chordates.

▶ Figure 13.1 **Three Characteristics of a Chordate** The structure of the lancelet *Amphioxus* (approximately 5 cm long) illustrates the three characteristics of a chordate: a notochord, a dorsal hollow nerve cord, and pharyngeal slits. Reproduced with permission from "The Lancelet" by M. Dale Stohes and N. D. Hallond. *American Scientist* 86: 552–560, Fig. 2. Copyright © 1998 American Scientist.

▶ **Figure 13.2** *Yunnanozoon lividum* Found in 525-million-year-old rocks in Yunnan province, China, *Yunnanozoon lividum*, a 5-cm-long animal, is one of the oldest known chordates.

▶ **Figure 13.3** **Cell Cleavage**

© 2013 Cengage Learning

© 2013 Cengage Learning

Spiral Cleavage

Radial Cleavage

(a) Arrangement of cells resulting from radial cleavage is characteristic of chordates and echinoderms. In this configuration, cells are directly above each other.

(b) Arrangement of cells resulting from spiral cleavage. In this arrangement, cells in successive rows are nested between each other. Spiral cleavage is characteristic of all invertebrates except echinoderms.

The ancestors and early members of the phylum Chordata were soft-bodied organisms that left few fossils (▶Figure 13.2). Consequently, we know little about the early evolutionary history of the chordates, and likewise little about the subphylum Vertebrata. Surprisingly, a close relationship exists between echinoderms (see Table 12.1) and chordates, and they may even have shared a common ancestor. This is because the development of the embryos of echinoderms and chordates are the same in both groups—that is, the cells divide by radial cleavage so that they are aligned directly above each other (▶Figure 13.3a). In all other invertebrates, cells undergo spiral cleavage, which results in having cells nested between each other in successive rows (Figure 13.3b). Furthermore, the biochemistry of muscle activity and blood proteins, as well as the larval stages, are similar in both echinoderms and chordates.

The evolutionary pathway to vertebrates thus appears to have taken place much earlier and more rapidly than many scientists have long thought. Based on fossil evidence and recent advances in molecular biology, one scenario suggests that vertebrates evolved shortly after an ancestral chordate, probably resembling *Yunnanozoon* (Figure 13.2), acquired a second set of genes. According to this hypothesis, a random mutation produced a duplicate set of genes, letting the ancestral vertebrate animal evolve entirely new body structures that proved to be evolutionarily advantageous. Not all scientists accept this hypothesis, and the origin of vertebrates is still hotly debated.

Fish

The most primitive vertebrates are fish, and some of the oldest fish remains are found in the Upper Cambrian Deadwood Formation in northeastern Wyoming (▶Figure 13.4). Here, phosphatic scales and plates of *Anatolepis*,

a primitive member of the class Agnatha (jawless fish), have been recovered from marine sediments. All known Cambrian and Ordovician fossil fish have been found in shallow, nearshore marine deposits, whereas the earliest nonmarine (freshwater) fish remains have been found in Silurian strata. This does not prove that fish originated in the oceans, but it does lend strong support to the idea.

As a group, fish range from the Late Cambrian to the present (▶Figure 13.5). The oldest and most primitive of the class Agnatha are the **ostracoderms**, whose name means "bony skin" (Table 13.1). These are armored, jawless fish that first evolved during the Late Cambrian, reached their zenith during the Silurian and Devonian, and then became extinct.

The majority of ostracoderms lived on the seafloor. *Hemicyclaspis* is a good example of a bottom-dwelling ostracoderm (▶Figure 13.6a). Vertical scales allowed *Hemicyclaspis* to wiggle sideways, propelling itself along the seafloor, and the eyes on the top of its head allowed it to see such predators as cephalopods and jawed fish approaching from above. While moving along the sea bottom, it probably sucked up small bits of food and sediments through its jawless mouth. Another type of ostracoderm, represented

▶ **Figure 13.4** **Ostracoderm Fish Plate** A fragment of a plate from *Anatolepis* cf. *A. heintzi* from the Upper Cambrian Deadwood Formation of Wyoming. *Anatolepis* is one of the oldest known fish.

John E. Repetski/USGS

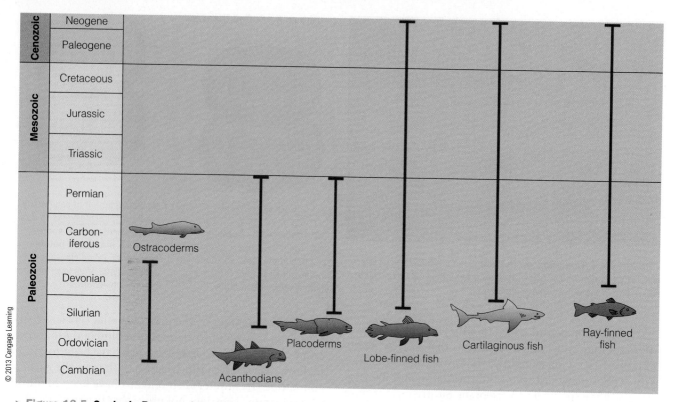

▶ **Figure 13.5** **Geologic Ranges of the Major Fish Groups** Ostracoderms are early members of the class Agnatha (jawless fish). Acanthodians (class Acanthodii) are the first fish with jaws. Placoderms (class Placodermii) are armored, jawed fish. Lobe-finned (subclass Sarcopterygii) and ray-finned (subclass Actinopterygii) fish are members of the class Osteichthyes (bony fish), whereas cartilaginous fish belong to the class Chondrichthyes.

by *Pteraspis,* was more elongated and probably an active swimmer, although it also seemingly fed on small pieces of food that it was able to suck up.

The evolution of jaws was a major evolutionary advance among primitive vertebrates. Although their jawless ancestors could feed only on detritus, jawed fish could chew food and become active predators, thus opening many new ecologic niches.

Indeed, the vertebrate jaw is an excellent example of evolutionary opportunism. Various studies suggest that the jaw originally evolved from the first two or three anterior gill arches of jawless fish. Because the gills are soft, they are supported by gill arches of bone or cartilage. The evolution of the jaw may thus have been related to respiration rather than to feeding (▶ Figure 13.7). By evolving joints in the forward gill arches, jawless fish could open their mouths

TABLE 13.1 **Brief Classification of Fish Groups Referred to in the Text**

Classification	Geologic Range	Living Example
Class Agnatha (jawless fish)	Late Cambrian–Recent	Lamprey, hagfish
Early members of the class are		
called ostracoderms	Late Cambrian Devonian	No living ostracoderms
Class Acanthodii (the first fish with jaws)	Early Silurian–Permian	None
Class Placodermii (armored jawed fish)	Late Silurian–Permian	None
Class Chondrichthyes (cartilagenous fish)	Devonian–Recent	Sharks, rays, skates
Class Osteichthyes (bony fish)	Devonian–Recent	Tuna, perch, bass, pike, catfish, trout, salmon, lungfish, *Latimeria*
Subclass Actinopterygii (ray-finned fish)	Devonian–Recent	Tuna, perch, bass, pike, catfish, trout, salmon
Subclass Sarcopterygii (lobe-finned fish)	Devonian–Recent	Lungfish, *Latimeria*
Order Coelacanthimorpha	Devonian–Recent	*Latimeria*
Order Dipnoi	Devonian–Recent	Lungfish
Order Crossopterygii	Devonian–Permian	None
Suborder Rhipidistia	Devonian–Permian	None

© 2013 Cengage Learning

▶ **Figure 13.6 Devonian Seafloor** Recreation of a Devonian seafloor showing (a) an ostracoderm, (b) a placoderm, (c) an acanthodian, and (d) a ray-finned fish.

wider. Every time a fish opened and closed its mouth, it would pump more water past the gills, thereby increasing the oxygen intake. The modification from rigid to hinged forward gill arches let fish increase both their food consumption and oxygen intake, and the evolution of the jaw as a feeding structure rapidly followed.

The fossil remains of the first jawed fish are found in Lower Silurian rocks and belong to the *acanthodians* (class Acanthodii), a group of small, enigmatic fish characterized by large spines, paired fins, scales covering much of the body, jaws, teeth, and greatly reduced body armor (Figure 13.6c, Table 13.1). Although their relationship to other fish is not yet well established, many scientists think the acanthodians included the ancestors of the present-day bony and cartilaginous fish groups. The acanthodians were most abundant during the Devonian, declined in importance through the Carboniferous, and became extinct in the Permian.

The other jawed fish, the **placoderms** (class Placodermii), whose name means "plate-skinned," evolved during the Silurian (Table 13.1). Placoderms were heavily armored, jawed fish that lived in both freshwater and the ocean. Like the acanthodians, they reached their peak of abundance and diversity during the Devonian.

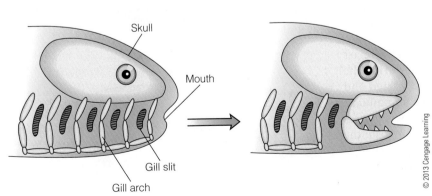

▶ **Figure 13.7 Evolution of the Vertebrate Jaw** The evolution of the vertebrate jaw is thought to have begun from the modification of the first two or three anterior gill arches. This theory is based on the comparative anatomy of living vertebrates.

Fish **257**

(d) *Cheirolepsis* (ray-finned)

(a) *Dunkleosteus* (placoderm)

(b) *Cladoselache* (cartilaginous)

(c) *Bothriolepsis* (placoderm)

▶ Figure 13.8 **Late Devonian Seascape** A Late Devonian marine scene from the midcontinent of North America. (a) The giant placoderm *Dunkleosteus* (length more than 12 m) is pursuing (b) the shark *Cladoselache* (length up to 1.2 m), a cartilaginous fish. Also shown are (c) the bottom-dwelling placoderm *Bothriolepsis* and (d) the swimming ray-finned fish *Cheirolepsis,* both of which reached a length of 40–50 cm.

The placoderms showed considerable variety, including small bottom dwellers (Figure 13.6b), as well as large major predators such as *Dunkleosteus,* a Late Devonian fish that lived in the mid-continental North American epeiric seas (▶Figure 13.8a). It was by far the largest fish of the time, reaching a length of more than 12 m. It had a heavily armored head and shoulder region, a huge jaw lined with razor-sharp bony teeth, and a flexible tail, all features consistent with its status as a ferocious predator.

Besides the abundant acanthodians, placoderms, and ostracoderms, other fish groups, such as the cartilaginous and bony fish, also evolved during the Devonian Period. Small wonder, then, that the Devonian is informally called the "Age of Fish," because all major fish groups were present during this time period.

The **cartilaginous fish** (class Chrondrichthyes) (Table 13.1), represented today by sharks, rays, and skates, first evolved during the Early Devonian, and by the Late Devonian, primitive marine sharks such as *Cladoselache* were abundant (Figure 13.8b). Cartilaginous fish have never been as numerous or as diverse as their cousins, the bony fish, but they were, and still are, important members of the marine vertebrate fauna.

Along with cartilaginous fish, the **bony fish** (class Osteichthyes) (Table 13.1) also first evolved during the Devonian. Because bony fish are the most varied and numerous of all the fishes, and because the amphibians evolved from them, their evolutionary history is particularly important. There are two groups of bony fish: the common *ray-finned fish* (subclass Actinopterygii) (Figure 13.8d) and the less familiar **lobe-finned fish** (subclass Sarcopterygii) (Table 13.1).

The term *ray-finned* refers to the way the fins are supported by thin bones that spread away from the body (▶Figure 13.9a). From a modest freshwater beginning during the Devonian, ray-finned fish, which include most of the familiar fish such as trout, bass, perch, salmon, and tuna, rapidly diversified to dominate the Mesozoic and Cenozoic seas.

▶ Figure 13.9 **Ray-Finned and Lobe-Finned Fish** Arrangement of fin bones for (a) a typical ray-finned fish and (b) a lobe-finned fish. The muscles extend into the fin of the lobe-finned fish, allowing greater flexibility of movement than that of the ray-finned fish.

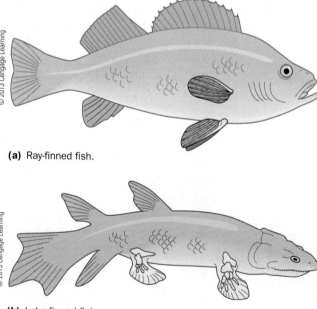

(a) Ray-finned fish.

(b) Lobe-finned fish.

Present-day lobe-finned fish are characterized by muscular fins. The fins do not have radiating bones but rather have articulating bones with the fin attached to the body by a fleshy shaft (Figure 13.9b). Such an arrangement allows for a powerful stroke of the fin, making the fish an effective swimmer. Three orders of lobe-finned fish are recognized: *coelacanths, lungfish,* and *crossopterygians* (Table 13.1).

Coelacanths (order Coelacanthimorpha) are marine lobe-finned fish that evolved during the Middle Devonian and were thought to have gone extinct at the end of the Cretaceous. In 1938, however, a fisherman caught a coelacanth in the deep waters off Madagascar (see Figure 7.13a), and since then, several dozen more have been caught, both there and in Indonesia.

Lungfish (order Dipnoi) were fairly abundant during the Devonian, but today only three freshwater genera exist, one each in South America, Africa, and Australia. Their present-day distribution presumably reflects the Mesozoic breakup of Gondwana (see Chapter 14).

The "lung" of a modern-day lungfish is actually a modified swim bladder that most fish use for buoyancy in swimming. In lungfish, this structure absorbs oxygen, allowing them to breathe air when the lakes or streams in which they live become stagnant and dry up. During such times, they burrow into the sediment to prevent dehydration and breathe through their swim bladder until the stream begins flowing or the lake that they were living in fills with water. When they are back in the water, lungfish then rely on gill respiration.

The **crossopterygians** (order Crossopterygii) are an important group of lobe-finned fish, because it is probably from them that amphibians evolved. However, the transition between crossopterygians and true amphibians is not as simple as it was once portrayed. Among the crossopterygians, the *rhipidistians* appear to be the ancestral group (Table 13.1). These fish, reaching lengths of over 2 m, were the dominant freshwater predators during the Late Paleozoic. *Eusthenopteron,* a good example of a rhipidistian crossopterygian and a classic example of the transitional form between fish and amphibians, had an elongated body that helped it move swiftly in the water, and paired, muscular fins that many scientists thought could be used for moving on land (▶ Figure 13.10). The structural similarity between crossopterygian fish and the earliest amphibians is striking and one of the most widely cited examples of a transition from one major group to another (▶ Figure 13.11). However, recent discoveries of older lobe-finned fish and newly published findings of tetrapod footprints and tetrapod-like fish are filling the gaps in the evolution from fish and tetrapods.

Before discussing this transition and the evolution of amphibians, it is useful to place the evolutionary history of Paleozoic fish in the larger context of Paleozoic evolutionary events. Certainly, the evolution and diversification of

▶ **Figure 13.11 Similarities Between Crossopterygians and Labyrinthodonts** Similarities between the crossopterygian lobe-finned fish and the labyrinthodont amphibians.

Labyrinthodont

Lobe fin

(a) Skeletal similarity.

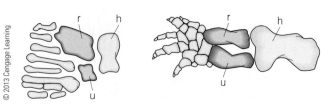

(b) Comparison of the limb bones of a crossopterygian (*left*) and amphibian (*right*); color identifies the bones (u = ulna, shown in blue; r = radius, mauve; h = humerus, gold) that the two groups have in common.

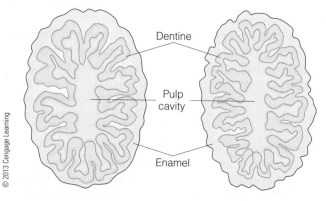

Dentine

Pulp cavity

Enamel

(c) Comparison of tooth cross sections shows the complex and distinctive structure found in both the crossopterygians (*left*) and labyrinthodont amphibians (*right*).

jawed fish, as well as eurypterids and ammonoid cephalopods, had a profound effect on the marine ecosystem. Previously defenseless organisms either evolved defensive mechanisms or suffered great losses, possibly even extinction.

Recall from Chapter 12 that trilobites experienced extinctions at the end of the Cambrian, recovered slightly during the Ordovician, and then declined greatly from the end of the Ordovician to final extinction at the end of the Permian. Perhaps their lightly calcified external covering made them easy prey for the rapidly evolving jawed fish and cephalopods.

Ostracoderms, although armored, would also have been easy prey for the swifter, jawed fishes. Ostracoderms became extinct by the end of the Devonian, a time that coincides with the rapid evolution of jawed fish. Placoderms, like acanthodians, greatly decreased in abundance after the Devonian and became extinct by the end of the Paleozoic Era. In contrast, cartilaginous and ray-finned bony fish expanded during the Late Paleozoic, as did the ammonoids (see Figure 12.16), the other major predators of the Late Paleozoic seas.

Amphibians—Vertebrates Invade the Land

Although amphibians were the first vertebrates to live on land, they were not the first land-living organisms. Land plants, which probably evolved from green algae, first evolved during the Ordovician. Furthermore, insects, millipedes, spiders, and even snails invaded the land before amphibians. Fossil evidence indicates that such land-dwelling arthropods as scorpions and flightless insects had evolved by at least the Devonian.

The transition from water to land required animals to surmount several barriers. The most critical were desiccation, reproduction, the effects of gravity, and the extraction of oxygen from the atmosphere by lungs rather than from water by gills. Until the 1990s, the traditional evolutionary sequence had a rhipidistian crossopterygian, like *Eusthenopteron,* evolving into a primitive amphibian such as *Ichthyostega.* At that time, fossils of those two genera were about all paleontologists had to work with, and although there were gaps in morphology, the link between crossopterygians and these earliest amphibians was easy to see (Figure 13.11). However, the recent discovery of fossilized tetrapod footprints in rocks dated at 397 million years old (earliest Middle Devonian) from the Holy Cross Mountains of Poland (see Introduction), as well as additional tetrapod fossil specimens, has forced paleontologist to rethink when animals emerged onto land.

Nevertheless, the question still remained as to when animals made the transition from water to land, and perhaps even more intriguing is the question of why limbs evolved in the first place. The answer to that question is that they probably were not for walking on land. The current thinking of many scientists is that aquatic limbs made it easier for animals to move around in streams, lakes, or swamps that were choked with water plants or other debris. The fossil evidence that began to emerge in the 1990s also seems to support this hypothesis. However, as pointed out in the Introduction, the discovery of fossilized tetrapod footprints from the mud of coral-reef lagoon deposits in

Poland is forcing paleontologists to rethink their long-held ideas of a terrestrial origin for tetrapods.

Fossils of *Acanthostega,* a tetrapod found in the 360-million-year-old Old Red Sandstone from Greenland, reveal an animal that had limbs, but was clearly unable to walk on land. Paleontologist Jennifer Clack, who has recovered and analyzed hundreds of specimens of *Acanthostega,* points out that *Acanthostega's* limbs were not strong enough to support its weight on land, and its rib cage was too small for the necessary muscles needed to hold its body off the ground. In addition, *Acanthostega* had gills and lungs, meaning that it could survive on land, but it was more suited for the water. Clack thinks that *Acanthostega* used its limbs to maneuver around in swampy, plant-filled waters, where swimming would be difficult and limbs would be an advantage.

Fragmentary fossils from other tetrapods living at about the same time as *Acanthostega* suggest, however, that some of these early tetrapods may have spent more time on dry land than in the water. These oldest amphibians, many of which are also found in the Upper Devonian Old Red Sandstone of eastern Greenland and belong to genera such as *Ichthyostega,* had streamlined bodies, long tails, and fins along their backs. In addition to four legs, they had a strong backbone, a rib cage, and pelvic and pectoral girdles, all of which were structural adaptations for walking on land (▶Figure 13.12).

These earliest amphibians thus appear to have inherited many characteristics from the crossopterygians with little modification (Figure 13.11). However, with the discovery of such fossils as *Acanthostega* and others like it, the transition between fish and amphibians involves a number of new genera that are intermediary between the two groups, and thus fills in some of the gaps between the earlier postulated rhipidistian crossopterygian–amphibian phylogeny.

In 2006, an exciting discovery of a 1.2–2.8-m-long, 375-million-year-old (Late Devonian) "fishapod" was announced. Discovered on Ellesmere Island, Canada, *Tiktaalik roseae,* from the Inuktitut meaning "large fish in a stream," was hailed as an intermediate form between the lobe-finned fish and the earliest tetrapod, *Acanthostega* (▶Figure 13.13a).

Tiktaalik roseae is truly a "fishapod" in that it has a mixture of both fish and tetrapod characteristics (Figure 13.13b). For example, it has gills and fish scales but also a broad skull, eyes on top of its head, a flexible neck and large rib cage that could support its body on land or in shallow water, and lungs, all of which are tetrapod features. What really excited scientists, however, was that *Tiktaalik roseae* has the beginnings of a true tetrapod forelimb, complete with functional wrist bones and five digits, as well as a modified ear region. Sedimentological evidence suggests *Tiktaalik roseae* lived in a shallow-water habitat associated with the Late Devonian floodplains of Laurasia.

As previously mentioned, the oldest known amphibian, *Ichthyostega,* had skeletal features that allowed it to spend its life on land. Because amphibians did not evolve until the Late Devonian, they were a minor element of the Devonian terrestrial ecosystem. Like other groups that moved into new and previously unoccupied niches, amphibians underwent rapid adaptive radiation and became abundant during the Carboniferous and Early Permian.

The Late Paleozoic amphibians did not at all resemble the familiar frogs, toads, newts, and salamanders that make up the modern amphibian fauna. Rather, they displayed a broad spectrum of sizes, shapes, and modes of life (▶Figure 13.14). One group of amphibians was the **labyrinthodonts**, so named for the labyrinthine wrinkling and folding of the chewing surface of their teeth (Figure 13.11c). Most labyrinthodonts were large animals, as much as 2 m in length. These typically sluggish creatures lived in

▶ Figure 13.12 **Late Devonian Landscape** Shown is *Ichthyostega,* both in the water and on land. *Ichthyostega* was an amphibian that grew to a length of approximately 1 m. The flora at this time was diverse, and consisted of a variety of small and large seedless vascular plants.

▶ Figure 13.13 ***Tiktaalik roseae*** *Tiktaalik roseae*, a "fishapod," has been hailed as an intermediary between lobe-finned fish and tetrapods because it has characteristics of both fish and tetrapods.

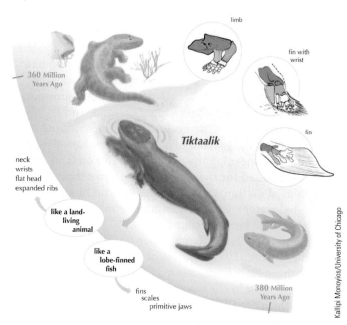

(a) Skeleton of *Taktaalik roseae*.

(b) Diagram illustrating how *Tiktaalik roseae* is a transitional species between lobe-finned fish and tetrapods.

▶ Figure 13.14 **Carboniferous Coal Swamp** The amphibian fauna during the Carboniferous was varied. Shown is the serpentlike *Dolichosoma* (foreground) and the large labyrinthodonts amphibian *Eryops*.

swamps and streams, eating fish, vegetation, insects, and other small amphibians (Figure 13.14).

Labyrinthodonts were abundant during the Carboniferous when swampy conditions were widespread (see Chapter 11) but soon declined in abundance during the Permian, perhaps in response to changing climatic conditions. Only a few species survived into the Triassic.

Evolution of the Reptiles— The Land Is Conquered

Amphibians were limited in colonizing the land because they had to return to water to lay their gelatinous eggs. The evolution of the **amniote egg** (▶ Figure 13.15) freed reptiles from this constraint. In such an egg, the developing embryo is surrounded by a liquid-filled sac called the *amnion* and provided with both a *yolk*, or food sac, and an *allantois*, or waste sac. In this way, the emerging reptile is in essence a miniature adult, bypassing the need for a larval stage in the water. The evolution of the amniote egg allowed vertebrates to colonize all parts of the land, because they no longer had to return to the water as part of their reproductive cycle.

Many of the differences between amphibians and reptiles are physiologic and are not preserved in the fossil record. Nevertheless, amphibians and reptiles differ sufficiently in skull structure, jawbones, ear location, and limb and vertebral construction to suggest that reptiles

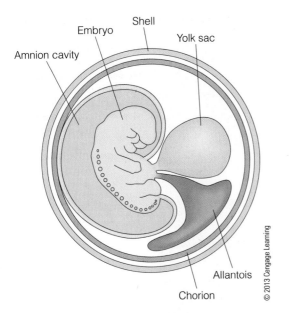

Figure 13.15 **Amniote Egg** In an amniote egg, the embryo is surrounded by a liquid sac (amnion cavity) and provided with a food source (yolk sac) and waste sac (allantois). The evolution of the amniote egg freed reptiles from having to return to the water for reproduction and let them inhabit all parts of the land.

▶ Figure 13.16 *Hylonomus lyelli* Reconstruction and skeleton of one of the oldest known reptiles, *Hylonomus lyelli*, from the Pennsylvanian Period. Fossils of this animal have been collected from sediments that filled tree stumps. *Hylonomus lyelli* was about 30 cm long.

▶ Figure 13.17 **Evolutionary Relationship Among the Paleozoic Reptiles**

evolved from labyrinthodont ancestors by the Late Mississippian. This assessment is based on the discovery of a well-preserved fossil skeleton of the oldest known reptile, *Westlothiana,* and other fossil reptile skeletons from Late Mississippian-age rocks in Scotland.

Other early reptile fossils occur in the Lower Pennsylvanian Joggins Formation in Nova Scotia, Canada. Here, remains of *Hylonomus* are found in the sediments filling in tree trunks (▶ Figure 13.16). These earliest reptiles from Scotland and Canada were small and agile and fed largely on grubs and insects. They are loosely grouped together as **protorothyrids**, whose members include the earliest known reptiles (▶ Figure 13.17). During the Permian Period, reptiles diversified and began displacing many amphibians. The reptiles succeeded partly because of their advanced method of reproduction and their more advanced jaws and teeth, as well as their tough skin and scales to prevent desiccation, and their ability to move rapidly on land.

The **pelycosaurs**, or finback reptiles, evolved from the protorothyrids during the Pennsylvanian and were the dominant reptile group by the Early Permian. They evolved into a diverse assemblage of herbivores, exemplified by the herbivore *Edaphosaurus* and carnivores such as *Dimetrodon* (▶ Figure 13.18). An interesting feature of most pelycosaurs is their sail. It was formed by vertebral spines that, in life, were covered with skin. The sail has been variously explained as a type of sexual display, a means of protection, and a display to look more ferocious. The current consensus seems to be that the sail served as some type of rudimentary thermoregulatory device, raising the reptile's

temperature by catching the sun's rays or cooling it by facing the wind. Because pelycosaurs are considered to be the group from which therapsids evolved, it is interesting that they may have had some sort of body-temperature control.

The pelycosaurs became extinct during the Permian and were succeeded by the **therapsids**, mammal-like reptiles that evolved from the carnivorous pelycosaur ancestry and rapidly diversified into herbivorous and carnivorous lineages (▶ Figure 13.19). Therapsids were small- to medium-sized animals that displayed the beginnings of many mammalian features: fewer bones in the skull, because many of the small skull bones were fused;

© 2013 Cengage Learning

▶ Figure 13.18 **Pelycosaurs** Most pelycosaurs, or finback reptiles, have a characteristic sail on their back. One hypothesis explains the sail as a rudimentary thermoregulatory device. Other hypotheses are that it was a type of sexual display or a device to make the reptile look more intimidating. Shown here are the carnivore *Dimetrodon* and the herbivore *Edaphosaurus*.

▶ Figure 13.19 **Therapsids** A Late Permian scene in southern Africa showing various therapsids including *Dicynodon* and *Moschops*. Many paleontologists think therapsids were endothermic and may have had a covering of fur, as shown here.

enlarged lower jawbone; differentiation of teeth for various functions such as nipping, tearing, and chewing food; and more vertically placed legs for greater flexibility, as opposed to the way the legs sprawled out to the side in primitive reptiles.

Furthermore, many paleontologists think therapsids were *endothermic*, or warm-blooded, enabling them to maintain a constant internal body temperature. This characteristic would have let them expand into a variety of habitats, and, indeed, the Permian rocks in which their fossil remains are found are distributed not only in low latitudes but also in middle and high latitudes as well.

As the Paleozoic Era came to an end, the therapsids constituted about 90 percent of the known reptile genera and occupied a wide range of ecologic niches. The mass extinctions that decimated the marine fauna at the close of the Paleozoic had an equally great effect on the terrestrial population (see Chapter 12). By the end of the Permian, about 96 percent of all marine invertebrate species became extinct, compared with more than two-thirds of all amphibians and reptiles. Plants, in contrast, apparently did not experience as great a turnover as animals.

Plant Evolution

When plants made the transition from water to land, they had to solve most of the same problems that animals did:

TABLE 13.2 Major Events in the Evolution of Land Plants

The Devonian Period was a time of rapid evolution for the land plants. Major events were the appearance of leaves, heterospory, secondary growth, and the emergence of seeds.

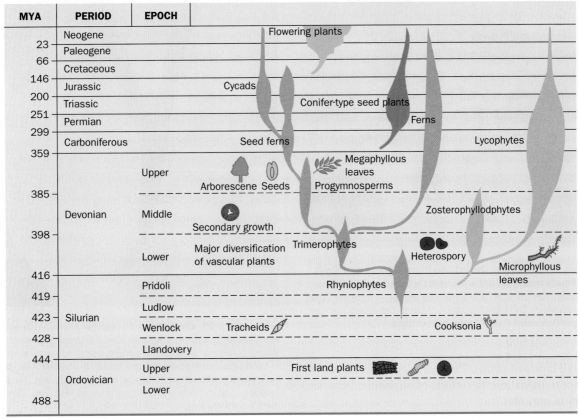

© 2013 Cengage Learning

desiccation, support, reproduction, and the effects of gravity. Plants did so by evolving a variety of structural adaptations that were fundamental to the subsequent radiations and diversification that occurred during the Silurian, Devonian, and later periods (Table 13.2). Most experts agree that the ancestors of land plants first evolved in a marine environment, then moved into a freshwater environment before finally progressing onto land. In this way, the differences in osmotic pressures between saltwater and freshwater were overcome while the plant was still in the water.

The higher land plants are divided into two major groups: nonvascular and vascular. Most land plants are **vascular**, meaning that they have a tissue system of specialized cells for the movement of water and nutrients. The *nonvascular* plants, such as bryophytes (liverworts, hornworts, and mosses), do not have these specialized cells and are typically small and usually live in low, moist areas.

The earliest land plants from the Middle to Late Ordovician were probably small and bryophyte-like in their overall organization (but not necessarily related to bryophytes). The evolution of vascular tissue in plants was an important step because it allowed transport of nutrients and water.

Discoveries of probable vascular plant megafossils and characteristic spores indicate to many paleontologists that vascular plants evolved well before the Middle Silurian.

Sheets of cuticle-like cells—that is, the cells that cover the surface of present-day land plants—and tetrahedral clusters that closely resemble the spore tetrahedrals of primitive land plants have been reported from Middle to Upper Ordovician rocks from western Libya and elsewhere (▶ Figure 13.20).

The ancestor of terrestrial vascular plants was probably some type of green alga. Although no fossil record of the transition from green algae to terrestrial vascular plants has been found, comparison of their physiology reveals a strong link. Primitive *seedless vascular plants* (discussed later in this chapter), such as ferns, resemble green algae in their pigmentation, important metabolic enzymes, and type of reproductive cycle. Furthermore, green algae are one of the few plant groups to have made the transition from saltwater to freshwater.

The evolution of terrestrial vascular plants from an aquatic, probable green algal ancestry was accompanied by various modifications that let them occupy this new, harsh environment. Besides the primary function of transporting water and nutrients throughout a plant, vascular tissue also provides some support for the plant body. Additional strength is derived from the organic compounds *lignin* and *cellulose,* found throughout a plant's walls.

The problem of desiccation was circumvented by the evolution of *cutin,* an organic compound found in the

Courtesy of Jane Gray, University of Oregon

▸ **Figure 13.20** **Upper Ordovician Plant Spores and Cells**
These fossils closely resemble the spore tetrahedrals of primitive
land plants. The sheet of cuticle-like cells (*center*) is from the Upper
Ordovician Melez Chograne Formation of Libya. The others are from
the Upper Ordovician Djeffara Formation of Libya.

Courtesy of Dianne Edwards, University of Wales College

▸ **Figure 13.21** ***Cooksonia*** The earliest known fertile land plant
was *Cooksonia,* seen in this fossil from the Upper Silurian of South
Wales. *Cooksonia* consisted of upright, branched stems terminating
in sporangia (spore-producing structures). It also had a resistant
cuticle and produced spores typical of a vascular plant. These plants
probably lived in moist environments such as mud flats. This specimen is 1.49 cm long.

outer-wall layers of plants. Cutin also provides additional
resistance to oxidation, the effects of ultraviolet light, and
the entry of parasites.

Roots evolved in response to the need to collect water
and nutrients from the soil and to help anchor the plant in
the ground. The evolution of *leaves* from tiny outgrowths
on the stem or from branch systems provided plants with
an efficient light-gathering system for photosynthesis.

Silurian and Devonian Floras

The earliest known vascular land plants are small Y-shaped
stems assigned to the genus *Cooksonia* from the Middle
Silurian of Wales and Ireland. Together with Upper Silurian and Lower Devonian species from Scotland, New
York State, and the Czech Republic, these earliest plants
were small, simple, leafless stalks with a spore-producing
structure at the tip (▸ Figure 13.21). They are known as
seedless vascular plants because they did not produce
seeds. They also did not have a true root system. A *rhi-
zome,* the underground part of the stem, transferred water
from the soil to the plant and anchored the plant in the
ground. The sedimentary rocks in which these plant fossils are found indicate that they lived in low, wet, marshy,
freshwater environments.

An interesting parallel can be seen between seedless
vascular plants and amphibians. When they made the
transition from water to land, both plants and animals
had to overcome the same problems that such a transition involved. Both groups, while successful, nevertheless

required a source of water in order to reproduce. In the
case of amphibians, their gelatinous egg had to remain
moist, whereas the seedless vascular plants required water
for the sperm to travel through to reach the egg.

From this simple beginning, the seedless vascular
plants evolved many of the major structural features characteristic of modern plants such as leaves, roots, and secondary growth. These features did not all evolve simultaneously but rather at different times, a pattern known as
mosaic evolution. This diversification and adaptive radiation took place during the Late Silurian and Early Devonian and resulted in a tremendous increase in diversity
(▸ Figure 13.22). During the Devonian, the number of
plant genera remained about the same, yet the composition of the flora changed. Whereas the Early Devonian
landscape was dominated by relatively small, low-growing,
bog-dwelling types of plants, the Late Devonian witnessed
forests of large, tree-sized plants up to 10 m tall.

In addition to the diverse seedless vascular plant flora
of the Late Devonian, another significant floral event took
place. The evolution of the seed at this time liberated land
plants from their dependence on moist conditions and
allowed them to spread throughout all terrestrial environments. Seedless vascular plants require moisture for successful fertilization because the sperm must travel to the

▶ **Figure 13.22 Early Devonian Landscape** Reconstruction of an Early Devonian landscape showing *Asteroxylon*, a common seed-less vascular plant with curling branches that grew up to 4 m tall, and various low-growing seedless vascular plants.

egg on the surface of the gamete-bearing plant (gameto-phyte) to produce a successful spore-generating plant (sporophyte). Without moisture, the sperm would dry out before reaching the egg (▶ Figure 13.23a). In the seed method of reproduction, the spores are not released to the environment as they are in the seedless vascular plants, but instead, they are retained on the spore-bearing plant, where they grow into the male and female forms of the gamete-bearing generation.

In the case of the **gymnosperms**, or flowerless seed plants, these are the male and female cones (Figure 13.23b). The male cone produces pollen, which contains the sperm and has a waxy coating to prevent desiccation, and the egg, or embryonic seed, is contained in the female cone. After fertilization, the seed then develops into a mature, cone-bearing plant. In this way, the need for a moist environment for the gametophyte generation is solved. The significance of this development is that seed plants, like reptiles, were no longer restricted to wet areas but were free to migrate into previously unoccupied dry environments.

Before seed plants evolved, an intermediate evolu-tionary step was necessary. This was the development

of *heterospory*, whereby a species produces two types of spores: a large one (megaspore) that gives rise to the female gamete-bearing plant and a small one (microspore) that produces the male gamete-bearing plant (Table 13.2). The earliest evidence of heterospory is found in the Early Devonian plant *Chaleuria cirrosa,* which produced spores of two distinct sizes (▶ Figure 13.24). The appearance of heterospory was followed several million years later by the emergence of progymnosperms—Middle and Late Devo-nian plants with fernlike reproductive habits and a gym-nosperm anatomy—which gave rise in the Late Devonian to such other gymnosperm groups as the seed ferns and conifer-type seed plants (Table 13.2).

Although the seedless vascular plants dominated the flora of the Carboniferous coal-forming swamps, the gymnosperms made up an important element of the Late Paleozoic flora, particularly in the nonswampy areas.

Late Carboniferous and Permian Floras

As discussed earlier, the rocks of the Pennsylvanian Period (Late Carboniferous) are the major source of the world's

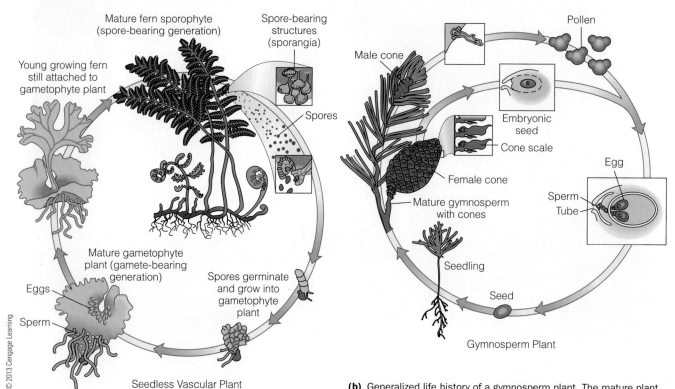

(a) Generalized life history of a seedless vascular plant. The mature sporophyte plant produces spores, which, upon germination, grow into small gametophyte plants that produce sperm and eggs. The fertilized eggs grow into the spore-producing mature plant, and the sporophyte–gametophyte life cycle begins again.

(b) Generalized life history of a gymnosperm plant. The mature plant bears both male cones that produce sperm-bearing pollen grains and female cones that contain embryonic seeds. Pollen grains are transported to the female cones by the wind. Fertilization occurs when the sperm moves through a moist tube growing from the pollen grain and unites with the embryonic seed, which then grows into a cone-bearing mature plant.

coal. Coal results from the alteration of plant remains accumulating in low, swampy areas. The geologic and geographic conditions of the Pennsylvanian were ideal for the growth of seedless vascular plants, and consequently these coal swamps had a very diverse flora (▸ Figure 13.25).

It is evident from the fossil record that whereas the Early Carboniferous (Mississippian Period) flora was

▸ Figure 13.24 **Chaleuria cirrosa** Specimen of the Early Devonian plant *Chaleuria cirrosa* from New Brunswick, Canada. This plant was heterosporous, meaning that it produced two sizes of spores.

similar to its Late Devonian counterpart, a great deal of evolutionary experimentation was taking place that would lead to the highly successful Late Paleozoic flora of the coal swamps and adjacent habitats. Among the seedless vascular plants, the *lycopsids* and *sphenopsids* were the most important coal-forming groups of the Pennsylvanian Period.

The lycopsids were present during the Devonian, chiefly as small plants, but by the Pennsylvanian, they were the dominant element of the coal swamps, achieving heights up to 30 m in such genera as *Lepidodendron* and *Sigillaria*. The Pennsylvanian lycopsid trees are interesting because they lacked branches except at their top, which had elongate leaves similar to the individual palm leaf of today. As the trees grew, the leaves were replaced from the top, leaving prominent and characteristic rows or spirals of scars on the trunk.

The sphenopsids, the other important coal-forming plant group, are characterized by being jointed and having horizontal underground stem-bearing roots. Many of these plants, such as *Calamites,* average 5 to 6 m tall. Small seedless vascular plants and seed ferns formed a thick undergrowth, or ground cover, beneath these treelike plants. Living sphenopsids include the horsetail (*Equisetum)* or scouring rushes (▸ Figure 13.26).

Not all plants were restricted to the coal-forming swamps. Among those plants that occupied higher and drier ground

▶ Figure 13.25 **Pennsylvanian Coal Swamp** Reconstruction of a Pennsylvanian coal swamp with its characteristic vegetation of seedless vascular plants.

© The Field Museum, #GEO75400

Reed Wicander

▶ Figure 13.26 **Horsetail (*Equisetum*)** Living sphenopsids include the horsetail (*Equisetum*).

© 2013 Cengage Learning

▶ Figure 13.27 **Late Carboniferous Cordaite Forest** Cordaites were a group of gymnosperm trees that grew up to 50 m tall.

were some of the *cordaites*, a group of tall gymnosperm trees that grew up to 50 m high and probably formed vast forests (▶ Figure 13.27). Another important nonswamp dweller was *Glossopteris,* the famous plant so abundant in Gondwana (see Figure 3.1), whose distribution is cited as critical evidence that the continents have moved through time.

The floras that were abundant during the Pennsylvanian persisted into the Permian, but because of climatic and geologic changes resulting from tectonic events (see Chapter 11), they declined in abundance and importance. By the end of the Permian, the cordaites became extinct, and the lycopsids and sphenopsids were reduced to mostly small, creeping forms. Gymnosperms, with lifestyles more suited to the warmer and drier Permian climates, diversified and came to dominate the Permian, Triassic, and Jurassic landscapes (see Perspective).

Perspective

Using Palynology to Address Geologic and Environmental Issues

Palynology is the study of organic micro-fossils called palynomorphs. These include acritarchs (organic-walled phytoplankton of unknown affinity) (see Figure 12.9), dinoflagellates (marine and fresh-water single-celled phytoplankton, some species of which in high concentrations make shellfish toxic to humans) (see Figure 15.4c), as well as chitinozoans (vase-shaped microfossils of unknown origin), scolecodonts (jaws of marine annelid worms), and microscopic colonial algae.

The palynomorphs that most people are probably familiar with are spores and pollen, in part, because both can cause allergies. Palynologists, however, study fossil spores and pollen preserved in sedimentary rocks because they reveal when plants colonized Earth's surface (Figure 13.20), which in turn influenced weathering and erosion rates, soil formation, and changes in the composition of atmospheric gases. Furthermore, because plants are not particularly common as fossils, spores and pollen, which are extremely resistant to decay and occur in tremendous numbers, are frequently the only fossils found in outcrops or cores.

One of the more important uses of fossil spores and pollen is determining the geologic age of sedimentary rocks. Because spores and pollen are micro-scopic, resistant to decay, deposited in both marine and terrestrial environments, extremely abundant, and part of the life cycle of plants (Figure 13.23), they are very useful for determining age. Further-more, many spore and pollen species have narrow time ranges that make them excellent guide fossils (see Chapter 5).

Fossil spores and pollen are also useful in determining the environment and climate in the past. Their presence in sedimentary rocks helps palynologists determine what plants and trees were living at the time, even if the fossils of those plants and trees are not preserved. Plants are very sensitive to climatic changes, and by plotting the abundance and types of vegetation present, based on their preserved spores and pollen, palynologists can determine past climates and changes in climates through time (see Figure 17.7).

An interesting study by C. B. Foster and S. A. Afonin published in 2005 related morphologic abnormalities in gymnosperm pollen grains to deteriorating atmospheric conditions around the Permian–Triassic boundary—that is, at the time of the global Permian extinction event (see Chapter 12). One cause of morphologic abnormalities in living gymnosperms and angiosperms

Photos courtesy of Clinton B. Foster, Petroleum and Marine Division, Geoscience Australia, and Sergey A. Afonin, formerly of the Palaeontological Institute, Moscow, Russia.

Figure 1 Normal and abnormal pollen grains of *Klausipollenites schaubergeri* from Nedubrovo, Russia (a–c) and the Junggar Basin, China (d–f). Normal pollen grains of *Klausipollenites schaubergeri* are bisaccate—that is, they have a central body with two air sacs (a and d)—whereas abnormal forms have three or four air sacs (b–c and e–f). The high percentage of abnormal pollen grains of *Klausipollenites schaubergeri* recovered from sedimentary rocks spanning the Permian–Trissic boundary in Russia and China have been used as evidence of increased atmospheric pollution related to the Permian mass extinctions. The scale bar in (c) represents 25 μm (25/1,000 mm).

is environmental stress on the parent plant. Plants are sensitive indicators of environmental change, and studies have shown that pollen-wall abnormalities are caused by atmospheric pollution, ultraviolet-B (UV-B) radiation, or a combination of both.

Samples from nonmarine sedimentary rock sequences that span the Permian–Triassic boundary from the Junggar Basin, Xinjiang Province, China, and Nedubrovo, Russia, yielded a diverse, abundant, and well-preserved pollen assemblage. Examination of the assemblage revealed that greater than 3 percent of the pollen showed morphologic abnormalities (Figure 1). Based on this finding, the authors concluded these abnormalities were the result of atmospheric pollution, including increased UV-B radiation, caused

by extensive volcanism. Other studies show similar pollen morphologic abnormalities in end-Permian sediments ranging from Greenland to Australia.

A 2007 publication discussed the role that the Siberian Traps flood basalts may have played in causing widespread ozone depletion, thus allowing an increase in UV-B radiation at the end of the Permian (see page 249 in Chapter 12). The authors of this study also noted the increase in morphologic abnormalities of gymnosperm pollen around the Permian–Triassic boundary.

Lastly, a 2008 study using microspectroscopy examined the biochemistry of the outer wall of the spores of *Lycopodium magellanicum* and *Lycopodium annotinum* from present-day high northern- and southern-latitude localities. This study indicates that the

concentrations of two UV-B-absorbing compounds reflects variations in UV-B radiation. As such, these compounds may prove valuable for reconstructing past variations in stratospheric ozone, which screens the surface biota from harmful UV-B radiation.

Recall from Chapter 12 that one of the possible causes of the Permian extinctions was increased global warming caused by higher atmospheric carbon dioxide levels and volcanic activity. One way of increasing atmospheric carbon dioxide levels is from volcanic and continental fissure eruptions. Although the cause of the end-Permian mass extinctions is still unresolved, the results of these three studies point out some of the exciting research being conducted on climate change and its impact on Earth's biota.

Summary

Table 13.3 summarizes the major evolutionary and geologic events of the Paleozoic Era and shows their relationships to each other.

▶ Chordates are characterized by a notochord, dorsal hollow nerve cord, and pharyngeal slits. The earliest chordates were soft-bodied organisms that were rarely fossilized. Vertebrates are a subphylum of the chordates.

▶ Fish are the earliest known vertebrates, with their first fossil occurrence in Upper Cambrian rocks. They have had a long and varied history, including jawless and jawed armored forms (ostracoderms and placoderms), cartilaginous forms, and bony forms. It is from the lobe-finned fish that amphibians evolved.

▶ The link between crossopterygian lobe-finned fish and the earliest amphibians is convincing and includes a close similarity of bone and tooth structures. However, new fossil discoveries show that the transition between the two groups is more complicated than originally hypothesized and includes several intermediate forms.

▶ Amphibians evolved during the Late Devonian, with labyrinthodont amphibians becoming the dominant terrestrial vertebrate animals during the Carboniferous.

▶ The Late Mississippian marks the earliest fossil record of reptiles. The evolution of an amniote egg was the critical factor that allowed reptiles to completely colonize the land.

▶ Pelycosaurs were the dominant reptiles during the Early Permian, whereas therapsids dominated the landscape for the rest of the Permian Period.

▶ In making the transition from water to land, plants had to overcome the same basic problems as animals—namely, desiccation, reproduction, and gravity.

▶ The earliest fossil record of land plants is from Middle to Upper Ordovician rocks. These plants were probably small and bryophyte-like in their overall organization.

▶ The evolution of vascular tissue was an important event in plant evolution as it allowed nutrients and water to be transported throughout the plant and provided the plant with additional support.

▶ The ancestor of terrestrial vascular plants was probably some type of green alga based on such similarities as pigmentation, metabolic enzymes, and the same type of reproductive cycle.

▶ The earliest seedless vascular plants were small, leafless stalks with spore-producing structures on their tips. From this simple beginning, plants evolved many of the major structural features characteristic of today's plants.

▶ By the end of the Devonian Period, forests with tree-sized plants up to 10 m had evolved. The Late Devonian also witnessed the evolution of the flowerless seed plants (gymnosperms), whose reproductive style freed them from having to stay near water.

▶ The Carboniferous Period was a time of vast coal swamps, where conditions were ideal for the seedless vascular plants. With the onset of more arid conditions during the Permian, the gymnosperms became the dominant element of the world's flora.

Important Terms

amniote egg, p. 262
bony fish, p. 258
cartilaginous fish, p. 258
chordate, p. 254
crossopterygian, p. 259
gymnosperm, p. 267

labyrinthodont, p. 261
lobe-finned fish, p. 258
ostracoderm, p. 255
pelycosaur, p. 263
placoderm, p. 257
protorothyrid, p. 263

seedless vascular plant, p. 266
therapsid, p. 263
vascular, p. 265
vertebrate, p. 254

Review Questions

1. Which of the following groups did amphibians evolve from?
   a. _____ Coelacanths; b. _____ Ray-finned fish;
   c. _____ Lobe-finned fish; d. _____ Pelycosaurs;
   e. _____ Therapsids.

2. An organism must possess which of the following during at least part of its life cycle to be classified a chordate?
   a. _____ Notochord, dorsal solid nerve cord, lungs;
   b. _____ Vertebrae, dorsal hollow nerve cord, pharyngeal slits; c. _____ Vertebrae, dorsal hollow nerve cord, lungs;
   d. _____ Notochord, ventral solid nerve cord, lungs;
   e. _____ Notochord, dorsal hollow nerve cord, pharyngeal slits.

3. Which plant group first successfully invaded land?
   a. _____ Seedless vascular; b. _____ Gymnosperms;
   c. _____ Naked seed bearing; d. _____ Angiosperms;
   e. _____ Flowering.

4. Labyrinthodonts are
   a. _____ plants; b. _____ fish; c. _____ amphibians;
   d. _____ reptiles; e. _____ none of the previous answers.

5. Which reptile group gave rise to the mammals?
   a. _____ Labyrinthodonts; b. _____ Acanthodians;
   c. _____ Pelycosaurs; d. _____ Protothyrids; e. _____ Therapsids.

6. Which evolutionary innovation allowed reptiles to colonize all of the land?
   a. _____ Tear ducts; b. _____ Additional bones in the jaw; c. _____ The middle-ear bones; d. _____ An egg that contained a food-and-waste sac and surrounded the embryo in a fluid-filled sac; e. _____ Limbs and a backbone capable of supporting the animals on land.

7. The first jawed fish belong to which group?
   a. _____ Cartilaginous; b. _____ Placoderms; c. _____ Acanthodians; d. _____ Ostracoderms; e. _____ Bony.

8. Based on similarity of embryo cell division, which invertebrate phylum is most closely allied with the chordates?
   a. _____ Mollusca; b. _____ Echinodermata; c. _____ Porifera; d. _____ Annelida; e. _____ Arthropoda.

9. Pelycosaurs are
   a. _____ jawless fish; b. _____ jawed armored fish;
   c. _____ reptiles; d. _____ amphibians; e. _____ plants.

10. The discovery of *Tiktaalik roseae* is significant because it is
    a. _____ the ancestor of modern reptiles; b. _____ an intermediate between lobe-finned fish and amphibians;
    c. _____ the first vascular land plant; d. _____ the "missing link" between amphibians and reptiles; e. _____ the oldest known fish.

11. Which algal group was the probable ancestor to vascular plants?
    a. _____ Yellow ; b . _____ Blue-green; c. _____ Red;
    d. _____ Brown; e. _____ Green.

12. The oldest and most primitive of the fish are
    a. _____ acanthodians; b. _____ cartilaginous; c. _____ placoderms; d. _____ ostracoderms; e. _____ bony.

13. Discuss how changing geologic conditions affected the evolution of plants and vertebrates.

14. Outline the evolutionary history of fish.

15. Based on what you know about Carboniferous geology (Chapter 12), why was this time period so advantageous to the evolution of both plants and amphibians?

16. Describe the problems that had to be overcome before organisms could inhabit and completely colonize the land.

17. Discuss the significance and possible advantages of the pelycosaur sail.

18. What are the major differences between the seedless vascular plants and the gymnosperms, and why are these differences significant in terms of exploiting the terrestrial environment?

19. Why were the reptiles so much more successful at extending their habitat than the amphibians?

20. The fossilized footprints shown in the accompanying image, which are more than 365 million years old, are part of a tetrapod trackway at Valentia Island, Ireland, and until recently, represented evidence of some of the oldest land-dwelling tetrapods. What evidence in this photo indicates the sediments were deposited in a wet environment, and how can you use the footprints to establish that the animal making them was a tetrapod?

Courtesy of Ken Higgs, Department of Geology, University College Cork, Ireland

TABLE 13.3 Major Evolutionary and Geologic Events of the Paleozoic Era

Age (millions of years)	Geologic Period		Invertebrates	Vertebrates	Plants	Major Geologic Events
251	Permian		Largest mass extinction event to affect the invertebrates	Acanthodians, placoderms, and pelycosaurs become extinct Therapsids and pelycosaurs the most abundant reptiles	Gymnosperms diverse and abundant	Formation of Pangaea Alleghenian orogeny Hercynian (Variscan) orogeny
299	Carboniferous	Pennsylvanian	Fusulinids diversify	Amphibians abundant and diverse	Coal swamps with flora of seedless vascular plants and gymnosperms	Coal-forming swamps common Formation of Ancestral Rockies Continental glaciation in Gondwana
318		Mississippian	Crinoids, lacy bryozoans, blastoids become abundant Renewed adaptive radiation following extinctions of many reef-builders	Reptiles evolve	Gymnosperms appear (may have evolved during Late Devonian)	Ouachita orogeny Widespread deposition of black shale
359	Devonian		Extinctions of many reef-building invertebrates near end of Devonian Reef building continues Eurypterids abundant	Amphibians evolve All major groups of fish present—Age of Fish	First seeds evolve Seedless vascular plants diversify	Widespread deposition of black shale Acadian orogeny Antler orogeny
416	Silurian		Major reef building Diversity of invertebrates remains high	Ostracoderms common Acanthodians, the first jawed fish, evolve	Early land plants— seedless vascular plants	Caledonian orogeny Extensive barrier reefs and evaporites
444	Ordovician		Extinctions of a variety of marine invertebrates near end of Ordovician Major adaptive radiation of all invertebrate groups Suspension feeders dominant	Ostracoderms diversify	Plants move to land?	Continental glaciation in Gondwana Taconic orogeny
488	Cambrian		Many trilobites become extinct near end of Cambrian Trilobites, brachiopods, and archaeocyathids are most abundant	Earliest vertebrates— jawless fish called ostracoderms		First Phanerozoic transgression (Sauk) onto North American craton
542						

© 2013 Cengage Learning

14

Mesozoic Earth History

Aerial view of the Sierra Nevada, showing Yosemite Valley and Half Dome, California. The rocks in this image are part of the Sierra Nevada batholith, a huge mass of granite and related rocks made up of many intrusive bodies. The massive plutons comprising the Sierra Nevada were emplaced during the Late Jurassic Nevadan orogeny, and later exposed by uplift along faults on the east side of the present-day mountain range. Yosemite Valley and Half Dome were formed by valley glaciers during the Pleistocene.

© 2009 Lars Oppermann/Flickr/Getty Images

▶ The Mesozoic breakup of Pangaea profoundly affected geologic and biologic events.

▶ During the Triassic Period most of North America was above sea level.

▶ During the Jurassic Period a seaway flooded the interior of western North America.

▶ A global rise in sea level during the Cretaceous Period resulted in an enormous interior seaway that extended

from the Gulf of Mexico to the Arctic Ocean and divided North America into two large landmasses.

▶ Western North America was affected by four interrelated orogenies that took place along an oceanic–continental plate boundary.

▶ Terrane accretion played an important role in the Mesozoic geologic history of western North America.

▶ Coal, petroleum, uranium, and copper deposits are major Mesozoic mineral resources.

Introduction

Approximately 150 to 210 million years after the emplacement of massive plutons created the Sierra Nevada (Nevadan orogeny), gold was discovered at Sutter's Mill on the South Fork of the American River at Coloma, California. On January 24, 1848, James Marshall, a carpenter building a sawmill for John Sutter, found bits of the glittering metal in the mill's tailrace. Soon settlements throughout the state were abandoned as word of the chance for instant riches spread throughout California.

Within a year after the news of the gold discovery reached the East Coast, the Sutter's Mill area was swarming with more than 80,000 prospectors, all hoping to make their fortune. At least 250,000 gold seekers worked the Sutter's Mill area, and although most were Americans, prospectors came from all over the world, even as far away as China.

Although many prospectors searched for the source of the gold, or the mother lode, the gold that they recovered was mostly in the form of placer deposits (deposits of sand and gravel containing gold particles large enough to be recovered by panning). Unfortunately, most thought the gold was simply waiting to be taken and didn't realize that prospecting was hard work. Life in the mining camps was extremely difficult and expensive, and the shop owners and traders frequently made more money than the prospectors. In reality, only a few prospectors ever hit it big or were even moderately successful. Most barely made a living wage working their claims before they finally abandoned their dream and went home. Nevertheless, during the five years from 1848 to 1853 that constituted the Gold Rush proper, more than $200 million in gold was extracted.

The Mesozoic Era (251 to 66 million years ago) was an important time in Earth's history. Just as the formation of Pangaea influenced geologic and biologic events during the Paleozoic Era, the breakup of this supercontinent had

profound geologic and biologic effects during the Mesozoic. As a result of plate movements during the Mesozoic, the Atlantic Ocean basin formed, as well as the start to the growth of the Rocky Mountains. The accumulation of vast salt deposits eventually formed salt domes adjacent to which oil and natural gas were trapped, and the emplacement of huge batholiths accounted for the origin of various mineral resources, including the gold that fueled the California Gold Rush of the mid-1800s.

A main emphasis of this book has been the systems approach to Earth and life history. In this chapter and the next, we will explore how the breakup of Pangaea affected the global climatic and oceanic regimes, tectonics and sedimentation patterns, and evolutionary changes in the world's biota.

The Breakup of Pangaea

Because of the magnetic anomalies preserved in the oceanic crust (see Figure 3.11), geologists have a very good record of the history of Pangaea's breakup, and the direction of movement of the various continents during the Mesozoic and Cenozoic eras.

Pangaea's breakup began with rifting between Laurasia and Gondwana during the Triassic (▶Figure 14.1a). By the end of the Triassic, the newly formed and expanding Atlantic Ocean separated North America from Africa, and sometime during the Late Triassic and Early Jurassic, North America rifted from South America.

Separation of the continents allowed water from the Tethys Sea to flow into the expanding central Atlantic Ocean, whereas Pacific Ocean waters flowed into the newly

(a) Triassic Period.

(b) Jurassic Period.

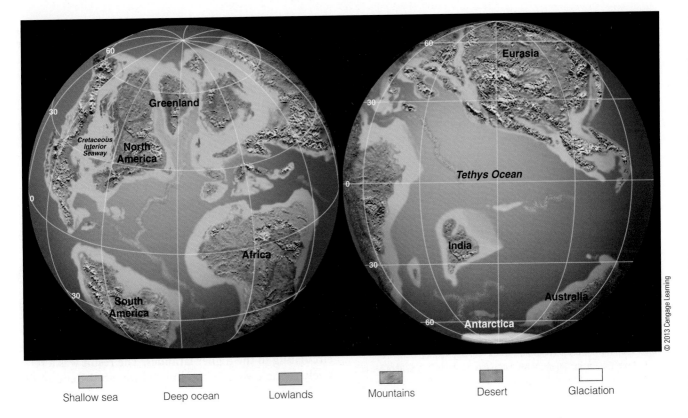

| Shallow sea | Deep ocean | Lowlands | Mountains | Desert | Glaciation |

(c) Late Cretaceous Period.

formed Gulf of Mexico, which then was little more than a restricted bay (▶ Figure 14.2). During that time, these areas were located in the low tropical latitudes, where high temperatures and high rates of evaporation were ideal for the formation of thick evaporite deposits.

The initial breakup of Gondwana took place during the Late Triassic and Jurassic periods. Antarctica and Australia, which remained sutured together, began separating from South America and Africa, whereas India began rifting from the Gondwana continent and moved northward.

South America and Africa began rifting apart during the Jurassic (Figure 14.1b) and the subsequent separation of these two continents formed a narrow basin where thick evaporite deposits accumulated from the evaporation of southern ocean waters (Figure 14.2). During this time, the eastern end of the Tethys Sea began closing as a result of the clockwise rotation of Laurasia and the northward movement of Africa. This narrow Late Jurassic and Cretaceous seaway between Africa and Europe was the forerunner of the present Mediterranean Sea.

By the end of the Cretaceous, Australia and Antarctica had detached from each other, and India had moved into the low southern latitudes and was nearly to the equator. South America and Africa were now widely separated, and Greenland was essentially an independent landmass with only a shallow sea between it and North America and Europe (Figure 14.1c).

A global rise in sea level during the Cretaceous resulted in worldwide transgressions onto the continents. Higher heat flow and rapid expansion of oceanic ridges were responsible for these transgressions. By the Middle Cretaceous, sea level was probably as high as at any time since the Ordovician, and about one-third of the present land area was inundated by epeiric seas (Figure 14.1c).

The final stage in Pangaea's breakup occurred during the Cenozoic. During this time, Australia continued moving northward, and Greenland was completely separated from Europe and North America and formed a separate landmass.

The Effects of the Breakup of Pangaea on Global Climates and Ocean Circulation Patterns

By the end of the Permian Period, Pangaea extended from pole to pole, covered about one-fourth of Earth's surface, and was surrounded by Panthalassa, a global ocean that encompassed approximately 300 degrees of longitude (see Figure 11.2b). Such a configuration exerted tremendous influence on the world's climate and resulted in generally arid conditions over large parts of Pangaea's interior.

The world's climates result from the complex interaction between wind and ocean currents and the location and topography of the continents. In general, dry climates

▶ **Figure 14.2 Gulf of Mexico Evaporites** Evaporites accumulated in shallow basins as Pangaea broke apart during the Early Mesozoic. Water from the Tethys Sea flowed into the central Atlantic Ocean, and water from the Pacific Ocean flowed into the newly formed Gulf of Mexico. Marine water from the south flowed into the southern Atlantic Ocean.

receive, whereas continents absorb only about 50 percent, even less if they are snow covered. The rest of the solar radiation is reflected back into space. Areas dominated by seas are warmer than those dominated by continents. By knowing the distribution of continents and ocean basins, geologists can generally estimate the average annual temperature for any region on Earth, as well as determine a temperature gradient.

The breakup of Pangaea during the Late Triassic caused the global temperature gradient to increase because the Northern Hemisphere continents moved farther northward, displacing higher-latitude ocean waters. Due to the steeper global temperature gradient, produced by a decrease in temperature in the high latitudes and the changing positions of the continents, oceanic and atmospheric circulation patterns greatly accelerated during the Mesozoic (▶ Figure 14.3).

Although the temperature gradient and seasonality on land were increasing during the Jurassic and Cretaceous, the middle- and higher-latitude oceans were still quite warm because warm waters from the Tethys Sea were circulating to the higher latitudes. The result was a relatively equable worldwide climate through the end of the Cretaceous.

occur on large landmasses in areas remote from sources of moisture and where barriers to moist air exist, such as mountain ranges. Wet climates occur near large bodies of water or where winds can carry moist air over land.

Past climatic conditions can be inferred from the distribution of climate-sensitive deposits. Evaporite deposits result when evaporation exceeds precipitation. Although sand dunes and red beds may form locally in humid regions, they are characteristic of arid regions. Coal forms in both warm and cool humid climates. Vegetation that is eventually converted into coal requires at least a good seasonal water supply; thus, coal deposits are indicative of humid conditions.

Widespread Triassic evaporites, red beds, and desert dunes in the low and middle latitudes of North and South America, Europe, and Africa indicate dry climates in those regions, whereas coal deposits are found mainly in the high latitudes, indicating humid conditions (Figure 14.1a). These high latitude coals are analogous to today's Scottish peat bog or Canadian muskeg. The lands bordering the Tethys Sea were probably dominated by seasonal monsoon rains resulting from the warm, moist winds and warm oceanic currents impinging against the east-facing coast of Pangaea.

The temperature gradient between the tropics and the poles also affects oceanic and atmospheric circulation. The greater the temperature difference between the tropics and the poles, the steeper the temperature gradient is, and thus the faster the circulation of the oceans and atmosphere. Oceans absorb about 90 percent of the solar radiation they

Mesozoic History of North America

The beginning of the Mesozoic Era was essentially the same in terms of tectonism and sedimentation as the preceding Permian Period in North America (see Figure 11.11). Terrestrial sedimentation continued over much of the craton, and block faulting and igneous activity began in the Appalachian region as North America and Africa began separating (▶ Figure 14.4). The newly forming Gulf of Mexico experienced extensive evaporite deposition during the Late Triassic and Jurassic as North America separated from South America (Figure 14.2 and ▶ Figure 14.5).

A global rise in sea level during the Cretaceous resulted in worldwide transgressions onto the continents such that marine deposition was continuous over much of the North American Cordilleran (▶ Figure 14.6).

A volcanic island arc system that formed off the western edge of the craton during the Permian was sutured to North America sometime during the Permian or Triassic. This event is referred to as the *Sonoma orogeny* and will be discussed later in the chapter. During the Jurassic, the entire Cordilleran area was involved in a series of major mountain-building episodes resulting in the formation of the Sierra Nevada, the Rocky Mountains, and other lesser mountain ranges. Although each orogenic episode has its own name, the entire mountain-building event is simply called the *Cordilleran orogeny* (also discussed later in this chapter). With this simplified overview of the Mesozoic

(a) Triassic Period.

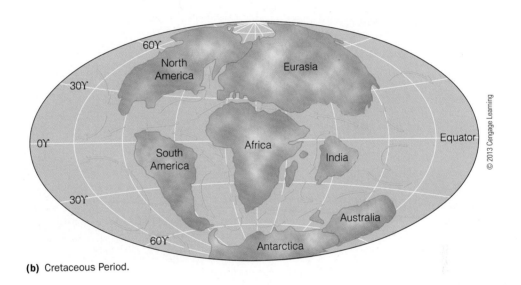

(b) Cretaceous Period.

history of North America in mind, we will now examine the specific regions of the continent.

Continental Interior

Recall that the history of the North American craton is divided into unconformity-bound sequences reflecting advances and retreats of epeiric seas over the craton (see Figure 10.3). Although these transgressions and regressions played a major role in the Paleozoic geologic history of the continent, they were not as important during the Mesozoic. Most of the continental interior during the Mesozoic was well above sea level and was not inundated by epeiric seas. Consequently, the two Mesozoic cratonic sequences, the *Absaroka Sequence* (Late Mississippian to Early Jurassic) and the *Zuni Sequence* (Early Jurassic to Early Paleocene) (see Figure 10.3), are not treated separately here; instead, we will examine the Mesozoic history of the three continental margin regions of North America.

Eastern Coastal Region

During the Early and Middle Triassic, coarse detrital sediments derived from erosion of the recently uplifted Appalachians (Alleghenian orogeny) filled the various intermontane basins and spread over the surrounding areas. As

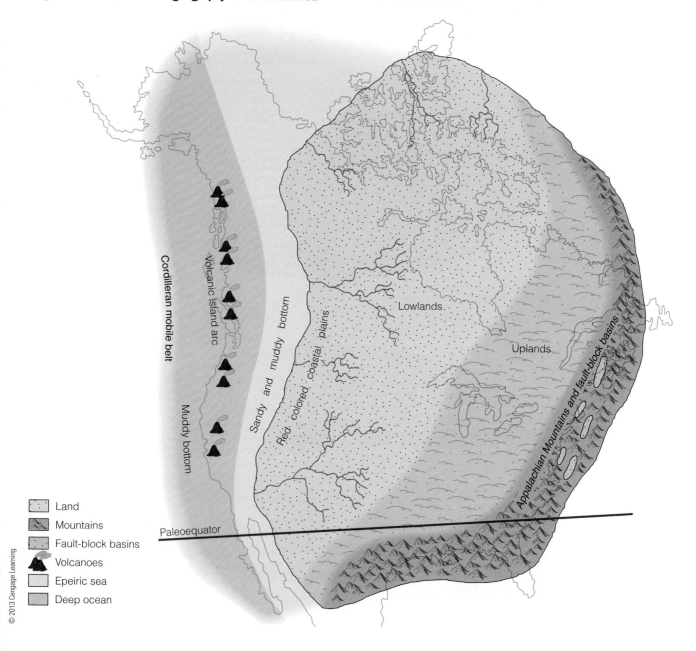

Land

Mountains

Fault-block basins

Volcanoes

Epeiric sea

Deep ocean

Cordilleran mobile belt

Volcanic island arc

Muddy bottom

Sandy and muddy bottom

Red colored coastal plains

Lowlands

Uplands

Appalachian Mountains and fault-block basins

Paleoequator

© 2013 Cengage Learning

weathering and erosion continued during the Mesozoic, this once lofty mountain system was reduced to a low-lying plain.

During the Late Triassic, the first stage in the breakup of Pangaea began with North America separating from Africa. Fault-block basins developed in response to upwelling magma beneath Pangaea in a zone stretching from present-day Nova Scotia to North Carolina (▶ Figure 14.7). Erosion of the fault-block mountains filled the adjacent basins with great quantities (up to 6,000 m) of poorly sorted red non-marine detrital sediments known as the *Newark Group*.

Reptiles roamed along the margins of the lakes and streams that formed in these basins, leaving their footprints and trackways in the soft sediments (▶ Figure 14.8). Although the Newark Group rocks contain numerous dinosaur foot-prints, they are almost completely devoid of dinosaur bones! The Newark Group is mostly Late Triassic in age, but in some areas deposition did not begin until the Early Jurassic.

Concurrent with sedimentation in the fault-block basins were extensive lava flows that blanketed the basin floors, as well as intrusions of numerous dikes and sills. The most famous intrusion is the prominent Palisades sill

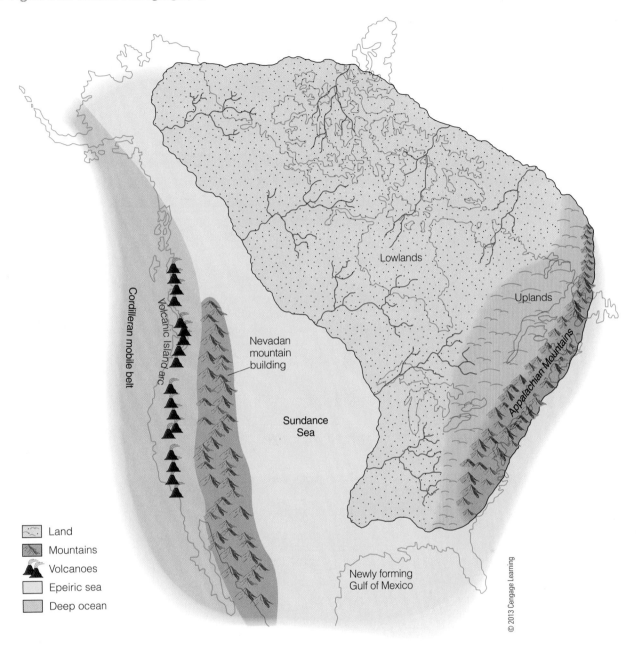

Cordilleran mobile belt

Volcanic Island arc

Nevadan mountain building

Sundance Sea

Lowlands

Uplands

Appalachian Mountains

Newly forming Gulf of Mexico

Land

Mountains

Volcanoes

Epeiric sea

Deep ocean

© 2013 Cengage Learning

along the Hudson River in the New York–New Jersey area (Figure 14.7d).

As the Atlantic Ocean grew, rifting ceased along the eastern margin of North America, and this once-active convergent plate margin became a passive, trailing continental margin. The fault-block mountains produced by this rifting continued to erode during the Jurassic and Early Cretaceous until all that was left was an area of low relief.

The sediments produced by this erosion contributed to the growing eastern continental shelf. During the Cretaceous Period, the Appalachian region was reelevated and once again shed sediments onto the continental shelf, forming a gently dipping, seaward-thickening wedge of rocks up to 3,000 m thick. These rocks are currently exposed in a belt extending from Long Island, New York, to Georgia.

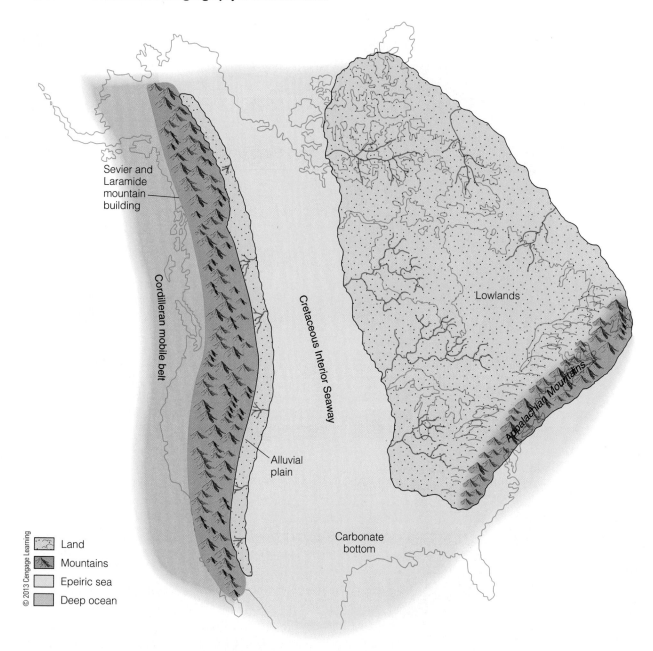

Land

Mountains

Epeiric sea

Deep ocean

© 2013 Cengage Learning

Gulf Coastal Region

The Gulf Coastal region was above sea level until the Late Triassic (Figure 14.4). As North America separated from South America during the Late Triassic and Early Jurassic, the Gulf of Mexico began to form (Figure 14.5). With oceanic waters flowing into this newly formed, shallow, restricted basin, conditions were ideal for evaporite formation. More than 1,000 m of evaporites were precipitated at this time, and most geologists think that these Jurassic evaporites are the source for the Cenozoic salt domes found today in the Gulf of Mexico and southern Louisiana. The history of these salt domes and their associated petroleum accumulations is discussed in Chapter 16.

By the Late Jurassic, circulation in the Gulf of Mexico was less restricted, and evaporite deposition ended. Normal marine conditions returned to the area with alternating transgressing and regressing seas. The resulting sediments were covered and buried by thousands of meters of Cretaceous and Cenozoic sediments.

(a) Areas where Triassic fault-block basin deposits crop out in eastern North America.

(b) After the Appalachians were eroded to a low-lying plain by the Middle Triassic, fault-block basins such as this one (shown in cross section) formed as a result of Late Triassic rifting between North America and Africa.

(c) These valleys accumulated tremendous thickness of sediments and were themselves broken by a complex of normal faults during rifting.

(d) Palisades of the Hudson River. This sill was one of many intruded into the Newark sediments during the Late Triassic rifting that marked the separation of North America from Africa.

▶ Figure 14.8 **Triassic Newark Group Dinosaur Footprints**
Dinosaur tracks in the Triassic Newark Group were uncovered in 1966 during the excavation for a new state building in Hartford, Connecticut. Because the tracks were so spectacular, the building site was moved, and the excavation site became Dinosaur State Park, a Registered National Landmark. Currently, a 5,100-square-meter geodesic dome covers some 500 of the tracks. The three-toed tracks, named *Eubronites*, are 25 to 41 cm long and are spaced 1.1 to 1.4 m apart. Although no bones of the dinosaur that made the tracks have yet been found, paleontologists surmise that the trackmaker was a dinosaur similar in size and shape to *Dilophosaurus*.

During the Cretaceous, the Gulf Coastal region, like the rest of the continental margin, was flooded by northward-transgressing seas (Figure 14.6). As a result, nearshore sandstones are overlain by finer sediments characteristic of deeper waters. Following an extensive regression at the end of the Early Cretaceous, a major transgression began, during which a wide seaway extended from the Arctic Ocean to the Gulf of Mexico (Figure 14.6). Sediments deposited in the Gulf Coastal region during the Cretaceous formed a seaward-thickening wedge.

Reefs were also widespread in the Gulf Coastal region during the Cretaceous and were composed primarily of bivalves called *rudists* (see Chapter 15). Because of their high porosity and permeability, rudistoid reefs make excellent petroleum reservoirs. A good example of a Cretaceous reef complex occurs in Texas where the reef trend strongly influenced the carbonate platform deposition of the region (▶ Figure 14.9). The facies patterns of these Cretaceous carbonate rocks are as complex as those in the major barrier-reef systems of the Paleozoic Era.

From D. G. Bebout and R. G. Loucks, "Lower Cretaceous Reefs, South Texas," *AAPG Memoir Series*, No. 33. AAPG (Figures 1 and 3, pp. 441–442). Copyright © 1983 by AAPG. Reprinted with permission of the AAPG, whose permission is required for future use.

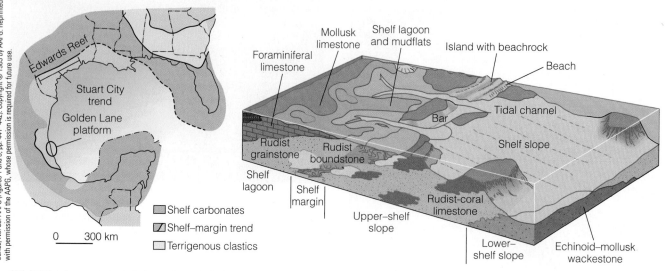

(a) Early Cretaceous shelf-margin facies around the Gulf of Mexico Basin. The reef trend is shown as a black line.

(b) Reconstruction of the depositional environment and facies changes across the Stuart City reef trend, South Texas.

Western Region

Except for the fault-block basins along the east coast of North America that developed in response to the separation between North America and Africa during the Late Triassic, tectonic activity and major mountain-building activity took place in the Cordilleran mobile belt during the Mesozoic. The western area of North America was also the site of two major seaways, one during the Jurassic (Figure 14.5) and the other, extending from the Arctic Ocean to the Gulf of Mexico, during the Cretaceous (Figure 14.6).

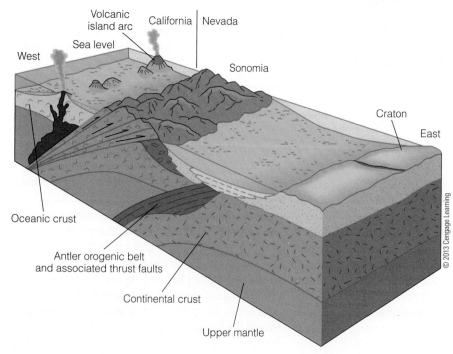

Mesozoic Tectonics

The Mesozoic geologic history of the North American Cordilleran mobile belt is very complex, involving the eastward subduction of the oceanic Farallon plate under the continental North American plate. Activity along this oceanic–continental convergent plate boundary resulted in an eastward movement of deformation. This orogenic activity progressively affected the trench and continental slope, the continental shelf, and the cratonic margin, causing a thickening of the continental crust. In addition, the accretion of terranes and microplates along the western margin of North America also played a significant role in the Mesozoic tectonic history of this area.

Except for the Late Devonian–Early Mississippian Antler orogeny (see Figure 11.15), the Cordilleran region of North America experienced little tectonism during the Paleozoic. However, an island arc and ocean basin formed off the western North American craton during the Permian (Figure 14.4), followed by subduction of an oceanic plate beneath the island arc and the thrusting of oceanic and island arc rocks eastward against the craton margin (▶ Figure 14.10). This event,

© 2013 Cengage Learning

▶ Figure 14.10 **Sonoma Orogeny** Tectonic activity that culminated in the Permian–Triassic Sonoma orogeny in western North America. The Sonoma orogeny was the result of a collision between the southwestern margin of North America and an island arc system.

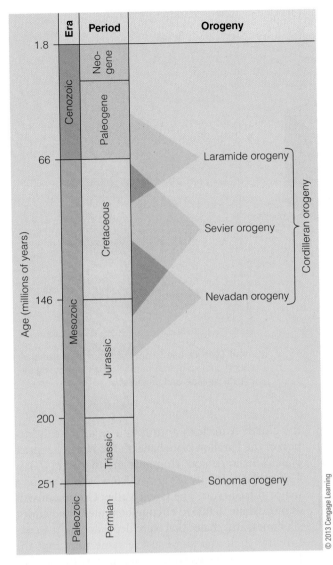

▶ **Figure 14.11** **Mesozoic Cordilleran Orogenies** Mesozoic orogenies occurring in the Cordilleran mobile belt.

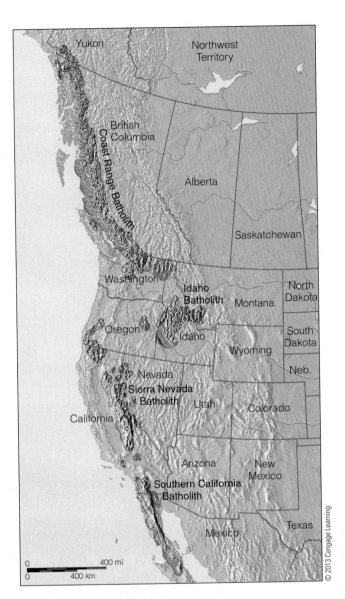

▶ **Figure 14.12** **Cordilleran Batholiths** Location of Jurassic and Cretaceous batholiths in western North America.

similar to the preceding Antler orogeny, and known as the **Sonoma orogeny**, occurred at or near the Permian–Triassic boundary. Like the Antler orogeny, it resulted in the suturing of island-arc terranes along the western edge of North America and the formation of a landmass called *Sonomia*.

Following the Late Paleozoic–Early Mesozoic destruction of the volcanic island arc during the Sonoma orogeny, the western margin of North America became an oceanic–continental convergent plate boundary. During the Late Triassic, a steeply dipping subduction zone developed along the western margin of North America in response to the westward movement of North America over the Farallon plate. This newly created oceanic–continental plate boundary controlled Cordilleran tectonics for the rest of the Mesozoic and for most of the Cenozoic Era; this subduction zone marks the beginning of the modern circum-Pacific orogenic system (see Chapter 16).

The general term **Cordilleran orogeny** is applied to the mountain-building activity that began during the Jurassic and continued into the Cenozoic (▶Figure 14.11). The Cordilleran orogeny consisted of a series of individual mountain-building events, or pulses, that occurred in different regions at different times but overlapped to some extent. Most of this Cordilleran orogenic activity is related to the continued westward movement of the North American plate as it overrode the Farallon plate, and its history is very complex.

The first pulse of the Cordilleran orogeny, the **Nevadan orogeny** (Figure 14.11), began during the Late Jurassic and continued into the Cretaceous as large volumes of granitic magma were generated at depth beneath the active western margin of North America. These granitic masses were emplaced as huge batholiths that are now recognized as the Sierra Nevada, Southern California, Idaho, and Coast Range batholiths (▶Figure 14.12). It was also during this

▶ Figure 14.13 **The Franciscan Complex**

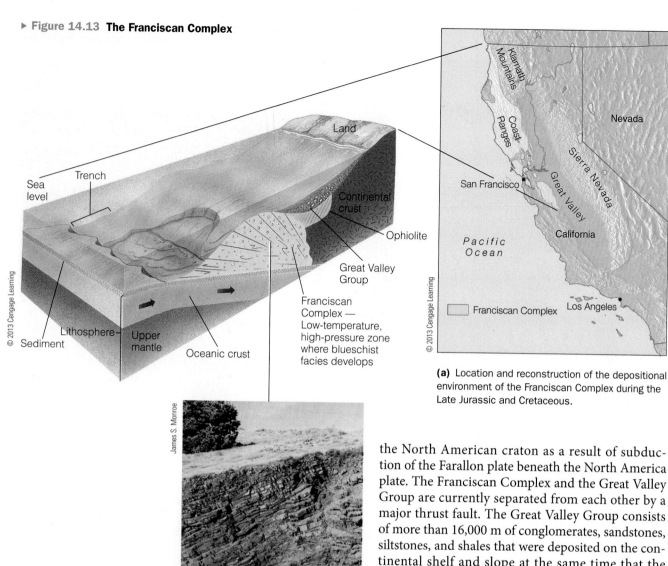

(a) Location and reconstruction of the depositional environment of the Franciscan Complex during the Late Jurassic and Cretaceous.

(b) Bedded chert exposed in Marin County, California. Most of the layers are about 5 cm thick.

time that the *Franciscan Complex* and *Great Valley Group* were deposited and deformed as part of the Nevadan orogeny within the Cordilleran mobile belt.

The Franciscan Complex, which is up to 7,000 m thick, is an unusual rock unit consisting of a chaotic mixture of rocks that accumulated during the Late Jurassic and Cretaceous. The various rock types—graywacke, volcanic breccia, siltstone, black shale, chert, pillow basalt, and blueschist metamorphic rocks—suggest that continental shelf, slope, and deep-sea environments were brought together in a submarine trench when North America overrode the subducting Farallon plate (▶ Figure 14.13).

The Franciscan Complex, and the Great Valley Group that lies east of it, were both squeezed against the edge of

the North American craton as a result of subduction of the Farallon plate beneath the North America plate. The Franciscan Complex and the Great Valley Group are currently separated from each other by a major thrust fault. The Great Valley Group consists of more than 16,000 m of conglomerates, sandstones, siltstones, and shales that were deposited on the continental shelf and slope at the same time that the Franciscan deposits were accumulating in the submarine trench (Figure 14.13).

By the Late Cretaceous, most of the volcanic and plutonic activity had migrated eastward into Nevada and Idaho. This migration was probably caused by a change from high-angle to low-angle subduction, resulting in the subducting oceanic plate reaching its melting depth farther east (▶ Figure 14.14). Thrusting occurred progressively farther east so that by the Late Cretaceous, it extended all the way to the Idaho–Washington border.

The second pulse of the Cordilleran orogeny, the **Sevier orogeny**, affected western North America from Alaska to Mexico, and was mostly a Cretaceous event, even though it began in the Late Jurassic and is associated with the tectonic activity of the earlier Nevadan orogeny (Figure 14.11). Subduction of the Farallon plate beneath the North American plate during this time caused numerous overlapping, low-angle thrust faults in which blocks of older Paleozoic strata were thrust eastward on top of younger strata (▶ Figure 14.15). This deformation resulted in crustal shortening and produced generally north–south-trending mountain ranges that stretch from Montana to western Canada.

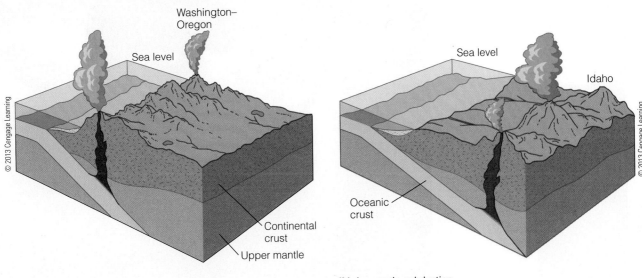

(a) High-angle subduction.

(b) Low-angle subduction.

During the Late Cretaceous to Early Cenozoic, the final pulse of the Cordilleran orogeny took place (Figure 14.11). The **Laramide orogeny** developed east of the Sevier orogenic belt in the present-day Rocky Mountain areas of New Mexico, Colorado, and Wyoming. Most features of the present-day Rocky Mountains resulted from the Cenozoic phase of the Laramide orogeny, and for that reason, it will be discussed in Chapter 16.

Mesozoic Sedimentation

Concurrent with the tectonism in the Cordilleran mobile belt, Early Triassic sedimentation on the western continental shelf consisted of shallow-water marine sandstones, shales, and limestones. During the Middle and Late Triassic, the western shallow seas regressed farther west, exposing large areas of former seafloor to erosion. Marginal marine and nonmarine Triassic rocks, particularly red

▶ Figure 14.15 **Sevier Orogeny**

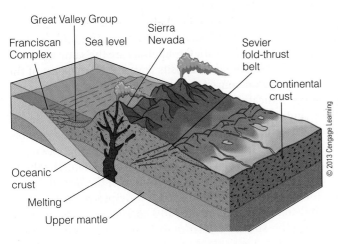

(a) Restoration showing the associated tectonic features of the Late Cretaceous Sevier orogeny caused by subduction of the Farallon plate under the North American plate.

(b) The Keystone thrust fault, a major fault in the Sevier overthrust belt, is exposed west of Las Vegas, Nevada. The sharp boundary between the light-colored Mesozoic rocks and the overlying dark-colored Paleozoic rocks marks the trace of the Keystone thrust fault.

(a) Stratigraphic column of Triassic and Lower Jurassic formations in the western United States.

(b) View of Chimney Rock, Capitol Reef National Park, Utah. The red beds comprising Chimney Rock are the Lower Triassic Moenkopi Formation. The light-colored and more resistant rocks forming the top of Chimney Rock are the Upper Triassic Shinarump Conglomerate.

beds, contribute to the spectacular and colorful scenery of the region.

The Lower Triassic *Moenkopi Formation* of the southwestern United States consists of a succession of brick red and chocolate-colored mudstones (► Figure 14.16). Sedimentary structures such as mud cracks and ripple marks, as well as fossil amphibians and reptiles and their tracks, indicate deposition in a variety of continental environments, including stream channels, floodplains, and both freshwater and brackish ponds. Thin tongues of marine limestones indicate brief incursions of the sea, whereas local beds with gypsum and halite crystal casts attest to a rather arid climate. Unconformably overlying the Moenkopi is the Upper Triassic *Shinarump Conglomerate*, a

widespread unit generally less than 50 m thick and usually forming ledges where it is exposed (Figure 14.16).

Above the Shinarump are the multicolored shales, siltstones, and sandstones of the Upper Triassic *Chinle Formation* (Figure 14.16a). This formation is widely exposed throughout the Colorado Plateau region and is probably most famous for its petrified wood, spectacularly exposed in Petrified Forest National Park, Arizona (see Perspective). Although best known for its petrified wood, the Chinle Formation has also yielded fossils of labyrinthodont amphibians, phytosaurs, and small dinosaurs (see Chapter 15 for a discussion of the latter two animal groups). In addition, palynologic studies show a similar assemblage of pollen from the Chinle Formation and the Lower Newark Group on the east coast, indicating that the Chinle Formation and Lower Newark Group units are the same age.

Early Jurassic-age deposits covering a large part of the western region consist mostly of clean, cross-bedded sandstones indicative of wind-blown deposits. The lowermost of these formations is the *Wingate Sandstone,* a desert dune deposit, which is overlain by the *Kayenta Formation,* a stream and lake deposit (Figure 14.16a). These two formations are well exposed in southwestern Utah. The thickest, best known, and most prominent of the Jurassic sandstones is the *Navajo Sandstone,* a widespread cross-bedded sandstone that accumulated in a coastal dune environment along the southwestern margin of the craton (► Figure

► Figure 14.17 **Navajo Sandstone** Large cross-beds of the Jurassic Navajo Sandstone exposed in Zion National Park, Utah.

(a) Panoramic view of the Jurassic Morrison Formation as seen from the visitors' center at Dinosaur National Monument, near Vernal, Utah.

(b) North wall of visitors' center showing dinosaur bones in bas relief, just as they were deposited 140 million years ago.

14.17). The sandstone's most distinguishing feature is its large-scale cross-beds, some of which are more than 25 m high (see Figure 6.20). The upper part of the Navajo contains smaller cross-beds as well as dinosaur and crocodilian fossils.

Marine conditions returned to the region during the Middle Jurassic when a seaway called the **Sundance Sea** twice flooded the interior of western North America (Figure 14.5). The resulting deposits, the *Sundance Formation,* were produced from erosion of tectonic highlands to the west that paralleled the shoreline. These highlands resulted from intrusive igneous activity and associated volcanism that began during the Triassic.

During the Late Jurassic, a mountain chain formed in Nevada, Utah, and Idaho as a result of the deformation produced by the Nevadan orogeny. As the mountain chain grew and shed sediments eastward, the Sundance Sea began retreating northward. A large part of the area formerly occupied by the Sundance Sea was then covered by multicolored sandstones, mudstones, shales, and occasional lenses of conglomerates that comprise the world-famous *Morrison Formation* (► Figure 14.18a).

The Morrison Formation contains one of the world's richest assemblages of Jurassic dinosaur remains. Although most of the skeletons are broken up, as many as 50 individual skeletons have been found together in a small area. Such a concentration indicates that the skeletons were brought together during times of flooding and deposited on sandbars in stream channels. Soils in the Morrison indicate that the climate was seasonably dry.

Although most major museums have either complete dinosaur skeletons or at least bones from the Morrison Formation, the best place to see the bones still embedded in the rocks is the visitors' center at Dinosaur National Monument near Vernal, Utah (Figure 14.18b).

Shortly before the end of the Early Cretaceous, Arctic waters spread southward over the craton, forming a large inland sea in the Cordilleran region. Mid-Cretaceous transgressions also occurred on other continents, and all were part of the global mid-Cretaceous rise in sea level that resulted from accelerated seafloor spreading as Pangaea continued to fragment. These Middle Cretaceous transgressions are marked by widespread black shale deposition within the oceanic areas, the shallow-sea shelf areas, and the continental regions that were inundated by the transgressions.

By the beginning of the Late Cretaceous, this incursion joined the northward-transgressing waters from the Gulf area to create an enormous **Cretaceous Interior Seaway** that occupied the area east of the Sevier orogenic belt. Extending from the Gulf of Mexico to the Arctic Ocean, and more than 1,500 km wide at its maximum extent, this seaway effectively divided North America into two large landmasses until just before the end of the Late Cretaceous (Figure 14.6).

Deposition in this seaway and the resulting sedimentary rock sequences are a result of complex interactions involving sea-level changes, sediment supply from the adjoining landmasses, tectonics, and climate. Cretaceous deposits less than 100 m thick indicate that the eastern margin of the Cretaceous Interior Seaway subsided slowly and received little sediment from the emergent, low-relief craton to the east. The western shoreline, however, shifted back and forth, primarily in response to fluctuations in the supply of sediment from the Cordilleran Sevier orogenic belt to the west. The facies relationships show lateral changes from conglomerate and coarse sandstone adjacent to the mountain belt through finer sandstones, siltstones, shales, and even limestones and chalks in the east (► Figure 14.19). During times of particularly active

Petrified Forest National Park, Arizona

Petrified Forest National Park is located in eastern Arizona about 42 km east of Holbrook (Figure 1). The park consists of two sections: the Painted Desert, which is north of Interstate 40, and the Petrified Forest, which is south of the Interstate.

The Painted Desert is a brilliantly colored landscape whose colors and hues change constantly throughout the day. The multicolored rocks of the Triassic Chinle Formation have been weathered and eroded to form a badlands topography of numerous gullies, valleys, ridges, mounds, and mesas (Figure 2). The Chinle Formation is composed predominantly of various-colored shale beds. These shales and associated volcanic ash layers are easily weathered and eroded. Interbedded locally with the shales are lenses of conglomerates, sandstones, and limestones, which are more resistant to weathering and erosion than the shales, and form resistant ledges.

The Petrified Forest was originally set aside as a national monument to protect the large number of petrified logs that lay exposed in what is now the southern part of the park (Figure 3). When the transcontinental railroad constructed a coaling and watering stop in Adamana, Arizona, passengers were encouraged to take excursions to "Chalcedony Park," as the area was then called, to see the petrified forests. In a short time, collectors and souvenir hunters hauled off tons of petrified wood, quartz crystals, and Native American relics. Not until a huge rock crusher was built to crush the logs for the manufacture of abrasives was the area declared a national monument and the petrified forests preserved and protected.

During the Triassic Period, the climate of the area was much wetter than

Figure 1 Petrified Forest National Park, Arizona, consists of two parts: the Painted Desert and the Petrified Forest. The Painted Desert is located north of Interstate 40, whereas the Petrified Forest stretches north of Highway 180 and south of Interstate 40.

it is today, and terrestrial sediments that became the Chinle Formation were deposited by rivers, streams, and in lakes. As just mentioned, the multicolored rocks of the Chinle Formation are what give the Painted Desert its name. However, the Chinle Formation is best known for the abundant and beautifully preserved conifer logs found throughout the Petrified Forest.

Figure 2 Panoramic view of the Painted Desert of Petrified Forest National Park, Arizona. Shown is the badland topography produced by weathering and erosion of the Upper Triassic Chinle Formation (Figure 14.16a). The weathering of the shales, siltstones, and sandstones is what gives the Painted Desert its constantly changing and brilliantly colored landscape.

Most of the petrified logs are conifers, some of which were more than 60 m tall and up to 4 m in diameter, and belong to the genus *Araucarioxylon* (Figure 3). Although most of the conifers apparently grew on higher ground or riverbanks, and were buried in place, many were uprooted and transported during floods and were deposited on sandbars and floodplains, where fossilization took place. Following burial, silica-rich groundwater percolated through the ash layers and silicified (permineralized) the wood.

During the ensuing Jurassic and Cretaceous periods, deposition continued in the Colorado Plateau region, further burying the Chinle Formation. During the Laramide orogeny, which began in the Cretaceous, the Colorado Plateau area was uplifted and eroded, exposing the Chinle Formation. Because the Chinle is composed mostly of shales, it was easily eroded, leaving the more resistant petrified logs and log fragments exposed on the surface—much as we see them today (Figure 3).

Figure 3 Petrified Forest National Park, Arizona. All of the logs here are *Araucarioxylon*, the most abundant tree in the park. The petrified logs have been weathered from the Chinle Formation, and they are mostly in the position in which they were buried some 200 million years ago.

▶ **Figure 14.19 Cretaceous Facies of the Western Cretaceous Interior Seaway** This restored west–east cross section of Cretaceous facies of the western Cretaceous Interior Seaway shows their relationship to the Sevier orogenic belt.

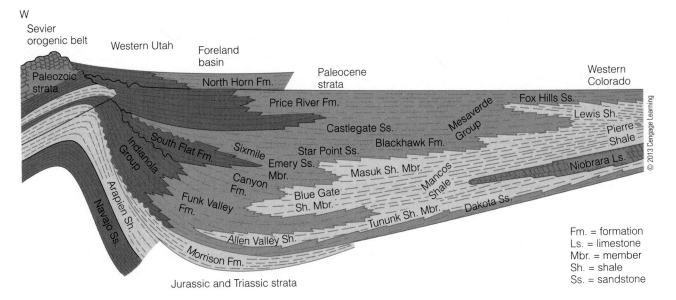

mountain building, these coarse clastic wedges of gravel and sand prograded even farther east.

As the Mesozoic Era ended, the Cretaceous Interior Seaway withdrew from the craton. During this regression, marine waters retreated to the north and south, and marginal marine and continental deposition formed widespread coal-bearing deposits on the coastal plain.

What Role Did Accretion of Terranes Play in the Growth of Western North America?

In the preceding sections, we have discussed orogenies along convergent plate boundaries resulting in continental accretion. Much of the material accreted to continents during such events is simply eroded older continental crust; however, a significant amount of new material was added to continents as well, such as igneous rocks that formed as a consequence of subduction and partial melting. Although subduction is the predominant influence on the tectonic history in many regions of orogenesis, other processes are also involved in mountain building and continental accretion, especially the accretion of terranes.

Geologists now know that portions of many mountain systems are composed of small, accreted lithospheric blocks that clearly formed elsewhere. These **terranes** differ completely in their fossil content, stratigraphy, structural trends, and paleomagnetic properties from the rocks of the surrounding mountain system and adjacent craton. In fact, these terranes are so different from adjacent rocks that most geologists think they formed elsewhere and were carried great distances as parts of other plates until they collided with other terranes or continents (see Chapter 2).

Geologic evidence indicates that more than 25 percent of the entire Pacific Coast from Alaska to Baja California consists of accreted terranes. These accreted terranes are composed of volcanic island arcs, oceanic ridges, seamounts, volcanic plateaus, and small fragments of continents that were scraped off and accreted to the continent's margin as the oceanic plate on which they were carried was subducted under the continent.

Geologists estimate that more than 100 different-sized terranes have been added to the western margin of North America during the last 200 million years (▶Figure 14.20). The Wrangellian terranes (Figure 14.1b) are a good example of terranes that have been accreted to North America's western margin (Figure 14.20).

The basic plate tectonic reconstruction of orogenies and continental accretion remains unchanged; however, the details of such reconstructions are decidedly different in view of terrane tectonics. For example, growth along active continental margins is faster than along passive continental

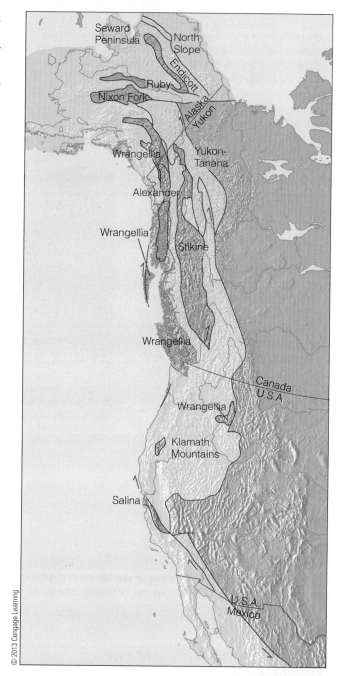

© 2013 Cengage Learning

▶ **Figure 14.20 Mesozoic Terranes** Some of the accreted lithospheric blocks, called terranes, that form the western margin of the North American craton. The dark brown blocks probably originated as terranes and were accreted to North America. The light green blocks are possibly displaced parts of North America. The North American craton is shown in dark green. See Figure 14.1b for the position of the Wrangellian terranes during the Jurassic.

margins because of the accretion of terranes. Furthermore, these accreted terranes are often new additions to a continent, rather than reworked older continental material.

So far, most Mesozoic terranes have been identified in mountains of the North American Pacific Coast region, but geologists suspect that a number of others are present in other mountain systems as well. They are more difficult to recognize in older mountain systems such as the

Appalachians because of greater deformation and erosion (see Chapter 11). As more of these terranes and microplates are recognized, they are providing a better understanding of the geologic history of the continents.

Mesozoic Mineral Resources

Although much of the coal in North America is Pennsylvanian or Paleogene in age, important Mesozoic coals occur in the Rocky Mountain states. These are mostly lignite and bituminous coals, but some local anthracites are present as well. Particularly widespread in western North America are coals of Cretaceous age. Mesozoic coals are also known from Australia, Russia, and China.

Large concentrations of petroleum occur in many areas of the world, but more than 50 percent of all proven reserves are in the Persian Gulf region. Conditions similar to those that led to the formation of petroleum in the Persian Gulf area also were present in what is now the Gulf Coast region of the United States and Central America. Here, petroleum and natural gas formed on a broad shelf over which transgressions and regressions occurred. In this region, the hydrocarbons are largely in reservoir rocks that were deposited as distributary channels on deltas and as barrier-island and beach sands. Some of these hydrocarbons are associated with structures formed adjacent to rising salt domes. The salt, called the *Louann Salt,* initially formed in a long, narrow sea when North America separated from Europe and North Africa during the fragmentation of Pangaea (Figure 14.2).

The richest uranium ores in the United States are widespread in Mesozoic rocks of the Colorado Plateau area of Colorado and adjoining parts of Wyoming, Utah, Arizona, and New Mexico. These ores, consisting of fairly pure masses of a complex potassium-, uranium-, vanadium-bearing mineral called *carnotite,* are associated with plant remains in sandstones that were deposited in ancient stream channels.

As noted in Chapter 9, Proterozoic banded iron formations are the main sources of iron ores. There are, however, some important exceptions. For example, the Jurassic-age "Minette" iron ores of western Europe, composed of oolitic limonite and hematite, are important ores in France, Germany, Belgium, and Luxembourg. In Great Britain, low-grade iron ores of Jurassic age consist of oolitic siderite, which is an iron carbonate. And in Spain, Cretaceous rocks are the host rocks for iron minerals.

South Africa, the world's leading producer of gem-quality diamonds and among the leaders in industrial diamond production, mines these minerals from kimberlite pipes, conical igneous intrusions of dark gray or blue igneous rock. Diamonds, which form at great depth where pressure and temperature are high, are brought to the surface during the explosive volcanism that forms kimberlite pipes. Although kimberlite pipes have formed throughout geologic time, the most intense episode of such activity in South Africa and adjacent countries was during the Cretaceous Period. Emplacement of Triassic and Jurassic diamond-bearing kimberlites also occurred in Siberia.

In the Introduction we noted that the mother lode, or source, for the placer deposits mined during the California Gold Rush is in Jurassic-age intrusive rocks of the Sierra Nevada. Gold placers are also known in Cretaceous-age conglomerates of the Klamath Mountains of California and Oregon.

Porphyry copper was originally named for copper deposits in the western United States mined from porphyritic granodiorite; however, the term now applies to large, low-grade copper deposits disseminated in a variety of rocks. These porphyry copper deposits are an excellent example of the relationship between convergent plate boundaries and the distribution, concentration, and exploitation of valuable metallic ores. Magma generated by partial melting of a subducting plate rises toward the surface, and as it cools, it precipitates and concentrates various metallic ores. The world's largest copper deposits were formed during the Mesozoic and Cenozoic in a belt along the western margins of North and South America (see Figure 3.26).

Summary

Table 14.1 summarizes the geologic history of North America, as well as global events and sea-level changes during the Mesozoic Era.

▶ We can summarize the breakup of Pangaea as follows:

1. During the Late Triassic, North America began separating from Africa. This was followed by the rifting of North America from South America.
2. During the Late Triassic and Jurassic periods, Antarctica and Australia—which remained sutured together—began separating from South America and Africa, and India began rifting from Gondwana.
3. South America and Africa began separating during the Jurassic, and Europe and Africa began converging during this time.
4. The final stage in Pangaea's breakup occurred during the Cenozoic, when Greenland completely separated from Europe and North America.

▶ The breakup of Pangaea influenced global climatic and atmospheric circulation patterns. Although the temperature gradient from the tropics to the poles gradually increased during the Mesozoic, overall global temperatures remained equable.

- An increased rate of seafloor spreading during the Cretaceous Period caused sea level to rise and transgressions to occur.
- Except for incursions along the continental margin and two major transgressions (the Sundance Sea and the Cretaceous Interior Seaway), the North American craton was above sea level during the Mesozoic Era.
- The Eastern Coastal Plain was the initial site of the separation of North America from Africa that began during the Late Triassic. During the Cretaceous Period, it was inundated by marine transgressions.
- The Gulf Coastal region was the site of major evaporite accumulation during the Jurassic as North America rifted from South America. During the Cretaceous, it was inundated by a transgressing sea, which, at its maximum, connected with a sea transgressing from the north to create the Cretaceous Interior Seaway.
- Mesozoic rocks of the western region of North America were deposited in a variety of continental and marine environments. One of the major controls of sediment distribution patterns was tectonism.
- Western North America was affected by four interrelated orogenies: the Sonoma, Nevadan, Sevier, and Laramide. Each involved igneous intrusions, as well as eastward thrust faulting and folding.
- The Sonoma orogeny was the result of subduction of an oceanic plate beneath an island arc system and the subsequent thrusting of oceanic and island arc rocks eastward against the southwestern cratonic margin of North America.
- The cause of the Nevadan, Sevier, and Laramide orogenies was the changing angle of subduction of the oceanic Farallon plate under the continental North American plate. The timing, rate, and, to some degree, the direction of plate movement were related to seafloor spreading and the opening of the Atlantic Ocean.
- Although the structural features of North America's western margin are associated with activity along an oceanic–continental convergent plate boundary, it is thought that more than 25 percent of the western margin originated from the accretion of terranes.
- Mesozoic rocks contain a variety of mineral resources, including coal, petroleum, uranium, gold, and copper.

Important Terms

Cordilleran orogeny, p. 285
Cretaceous Interior Seaway, p. 289
Laramide orogeny, p. 287

Nevadan orogeny, p. 285
Sevier orogeny, p. 286
Sonoma orogeny, p. 285

Sundance Sea, p. 289
terrane, p. 292

Review Questions

1. During which period of the Mesozoic Era was the global climate most equable?
   a. _____ Permian; b. _____ Triassic; c. _____ Jurassic; d. __✓__ Cretaceous; e. _____ Paleogene.
2. Triassic rifting between two continental landmasses initiated the breakup of Pangaea. Which two landmasses were these?
   a. _____ India and Australia; b. _____ Antarctica and India; c. _____ South America and Africa; d. _____ North America and Eurasia; e. __✓__ Laurasia and Gondwana.
3. The time of greatest post-Paleozoic inundation of the craton occurred during which geologic period?
   a. _____ Triassic; b. _____ Jurassic; c. __✓__ Cretaceous; d. _____ Paleogene; e. _____ Neogene.
4. A possible cause for the eastward migration of igneous activity in the Cordilleran region during the Cretaceous was a change from
   a. _____ high-angle to low-angle subduction; b. _____ divergent plate margin activity to subduction; c. _____ subduction to divergent plate margin activity; d. _____ oceanic–oceanic convergence to oceanic–continental convergence; e. _____ divergent to convergent plate margin activity.

5. During the Jurassic, the newly forming Gulf of Mexico was the site of primarily what type of deposition?
   a. __✓__ Evaporites; b. _____ Siliciclastics; c. _____ Volcaniclastics; d. _____ Detrital; e. _____ Answers b and c.
6. The second major Mesozoic seaway to flood North America was the
   a. _____ Zuni; b. _____ Sundance; c. _____ Newark; d. _____ Cordillera; e. __✓__ Cretaceous Interior Seaway.
7. Which formation contains one of the world's richest assemblages of Jurassic dinosaur fossils?
   a. _____ Franciscan; b. _____ Shinarump; c. _____ Chinle; d. _____ Newark; e. __✓__ Morrison.
8. The orogeny responsible for the present-day Rocky Mountains is the
   a. _____ Sevier; b. _____ Nevadan; c. _____ Antler; d. _____ Sonoma; e. __✓__ Laramide.
9. More than 50 percent of the world's proven reserves of petroleum occur in
   a. _____ the Gulf of Mexico; b. __✓__ Russia; c. _____ Canada; d. _____ Alaska; e. __✓__ the Persian Gulf.
10. Which formation or group filled the Late Triassic-age fault-block basins of the east coast of North America with red nonmarine sediments?
    a. _____ Morrison; b. _____ Chinle; c. _____ Navajo; d. _____ Franciscan; e. __✓__ Newark.

11. Briefly explain some of the evidence geologists use to interpret climatic conditions during the Mesozoic Era.

12. How does terrane accretion relate to the Mesozoic orogenies that took place on the western margins of North America?

13. Compare the tectonic setting and depositional environment of the Gulf of Mexico evaporites with the evaporite sequences of the Paleozoic Era.

14. What effect did the breakup of Pangaea have on oceanic and climatic circulation patterns? Compare the oceanic circulation pattern during the Triassic with that during the Cretaceous.

15. Why are Mesozoic-age coals mostly lignite and bituminous, whereas Paleozoic-age coals tend to be high-grade bituminous and anthracite?

16. Why is the paleogeography of the Mesozoic Era in some ways easier to reconstruct and more accurate than that of the Paleozoic Era?

17. From a plate tectonic perspective, how does the orogenic activity that occurred in the Cordilleran mobile belt during the Mesozoic Era differ from that which took place in the Appalachian mobile belt during the Paleozoic?

18. How did the Mesozoic rifting that took place on the east coast of North America affect the tectonics in the Cordilleran mobile belt?

19. Explain and diagram how increased seafloor spreading can cause a rise in sea level along the continental margins.

20. The gold discovered at Sutter's Mill, California, in 1848 sparked the California Gold Rush. This gold is widely disseminated throughout the granitic rocks of the Sierra Nevada batholith and was concentrated in placer deposits. During the Nevadan orogeny, several other large granitic batholiths were intruded in the Cordillera region of North America. Have there been any gold discoveries associated with these intrusions? Why?

TABLE 14.1 Summary of Mesozoic Geologic Events

Geologic Period	Sequence	Relative Changes in Sea Level (Rising ← → Falling)	Cordilleran Mobile Belt	North American Interior	Gulf Coastal Region	Eastern Coastal Region	Global Plate Tectonic Events
Cretaceous (66)	Zuni		Laramide orogeny / Sevier orogeny / Nevadan orogeny (Cordilleran orogeny)		Major Late Cretaceous transgression; Reefs particularly abundant; Regression at end of Early Cretaceous; Early Cretaceous transgression and marine sedimentation; Sandstones, shales, and limestones are deposited in transgressing and regressing seas; Thick evaporites are deposited in newly formed Gulf of Mexico	Appalachian region uplifted; Erosion of fault-block mountains formed during the Late Triassic to Early Jurassic	South America and Africa are widely separated; Greenland begins separating from Europe
Jurassic (146)			Jurassic and Cretaceous tectonism controlled by eastward subduction of the Pacific plate beneath North America and accretion of terranes.				South America and Africa begin separating in the Late Jurassic
Triassic (200)	Absaroka		Subduction zone develops as a result of westward movement of North America; Sonoma orogeny		Gulf of Mexico begins forming during Late Triassic	Fault-block mountains and basins develop in eastern North America; Deposition of Newark Group; lava flows, sills, and dikes	Breakup of Pangaea begins with rifting between Laurasia and Gondwana; Supercontinent Pangaea still in existence
(251)							

Age (millions of years)

© 2013 Cengage Learning

Camarasaurus lentus belongs to a group of dinosaurs known as sauropods, which were very large, long-necked herbivores that walked on all four limbs. This species of *Camarasaurus* was about 15 m long and is the most common sauropod found in Jurassic-age rocks. Although *Camarasaurus* was large, much larger sauropods existed. A good place to see *Camarasaurus* is Dinosaur National Monument on the Utah–Colorado border.

Photo by Daderot

At the end of this chapter, you will have learned that

▶ Marine invertebrates that survived the Paleozoic extinctions diversified and repopulated the seas.

▶ Land-plant communities changed considerably when flowering plants evolved during the Cretaceous.

▶ Reptile diversification began during the Mississippian and continued throughout the Mesozoic Era.

▶ Among the Mesozoic reptiles, dinosaurs had evolved by the Late Triassic and soon became the dominant land-dwelling vertebrate animals.

▶ In addition to dinosaurs, the Mesozoic was also the time of flying reptiles and marine reptiles, as well as turtles, lizards, snakes, and crocodiles.

▶ Birds evolved from reptiles, probably from some small, carnivorous dinosaur.

▶ Mammals evolved from reptiles only distantly related to dinosaurs, and they existed as contemporaries with dinosaurs.

▶ The transition from reptiles to mammals is very well supported by fossil evidence.

▶ Several varieties of Mesozoic mammals are known, all of which were small, and their diversity remained low.

▶ The proximity of continents and generally mild Mesozoic climates allowed many plants and animals to spread over extensive geographic areas.

▶ Extinctions at the end of the Mesozoic Era were second in magnitude only to the Paleozoic extinctions. These extinctions have received more attention than any others because dinosaurs were among the victims.

Introduction

Restorations of prehistoric floras and faunas—especially those of Mesozoic dinosaurs, flying reptiles, and marine reptiles—fascinate many people. Indeed, ever since 1842 when Sir Richard Owen coined the term *dinosaur*, these animals have been the object of intense curiosity, the subject of countless articles and books, many TV specials, and movies, including *Jurassic Park* (1993) and its two sequels, and several others. The popularity of dinosaurs and their relatives is no doubt related to the tremendous increase in knowledge that scientists have gained about them in the last few decades. For example, scientists now even know the color of some dinosaurs.

The Mesozoic Era is commonly called the "Age of Reptiles," calling attention to the fact that reptiles predominated among land-dwelling vertebrate animals. Keep in mind, though, that these "Age of" designations simply reflect our personal preferences; there were far more species of plants, fish, and insects at this time than there were reptiles. Nevertheless, dinosaurs and other Mesozoic reptiles have thoroughly captured the public imagination to such an extent that many people are not as aware of other equally important Mesozoic life events. For example, mammals evolved from mammal-like reptiles during the Triassic, and by Jurassic time birds had evolved from small, carnivorous dinosaurs. Crocodiles, turtles, lizards, and snakes were also present, although the fossil record of the last two is not so good.

Important changes also took place in plant communities when the flowering plants (angiosperms) evolved and soon became the most common plants on land. Marine invertebrates such as bivalves, snails, and cephalopods made a remarkable resurgence following the Permian mass extinctions. In short, biotic diversity once again increased in all realms of the organic world, only to decrease again at the end of the Mesozoic. These end-of-Mesozoic extinctions were second in magnitude only to those at the end of the Paleozoic, but they are much better known because dinosaurs were among the casualties.

Throughout this book, we have emphasized the systems approach to Earth and life history, and we do so again in this chapter. Remember that the distribution of land and sea has a profound influence on oceanic circulation patterns, which in turn partly control climate. Climate, as well as the proximity of continents, have a direct effect on the distribution of organisms and their evolution. In Chapter 14, we discussed the breakup of the supercontinent Pangaea, which began during the Triassic and continues even now. As a consequence, intercontinental migrations became more and more difficult for most organisms. In fact, South America and Australia became isolated from the other continents, and their faunas, evolving in isolation, became increasingly different from those elsewhere (see Figure 18.19).

Marine Invertebrates and Phytoplankton

Following the Permian mass extinctions, the Mesozoic was a time when marine invertebrates repopulated the seas. The Early Triassic invertebrate fauna was not very diverse, but by the Late Triassic, the seas were once again swarming with these animals—from planktonic foraminifera to cephalopods. The brachiopods that had been so abundant during the Paleozoic never completely recovered from their near extinction, and although they still exist today, the bivalves have largely taken over their ecologic niche.

Mollusks such as cephalopods, bivalves, and gastropods were the most important elements of the Mesozoic marine invertebrate fauna. Their rapid evolution and the fact that many cephalopods were nektonic make them excellent guide fossils (▸ Figure 15.1). The Ammonoidea, cephalopods with wrinkled sutures, constitute three groups: the goniatites, ceratites, and ammonites. The latter,

though present during the entire Mesozoic, were most prolific during the Jurassic and Cretaceous. Most ammonites were coiled, with some attaining diameters of 2 m, whereas others were uncoiled and led a near benthonic existence. Ammonites became extinct at the end of the Cretaceous, but two related groups of cephalopods survived into the Cenzoic: the *nautiloids*, including the living pearly nautilus, and the *coleoids*, represented by the extinct belemnoids (▸ Figure 15.2), which are good Jurassic and Cretaceous guide fossils, as well as the living squid and octopus.

Mesozoic bivalves diversified to inhabit many epifaunal and infaunal niches. Oysters and clams (epifaunal suspension feeders) became particularly diverse and abundant, and despite a reduction in diversity at the end of the Cretaceous, remain important animals in the marine fauna today.

As is true now, where shallow marine waters were warm and clear, coral reefs proliferated. However, reefs did not rebound from the Permian extinctions until the Middle Triassic. An important reef-builder throughout the Mesozoic was a group of bivalves known as *rudists*. Rudists are important because they displaced corals as the main reef builders during the later Mesozoic and are good guide fossils for the Late Jurassic and Cretaceous.

A new and familiar type of coral also appeared during the Triassic, the *scleractinians*. Whether scleractinians evolved from rugose corals or from an as yet unknown soft-bodied ancestor with no known fossil record is still unresolved. In addition to the familiar present-day reef-building colonial scleractinian corals, solitary or individual scleractinian corals also inhabited relatively deep waters during the Mesozoic.

Another invertebrate group that prospered during the Mesozoic was the echinoids. Echinoids were exclusively epifaunal during the Paleozoic but branched out into the

▸ Figure 15.1 **Cretaceous Seascape** Cephalopods such as the Late Cretaceous ammonoids *Baculites* (foreground) and *Helioceros* (background) were present throughout the Mesozoic, but they were most abundant during the Jurassic and Cretaceous. They were important predators, and they are excellent guide fossils.

© Tom McHugh/Photo Researchers, Inc.

▸ Figure 15.2 **Belemnoids** Belemnoids are extinct, squidlike cephalopods that were particularly abundant during the Cretaceous. They are very good guide fossils for the Jurassic and Cretaceous. Shown here are several belemnoids swimming in a Cretaceous sea.

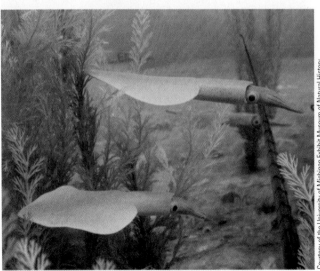

Courtesy of the University of Michigan Exhibit Museum of Natural History

infaunal habitat during the Mesozoic. Both groups began a major adaptive radiation during the Late Triassic that continued throughout the remainder of the Mesozoic and into the Cenozoic.

A major difference between Paleozoic and Mesozoic marine invertebrate faunas was the increased abundance and diversity of burrowing organisms. With few exceptions, Paleozoic burrowers were soft-bodied animals such as worms. The bivalves and echinoids, which were epifaunal animals during the Paleozoic, evolved various means of entering infaunal habitats. This trend toward an infaunal existence may have been an adaptive response to increasing predation from the rapidly evolving fish and cephalopods. Bivalves, for instance, expanded into the infaunal niche during the Mesozoic and, by burrowing, they escaped predators.

The foraminifera (single-celled consumers) diversified rapidly during the Jurassic and Cretaceous and continued to be diverse and abundant to the present. The planktonic forms (▶Figure 15.3), in particular, diversified rapidly, but most genera became extinct at the end of the Cretaceous. The planktonic foraminifera are excellent guide fossils for the Cretaceous.

The primary producers in the Mesozoic seas were various types of microorganisms. *Coccolithophores* are an important group of calcareous phytoplankton (▶Figure 15.4a) that first evolved during the Jurassic and became extremely common during the Cretaceous. *Diatoms* (Figure 15.4b), which build skeletons of silica, made their appearance during the Cretaceous, but they are more important as primary producers during the Cenozoic. Diatoms are presently most abundant in cooler oceanic waters, and some species inhabit freshwater lakes. *Dinoflagellates*, which are organic-walled phytoplankton, were common during the Mesozoic and today are the major primary producers in warm water (Figure 15.4c).

In general terms, we can think of the Mesozoic as a time of increasing complexity among the marine invertebrate fauna. At the beginning of the Triassic, diversity was low and food chains were short. Near the end of the Cretaceous, though, the marine invertebrate fauna was highly complex, with interrelated food chains. This evolutionary history reflects changing geologic conditions influenced by plate tectonic activity, as discussed in Chapter 3.

Aquatic and Semiaquatic Vertebrates

Remember that fishes evolved by Cambrian time and then diversified, especially during the Devonian Period, when several major groups appeared (see Figure 13.5). Amphibians evolved during the Devonian and persist to the present, but their greatest abundance and diversity occurred during the Pennsylvanian.

▶ **Figure 15.4 Primary Producers** Coccolithophores, diatoms, and dinoflagellates were primary producers in the Mesozoic and Cenozoic oceans. (a) A Miocene coccolith from the Gulf of Mexico (*left*) and a Miocene–Pliocene coccolith from the Gulf of Mexico (*right*). Coccoliths are the calcareous plates covering the living coccolithosphere. (b) Upper Miocene diatoms from Java (*left* and *right*). (c) Miocene dinoflagellates from Sale, Morocco (*left* and *right*).

▶ **Figure 15.3 Planktonic Foraminifera** Planktonic foraminifera diversified and became abundant during the Jurassic and Cretaceous, and continued to be diverse and abundant throughout the Cenozoic. Many planktonic foraminifera are excellent guide fossils for the Cretaceous, such as species of the genus *Globotruncana,* which is restricted to the Upper Cretaceous. Shown here are three views of the holotype (the specimen that defines the species) of *Globotruncana loeblichi,* a Late Cretaceous planktonic foraminifera.

Fishes

Recall from Chapter 13 that the bony fish (class Osteichthyes) evolved by the Devonian Period, and among the bony fish there are two subclasses: the ray-finned fish (Actinoperygii) and the lobe-finned fish (Sarcopterygii) (see Figure 13.9). Only a few of the lobe-finned fishes, represented by lungfish and crossopterygians, persisted into the Mesozoic, but the latter declined and with one exception went extinct by the end of that era (see Figure 7.13a). The lungfish have done only slightly better, with only a few species living today in Africa, South America, and Australia.

The sharks and their relatives (class Chondrichthyes), or cartilaginous fish, became more abundant during the Mesozoic, but never came close to matching the diversity of bony fish; about 930 species exist now. Nevertheless, sharks, an evolutionarily conservative group, remain important in the marine fauna, especially as predators, although a few strain plankton from seawater instead of eating their prey.

The story for the ray-finned fishes is much different; today they account for almost all of the 24,000 or so species of bony fish. The most primitive bony fish existed mostly during the Paleozoic but gave rise to an intermediate group that was common during the Middle Mesozoic (▶ Figure 15.5). In both of these groups, the skeleton remained partly cartilaginous, but one trend was the replacement of cartilage by bone. Furthermore, the most primitive bony fish had a tail with an upturned lobe much like the sharks, whereas the intermediate Mesozoic fish had a nearly symmetrical tail.

The most advanced bony fish, formally known as telosts, continued the evolutionary trends established much earlier. For instance, they had no cartilage in their skeletons, their tails were symmetrical, their jaw mechanism worked differently, and they continued a general reduction in the number of bones. By Cretaceous time, they were by far the most common fish in both fresh and saltwater habitats and now are the most varied and numerous of all vertebrate animals.

Amphibians

The labyrinthodont amphibians were common during the latter part of the Paleozoic, but the few surviving Mesozoic species died out by the end of the Triassic. Since their greatest diversity and abundance during the Pennsylvanian, amphibians have made up only a small part of the total vertebrate fauna. A recently analyzed fossil dubbed a *frogmander* from the Permian of Texas, coupled with molecular evidence from living amphibians, leads some investigators to conclude that frogs and salamanders diverged during the Late Permian or Early Triassic. Frogs and salamanders were certainly present by the Mesozoic, but their fossil records are poor.

Plants—Primary Producers on Land

Just as they did during the Late Paleozoic, seedless vascular plants and gymnosperms dominated Triassic and Jurassic land-plant communities, and, in fact, representatives of both groups are still common (▶ Figure 15.6). Among the gymnosperms, the large seed ferns became extinct by the end of the Triassic, but *ginkgos* (see Figure 7.13b) remained abundant and still exist in isolated regions, and *conifers* continued to diversify and are now widespread, particularly at high elevations and high latitudes. A new group of gymnosperms known as *cycads* made their appearance during the Triassic. They became widespread and now exist in tropical and semitropical areas (Figure 15.6b).

The long dominance of seedless vascular plants and gymnosperms ended during the Early Cretaceous, or perhaps the Late Jurassic, when many of them were replaced by the flowering plants, or **angiosperms** (▶ Figure 15.7). Unfortunately, the fossil record of the earliest angiosperms is sparse, so their precise ancestors remain obscure. Nevertheless, studies of living plants and the fossils that are available indicate close relationships to gymnosperms.

Since they first evolved, angiosperms have adapted to nearly every terrestrial habitat, from high mountains to deserts, marshes and swamps, and some have even adapted to shallow coastal waters. Several features account for their phenomenal evolutionary success, including enclosed seeds and, above all, the origin of flowers, which attract animal pollinators, especially insects. All organisms interact in some ways and influence the evolution of others, but the interrelationships among flowering plants and insects are so close that biologists refer to changes in one induced by the other as *coevolution*.

The 250,000 to 300,000 species of angiosperms that now exist, accounting for more than 90 percent of all land plant species, and the fact that they inhabit some

© Dmitry Bogdanov

▶ **Figure 15.5 The Mesozoic Fish *Leedsichthys* (background) and the Short-Necked Plesiosaur *Liopluerodon*** This fish, from the intermediate group of fishes, was one of the largest ever, but because some of its spine is missing, its total length is not known; estimates range from 9 to 30 m. It was probably a plankton feeder.

(a) This Jurassic landscape was dominated by seed-less vascular plants, especially ferns, as well as gymnosperms, including the conifers and cycads.

(b) These living cycads look much like some of the vegetation shown in (a).

▸ Figure 15.7 **The Angiosperms (Flowering Plants)**

(b) *Archaefructus sinensis,* from Lower Cretaceous rocks in China, is among the oldest known angiosperms.

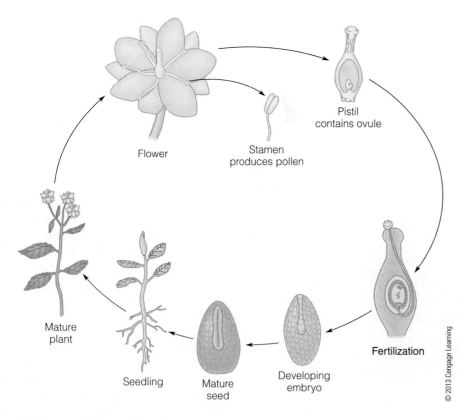

Flower

Stamen produces pollen

Pistil contains ovule

Fertilization

Developing embryo

Mature seed

Seedling

Mature plant

(a) The reproductive cycle in angiosperms.

(c) Restoration of *Archaefructus sinensis.*

environments hostile to other plants, is a testimony to their success. Nevertheless, seedless vascular plants and gymnosperms remain important in the world's flora.

The Diversification of Reptiles

Reptiles, birds, and mammals are collectively called **amniotes** because they all produce amniote eggs, and they all evolved from a common ancestor that lived during the Late Paleozoic. The first of these ancestors, the so-called protorothyrids, evolved during the Mississippian Period (see Figure 13.17). From this basal stock of reptiles two main branches emerged, one leading to the pelycosaurs and therapsids and eventually to mammals, and the other to all other reptiles and birds (▶Figure 15.8).

Dinosaurs are included among the reptiles, but they possessed several characteristics that set them apart. They had teeth set in individual sockets, a reduced lower leg bone (fibula), a pelvis anchored to the vertebral column by three or more vertebrae, a ball-like head on the upper leg bone (femur), and elongate bones in the palate. In addition, dinosaurs had a fully upright posture, with the limbs directly beneath their bodies, rather than the sprawling stance or semi-erect posture of other reptiles. In fact, their upright posture and other limb modifications may have been partly responsible for their incredible success.

Contrary to popular belief, there were no flying dinosaurs, nor any fully aquatic ones, although there were Mesozoic reptiles that filled these niches. Neither were all dinosaurs large, even though some certainly were. Also, dinosaurs lived only during the Mesozoic Era, unless we consider the birds, which evolved from one specific group of dinosaurs.

Archosaurs and the Origin of Dinosaurs

Reptiles called **archosaurs** (*archo* meaning "ruling" and *sauros* meaning "lizard") include such diverse animals as crocodiles, pterosaurs (flying reptiles), dinosaurs, birds, and several other extinct groups. In fact, following the Paleozoic extinctions, archosaurs were the dominant land-dwelling vertebrate animals. All archosaurs are united by several shared characteristics, but most notably all have teeth set in individual sockets—except for today's birds. However, the earliest birds had this feature in common with other archosaurs.

▶ **Figure 15.8 Evolution of the Amniotes** All of the organisms shown here are amniotes—that is, animals that lay amniote eggs. The living amniotes are snakes, lizards, tuataras, birds, turtles, crocodiles, and mammals.

© 2013 Cengage Learning

De Agostini Picture Library/De Agostini/Getty Images

▶ **Figure 15.9** **Eoraptor** This small, carnivorous biped was one of the first dinosaurs. Populations of these animals, or ones very much like them, are the probable ancestors of all later dinosaurs. *Eoraptor* lived during the Late Triassic Period (about 230 million years ago) in what is now Argentina. It was only about a meter long and likely weighed no more than 10 kg.

The probable ancestors of dinosaurs were small (about 1 m long), long-legged archosaurs much like those from the Middle and Late Triassic-age rocks in Argentina and Africa. They walked and ran on their hind limbs, so they were **bipedal** (▶ Figure 15.9), as opposed to **quadrupedal**, which refers to animals that move on all four limbs. Dinosaurs had evolved by the Late Triassic, but, remember, there really was no single, first dinosaur, because it is populations that evolve, not individuals.

Dinosaurs share many characteristics, and yet they differ enough for paleontologists to recognize two distinct orders based primarily on their type of pelvis: the **Saurischia** and the **Ornithischia**. Saurischian dinosaurs have a lizard-like pelvis and are thus informally called lizard-hipped dinosaurs, whereas ornithischians have a birdlike pelvis and are called bird-hipped dinosaurs (▶ Figure 15.10). For decades,

© 2013 Cengage Learning

▶ **Figure 15.10** **Cladogram Showing Relationships Among Dinosaurs** Pelvises of ornithischian and saurischian dinosaurs are shown for comparison. All the dinosaurs shown here were herbivores except for theropods. Note that bidpedal and quadrupedal dinosaurs are found among both ornithischians and saurischians.

paleontologists thought that each order evolved independently during the Late Triassic, but it is now clear that they had a common ancestor, much like the archosaurs known from Middle Triassic rocks in Argentina (Figure 15.9).

Dinosaurs

The mistaken belief that dinosaurs were dim-witted, lethargic beasts persists even today, although the media now portray them as more active animals than was previously thought. Actually, some of the predator dinosaurs had fairly large brains, at least by reptile standards. Moreover, it is possible that some dinosaurs were warm-blooded, and we now know that a few species tended their young long after they hatched, a behavioral characteristic more typical of birds and mammals than reptiles. Many questions about dinosaurs remain unanswered, but their fossils and the rocks containing them are revealing more and more about their evolution and behavior.

The term **dinosaur** refers to any Mesozoic-age reptile belonging to the orders Saurischia and Ornithischia. When proposed in 1842, *dinosaur* meant "fearfully great lizard"—hence the characterization of dinosaurs as "terrible lizards." We have no reason to think that they were any more terrible than animals living today, and they were not lizards. Nevertheless, mistaken ideas about them persist, such as that all dinosaurs were large and that they must have been poorly adapted because they eventually became extinct. Some, it is true, were the largest land-dwelling animals ever, but among dinosaurs were species that weighed only 2 or 3 kg. And to consider them poorly adapted is to ignore the fact that, as a group, they diversified and became widespread for more than 160 million years.

Saurischian Dinosaurs Paleontologists recognize two groups of saurischians, known as *theropods* and *sauropods* (Figure 15.10). Theropods were bipedal carnivores* that varied from tiny *Compsognathuis* to giants such as *Tyrannosaurus* and similar, but even larger, genera (▶Figure 15.11, Table 15.1). Beginning in 1996, Chinese scientists have found several genera of theropods with feathers.** No one doubts that these dinosaurs had feathers, and molecular evidence indicates that their feathers were composed of the same material as bird feathers.

The movie *Jurassic Park* popularized some of the smaller theropods, especially *Velociraptor*, a 1.8-m-long predator with large sickle-like claws on its back feet. This carnivore and its somewhat larger relative, *Deinonychus*, likely used these claws in a slashing type of attack (Figure

---

*Some studies indicate that a number of theropods also ate vegetation.

**One feathered specimen purchased in Utah, that was reported to have come from China, turned out to be a forgery. It even appeared in *National Geographic* before scientists exposed it as a fraud.

15.11b). Despite what you might see in movies and on TV, theropods, like predators today, probably avoided large, dangerous prey and went for the easy kill—the young, old, or disabled—or they dined on carrion. No doubt the larger theropods simply chased smaller predators away from their kill. From the evidence available, some theropods, such as diminutive *Coelophysis* and medium-sized *Deinonychus*, hunted in packs.

The second group of saurischians, the sauropods, includes the truly giant, quadrupedal, herbivorous dinosaurs such as *Apatosaurus*, *Diplodocus*, and *Brachiosaurus*. Among the sauropods were the largest land animals ever (see the chapter opening photo, Table 15.1), and *Brachiosaurus* was a giant even by sauropod standards—it may have weighed 75 metric tons, and partial remains indicate that even larger ones existed. Trackways show that sauropods moved in herds.

Sauropods were preceded in the fossil record by the smaller, Late Triassic to Early Jurassic prosauropods, which were undoubtedly related to sauropods. In fact, a 6-m-long prosauropod named *Aardonyx* that was discovered in Africa in 2009 is likely the closest relative of sauropods found so far. In any case, once sauropods evolved, they were most common during the Jurassic; only a few genera existed during the Cretaceous.

Ornithischian Dinosaurs Recall that the distinguishing features of ornithischians is their birdlike pelvis. However, they differ from saurischians in other ways, too. For instance, ornithischians have no teeth in the front of the mouth, whereas saurischians do, and they also have ossified (bonelike) tendons in the back region. Scientists identify five groups of ornithischians: ornithopdods, pachycephalosaurs, ankylosaurs, stegosaurs, and ceratopsians (Figure 15.10, Table 15.1). Although you might not know the names of these groups, you have probably seen examples of them all.

In 1822, Mary Ann Mantell discovered some fossil teeth that her husband, Gideon Mantell, later assigned to the genus *Iguanodon*, which proved to be a member of the ornithischian subgroup known as ornithopods. Another ornithopod known as *Hadrosaurus* was discovered in North America in 1858, and it was the first dinosaur to be assembled and displayed in a museum.

Among the several varieties of ornithopods, the duck-billed dinosaurs, or *hadrosaurs*, were especially numerous during the Cretaceous, and several had crests on their heads that may have been used to amplify bellowing, for sexual display, or for species recognition (▶Figure 15.12). All ornithopods were herbivores and primarily bipedal, but they had well-developed forelimbs and could also walk in a quadrupedal manner.

Another duck-billed dinosaur of some interest is *Maiasaura* ("good mother dinosaur") that nested in colonies and used the same nesting area repeatedly. Furthermore, their 2-m-diameter nests were spaced 7 m apart, or about the length of an adult. Some nests contain the remains of

TABLE 15.1 **Summary Chart for the Orders and Suborders of Dinosaurs**

Lengths and weights approximate, from several sources.

Order	Suborder	Familiar Genera	Comments
Saurischia	Theropoda	Allosaurus, Coelophysis, Compsognathus, Deinonychus, Tyrannosaurus,* Velociraptor	Bipedal carnivores. Late Triassic to end of the Cretaceous. Size from 0.6 to 15 m long, 2 or 3 kg to 7.3 metric tons. Some smaller genera may have hunted in packs.
	Sauropoda	Apatosaurus, Brachiosaurus, Camarasaurus, Diplodocus, Titanosaurus	Giant quadrupedal herbivores. Late Triassic to Cretaceous, but most common during Jurassic. Size up to 27 m long, 75 metric tons.** Trackways indicate sauropods lived in herds. Preceded in fossil record by the smaller prosauropods.
Ornithischia	Ornithopoda	Anatosaurus, Camptosaurus, Hypsilophodon, Iguanodon, Parasaurolophus	Some ornithopods, such as Anatosaurus, had a flattened bill-like mouth and are called duck-billed dinosaurs. Size from a few meters long up to 13 m and 3.6 metric tons. Especially diverse and common during the Cretaceous. Primarily bipedal herbivores, but could also walk on all fours.
	Pachycephalosauria	Stegoceras	Stegoceras only 2 m long and 55 kg, but larger species known. Thick bones of skull cap might have aided in butting contests for dominance and mates. Bipedal herbivores of the Cretaceous.
	Ankylosauria	Ankylosaurus	Ankylosaurus more than 7 m long and about 4.5 metric tons. Heavily armored with bony plates on top of head, back, and sides. Quadrupedal herbivore.
	Stegosauria	Stegosaurus	A variety of stegosaurs are known, but Stegosaurus, with bony plates on its back and a spiked tail, is best known. Plates probably were for absorbing and dissipating heat. Quadrupedal herbivores that were most common during the Jurassic. Size was 9 m long, 1.8 metric tons.
	Ceratopsia	Triceratops	Numerous genera known. Some early ones bipedal, but later large animals were quadrupedal herbivores. Much variation in size; size was up to 7.6 m long and 5.4 metric tons, with large bony frill over top of neck, three horns on skull, and beaklike mouth. Especially common during the Cretaceous.

*Tyrannosaurus, at 4.5 metric tons, was the largest known theropod, but now similar and larger animals are known from Argentina and Africa.
** Partial remains indicate even larger brachiosaurs existed, perhaps measuring 30 m long and weighing 100 metric tons.

© 2013 Cengage Learning

juveniles up to 1 m long, which is much larger than when they hatched, so they must have stayed in the nest area where adults protected and perhaps fed them. Another interesting fact about these dinosaurs is that a bone bed in Montana has the remains of an estimated 10,000 individuals. The evidence indicates that they were overcome by volcanic gases and later buried by flood deposits.

The most distinctive feature of the bipedal, herbivorous pachycephalosaurs is their thick-boned, domed skulls, although the earliest ones were flat headed (Figure 15.10, Table 15.1). They varied from 1.0 to 4.5 m long and have been found only in Cretaceous-age rocks and only on the Northern Hemisphere continents. Although not accepted by all paleontologists, the traditional view is that these animals butted heads for dominance or competition for mates.

The fossil record of ceratopsians (horned dinosaurs) (Figure 15.10), indicates that small Late Jurassic bipeds were the ancestors of large Late Cretaceous quadrupeds such as Triceratops. Triceratops, with three horns on its skull, and related genera had a huge head and a bony frill over the top of the neck; they were especially common in North America during the Cretaceous. Fossil trackways show that these large herbivores moved in herds. Indeed, bone beds with dozens of individuals of a single species indicate that large numbers of animals perished quickly, probably during river crossings.

Stegosaurus, as a representative of the Stegosauria, which is well-known for plates along its back and spikes at the end of its tail for defense (▶Figure 15.13, Table 15.1). The arrangement of the plates is not precisely known, but most restorations show two rows, with the plates on one

(a) A Late Cretaceous scene showing the heavily armored dinosaur *Ankylosaurus* defending itself from *Tyrannosaurus*.

(b) Life-like restoration of *Deinonychus*. It was 3 m long and may have weighed 80 kg. This dinosaur had a huge curved claw on each back foot.

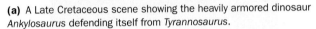

▶ Figure 15.12 **Hadrosaurs (Duck-Billed Dinosaurs)** This group of ornithopods had flattened, bill-like mouths, and some had crests or other ornamentation on their heads.

(a) *Shantungosaurus* had the typical flattened bill-like mouth of hadrosaurs, but it had no crest.

(b) *Parasaurolophus* was a crested hadrosaur.

side offset from those on the other. Regardless of the plates' arrangement, most paleontologists think they functioned to absorb and dissipate heat. These medium-size, quadrupedal herbivores lived during the Jurassic. In addition to *Stegosaurus*, there were several other genera that did not have broad plates on their back but rather had spikes.

Ankylosaurs were quadrupedal herbivores that were more heavily armored than any other dinosaur (Figure 15.11a, Table 15.1), and as a result were not very fast; one estimate has their top speed at 10 km/hr. The animal's back, flanks, and top of its head were protected by bony armor, and the tail of some species, such as *Ankylosaurus*, ended in a huge bony club that could no doubt deliver a crippling blow to an attacking predator (Figure 15.11a). If the tail proved inadequate, the animal probably simply hunkered

down; it would have been difficult even for *Tyrannosaurus* to flip over a 9-m-long, 4.5 metric ton *Ankylosaurus*.

Typically, dinosaurs have been depicted as aggressive, dangerous beasts, but we have every reason to think that they behaved much as land animals do now. Certainly, some lived in herds and no doubt interacted by bellowing, snorting, grunting, and foot stomping in defensive actions, territorial disputes, and attempts to attract mates.

Warm-Blooded Dinosaurs?

Were dinosaurs **endotherms** (warm-blooded) like today's mammals and birds, or were they **ectotherms** (cold-blooded) like all of today's reptiles? Almost everyone now agrees some compelling evidence exists for dinosaur

▶ **Figure 15.13 *Stegosaurus*** *Stegosaurus* is noted for the rows of plates on its back and the bony spikes at the end of its tail. Stegosaurs died out by the end of the Jurassic Period.

▶ **Figure 15.14 *Sinosauropteryx*** This 1-m-long genus of therapod from Lower Cretaceous Rocks in China was the first dinosaur with evidence of feathers. Recent studies of preserved structures called melanosomes, that partly control feather color in today's birds, indicate this animal was probably reddish brown and had stripes on its tail.

endothermy, but opinion is divided among (1) those holding that all dinosaurs were endotherms, (2) those who think only some were endotherms, and (3) those proposing that dinosaur metabolism, and thus the ability to regulate body temperature, changed as they matured.

Bones of endotherms typically have numerous passageways that, when the animals are alive, contain blood vessels, but the bones of ectotherms have considerably fewer passageways. Proponents of dinosaur endothermy note that dinosaur bones are more similar to the bones of living endotherms than living ectotherms. However, crocodiles and turtles have this so-called endothermic bone, but they are ectotherms, and some small mammals have bones more typical of ectotherms. Perhaps bone structure is related more to body size and growth patterns than to endothermy, so this evidence is not conclusive.

Endotherms must eat more than comparably sized ectotherms because their metabolic rates are so much higher. Consequently, endothermic predators require large prey populations and thus constitute a much smaller proportion of the total animal population than their prey, usually only a few percent. In contrast, the proportion of ectothermic predators to prey may be as high as 50 percent. Where data are sufficient to allow an estimate, dinosaur predators made up 3 percent to 5 percent of the total population. Nevertheless, uncertainties in the data make this argument less than convincing for many paleontologists.

A large brain in comparison to body size requires a rather constant body temperature and thus implies endothermy. And some dinosaurs were indeed brainy, especially the small- and medium-sized theropods. So brain size might be a convincing argument for these dinosaurs. Even more compelling evidence for theropod endothermy comes from their relationship to birds and from the recent discoveries in China of dinosaurs with feathers or a featherlike covering, and for some of these even the color is known (▶ Figure 15.14). Today, only endotherms have hair, fur, or feathers for insulation.

Some scientists point out that certain duck-billed dinosaurs grew and reached maturity much more quickly than would be expected for ectotherms and conclude that they must have been warm-blooded. Furthermore, a fossil ornithopod discovered in 1993 has a preserved four-chambered heart much like that of living mammals and birds. Three-dimensional imaging of this structure, now on display at the North Carolina Museum of Natural Sciences, has convinced many scientists that this animal was an endotherm.

Good arguments for endothermy exist for several types of dinosaurs, particularly theropods. The large sauropods were probably not endothermic, but nevertheless were capable of maintaining a rather constant body temperature. Large animals heat up and cool down more slowly than smaller ones because they have a small surface area compared to their volume. With their comparatively smaller surface area for heat loss, sauropods probably retained heat more effectively than their smaller relatives.

Flying Reptiles

Paleozoic insects were the first animals to achieve flight, but the first among vertebrates were **pterosaurs**, or flying reptiles, which were common in the skies from the Late Triassic until their extinction at the end of the Cretaceous (▶ Figure 15.15). Adaptations for flight include a wing membrane supported by an elongated fourth finger, light, hollow bones; and development of those parts of the brain that controlled muscular coordination and sight.

The evolution of some Mesozoic reptiles is fairly well known, but pterosaurs are not among them. Pterosaurs

(b) Later pterosaurs such as *Pteranodon* of the Late Cretaceous were much larger, tailless, and had toothless beaks. *Pteranodon's* wingspan was about 6 m.

(a) *Pterodactylus* from the Late Jurassic is a good example of the earlier pterosaurs. It has a long tail, numerous teeth, and a wingspan of about 1.5 m.

At least one small pterosaur, called *Sordes pilosus* (hairy devil), found in 1971 in what is now Kazahkstan, had a coat of hair, or hairlike feathers. This outer covering and the fact that wing flapping requires a high metabolic rate and efficient respiratory and circulatory systems as in present-day birds, indicates that some, or perhaps all, pterosaurs were warm-blooded.

Mesozoic Marine Reptiles

Several types of Mesozoic reptiles adapted to a marine environment, including turtles and some crocodiles, as well as the Triassic mollusk-crushing placodonts. Here, though, we concentrate on the ichthyosaurs and plesiosaurs and the less familiar mosasaurs. All were thoroughly aquatic marine predators, but other than all being reptiles, they were not closely related to one another. Furthermore, none were dinosaurs, although some popular media depict them as such.

The streamlined, rather porpoise-like **ichthyosaurs** varied from species measuring only 0.7 m long to giants more than 15 m long (▶ Figure 15.16a). Details of their evolution are still not clear, but fossil ichthyosaurs from Japan prompted researcher Ryosuke Motani to say, "I knew *Utatsusaurus* was exactly what paleontologists had been expecting to find for years: an ichthyosaur that looked like a lizard with legs."[*]

Ichthyosaurs used their powerful tail for propulsion and maneuvered with their flipperlike forelimbs. They had numerous sharp teeth, and preserved stomach contents reveal a diet of fish, cephalopods, and other marine

belong to the broader group known as archosaurs, so their relatives no doubt are in that group, but the question of exactly which relatives has not been resolved. Here we should clear up some misconceptions about these remarkable animals: they were not dinosaurs; they were not the ancestors of birds; and even though movies commonly depict them as large, aggressive creatures, some were no larger than today's sparrows, robins, and crows. There were some Cretaceous species with wingspans of at least 12 m, but even these probably weighed no more than a few tens of kilograms.

Experiments and studies of fossils indicate that the wing bones of large pterosaurs such as *Pteranodon* (Figure 15.15b) were too weak for sustained flapping. These comparatively large animals probably took advantage of rising air currents to stay airborne, mostly by soaring, but occasionally flapping their wings for maneuvering. In contrast, smaller pterosaurs probably stayed aloft by vigorously flapping their wings just as present-day small birds do.

Recently discovered fossils of a pterosaur called *Darwinopterus* indicate that this animal laid soft-shelled eggs much like those of today's turtles. Furthermore, at least in this genus, it seems that only males had crests on their heads.

---

[*]Ryosuke Motani. 2004. Rulers of the Jurassic Seas. *Scientific American*. v. 14, no. 2. p. 7.

(a) Ichthyosaurs.

(b) A long-necked plesiosaur.

(c) Mosasaur.

Mosasaurs were Late Cretaceous marine lizards related to the present-day Komodo dragon, or monitor lizard (Figure 15.16c). Some species measured no more than 2.5 m long, but a few, such as *Tylosaurus*, were large, measuring up to 9 m. Mosasaur limbs resemble paddles and were used mostly for maneuvering, whereas the long tail provided propulsion. All were predators, and preserved stomach contents indicate they ate fish, birds, smaller mosasaurs, and a variety of invertebrates, including ammonoids.

A possible mosasaur ancestor was found in Texas in 1989, although it was several years before it was examined and the results published. In any case, *Dallasaurus* from Upper Cretaceous rocks has a mosasaur skeleton, but rather than paddle-like flippers, it retained complete limbs for walking on land.

Crocodiles, Turtles, Lizards, and Snakes

All crocodiles are amphibious, spending much of their time in water, but they are well equipped for walking on land. By Jurassic time, crocodiles had become the most common freshwater predators. Overall, crocodile evolution has been conservative, involving changes mostly in size from a meter or so in Jurassic forms to 15 m in some Cretaceous species.

Turtles, too, have been evolutionarily conservative since their appearance during the Triassic. The most remarkable feature of turtles is their heavy, bony armor; they are more thoroughly armored than any other vertebrate animal, living or fossil. Turtle ancestry is uncertain. One Permian animal had eight broadly expanded ribs, which may represent the first stages in the development of turtle armor.

organisms. It is doubtful that ichthyosaurs could come onto land, so females must have retained eggs within their bodies and given birth to live young. A few fossils with small ichthyosaurs in the appropriate part of the body cavity support this interpretation.

An interesting side note in the history of paleontology is the story of Mary Anning (see Perspective), who, when she was only about 11 years old, discovered and directed the excavation of a nearly complete ichthyosaur in southern England.

The **plesiosaurs** belonged to one of two subgroups: short-necked and long-necked (Figures 15.5 and 15.16b). Most were modest-sized animals 3.6 to 6 m long, but one species found in Antarctica measures 15 m. Short-necked plesiosaurs might have been bottom feeders, but their long-necked cousins may have used their necks in a snake-like fashion, along with their numerous sharp teeth, to capture fish. These animals may have come ashore to lay their eggs, but one fossil plesiosaur with a fetus within the skeleton indicates that they gave birth to live young.

Illustration by John Sibbick, © The Natural History Museum/The Image Works

Roger Harris/Photo Researchers, Inc.

© Carl Buell

Mary Anning and Her Contributions to Paleontology

Men and women from many countries contribute to our understanding of prehistoric life, but this has not always been the case. The early history of paleontology was dominated by Western European males, but there was a notable exception: Mary Anning (1799–1847), who began a remarkable career as a fossil collector when she was only 11 years old (Figure 1).

Mary Anning was born in Lyme Regis on England's south coast. When only 15 months old, she survived a lightning strike that, according to one report, killed three girls, and according to another, killed a nurse tending her. Of the 10 or so children in the Anning family, only Mary and her brother Joseph reached maturity.

In 1810 Mary's father, a cabinetmaker who also sold fossils part time, died, leaving the family nearly destitute. Mary Anning expanded the fossil business and became a professional fossil collector known to the paleontologists of her time, some of whom visited her shop to buy fossils or gather information. She collected fossils from the Dorset coast near Lyme Regis and is reported to have been the inspiration for the tongue twister "She sells seashells by the seashore."

Soon after her father's death, she made her first important discovery: a nearly complete skeleton of a Jurassic ichthyosaur, which was described in 1814 by Sir Everard Home. The sale of this fossil specimen provided considerable financial relief for her family. In 1821 she made a second major discovery when she excavated the remains of another Mesozoic marine reptile, a plesiosaur. And in 1818 she excavated the remains of the first Mesozoic flying reptile (pterosaur) found in England, which was sent to the eminent geologist William Buckland of Oxford University.

By 1830 Mary Anning's fortunes began declining as collectors and museums had less money to spend on fossils. Indeed, she might have once again become destitute if it had not been for her friend Henry Thomas de la Beche, also a resident of Lyme Regis. De la Beche drew a fanciful scene called *Duria antiquior*, meaning "An earlier Dorset," in which he brought to life the fossils Mary Anning had collected. The scene was printed and sold widely, the proceeds of which went directly to Mary Anning.

Mary Anning died of cancer in 1847, and although only 48 years old, she had a fossil-collecting career that spanned 36 years. Her contributions to paleontology are now widely recognized, but, unfortunately, soon after her death she was mostly forgotten. Apparently, people who purchased her fossils were credited with finding them. So even though Mary Anning became a respected fossil collector, many scientists of that time could not accept that an untutored girl could possess such knowledge and skill. "It didn't occur to them to credit a woman from the lower classes with such astonishing work. So an uneducated little girl, with a quick mind and an accurate eye, played a key role in setting the course of the 19th-century geologic revolution. Then we simply forgot about her."[*]

Although Mary Anning's contributions to paleontology were mostly forgotten, beginning in 2002, the Palaeontological Association has presented the Mary Anning Award to someone who is "not professionally employed within palaeontology but who made an outstanding contribution to the subject."[**]

---

[*]John Lienhard, University of Houston.
[**]The Palaeontological Association (http://www.palass.org/modules.php?name=palaeo&sec=Awards&page=122)

English School (19th Century). Private Collection/The Bridgeman Art Library International

Figure 1 Mary Anning made several important fossil discoveries in England during the early 1800s, but she was largely forgotten after her death in 1847.

Lizards and snakes are closely related, and lizards were in fact ancestral to snakes. The limbless condition in snakes (some lizards are limbless, too) and skull modifications that allow snakes to open their mouths very wide are the main differences between these two groups. Lizards are known from Upper Permian strata, but they were not abundant until the Late Cretaceous. Snakes first appear during the Cretaceous, but the families to which most living snakes belong have differentiated since the Early Miocene. One Early Cretaceous genus from Israel shows characteristics intermediate between snakes and their lizard ancestors.

▶ Figure 15.17 **Restoration of *Archaeopteryx*** Several speci-mens of *Archaeopteryx* have been found in the Jurassic-age Solnhofen Limestone in Germany. This animal had feathers and a wishbone, so it is classified as a bird, but in most details of its anatomy, it more closely resembles small theropod dinosaurs—it had reptile-like teeth, a long tail, and claws on its wings.

From Reptiles to Birds

Birds have feathers, whereas reptiles have scales or a tough, beaded skin, and birds do not closely resemble any living reptile. So why do scientists think that birds evolved from reptiles? Long ago, scientists were aware of the probable relationships between these two groups of animals. Birds and reptiles both lay shelled, yolked eggs, and both share several skeletal characteristics, such as the way the jaw attaches to the skull. Furthermore, since 1860, approxi-mately 10 fossils have been recovered from the Solnhofen Limestone of Germany that provide evidence for reptile–bird relationships. The fossils definitely have feathers and a wishbone, consisting of the fused clavicle bones so typi-cal of birds, and yet, in most other physical characteristics, they most closely resemble small theropod dinosaurs.

These remarkable fossils, known as **Archaeop-teryx** (from Greek *archaios* for "ancient" and *pteryx* for "feather") are birds by definition, but their numerous rep-tilian features convince scientists that their ancestors were among the theropods (▶ Figure 15.17, and Chapter 7 open-ing photo).

Even the fused clavicles (wishbone) are found in sev-eral theropods, and during the last several years, paleon-tologists in China have discovered theropods with feath-ers, providing more evidence for this relationship. The few that oppose the theropod–bird view note that theropods are found in Cretaceous-age rocks, but *Archaeopteryx* is Jurassic. However, some of the fossils from China are about the same age as *Archaeopteryx*, narrowing the gap between presumed ancestor and descendant.

Another fossil bird from China that is slightly younger than *Archaeopteryx* retains ribs in the abdominal region just as *Archaeopteryx* and small theropods do, but it has a reduced tail more like present-day birds. More fossils found in China in 2004 and 2005 of five specimens of an Early Cretaceous bird indicate that today's birds may have had an aquatic ancestor. With few exceptions, the bones of these birds, known as *Gansus yumenesis*, are much like those of living birds.

The fossils of *Archaeopteryx* are significant, but there are not enough of them or of other early birds to resolve whether *Archaeopteryx* was the ancestor of today's birds or if it was an early bird that died out without leaving

descendants. Of course, this does not diminish the fact that *Archaeopteryx* had both reptile and bird features (recall the concept of *mosaic evolution* from Chapter 7). However, there is another candidate for earliest bird. Some claim that fossils of two crow-sized individuals known as *Protoavis*, from Upper Triassic rocks in Texas, are birds. *Protoavis* did have hollow bones and a wishbone as today's birds do, but because the specimens are fragmentary and no feather impressions were found, most paleontologists think that they were small theropods.

One hypothesis for the origin of bird flight—*from the ground up*—holds that the ancestors of birds were bipedal, fleet-footed ground dwellers that used their wings to leap into the air, at least for short distances, to catch insects or to escape predators. The *from-the-trees-down* hypothesis holds that bird ancestors were bipeds that climbed trees and used their wings for gliding or parachuting. The from-the-ground-up hypothesis is probably better supported in that a bipedal theropod ancestor is reasonable because small theropods had forelimbs much like those of *Archaeopteryx*. However, from-the-trees-down has an advantage because takeoff from an elevated position is easier, although landing is a challenge.

Origin and Evolution of Mammals

Recall from Chapter 13 that mammal-like reptiles called **therapsids** diversified into many species of herbivores and carnivores during the Permian Period. In fact, they were the most numerous and diverse land-dwelling vertebrates at that time. Among the therapsids, one group, known as **cynodonts**, was the most mammal-like of all, and by Late Triassic time true mammals evolved from them.

TABLE 15.2 Summary Chart Showing Some Characteristics and How They Changed During the Transition from Reptiles to Mammals

Features	Typical Reptile	Cynodont	Mammal
Lower Jaw	Dentary and several other bones	Dentary enlarged, other bones reduced	Dentary bone only, except in earliest mammals
Jaw–Skull Joint	Articular–quadrate	Articular–quadrate; some advanced cynodonts had both the reptile jaw–skull joint and the mammal jaw–skull joint	Dentary–squamosal
Middle-Ear Bones	Stapes	Stapes	Stapes, incus, malleus
Secondary Palate	Absent	Partly developed	Well developed
Teeth	No differentiation; chewing teeth single-rooted	Some differentiation; chewing teeth partly double-rooted	Fully differentiated into incisors, canines, and chewing teeth; chewing teeth double-rooted
Tooth Replacement	Teeth replaced continuously	Only two sets of teeth in some advanced cynodonts	Two sets of teeth
Occipital Condyle	Single	Partly divided	Double
Occlusion (chewing teeth meet surface to surface to allow grinding)	No occlusion	Occlusion in some advanced cynodonts	Occlusion
Endothermic vs. Ectothermic	Ectothermic	Probably endothermic	Endothermic
Body Covering	Scales	One fossil shows it had skin similar to that of mammals	Skin with hair or fur

© 2013 Cengage Learning

Cynodonts and the Origin of Mammals

We can easily recognize living mammals as warm-blooded animals that have hair or fur and mammary glands and, except for the platypus and spiny anteater, give birth to live young. However, these criteria are not sufficient for recognizing fossil mammals; for them, we must rely on skeletal structure only. Several skeletal modifications took place during the transition from mammal-like reptiles to mammals, but distinctions between the two groups are based mostly on details of the middle ear, the lower jaw, and the teeth (Table 15.2). The evolution of mammals from cynodonts is so well documented by fossils that classification of some fossils as reptile or mammal is difficult.

Reptiles have one small bone in the middle ear (the stapes), whereas mammals have three: the incus, the malleus, and the stapes. Also, the lower jaw of a mammal is composed of a single bone called the dentary, but a reptile's jaw is composed of several bones (▶ Figure 15.18). In addition, a reptile's jaw is hinged to the skull at a contact between the articular (a jaw bone) and quadrate bones of the skull, whereas in mammals the dentary contacts the squamosal bone (Figure 15.18).

During the transition from cynodonts to mammals, the quadrate and articular bones that had formed the joint between the jaw and skull in reptiles were modified into the incus and malleus of the mammalian middle ear (Figure 15.18, Table 15.2). Fossils document the progressive enlargement of the dentary until it became the only element in the mammalian jaw. Likewise, a progressive change from the reptile to mammal jaw joint is documented by fossil evidence. In fact, some of the most advanced cynodonts were truly transitional, because they had a compound jaw joint consisting of (1) the articular and quadrate bones typical of reptiles and (2) the dentary and squamosal bones as in mammals (Table 15.2). A recently discovered fossil mammal known as *Liaconodon hui* from China has the rarely preserved tiny jaw bones that became the middle ear bones in mammals. In fact, these bones have separated from the lower jaw, but are "still held in place by a piece of ossified [bony] cartilage that rested in a groove in the lower jaw."[*]

Several other features of cynodonts also indicate they were ancestors of mammals. Their teeth were becoming double-rooted as they are in mammals, and they were somewhat differentiated into distinct types that performed specific functions. In mammals the teeth are fully differentiated into incisors, canines, and chewing teeth, but typical reptiles do not have differentiated teeth (Figure 15.18). In addition, mammals have only two sets of teeth

Earth, v. 56, no. 12 (December 2011), p. 50.

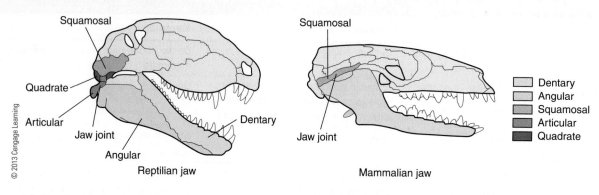

Squamosal
Quadrate
Articular
Jaw joint
Angular
Dentary
Reptilian jaw

© 2013 Cengage Learning

Squamosal
Jaw joint
Mammalian jaw

Dentary
Angular
Squamosal
Articular
Quadrate

(a) Cynodonts had several bones in the jaw and a jaw–skull joint between the articular and quadrate bones.

(b) Mammals have one bone in the jaw, the dentary, and a jaw–skull joint between the dentary and squamosal bones.

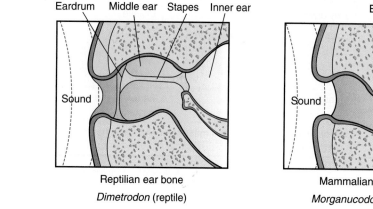

Eardrum   Middle ear   Stapes   Inner ear
Sound
Reptilian ear bone
Dimetrodon (reptile)

© 2013 Cengage Learning

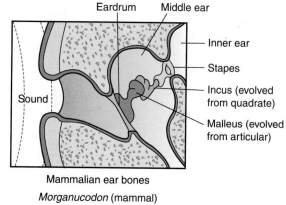

Eardrum   Middle ear
Inner ear
Stapes
Incus (evolved from quadrate)
Malleus (evolved from articular)
Sound
Mammalian ear bones
Morganucodon (mammal)

(c) In reptiles, sound is carried from the eardrum through the stapes, the only bone in the middle ear.

(d) Mammals have three middle-ear bones: the stapes, malleus, and incus. The malleus and incus were derived from the articular and quadrate bones.

during their lifetimes—a set of baby teeth and the permanent adult teeth. Reptiles, with the exception of some cynodonts, have teeth replaced continuously throughout their lives, but the cynodonts, in mammal fashion, had only two sets of teeth. Another important feature of mammal teeth is occlusion; that is, the chewing teeth meet surface to surface to allow grinding. Thus, mammals chew their food, but reptiles, amphibians, and fish do not. Tooth occlusion is known in some advanced cynodonts (Table 15.2).

Reptiles and mammals have a bony protuberance from the skull that fits into a socket in the first vertebra: the atlas. This structure, called the *occipital condyle*, is a single feature in typical reptiles, but in cynodonts it is partly divided into a double structure typical of mammals (Table 15.2). Another mammalian feature, the secondary palate, was partially developed in advanced cynodonts. This bony shelf separating the nasal passages from the mouth cavity is an adaptation for eating and breathing at the same time, a necessary requirement for endotherms with their high demands for oxygen.

In 1837, the German scientist Karl Reichert discovered that the embryos of mammals have an extra bone,

the articular, in the lower jaw, whereas adult mammals have one bone, the dentary. He also found an extra bone in the upper jaw, called the quadrate. Furthermore, these two bones formed the jaw–skull joint just as they do in reptiles. However, as the embryo matured, the articular and quadrate moved to the middle ear, where they became the incus and malleus, and the jaw–skull joint typical of mammals developed. In fact, when opossums are born they have this reptile-type jaw–skull joint between the articular and quadrate, but as they develop further, these bones move to the middle ear and a typical mammal-type jaw–skull joint forms.

Mesozoic Mammals

Mammals evolved during the Late Triassic not long after the first dinosaurs appeared, but for the rest of the Mesozoic Era most of them were small. There were, however, a few exceptions. One, a Middle Jurassic-age aquatic mammal found in China, measures about 50 cm long, and it also has the distinction of being the oldest known fossil with fur. The other is an Early Cretaceous-age mammal called *Repenomamus giganticus*, also from China, that was about 1 m long, weighed 12 to 14 kg, and had the remains of a juvenile dinosaur in its stomach. Most other Mesozoic mammals were about the size of mice and rats, and they were not very diverse—certainly not as diverse as they were during the Cenozoic Era. Furthermore, they retained reptile characteristics but had mammalian features, too. The Triassic triconodonts, for instance, had the fully differentiated teeth typical of mammals, but they also had both the reptile and the mammal types of jaw joints. In short, some mammal features appeared sooner than others (remember the concept of *mosaic evolution* from Chapter 7).

Mammals were certainly present by the Late Triassic and later diverged into two distinct groups. One group, the **monotremes**, or egg-laying mammals such as the spiny anteater and the platypus of the Australian region, may have diverged from other mammals by Middle Jurassic time. The other main mammal group consists of all other mammals, including the **marsupial** (pouched) **mammals** and the **placental mammals**. The history of monotremes is uncertain, but fossils tell us that marsupials and placentals diverged from a common ancestor during the Early Cretaceous (▸Figure 15.19).

Mesozoic Climates and Paleobiogeography

Fragmentation of the supercontinent Pangaea began by the Late Triassic, but during much of the Mesozoic, close connections existed between the various landmasses. The proximity of these landmasses alone, however, is not enough to explain Mesozoic biogeographic distributions, because climates are also effective barriers to wide dispersal. During much of the Mesozoic, though, climates were more equable and lacked the strong north and south zonation characteristic of the present. In short, Mesozoic plants and animals had greater opportunities to occupy much more extensive geographic ranges.

Pangaea persisted as a supercontinent through most of the Triassic (see Figure 14.1a), and the Triassic climate was warm-temperate to tropical, although some areas, such as the present southwestern United States, were arid. Mild temperatures extended 50 degrees north and south of the equator, and even the polar regions may have been temperate. The fauna had a truly worldwide distribution. Some dinosaurs had continuous ranges across Laurasia and Gondwana; the peculiar gliding lizards lived in New Jersey and England; and reptiles known as phytosaurs lived in North America, Europe, and Madagascar.

By the Late Jurassic, Laurasia had become partly fragmented by the opening North Atlantic, but a connection still existed. The South Atlantic had begun to open so that a long, narrow sea separated the southern parts of Africa and South America. Otherwise the southern continents were still close together.

▸ **Figure 15.19** **Restorations of the Oldest Known Marsupial and Placental Mammals** Both of these restorations are based on fossils that were found in Lower Cretaceous rocks in China.

(a) *Sinodelphys,* the oldest known marsupial, was only about 15 cm long.

(b) This restoration shows *Eomaia,* which was only 12 or 13 cm long. It is the oldest known placental mammal.

The mild Triassic climate persisted into the Jurassic. Ferns whose living relatives are now restricted to the tropics of southeast Asia lived as far as 63 degrees south and 75 degrees north. Dinosaurs roamed widely across Laurasia and Gondwana. For example, the giant sauropod *Brachiosaurus* is found in western North America and eastern Africa. Stegosaurs and some families of carnivorous dinosaurs lived throughout Laurasia and in Africa.

By the Late Cretaceous, the North Atlantic had opened further, and Africa and South America were completely separated (see Figure 14.1c). South America remained an island continent until late in the Cenozoic, and its fauna became increasingly different from faunas of the other continents (see Chapter 18). Marsupial mammals reached Australia from South America via Antarctica, but the South American connection was eventually severed. Placentals, other than bats and a few rodents, never reached Australia, explaining why marsupials continue to dominate the continent's fauna even today.

Cretaceous climates were more strongly zoned by latitude, but they remained warm and equable until the close of that period. Climates then became more seasonal and cooler, a trend that persisted into the Cenozoic. Dinosaur and mammal fossils demonstrate that interchange was still possible, especially between the various components of Laurasia.

Mass Extinctions—A Crisis in Life History

During the Phanerozoic Eon, there were at least five mass extinctions and several extinctions of lesser magnitude that drastically reduced Earth's biotic diversity (see Figure 12.20). The greatest of these extinctions took place at the end of the Paleozoic Era, but the one of interest here was at the end of the Mesozoic. It has certainly attracted much more attention than any other extinction because dinosaurs, marine reptiles, and pterosaurs died out, but it was equally devastating for several types of marine invertebrates, including ammonites, rudist bivalves, and some planktonic organisms.

Many hypotheses have been proposed to account for Mesozoic extinctions, but most have been dismissed as improbable or inconsistent with the available data. In 1980, however, a proposal was made that has gained wide acceptance. It was based on a discovery at the Cretaceous–Paleogene boundary in Italy of a 2.5-cm-thick clay layer with a notable concentration of the platinum-group element iridium. High iridium concentrations are now known from many other Cretaceous–Paleogene boundary sites around the world (▶ Figure 15.20a).

The significance of this **iridium anomaly** is that iridium is rare in crustal rocks but is found in much higher concentrations in some meteorites. Accordingly, some investigators propose a meteorite impact to explain the anomaly and further postulate that the meteorite, perhaps 10 km in diameter, set in motion a chain of events leading to extinctions. Some Cretaceous–Paleogene boundary sites also contain soot and shock-metamorphosed quartz grains, both of which are cited as additional evidence of an impact.

According to the impact hypothesis, about 60 times the mass of the meteorite was blasted from the crust high into the atmosphere, and the heat generated at impact started raging forest fires that added more particulate matter to the atmosphere. Sunlight was blocked for several months, temporarily halting photosynthesis; food chains collapsed; and extinctions followed. Furthermore, with sunlight greatly diminished, Earth's surface temperatures were drastically reduced, adding to the biologic stress. Another consequence of the impact was that vaporized rock and atmospheric gases produced sulfuric acid (H_2SO_4) and

▶ Figure 15.20 **End of Mesozoic Extinctions**

(a) Closeup view of the iridium-enriched, Cretaceous–Paleogene boundary clay (the white layer) in the Raton Basin, New Mexico.

(b) Proposed meteorite impact site centered on Chicxulub on the Yucatan Peninsula of Mexico.

nitric acid (HNO_3). Both would have contributed to strongly acid rain that may have had devastating effects on vegetation and marine organisms.

Now some geologists point to a probable impact site centered on the town of Chicxulub on the Yucatán Peninsula of Mexico (Figure 15.20b). The 170-km-diameter structure lies beneath layers of sedimentary rock and appears to be the right age. Evidence that supports the conclusion that the Chicxulub structure is an impact crater includes shocked quartz, the deposits of huge waves, and *tektites*—small pieces of rock melted during the impact and hurled into the atmosphere.

Even if a meteorite did hit Earth, did it lead to these extinctions? If so, both terrestrial and marine extinctions must have occurred at the same time. To date, strict time equivalence between terrestrial and marine extinctions has not been demonstrated. The selective nature of the extinctions is also a problem. In the terrestrial realm, large animals were the most affected, but not all dinosaurs were large, and crocodiles, close relatives of dinosaurs, survived, although some species died out. Some paleontologists think dinosaurs, some marine invertebrates, and many plants were already on the decline and headed for extinction before the end of the Cretaceous. A meteorite impact may have simply hastened the process.

In the final analysis, Mesozoic extinctions have not been explained to everyone's satisfaction. Most geologists now concede that a large meteorite impact occurred, but we also know that vast outpourings of lava were taking place in what is now India. Perhaps these brought about detrimental atmospheric changes. Furthermore, the vast, shallow seas that covered large parts of the continents had mostly withdrawn by the end of the Cretaceous, and the mild, equable Mesozoic climates became harsher and more seasonal by the end of that era. Nevertheless, these extinctions were selective, and no single explanation accounts for all aspects of this crisis in life history.

Summary

Table 15.3 summarizes the major Mesozoic evolutionary and climatic events.

▶ Invertebrate survivors of the Paleozoic extinctions diversified and gave rise to increasingly diverse marine communities.

▶ Some of the most abundant invertebrates were cephalopods, especially ammonoids, foraminifera, and the reef-building rudists.

▶ Land-plant communities of the Triassic and Jurassic consisted of seedless vascular plants and gymnosperms. The angiosperms, or flowering plants, evolved during the Late Jurassic or Early Cretaceous, diversified rapidly, and were soon the most abundant land plants.

▶ Dinosaurs evolved from small, bipedal archosaurs during the Late Triassic, but they were most common during the Jurassic and Cretaceous periods.

▶ All dinosaurs evolved from a common ancestor but differ enough that two distinct orders are recognized: the Saurischia and the Ornithischia.

▶ Bone structure, predator–prey relationships, and other features have been cited as evidence of dinosaur endothermy. Although there is still no solid consensus, many paleontologists think some dinosaurs were indeed endotherms.

▶ That some theropods had feathers indicates they were warm-blooded and provides further evidence of their relationship to birds.

▶ Pterosaurs, the first flying vertebrates, varied from sparrow size to comparative giants. The larger pterosaurs probably depended on soaring to stay aloft, whereas smaller ones flapped their wings. At least one species had hair or hairlike feathers.

▶ The fish-eating, porpoise-like ichthyosaurs were thoroughly adapted to an aquatic environment, whereas the plesiosaurs, with their paddle-like limbs, could most likely come out of the water to lay their eggs. The marine reptiles known as mosasaurs were most closely related to lizards.

▶ During the Jurassic, crocodiles became the dominant freshwater predators. Turtles and lizards were present during most of the Mesozoic. By the Cretaceous, snakes had evolved from lizards.

▶ Jurassic-age *Archaeopteryx*, the oldest known bird, possesses so many theropod characteristics that it has convinced most paleontologists that the two are closely related.

▶ Mammals evolved by the Late Triassic, but they differed little from their ancestors among the cynodonts. Minor differences in the lower jaw, teeth, and middle ear differentiate one group of fossils from the other.

▶ Several types of Mesozoic mammals existed, but most were small, and their diversity was low. Marsupial and placental mammals diverged from a common ancestor during the Cretaceous.

▶ Because during much of the Mesozoic the continents were close together and climates were mild, plants and animals occupied much larger geographic ranges than they do now.

▶ Among the victims of the Mesozoic mass extinction were dinosaurs, flying reptiles, marine reptiles, and several groups of marine invertebrates. A huge meteorite impact may have caused these extinctions, but some paleontologists think other factors were important, too.

Important Terms

amniote, p. 303
angiosperm, p. 301
Archaeopteryx, p. 312
archosaur, p. 303
bipedal, p. 304
cynodont, p. 312
dinosaur, p. 305

ectotherm, p. 307
endotherm, p. 307
ichthyosaur, p. 309
iridium anomaly, p. 316
marsupial mammal, p. 315
monotreme, p. 315
mosasaur, p. 310

Ornithischia, p. 304
placental mammal, p. 315
plesiosaur, p. 310
pterosaur, p. 308
quadrupedal, p. 304
Saurischia, p. 304
therapsid, p. 312

Review Questions

1. The flowering plants, otherwise known as _____, evolved during the _____ Period.
   a. _____ gymnosperms/Triassic; b. _____ seedless plants/Permian; c. _____ horsetails/Cenozoic; d. _____ angiosperms/Cretaceous; e. _____ cycads/Jurassic.

2. The only mammals that lay shelled eggs belong to the group known as the
   a. _____ therapsids; b. _____ monotremes; c. _____ pterosaurs; d. _____ endotherms; e. _____ bipeds.

3. All known carnivorous dinosaurs were
   a. _____ quadrupeds; b. _____ stegosaurs; c. _____ theropods; d. _____ ichthyosaurs; e. _____ marsupials.

4. One specialization in pterosaurs was
   a. _____ development of a placenta to nourish embryos; b. _____ flipper-like limbs for steering and maneuvering; c. _____ huge legs to support a massive body; d. _____ teeth differentiated into specific types; e. _____ modified hand with a long finger to support a wing.

5. Which one of the following statements is correct?
   a. _____ The first flying vertebrates were plesiosaurs; b. _____ Icthyosaurs look much like living opossums; c. _____ Most Triassic plants were angiosperms; d. _____ Rudists were important Mesozoic reef-building animals; e. _____ The lower jaw of archosaurs was made up of a single bone.

6. Because of their rapid evolution and nektonic lifestyle, the _____ are good guide fossils.
   a. _____ ammonites; b. _____ saurischians; c. _____ amphibians; d. _____ bivalves; e. _____ trilobites.

7. Which one of the following groups was not only common during the Mesozoic, but are now the major primary producers in today's warm seas?
   a. _____ Nautiloids; b. _____ Rudists; c. _____ Dinoflagellates; d. _____ Coleoids; e. _____ Placentals.

8. The most mammal-like of all mammal-like reptiles were the
   a. _____ cynodonts; b. _____ ornithopods; c. _____ angiosperms; d. _____ ophiolites; e. _____ saurischians.

9. *Archaeopteryx* is considered by most paleontologists to be the first known
   a. _____ dinosaur; b. _____ marine reptile; c. _____ bird; d. _____ mammal; e. _____ cynodont.

10. The middle-ear bones of mammals evolved from which of these bones in the mammal-like reptiles?
    a. _____ Palatine and vomer; b. _____ Articular and quadrate; c. _____ Prefrontal and parietal; d. _____ Dentary and incus; e. _____ Tibia and fibula.

11. Living crocodiles have a fairly well developed secondary palate and a four-chambered heart much like that in mammals. So why are crocodiles classified as reptiles rather than mammals?

12. Construct a cladogram showing the relationships among birds, saurischians, ornithischians, cynodonts, and mammals. (Hint: see Figure 15.8.)

13. Summarize the change that occurred in the lower jaw and middle ear during the transition from mammal-like reptiles to mammals.

14. Why classify *Archaeopteryx* as a bird? After all, we now know of several dinosaurs that had feathers.

15. What were the main groups of plants during the Triassic, Jurassic, and Cretaceous? Which group predominates now, and why has it been so successful?

16. How do dinosaurs differ from most other reptiles, and what criteria are used to distinguish between the two main groups of dinosaurs?

17. Summarize the evidence for endothermy in dinosaurs.

18. What happened in the marine invertebrate community following the extinctions at the end of the Paleozoic Era?

19. Summarize the evidence that convinces many geologists that an asteroid impact took place at the end of the Mesozoic Era.

20. Explain how plate positions and climate influence the geographic distribution of Mesozoic plants and animals.

TABLE 15.3 Major Evolutionary and Climatic Events of the Mesozoic Era

Age (millions of years)	Geologic Period	Invertebrates	Vertebrates	Plants	Climate
66	Cretaceous	Extinction of ammonites, rudists, and most planktonic foraminifera at end of Cretaceous Continued diversification of ammonites and belemnoids Rudists become major reef builders	Extinctions of dinosaurs, flying reptiles, marine reptiles, and some marine invertebrates Placental and marsupial mammals diverge	Angiosperms evolve and diversify rapidly Seedless plants and gymnosperms still common but less varied and abundant	Climate becomes more seasonal and cooler at end of Cretaceous North–south zonation of climates more marked but remains equable
146	Jurassic	Ammonites and belemnoid cephalopods increase in diversity Scleractinian coral reefs common Appearance of rudist bivalves	*Greatest Diversity of Dinosaurs* First birds (may have evolved in Late Triassic) Time of giant sauropod dinosaurs	Seedless vascular plants and gymnosperms only	Much like Triassic Ferns with living relatives restricted to tropics live at high latitudes, indicating mild climates
200	Triassic	The seas are repopulated by invertebrates that survived the Permian extinction event Bivalves and echinoids expand into the infaunal niche	Cynodonts become extinct Mammals evolve from cynodonts Ancestral archosaur gives rise to dinosaurs Flying reptiles and marine reptiles evolve	Land flora of seedless vascular plants and gymnosperms as in Late Paleozoic	Warm-temperate to tropical Mild temperatures extend to high latitudes; polar regions may have been temperate Local areas of aridity
251					

© 2013 Cengage Learning

16

Cenozoic Earth History
The Paleogene and Neogene Periods

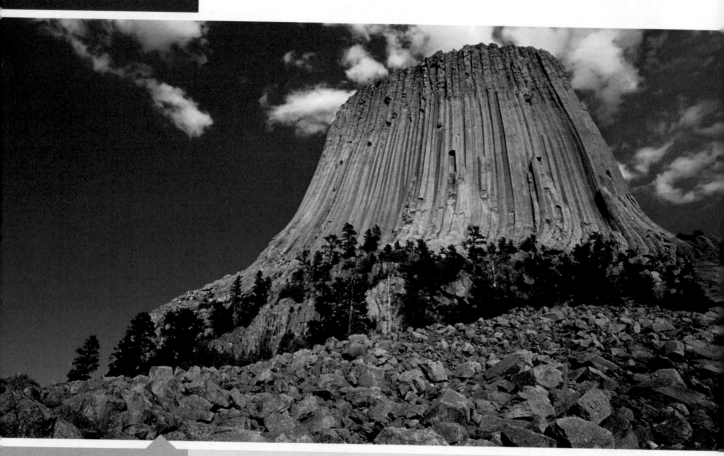

Devil's Tower in northeastern Wyoming is a small igneous intrusion that was emplaced in Earth's crust 45 to 50 million years ago. It shows well-developed columnar jointing, it stands 650 m above its base, and it can be seen from 48 km away. According to one Native American legend, Devil's Tower formed when the Great Spirit caused it to rise from the ground, carrying several children with it who were trying to escape from a gigantic grizzly bear. In its attempts to reach the children, the bear left deep scratch marks in the tower's rocks.

Jason Patrick Ross/Shutterstock.com

320

CHAPTER OBJECTIVES

At the end of this chapter, you will have learned that

▶ Geologists divide the 66-million-year-long Cenozoic Era into three periods: the Paleogene, Neogene, and Quaternary, each of which consists of several epochs.

▶ The breakup of Pangaea began during the Triassic Period and continues to the present, giving rise to the present distribution of land and sea.

▶ Cenozoic orogenies were concentrated in two belts, one that nearly encircles the Pacific Ocean basin, and another that trends east–west through the Mediterranean basin and on into southeast Asia.

▶ The Late Cretaceous to Eocene Laramide orogeny resulted in deformation of a large area in the west called the North American Cordillera, which extends from Alaska to Mexico.

▶ Following the Laramide orogeny, the North American Cordillera continued to evolve as it experienced volcanism, uplift of broad plateaus, large-scale block faulting, and deep erosion.

▶ The Great Plains consist of huge quantities of sediments that were eroded from the Rocky Mountains and transported eastward.

▶ A subduction zone was present along the western margin of the North American plate until the plate collided with a spreading ridge, producing the San Andreas and Queen Charlotte transform faults.

▶ An epeiric sea briefly occupied North America's continental interior during the Paleogene.

▶ Thick deposits of sediment accumulated along the Gulf and Atlantic Coastal plains.

▶ Renewed uplift and erosion account for the present-day Appalachian Mountains.

▶ Paleogene and Neogene rocks contain mineral resources such as oil, oil shale, coal, phosphorus, and gold.

Introduction

In Chapter 1, we noted that Earth is a complex planet that has changed continuously since its origin 4.6 billion years ago. We also mentioned that three interrelated subthemes run throughout our consideration of Earth history: plate tectonics, organic evolution, and geologic time. Here, we continue with these themes, but in this chapter we review the Cenozoic Era, representing the most recent 66 million years of all geologic time. In the context of geologic time, the Cenozoic is only 1.4 percent of all geologic time, or 20 minutes on a hypothetical 24-hour clock (see Figure 8.1). Of course, by any other measure, it was an incredibly long interval, certainly long enough for plates to change position, for landscapes to evolve, and for significant evolution of the biota to occur.

▶ Figure 16.1 shows the most recent time scale for the Cenozoic Era (see also Figure 1.11), which shows that the era consists of three periods of unequal duration. The *Paleogene Period* (66 to 23 million years ago) includes the Paleocene, Eocene, and Oligocene epochs, whereas the *Neogene Period* (23 to 2.6 million years ago) consists of the Miocene and Pliocene epochs, and the *Quaternary Period* (2.6 million years ago to the present) includes the Pleistocene and Recent (or Holocene) epochs. An older subdivision of the Cenozoic Era, consisting of the Tertiary and Quaternary periods, is still used by some geologists, but these designations are no longer recommended.

▶ Figure 16.1 **Cenozoic Geologic Time Scale** The International Commission on Stratigraphy recommends the terms used here for Cenozoic time. See Figure 1.11 for the complete geologic time scale. Dates are from the 2009 International Stratigraphic Chart. © 2009 by the International Commission of Stratigraphy.

Holocene/Recent (.0117)

Era	Period	Epoch	Duration, millions of years (approx.)	Millions of years ago (approx.)
Cenozoic	Quaternary	Pleistocene	2.59	
				— 2.6 —
	Neogene	Pliocene	3.2	
				— 5.0 —
		Miocene	18.0	
				— 23 —
	Paleogene	Oligocene	11	
				— 34 —
		Eocene	22	
				— 56 —
		Paleocene	10	
				66

(a) The Paleocene Cannonball Formation in Montana.

(b) The Eocene Ione Formation (light colored) and the overlying Miocene-age Lovejoy Basalt near Cherokee, California.

Geologists know more about Cenozoic Earth and life history than they do about previous eras because (1) Cenozoic-age rocks are the most accessible, being at or near the surface; (2) most have been little altered; and (3) many contain fossils. Vast exposures of Cenozoic sedimentary and igneous rocks present in western North America record the presence of a shallow sea in the continental interior, deposition in continental environments, the emplacement of plutons, and lava flows and volcanism on a huge scale in the Pacific Northwest (▶ Figure 16.2, and chapter opening photo).

Cenozoic-age marine sedimentary rocks are also present along the Gulf Coast and along the eastern margin of the United States, but deposits in continental environments are limited, except for a few in Maryland, South Carolina, and, most notably, in Florida. And whereas volcanism was widespread in the West, only a few areas of Cenozoic volcanic rocks are known in the East—mostly

dikes, sills, and volcanic plugs in West Virginia that date from the Eocene Epoch.

Why should you study Cenozoic Earth and life history? One reason is that the present distribution of land and sea, climatic and oceanic circulation patterns, and Earth's present distinctive topography and its biota developed during this time. In short, it was a continuation of processes that began much earlier. In this chapter, our main concern is Earth history of the Paleogene and Neogene periods. The Quaternary Period, which includes the Pleistocene and Recent (Holocene) epochs, was unusual because the Pleistocene was one of the few times in Earth history when widespread glaciers were present, and, accordingly, we cover it in Chapter 17. Evolution of the Cenozoic biota is considered in Chapters 18 and 19.

Cenozoic Plate Tectonics—An Overview

The ongoing fragmentation of Pangaea (see Figure 14.1), the supercontinent that existed at the end of the Paleozoic, accounts for the present distribution of Earth's landmasses. Moving plates also directly affect the biosphere because the geographic locations of continents profoundly influence the atmosphere and hydrosphere. As we examine Cenozoic life history, you will see that some important biologic events are related to isolation and/or connections between landmasses (see Chapter 18).

Notice in Figure 14.1 that by Cretaceous time the Americas had broken away from Europe and Africa, and by the Eocene (▶ Figure 16.3a) they were widely separated by the evolving Atlantic Ocean basin, a trend that continues even now. Remember that spreading ridges such as the Mid-Atlantic Ridge and East Pacific Rise became established, along which new oceanic crust forms. In the Pacific, however, the age of the oceanic crust is asymmetric across the spreading ridge because much of the crust that formed in the eastern Pacific has been subducted beneath the westerly moving North and South American plates (see Figure 3.11).

Another important plate tectonic event was the northward movement of the Indian plate and its eventual collision with Asia (Figure 16.3b). Simultaneous northward movement of the African plate caused the closure of the Tethys Sea and initiated the tectonic activity that currently takes place throughout an east–west zone from the Mediterranean through northern India. Erupting volcanoes in Italy and Greece as well as seismic activity in Italy, Turkey, Greece, Pakistan, and India remind us of the continuing plate interactions in this part of the world.

Neogene rifting began in East Africa, the Red Sea, and the Gulf of Aden (see Figure 3.15). Rifting in East Africa is in its early stages, because the continental crust has not yet stretched and thinned enough for new oceanic crust to form from below. Nevertheless, this area is seismically active and has many active volcanoes. In the Red Sea,

(a) The Eocene Epoch.

(b) The Miocene Epoch.

rifting and the Late Pliocene origin of oceanic crust followed vast eruptions of basalt, and in the Gulf of Aden, Earth's crust had stretched and thinned enough by Late Miocene time for upwelling basaltic magma to form new oceanic crust. Notice in Figure 3.15 that the Arabian plate is moving north, so it too causes some of the deformation taking place from the Mediterranean through India.

In the meantime, the North and South American plates continued their westerly movement as the Atlantic Ocean basin widened. Subduction zones bounded both continents on their western margins, but the situation changed in North America as it moved over the northerly extension

of the East Pacific Rise and it now has a transform plate boundary—a topic we discuss more fully in a later section—although subduction continues in the Pacific Northwest.

Cenozoic Orogenic Belts

Remember that an *orogeny* is an episode of mountain building during which deformation takes place over an elongate area. In addition, most orogenies involve volcanism, the emplacement of plutons, and regional metamorphism as Earth's crust is locally thickened and stands higher than

▶ **Figure 16.4**
Phanerozoic Orogenic Belts Most of Earth's geologically recent and continuing orogenic activity takes place in the circum-Pacific orogenic belt and the Alpine–Himalayan orogenic belt. Each belt is made up of smaller segments known as orogens.

Courtesy of Eric H. Christiansen

Map legend:
- ⌇ Paleozoic orogenic belts
- ⌇ Mesozoic and Cenozoic orogenic belts
- ⊢⊢ Oceanic ridges
- ⌇ Trenches
- ▢ Ocean basins
- ▢ Stable platforms
- ▨ Shield

adjacent areas. Cenozoic orogenic activity took place largely in two major zones or belts: the *Alpine–Himalayan orogenic belt* and the *circum-Pacific orogenic belt* (▶ Figure 16.4). Both belts are made up of smaller segments known as **orogens**, each of which shows the characteristics of an orogeny.

Alpine–Himalayan Orogenic Belt

The **Alpine–Himalayan orogenic belt** extends eastward from Spain through the Mediterranean region as well as the Middle East and India and on into Southeast Asia (Figure 16.4). Remember that during Mesozoic time the Tethys Sea separated much of Gondwana from Eurasia. Closure of this sea took place during the Cenozoic as the African and Indian plates collided with the huge landmass to the north and northwest (Figure 16.3). Volcanism, seismicity, and deformation indicate that the Alpine–Himalayan orogenic belt remains active.

The Alps During the **Alpine orogeny**, deformation occurred in a linear zone in southern Europe extending from Spain eastward through Greece and Turkey. Concurrent deformation also occurred along Africa's northwest margin (Figure 16.4). Many details of this long, complex event are poorly understood, but the overall picture is now becoming clear.

Events leading to Alpine deformation began during the Mesozoic, yet Eocene to Late Miocene deformation was also important. Northward movements of the African and Arabian plates against Eurasia caused compression and deformation, but the overall picture is complicated by the collision of several smaller plates with Europe. These

small plates were also deformed and are now found in the mountains in the Alpine orogen.

Mountain building produced the Pyrenees between Spain and France, the Apennines of Italy, as well as the Alps of mainland Europe (Figure 16.4). Indeed, the compressional forces generated by colliding plates resulted in complex thrust faults and huge overturned folds known as *nappes* (▶ Figure 16.5). As a result, the geology of the Alps in France, Switzerland, and Austria is extremely complex. Plate convergence also produced an almost totally isolated sea in the Mediterranean basin, which had previously been part of the Tethys Sea. Late Miocene deposition in this sea, which was then in an arid environment, accounts for evaporite deposits up to 2 km thick.

The collision of the African plate with Eurasia also accounts for the Atlas Mountains of northwest Africa, and farther to the east, in the Mediterranean basin, Africa continues to force oceanic lithosphere northward beneath Greece and Turkey. Active volcanoes in Italy and Greece as well as seismic activity throughout this region indicate that southern Europe and the Middle East remain geologically active. In 2005, for instance, an earthquake with a Richter magnitude of 7.6 killed more than 86,000 people in Pakistan, and Mount Vesuvius in Italy has erupted 80 times since it destroyed Pompeii in A.D. 79; its most recent eruption was in 1944.

The Himalayas—Roof of the World During the Early Cretaceous, India broke away from Gondwana and began moving north, and oceanic lithosphere was consumed at a subduction zone along the southern margin of Asia (▶ Figure 16.6a). The descending plate partially melted, forming

(a) View of the Alps near Interlaken, Switzerland. The peaks in this view from left to right are the Eiger (Ogre), Mönch (Monk), and the Jungfrau (Virgin).

(b) Folded rocks in the Alps near Lütschental, Switzerland.

▸ Figure 16.6 **Plate Tectonics and the Himalayan Orogen**

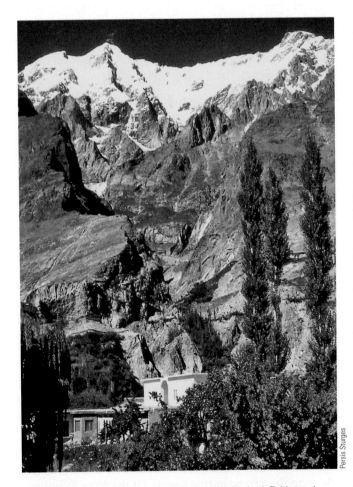

(a) The Indian plate moved northward for millions of years until it collided with Eurasia, causing crustal thickening and uplift of the Himalayas.

(b) The Karakoram Range, seen here from Karimabad, Pakistan, is within the Himalayan origin. The range lies on the border of Pakistan, China, and India.

magma that rose to form a volcanic chain and large granitic plutons in what is now Tibet. The Indian plate eventually approached these volcanoes and destroyed them as it collided with Asia. As a result, two continental plates were sutured along a zone now recognized as the **Himalayan orogeny** (Figure 16.6b).

Sometime between 40 and 50 million years ago, India's drift rate decreased abruptly from 15 to 20 cm/year to about 5 cm/year. Because continental lithosphere is not dense enough to be subducted, this decrease most likely marks the time of collision and India's resistance to subduction. As a result of India's low density and resistance to subduction, it was underthrust about 2,000 km beneath Asia, causing crustal thickening and uplift, a process that continues at about 5 cm/year. Furthermore, sedimentary rocks formed in the sea south of Asia were thrust northward into Tibet, and two huge thrust faults carried Paleozoic and Mesozoic rocks of Asian origin onto the Indian plate.

In the Himalayan origin there is no volcanism because the Indian plate does not penetrate deeply enough to generate magma, but seismic activity continues. Indeed, the entire Himalayan region, including the Tibetan plateau and well into China is seismically active. The May 12, 2008, Sichuan earthquake in China, in which about 68,000 people perished, was a result of this collision between India and Asia.

Circum-Pacific Orogenic Belt

The **circum-Pacific orogenic belt** consists of orogens along the western margins of South, Central, and North America as well as the eastern margin of Asia and the islands north of Australia and New Zealand (Figure 16.4). Subduction of oceanic lithosphere accompanied by deformation and igneous activity characterize the orogens in the western and northern Pacific. Japan, for instance, is bounded on the east by the Japan Trench, where the Pacific plate is subducted, and the Sea of Japan, a **back-arc basin**, lies between Japan and mainland Asia. According to some geologists, Japan was once part of mainland Asia and was separated when back-arc spreading took place (▶Figure 16.7). Separation began during the Cretaceous as Japan moved eastward over the Pacific plate and oceanic crust formed in the Sea of Japan. Japan's geology is complex, and much of its deformation predates the Cenozoic, but considerable deformation, metamorphism, and volcanism occurred during the Cenozoic and continues to the present. Indeed, the magnitude 9.0 earthquake and resulting tsunami off of Japan's west coast in March 2011, in which more than 13,000 people died, is an indication of the continuing tectonic activity in this region.

In the northern part of the Pacific Ocean, basin subduction of the Pacific plate at the Aleutian trench accounts for the tectonic activity in that region. Of the 80 or so potentially active volcanoes in Alaska, at least half have erupted since 1760, and, of course, seismic activity is ongoing.

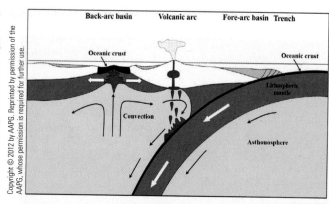

▶ **Figure 16.7** **Back-Arc Basin** Seafloor spreading on the landward side of a volcanic arc and the origin of a back-arc basin. The Sea of Japan probably formed in this way.

In the eastern part of the Pacific, the Cocos and Nazca plates moved east from the East Pacific Rise only to be consumed at subduction zones along the west coasts of Central and South America (Figure 16.3). Volcanism and seismic activity indicate these orogens in both Central and South America are active. One manifestation of ongoing tectonic activity in South America is the Andes Mountains, with more than 49 peaks higher than 6,000 m. The Andes formed, and continue to do so, as Mesozoic–Cenozoic plate convergence resulted in crustal thickening as sedimentary rocks were deformed, uplifted, and intruded by huge granitic plutons (▶Figure 16.8).

North American Cordillera

The **North American Cordillera**, a complex mountainous region in western North America, is a large segment of the circum-Pacific orogenic belt extending from Alaska to central Mexico. In the United States it widens to 1,200 km, stretching east–west from the eastern flank of the Rocky Mountains to the Pacific Ocean (▶Figure 16.9).

The geologic evolution of the North American Cordillera began during the Neoproterozoic when huge quantities of sediment accumulated along a westward-facing continental margin (see Figure 9.7a). Deposition continued into the Paleozoic, and during the Devonian, part of the region was deformed at the time of the Antler orogeny (see Chapter 11). A protracted episode of deformation known as the *Cordilleran orogeny* began during the Late Jurassic as the Nevadan, Sevier, and Laramide orogenies progressively affected areas from west to east (see Figure 14.11). The first two of these orogenies were discussed in Chapter 14; we discuss the Laramide orogeny, an episode of deformation occurring from the Late Cretaceous to the Eocene, in the following section.

After Laramide deformation ceased during Eocene time, the North American Cordillera continued to evolve

▶ Figure 16.8 **The Andes Mountains in South America**

(a) Prior to 200 million years ago, the western margin of South America was a passive continental margin.

(b) Orogeny began when this area became an active continental margin as the South American plate moved to the west and collided with oceanic lithosphere.

(c) Continued deformation, plutonism, and volcanism.

© 2013 Cengage Learning

as parts of it experienced large-scale block faulting, extensive volcanism, and vertical uplift and deep erosion. Furthermore, during about the first half of the Cenozoic Era a subduction zone was present along the entire western margin of the Cordillera, but now most of it is a transform plate boundary. Seismic activity and volcanism indicate plate interactions continue in the Cordillera, especially near its western margin.

Laramide Orogeny

We already mentioned that the **Laramide orogeny** was the third in a series of deformational events in the Cordillera beginning during the Late Jurassic (see Chapter 14). However, this orogeny was Late Cretaceous to Eocene, and it

differed from the previous orogenies in important ways. First, it occurred much farther inland from a convergent plate boundary, and neither volcanism nor emplacement of plutons was very common. In addition, deformation mostly took the form of vertical, fault-bounded uplifts rather than the compression-induced folding and thrust faulting typical of most orogenies. To account for these differences, geologists modified their model for orogenies at convergent plate boundaries. During the preceding Nevadan and Sevier orogenies, the Farallon plate was subducted at about a 50-degree angle along the western margin of North America. Volcanism and plutonism took place 150 to 200 km inland from the oceanic trench, and sediments of the continental margin were compressed and deformed. Most geologists agree that by Late Cretaceous–Early Paleogene

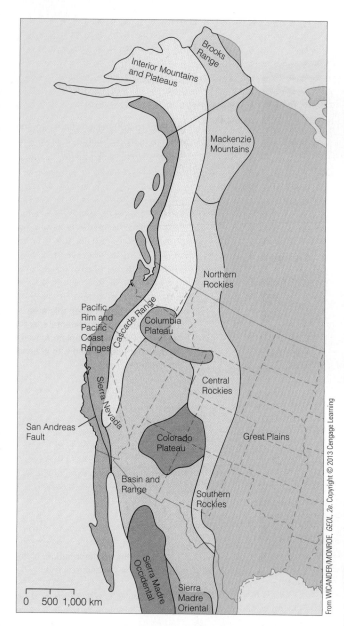

From WICANDER/MONROE, *GEOL*, 2e. Copyright © 2013 Cengage Learning

▶ Figure 16.9 **The North American Cordillera** The North American Cordillera is a complex mountainous area extending from the eastern margin of the Rocky Mountains to the Pacific coast.

time there was a change in the subduction angle from steep to gentle, and the Farallon plate moved nearly horizontally beneath North America, but they disagree on what caused the change in the angle of subduction.

According to one hypothesis, a buoyant oceanic plateau that was part of the Farallon plate that descended beneath North America resulted in shallow subduction. Another hypothesis holds that North America overrode the Farallon plate, beneath which was the deflected head of a mantle plume (▶Figure 16.10). The lithosphere above the mantle plume was buoyed up, accounting for the change from steep to shallow subduction. As a result, igneous activity shifted farther inland and finally ceased because the descending plate no longer penetrated to the mantle.

This changing angle of subduction also caused a change in the type of deformation—the fold-thrust deformation of the Sevier orogeny gave way to large-scale buckling and fracturing, which yielded fault-bounded vertical uplifts. Erosion of the uplifted blocks yielded rugged, mountainous topography and supplied sediments to the intervening basins.

The Laramide orogen is centered in the middle and southern parts of the Rocky Mountains of Wyoming and Colorado, but deformation also took place far to the north and south. In the northern Rocky Mountains of Montana and Alberta, Canada, huge slabs of pre-Laramide strata moved eastward along overthrust faults.* On the Lewis overthrust in Montana, a slab of Precambrian rocks was displaced eastward about 75 km (▶Figure 16.11), and similar deformation occurred in the Canadian Rocky Mountains. Far to the south of the main Laramide orogen, sedimentary rocks in the Sierra Madre Oriental of east-central Mexico are now part of a major fold-thrust belt.

By Middle Eocene time, Laramide deformation ceased and igneous activity resumed in the Cordillera when the mantle plume beneath the lithosphere disrupted the overlying oceanic plate (Figure 16.10c). The uplifted blocks of the Laramide orogen continued to erode, and by the Neogene, the rugged mountains had been nearly buried in their own debris, forming a vast plain across which streams flowed. During a renewed cycle of erosion, these streams removed much of the basin fill sediments and incised their valleys into the uplifted blocks. Late Neogene uplift accounts for the present ranges, and uplift continues in some areas, such as the Teton Range in Wyoming.

Cordilleran Igneous Activity

The enormous batholiths in Idaho, British Columbia, Canada, and the Sierra Nevada of California were emplaced during the Mesozoic (see Chapter 14), but intrusive activity continued into the Paleogene Period. Numerous small plutons formed, including copper- and molybdenum-bearing stocks in Utah, Nevada, Arizona, and New Mexico.

Volcanism was common in the Cordillera, but it varied in location, intensity, and eruptive style, and it ceased temporarily in the area of the Laramide orogen (▶Figure 16.12a). In the Pacific Northwest, the Columbia Plateau is underlain by 200,000 km³ of Miocene lava flows of the Columbia River basalts that have an aggregate thickness of about 2,500 m. These vast lava flows are now well exposed in the walls of the canyons eroded by the Columbia and Snake rivers and their tributaries (Figure 16.12b and c). The relationship of this huge outpouring of lava to plate tectonics remains unclear, but some geologists think it resulted from a mantle plume beneath western North America.

The Snake River Plain (Figure 16.12a), which is mostly in Idaho, is actually a depression in the crust that was filled

---

*An overthrust fault is a large-scale, low-angle thrust fault with displacement measured in kilometers.

▶ Figure 16.10 **Laramide Orogeny** The Late Cretaceous to Eocene Laramide orogeny took place as the Farallon plate was subducted beneath North America.

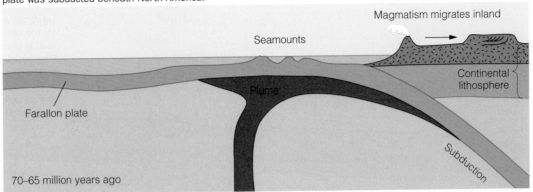

(a) As North America moved westward over the Farallon plate, beneath which was the deflected head of a mantle plume, the angle of subduction decreased and igneous activity shifted inland.

(b) With nearly horizontal subduction, igneous activity ceased and the continental crust was deformed mostly by vertical uplifts.

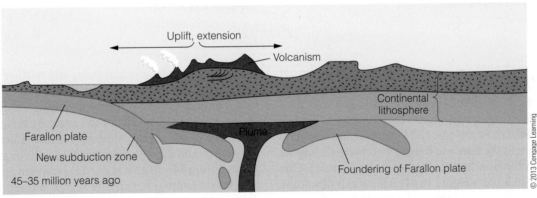

(c) Disruption of the oceanic plate by the mantle plume marked the onset of renewed igneous activity.

by Miocene and younger rhyolite, volcanic ash, and basalt (Figure 16.12c). These rocks are oldest in the southwest part of the area and become younger toward the northeast, leading some geologists to propose that North America has migrated over a mantle plume that now lies beneath Yellowstone National Park in Wyoming. Other geologists disagree, thinking that these volcanic rocks erupted along an intracontinental rift zone.

Bordering the Snake River Plain on the northeast is the Yellowstone Plateau, an area of Pliocene and Pleistocene volcanism mostly in Wyoming (Figure 16.12a). Perhaps a mantle plume lies beneath the area, as just noted, that accounts for the ongoing hydrothermal activity there, but the heat may come from an intruded body of magma that has not yet completely cooled.

Elsewhere in the Cordillera, andesite, volcanic breccia, and welded tuffs (ignimbrites), mostly of Oligocene age, cover more than 25,000 km^2 in the San Juan volcanic field in Colorado. Eruptions in the Cima volcanic field in California began during the Pliocene and continued until

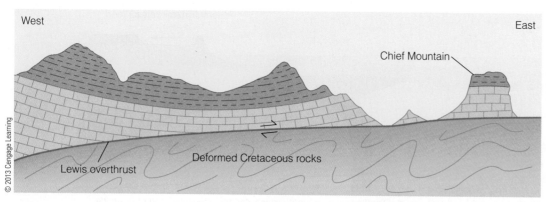

West   East

Chief Mountain

(a) Cross section showing the Lewis overthrust fault in Glacier National Park, Montana. Meso- and Neoproterozoic rocks of the Belt Supergroup rest on deformed Cretaceous rocks.

Deformed Cretaceous rocks

Lewis overthrust

© 2013 Cengage Learning

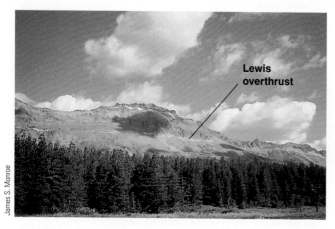

Lewis overthrust

James S. Monroe

(b) The trace of the Lewis overthrust is visible in this image as a light-colored line on the mountainside.

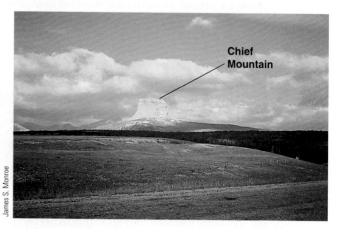

Chief Mountain

James S. Monroe

(c) Erosion has isolated Chief Mountain from the rest of the slab of overthrust rock.

only a few thousand years ago (▶Figure 16.13a); the San Francisco volcanic field of Arizona was active during the Pliocene and Pleistocene (Figure16.13b), and volcanism was also occurring along Oregon's coast.

Some of the most majestic and highest mountains in the Cordillera are in the **Cascade Range** of northern California, Oregon, Washington, and southern British Columbia, Canada (Figure 16.9 and ▶Figure 16.14). Thousands of volcanic vents are present, the most impressive of which are the dozen or so large composite volcanoes and Lassen Peak in California, the world's largest lava dome. Volcanism in this region is related to subduction of the Juan de Fuca plate beneath North America. Volcanism in the Cascade Range goes back at least to the Oligocene, but the most recent episode began during the Late Miocene or Early Pliocene. The eruption of Lassen Peak in California from 1914 to 1917, and the eruptions of Mount St. Helens in Washington in 1980 and again in 2004, indicate that Cascade volcanoes remain active.

Basin and Range Province

Earth's crust in the **Basin and Range Province** (Figure 16.9)—an area of nearly 780,000 km^2 centered on Nevada

but extending into adjacent states and northern Mexico—has been stretched and thinned, yielding north–south oriented mountain ranges with intervening valleys or basins (▶Figure 16.15a). The 400 or so ranges are bounded on one or both sides by steeply dipping normal faults that probably curve and dip less steeply with depth. In any case, the faults outline blocks that show displacement and rotation.

Before faulting began, the region was deformed during the Nevadan, Sevier, and Laramide orogenies. Then, during the Paleogene, the entire area consisted of highlands undergoing extensive erosion, but Early Miocene eruptions of rhyolitic lava flows and pyroclastic materials covered large areas. By the Late Miocene, large-scale faulting had begun, forming the basins and ranges. Sediment derived from the ranges was transported into the adjacent basins and accumulated as alluvial fan and playa lake deposits.

At its western margin, the Basin and Range Province is bounded by normal faults along the east flank of the Sierra Nevada (Figure 16.15b). Pliocene and Pleistocene uplift tilted the Sierra Nevada toward the west, and its crest now stands 3,000 m above the basins to the east. Before this uplift took place, the Basin and Range had a subtropical

Predominantly andesitic

Basalt and andesite

Predominantly basaltic

© 2013 Cengage Learning

(a) Distribution of Cenozoic volcanic rocks in the western United States.

(b) Outcrop of the Columbia River Basalt in Oregon. Huge accumulations of overlapping lava flows cover a large area mostly in Washington but extending into parts of Oregon and Idaho.

© Lyn Topinka

(c) Basalt lava flows of the Snake River Plain at Malad Gorge State Park, Idaho.

© Sue Monroe

▶ Figure 16.13 **Cenozoic Volcanism in California and Arizona**

Sue Monroe

Photo by Tyler finvold

(a) View of the Cima volcanic field in Mojave National Preserve in California, which was active between 7.6 million years (Late Miocene) and 10,000 years ago (Late Pleistocene). Basalt lava flows and 52 cinder cones are present here.

(b) The San Francisco Mountains in northern Arizona are part of the San Francisco volcanic field where 600 volcanoes formed between 6 million years ago (Late Miocene) and about 1,000 years ago.

climate, but the rising mountains created a rain shadow, and the climate became increasingly arid.

Geologists have proposed several models to account for basin-and-range structure but have not reached a consensus. Among these are back-arc spreading; spreading at the East Pacific Rise, the northern part of which is thought to now lie beneath this region; spreading above a mantle plume; and deformation related to movements along the San Andreas Fault.

Colorado Plateau

The vast, elevated region in Colorado, Utah, Arizona, and New Mexico known as the **Colorado Plateau** (Figure 16.9) has volcanic mountains rising above it, brilliantly colored rocks, and deep canyons. In Chapters 11 and 14, we noted that during the Permian and Triassic, the Colorado Plateau region was the site of extensive red bed deposition; many of these rocks are now exposed in the uplifts and canyons (▶Figure 16.16a). Deposition of detrital sediment as well as igneous activity continued into the Cenozoic in many areas of the Colorado Plateau (Figures 16.16b and 16.16c).

Cretaceous-age marine sedimentary rocks indicate the Colorado Plateau was below sea level, but during the Paleogene Period, Laramide deformation yielded broad anticlines and arches and basins, and a number of large normal faults. However, deformation was far less intense here than elsewhere in the Cordillera. Neogene uplift elevated the region from near sea level to the 1,200 to 1,800 m elevations seen today, and as uplift proceeded, streams and rivers began eroding deep canyons.

Geologists disagree on the details of just how the deep canyons so typical of the region developed—such as the Grand Canyon. Some think the streams were *antecedent*, meaning they existed before the present topography developed, in which case they simply eroded downward as uplift proceeded. Others think the streams were *superposed*, implying that younger strata covered the area, on which streams were established. During uplift, the streams stripped away these younger rocks and eroded down into the underlying strata. In either case, the landscape continues to evolve as erosion of the canyons and their tributaries deepens and widens them.

Rio Grande Rift

The **Rio Grande rift** extends north to south about 1,000 km from central Colorado, through New Mexico, and into northern Mexico. Recall our discussions of the Mesoproterozoic Midcontinent rift in the Great Lakes region (see Chapter 9) and the present-day rifting in the Gulf of Aden, the Red Sea, and East Africa (see Figure

Photo by Lyn Topinka/USGS

▶ Figure 16.14 **Cascade Range Volcanoes** Volcanism in the Cascade Range dates back to the Oligocene, but the large volcanoes of the range formed more recently. This view shows Mount Rainier in Washington State, which last erupted during the 1840s. Because of its elevation (4,329 m) and its northerly location, Mount Rainier has been deeply eroded by glaciers, several of which remain active.

(a) The Basin and Range Province is mostly in Nevada, but it extends into adjacent states and northern Mexico. The ranges and the intervening basins trend north–south. Each range is bounded on one or both sides by normal faults.

(b) The Sierra Nevada at the western margin of the Basin and Range Province has risen along normal faults so that it is more than 3,000 m above the valley to the east.

3.15). The Rio Grande rift is similar in that Earth's crust has been stretched and thinned, the rift is bounded on both sides by normal faults, seismic activity continues, and volcanoes and calderas are present (►Figure 16.17). Actually, the Rio Grande rift consists of several basins through which the present-day Rio Grande flows, although the river simply exploited an easy route to the sea and was not responsible for the rift itself.

Rifting in this area began about 29 million years ago and persisted for 10 to 12 million years, from the Late Oligocene into the Early Miocene. A second period of rifting began during the Middle Miocene, about 17 million years ago, and it continues to the present. The displacement on some of the faults is as much as 8,000 m, but concurrent with faulting, the basins within the rift filled with huge quantities of sediments and volcanic rocks. Some of the volcanic features, such as Valles Caldera, which measures 19 by 24 km, and the Bandelier Tuff, are prominent features in New Mexico (Figure 16.17b). Rifting continues, but very slowly—only 2 mm or less per year. So even though ongoing rifting may eventually split the area so that it resembles the Red Sea, it will be in the far distant future.

Pacific Coast

Before the Eocene, the entire Pacific Coast was a convergent plate boundary where the **Farallon plate** was consumed at a subduction zone that stretched from Mexico to Alaska. Now only two small remnants of the Farallon plate remain—the Juan de Fuca and Cocos plates (►Figure 16.18). Continuing subduction of these small plates accounts for the present seismic activity and volcanism in the Pacific Northwest and Central America, respectively.

Another consequence of these plate interactions was the westward movement of the North American plate and its collision with the **Pacific–Farallon Ridge**. Because the Pacific–Farallon Ridge was at an angle to the margin of North America, the continent–ridge collision took place first during the Eocene in northern Canada and only later during the Oligocene in southern California (Figure 16.18). In southern California, two triple junctions formed, one at the intersection of the North American, Juan de Fuca, and Pacific plates, the other at the intersection of the North American, Cocos, and Pacific plates. Continued westward movement of the North American plate over the Pacific plate caused the triple junctions to migrate, one to the north and the other to the south, giving rise to the **San Andreas transform fault** (Figure 16.18). A similar occurrence along Canada's west coast produced the *Queen Charlotte transform fault*, which continues into Alaska as the *Fairweather Fault*.

Seismic activity on the San Andreas Fault results from continuing movements of the Pacific and North American plates along this complex zone of shattered rocks. Indeed,

▶ Figure 16.16 **The Colorado Plateau** Cenozoic uplift has elevated the plateau to its present elevation, and erosion has exposed rocks from Proterozoic to Cenozoic age.

(a) Permian-age sedimentary rocks exposed in Monument Valley Navajo Tribal Park in Arizona.

(b) These sedimentary rocks at Bryce Canyon National Park in Utah date from the Paleocene.

(c) The Henry Mountains in Utah have a core of igneous intrusions of Oligocene age. In fact, the mountain on the left is an eroded laccolith.

▶ Figure 16.17 **The Rio Grande Rift**

(a) Location of the basins making up the Rio Grande rift. A complex of normal faults is present on both sides of the rift.

(b) The Bandelier Tuff in Bandelier National Monument, New Mexico, erupted in the Jemez volcanic field 1.14 million years ago.

where the fault cuts through coastal California it is actually a zone as much as 2 km wide, and it has numerous branches. Movements on such complex fault systems subject blocks of rocks adjacent to and within the fault zone to extensional and compressive stresses forming basins and elevated areas, the latter supplying sediments to the former. Many of the fault-bounded basins in the southern California area have subsided below sea level and soon filled with turbidites and other deposits. A number of these basins are areas of prolific oil and gas production.

© 2013 Cengage Learning

(a) Three stages in the westward movement of North America and its collision with the Pacific–Farallon Ridge. As North America overrode the ridge, its margin became bounded by transform faults except in the Pacific Northwest.

USGS

(b) Aerial view of the San Andreas Fault today. On land, we call it a right-lateral strike-slip fault. The creek has been offset nearly 100 m.

The Continental Interior

Much of central North America is a vast area called the **continental interior**, which in turn is made up of the *Great Plains* and the *Central Lowlands*. During the Cretaceous, the Great Plains were covered by the **Zuni epeiric sea**, but by Early Paleogene time, this sea had largely withdrawn except for a sizable remnant in North Dakota. Sediments eroded from the Laramide highlands were transported to this sea and deposited in transitional and marine environments. Following this brief episode of marine deposition, all other sedimentation in the Great Plains took place in

▶ **Figure 16.19** **Cenozoic Sedimentary and Igneous Rocks in the Great Plains**

(a) The Oligocene Brule Formation in Badlands National Park in South Dakota was deposited mostly in stream channels and on their floodplains.

(b) Sierra Grande is a shield volcano in northeastern New Mexico. It stands 600 m high and is 15 km in diameter. It was active from 4.0 to 2.6 million years ago.

continental environments, especially fluvial systems (see Perspective). These formed eastward-thinning wedges of sediment that now underlie the entire region.

The only local sediment source within the Great Plains was the Black Hills in South Dakota. The Great Plains have a history of marine deposition during the Cretaceous followed by the origin of terrestrial deposits derived from the Black Hills that are now well exposed in Badlands National Park, South Dakota (▶Figure 16.19a). Judging from the sedimentary rocks and their numerous fossil mammals and other animals, the area was initially covered by semitropical forest, but grasslands replaced the forests as the climate became more arid (see Chapter 18).

Igneous activity was not widespread in the continental interior but was significant in some parts of the Great Plains. For instance, igneous activity in northeastern New Mexico was responsible for volcanoes and numerous lava flows (▶Figure 16.19b) and several small plutons were emplaced in Colorado, Wyoming, Montana, South Dakota, and New Mexico. Indeed, one of the most widely recognized igneous bodies in the entire continent, Devil's Tower in northeastern Wyoming, is probably an Eocene volcanic neck, although some geologists think it is an eroded laccolith (see the chapter opening photo).

Our discussion thus far has focused on the Great Plains, but what about the Central Lowlands to the east? Pleistocene glacial deposits are present in the northern part of this region, as well as in the northern Great Plains (see Chapter 17), but during most of the Cenozoic Era nearly all the Central Lowlands was an area of active erosion rather than deposition. Of course, the eroded materials had to be deposited somewhere, and that was on the Gulf Coastal Plain (Figure 16.9).

Cenozoic History of the Appalachian Mountains

Deformation and mountain building in the area of the present Appalachian Mountains began during the Neoproterozoic with the Grenville orogeny (see Figure 9.2c). The area was deformed again during the Taconic and Acadian orogenies, and during the Late Paleozoic closure of the Iapetus Ocean, which resulted in the Hercynian (Variscan)–Alleghenian orogeny (see Chapters 10 and 11). Then, during Late Triassic time, the entire region experienced block faulting as Pangaea fragmented (see Figure 14.7). By the end of the Mesozoic, though, erosion had reduced the mountains to a plain across which streams flowed eastward to the ocean.

The present distinctive aspect of the Appalachian Mountains developed as a result of Cenozoic uplift and erosion (Figure 16.20). As uplift proceeded, upturned resistant rocks formed northeast–southwest trending ridges, with intervening valleys eroding into less resistant rocks. The preexisting streams eroded downward while uplift took place, were superposed on resistant rocks, and cut large canyons across the ridges, forming *water gaps* (▶Figure 16.21), deep passes through which streams flow, and *wind gaps*, which are water gaps no longer containing streams.

Erosion surfaces at different elevations in the Appalachians are a source of continuing debate among geologists. Some are convinced these more or less planar surfaces show evidence of uplift followed by extensive erosion and then renewed uplift and another cycle of erosion. Others think that each surface represents differential response to weathering and erosion. According to this view, a low-elevation erosion surface developed on softer strata that eroded more or less uniformly, whereas higher surfaces represent weathering and erosion of more resistant rocks.

Geology Along the Oregon Trail in Nebraska

The Oregon Trial flourished as a 3,200 km avenue for migration from Independence, Missouri, to Oregon, a journey that would take four to six months. When it was first established in the early 1830s, the trail crossed what would become Kansas, Nebraska, Wyoming, Idaho, and Oregon, and branches from the trail led to California and Utah. As part of the historical heritage of the United States, the trail is interesting in its own right; some estimate that 400,000 people migrated west along the trail from 1841 until 1869, when the Transcontinental Railroad finally linked east and west.

You can see many interesting geologic features along the Oregon Trail, but here we concentrate on a segment in Nebraska, particularly in the western part of the state. We should also note, though, that the Oregon Trail followed the Platte and North Platte rivers across Nebraska and into Wyoming. In Wyoming, the river is swift flowing, but when it reaches Nebraska it is wide, shallow, and muddy, or, as it is more popularly described, "too thin to plow, too thick to drink."

Following the Laramide orogeny, which ended during the Eocene, the area we now call Nebraska and adjacent areas in the Great Plains (Figure 16.9) was a site of detrital deposition in stream channels and their floodplains as well as lakes and swamps. Some of the sediment from the mountains to the west was carried across the Great Plains and eventually deposited on the Gulf Coastal Plain or in the Gulf of Mexico. Our interest here, however, is the Paleogene and Neogene sedimentary rocks at Chimney Rock National Historic Site and Scotts Bluff National Monument, Nebraska.

Chimney Rock is one of several landmarks along the Oregon Trail where it crossed the Great Plains, or what the immigrants called the Great American Desert. Indeed, Chimney Rock, an 87-m-high erosional remnant with a spire about 36 m high was, according to one traveler in 1832, visible from about 48 km (Figure 1). It is made up of rocks of the Eocene- to Oligocene-age Arikaree Group, which is mostly siltstone but also contains layers of sandstone, claystone, volcanic ash, and some limestone. In fact, hard layers of sandstone in Chimney Rock's spire protect the softer rocks below. During the time that wagon trains followed the Oregon Trail, many people climbed the lower slopes of Chimney Rock to carve their names on the spire.

After passing Chimney Rock, wagon trains heading west passed Scotts Bluff, another erosional remnant of Great Plains rocks that were once much more widespread (Figure 2). In fact, you can still see ruts made by wagons as they crossed Mitchell Pass, and the Visitors' Center at Scotts Bluff National Monument has displays of historical significance as well as literature on the monument and the Oregon Trail.

Scotts Bluff is an imposing edifice rising more than 150 m above the surrounding plain. It is composed of Oligocene- to Miocene-age sedimentary rocks, mostly siltstone and sandstone, as well as volcanic ash layers. Indeed, two prominent layers of volcanic ash are visible that must have come from volcanoes that were active far to the west. One formation at Scotts Bluff and in adjacent areas, the 30- to 40-million-year-old Brule Clay, has yielded many fossil mammals, including the now extinct piglike oreodonts and the rhinoceros-sized titanotheres, as well as three-toed horses, camels, dogs, cats, and rodents (see Chapter 18).

Figure 1 Chimney Rock is a prominent landmark along the Oregon Trail in western Nebraska. This erosional remnant rises 87 m above the surrounding plain. It is made up of several types of Eocene to Oligocene sedimentary rocks.

Caitlin Mirra/Shutterstock.com

Figure 2 Scotts Bluff in Nebraska is made up mostly of Oligocene- to Miocene-age siltstone and sandstone, but it also has two prominent layers of volcanic ash. It is about 150 m high, and, like Chimney Rock, it was a landmark along the Oregon Trail.

John Brueske/Shutterstock.com

▶ **Figure 16.20 Landsat Image of the Appalachian Mountains** This view of the Appalachians shows the central part of Pennsylvania. Notice the long ridges with intervening valleys.

NASA

North America's Southern and Eastern Continental Margins

In a previous section, we mentioned that much of the Central Lowlands eroded during the Cenozoic. Even in the Great Plains, where vast deposits of Cenozoic rocks are present, sediment was carried across the region and into the drainage systems that emptied into the Gulf of Mexico. Likewise, sediment that eroded from the western margin of the Appalachian Mountains ended up in the Gulf, but

▶ **Figure 16.21 Evolution of the Present Topography of the Appalachian Mountains** Although these mountains have a long history, their present topographic expression resulted mainly from Cenozoic uplift and erosion.

Cretaceous

Late Triassic

Recent

Cenozoic

► Figure 16.22 **Cenozoic Deposition on the Gulf Coastal Plain and Atlantic Costal Plain** Shown here are areas of deposition in the Gulf of Mexico and surface geology of the Gulf and Atlantic Coastal plains.

Legend:
- Quaternary
- Neogene
- Paleogene
- Cretaceous
- Detrital shelf sediments
- Carbonate shelf sediments
- Deeper water sediments of continental slope and Sigsbee Deep

© 2013 Cengage Learning

these mountains also shed huge quantities of sediment eastward that was deposited along the Atlantic Coastal Plain. Notice in ▶Figure 16.22 that the **Atlantic Coastal Plain** and the **Gulf Coastal Plain** form a continuous belt extending from the northeastern United States to Texas. Both areas have horizontal or gently seaward-dipping strata deposited mostly by streams. Seaward of the coastal plains lie the continental shelf, slope, and rise, also areas of notable Mesozoic and Cenozoic deposition.

Gulf Coastal Plain

After the withdrawal of the Cretaceous to Early Paleogene Zuni Sea, the Cenozoic **Tejas epeiric sea** made a brief appearance on the continent. But even at its maximum extent it was largely restricted to the Atlantic and Gulf Coastal plains and parts of coastal California. It did, however, extend up the Mississippi River Valley, where it reached as far north as southern Illinois.

The overall Gulf Coast sedimentation pattern was established during the Jurassic and persisted throughout the Cenozoic. Sediments derived from the Cordillera, western Appalachians, and the Central Lowlands were transported toward the Gulf of Mexico, where they were deposited in continental, transitional, and marine environments. In general, the sediments form seaward-thickening wedges grading from terrestrial facies in the north to marine facies in the south (Figure 16.22).

Sedimentary facies development was controlled mostly by regression of the Tejas epeiric sea. After its maximum extent onto the continent during the Paleogene, this sea began its long withdrawal toward the Gulf of Mexico. Its regression, however, was periodically reversed by minor transgressions—eight transgressive–regressive episodes are recorded in Gulf Coastal Plain sedimentary rocks, accounting for the intertonguing among the various facies.

Many sedimentary rocks in the Gulf Coastal Plain are either source rocks or reservoirs for hydrocarbons, a topic we discuss more fully in the section on Paleogene and Neogene mineral resources.

Most of the Gulf Coastal Plain was dominated by detrital deposition, but in the Florida section of the region and the Gulf Coast of Mexico significant carbonate deposition took place. Florida was a carbonate platform during the Cretaceous and continued as an area of carbonate

▸ Figure 16.23 **Deposits on the Eastern Continental Margin and the Atlantic Coastal Plain**

WEST

New Jersey   Shore   Shelf break

Coastal plain

Continental shelf

Continental slope and rise

Atlantic Ocean

EAST

Abyssal plain

Cenozoic

Cretaceous

reefs

Triassic basin

Jurassic

3,050 meters

80 km

salt

CONTINENTAL CRUST
(Precambrian and Paleozoic sedimentary and metamorphic rocks and intrusive igneous rocks)

OCEANIC CRUST
(Early Mesozoic mafic rocks & serpentinite)

(a) The coastal plain and continental margin of New Jersey are covered mostly by Cenozoic sandstones and shales. Beneath these rocks lie Cretaceous and probably Jurassic sedimentary rocks.

© 2013 Cengage Learning

Dr. James P. Reger/Maryland Geological Survey, Baltimore

(b) These Miocene- and Pliocene-age sedimentary rocks at Rocky Point in the Calvert Cliffs of Maryland were deposited in marginal marine environments. In addition to fossils of marine microorganisms, the rocks also contain fossil invertebrates, sharks, and marine mammals.

deposition into the Early Paleogene; carbonate deposition continues even now in Florida Bay and the Florida Keys. Southeast of Florida, across the 85-km-wide Florida Strait, lies the Great Bahama Bank, an area of carbonate deposition from the Cretaceous to the present.

Atlantic Continental Margin

The eastern coast of North America includes the Atlantic Coastal Plain and extends seaward across the continental shelf, slope, and rise (Figure 16.22 and ▸Figure 16.23a). When Pangaea began fragmenting during the Triassic, continental crust rifted, and a new ocean basin began to form. Remember that the North American plate moved westerly, so its eastern margin was within the plate, where a passive continental margin developed.

The Atlantic continental margin has a number of Mesozoic and Cenozoic basins, formed as a result of rifting, in which sedimentation began by Jurassic time. Even though Jurassic-age rocks have been detected in only a few deep wells, geologists assume they underlie the entire continental margin. The distribution of Cretaceous and Cenozoic rocks is better known, because both are exposed on the Atlantic Coastal Plain, and both have been penetrated by wells on the continental shelf.

Sedimentary rocks on the broad Atlantic Coastal Plain, as well as those underlying the continental shelf, slope, and rise, were derived from the Appalachian Mountains. Numerous rivers and streams transported sediments toward the east, where they were deposited in seaward-thickening wedges (up to 14 km thick) that grade from terrestrial deposits on the west to marine deposits farther east. For instance, the Calvert Cliffs in Maryland consist of rocks deposited in marginal marine environments (Figure 16.23b).

An interesting note regarding the geologic evolution of the Atlantic Coastal Plain is the evidence indicating that a 3- to 5-km-diameter comet or asteroid impact occurred in the present-day area of Chesapeake Bay. This event took place about 35 million years ago, during the Late Eocene, and left an impact crater measuring 85 km in diameter and 1.3 km deep. Now buried beneath 300 to 500 m of younger sedimentary rocks, it has been detected by drilling and geophysical surveys.

Paleogene and Neogene Mineral Resources

The Eocene Green River Formation of Wyoming, Utah, and Colorado, well known for fossils, also contains huge quantities of oil shale and evaporites of economic interest.

Oil shale consists of clay particles, carbonate minerals, and an organic compound called *kerogen* from which liquid oil and combustible gases can be extracted. No oil is currently being extracted from these rocks, but according to one estimate, 80 billion barrels of oil could be recovered with present technology. The evaporite mineral *trona* is mined from Green River rocks for sodium compounds.

Mining of phosphorus-rich sedimentary rocks in Central Florida accounts for more than half of that state's mineral production. The phosphorus from these rocks has a variety of uses in metallurgy, preserved foods, ceramics, matches, fertilizers, and animal feed supplements. Some of these phosphate rocks also contain interesting assemblages of fossil mammals (see Chapter 18).

Diatomite is a soft, low-density sedimentary rock made up of microscopic shells of diatoms, single-celled marine and freshwater plants with skeletons of silicon dioxide (SiO_2) (see Figure 15.4b). In fact, diatomite is so porous and light that when dry it will float. Diatomite is used mostly to purify gas and to filter liquids such as molasses, fruit juices, and sewage. The United States leads the world in diatomite production, mostly from Cenozoic deposits in California, Oregon, and Washington.

Historically, most coal mined in the United States (Canada has very little coal) has been Pennsylvanian-age bituminous coal from mines in Pennsylvania, West Virginia, Kentucky, and Ohio (see Chapter 11). Now, though, huge deposits of lignite and subbituminous coal in the Northern Great Plains are becoming important resources. These Late Cretaceous to Early Paleogene-age coal deposits are most abundant in the Williston and Powder River basins of North Dakota, Montana, and Wyoming. Besides having a low sulfur content, which makes them desirable, some of these coal beds are more than 30 m thick!

Gold from the Pacific Coast states, particularly California, comes largely from stream gravels in which placer deposits are found. A *placer* is an accumulation resulting from the separation and concentration of minerals of greater density from those of lesser density in streams or on beaches. The gold in these placers was weathered and eroded from Mesozoic-age quartz veins in the Sierra Nevada batholith and adjacent rocks (see Chapter 14 Introduction).

Hydrocarbons are recovered from the Cenozoic fault-bounded basins in Southern California and from many rocks of the Gulf Coastal Plain. Many rocks in the latter region form reservoirs for petroleum and natural gas because of different physical properties of the strata and are thus called *stratigraphic traps*. Hydrocarbons are also found in geologic structures, such as folds, particularly those adjacent to salt domes, and such reservoirs are accordingly called *structural traps*. Because rock salt is a low-density sedimentary rock, when deeply buried and under pressure it rises toward the surface, and in doing so it penetrates and deforms the overlying rocks.

Another potential resource is methane hydrate, which consists of single methane molecules bound up in networks formed by frozen water. Huge deposits of methane hydrate are present along the eastern continental margin of North America, but so far it is not known whether they can be effectively recovered and used as an energy source. According to one estimate, the amount of carbon in methane hydrates worldwide is double that in all coal, oil, and conventional natural gas reserves.

Summary

▶ The Late Triassic rifting of Pangaea continued through the Cenozoic and accounts for the present distribution of continents and oceans.

▶ Cenozoic orogenic activity was concentrated in two major belts: the Alpine–Himalayan orogenic belt and the circum-Pacific orogenic belt. Each belt is composed of smaller units called orogens.

▶ The Alpine orogeny resulted from convergence of the African and Eurasian plates. Mountain building took place in southern Europe, the Middle East, and North Africa. Plate motions also caused the closure of the Mediterranean basin, which became a site of evaporite deposition.

▶ India separated from Gondwana, moved north, and eventually collided with Asia, causing deformation and uplift of the Himalayas.

▶ Orogens characterized by subduction of oceanic lithosphere and volcanism took place in the western and northern Pacific Ocean basin. Back-arc spreading produced back-arc basins such as the Sea of Japan.

▶ Subduction of oceanic lithosphere occurred along the western margins of the Americas during much of the Cenozoic.

▶ Subduction continues beneath Central and South America, but the North American plate is now bounded mostly by transform faults, except in the Pacific Northwest.

▶ The North American Cordillera is a complex mountainous region extending from Alaska into Mexico. Its Cenozoic evolution included deformation during the Laramide orogeny, extensional tectonics that formed the Basin and Range structures, intrusive and extrusive igneous activity, and uplift and erosion.

▶ Shallow-angle subduction of the Farallon plate beneath North America resulted in the vertical uplifts of the Laramide orogeny. The Laramide orogeny is centered in the middle and southern Rockies, but deformation occurred from Alaska to Mexico.

▶ Cordilleran volcanism was more or less continuous in the Cordillera through the Cenozoic. The Columbia River basalts represent one of the world's greatest eruptive events. Volcanism continues in the Cascade Range of the Pacific Northwest.

▶ Crustal extension in the Basin and Range Province yielded north–south oriented, normal faults. Differential movement on these faults produced uplifted ranges separated by broad, sediment-filled basins.

▶ The Colorado Plateau was deformed less than other areas in the Cordillera. Late Neogene uplift and erosion were responsible for the present topography of the region.

▶ The Rio Grande rift formed as north–south oriented rifting took place in an area extending from Colorado into Mexico. The basins within this rift filled with sediments and volcanic rocks.

▶ The westward drift of North America resulted in its collision with the Pacific–Farallon Ridge. Subduction ceased, and the continental margin became bounded by major transform faults, except where the Juan de Fuca plate continues to collide with North America.

▶ Sediments eroded from Laramide uplifts were deposited in intermontane basins, on the Great Plains, and in a remnant of the Cretaceous epeiric sea in North Dakota.

▶ Deposition on the Gulf Coastal Plain and Atlantic Coastal Plain took place throughout the Cenozoic, resulting in seaward-thickening wedges of rocks grading from terrestrial facies to marine facies.

▶ Cenozoic uplift and erosion were responsible for the present topography of the Appalachian Mountains. Much of the sediment eroded from the Appalachians was deposited on the Atlantic Coastal Plain.

▶ Paleogene and Neogene mineral resources include oil and natural gas, gold, and phosphorus-rich sedimentary rocks.

Important Terms

Alpine–Himalayan orogenic belt, p. 324
Alpine orogeny, p. 324
Atlantic Coastal Plain, p. 339
back-arc basin, p. 326
Basin and Range Province, p. 330
Cascade Range, p. 330
circum-Pacific orogenic belt, p. 326

Colorado Plateau, p. 332
continental interior, p. 335
Farallon plate, p. 333
Gulf Coastal Plain, p. 339
Himalayan orogeny, p. 326
Laramide orogeny, p. 327
North American Cordillera, p. 326

orogen, p. 324
Pacific–Farallon Ridge, p. 333
Rio Grande rift, p. 332
San Andreas transform fault, p. 333
Tejas epeiric sea, p. 339
Zuni epeiric sea, p. 335

Review Questions

1. A large area of the North American Cordillera that has been uplifted but little deformed is the
a. _____ Farallon plate; b. _____ Cascade Range; c. ✓ Colorado Plateau; d. _____ Zuni epeiric sea; e. _____ Atlantic Coastal Plain.

2. The Basin and Range Province in the United States is
a. ✓ a huge area of block faulting; b. _____ mainly in Kansas and Nebraska; c. _____ made up mostly of volcanic mountains; d. _____ characterized by compression and crustal thickening; e. _____ bordered on the east and west by the Appalachians and the Great Plains, respectively.

3. The San Andreas Fault formed when North America moved westward and collided with the
a. _____ Rio Grande rift; b. _____ Gulf Coastal Plain; c. _____ continental interior; d. ✓ Pacific–Farallon Ridge; e. _____ Alpine orogeny.

4. The rocks in the Snake River Plain are mostly
a. _____ limestone; b. _____ siltstone; c. _____ metamorphic; d. ✓ igneous; e. _____ sedimentary.

5. The Himalayas formed when the _____ plate collided with the _____ plate.
a. _____ Farallon/Pacific; b. _____ Nazca/Cacos; c. _____ African/European; d. ✓ Indian/Asian; e. _____ Australian/South American.

6. Japan is separated from mainland Asia by a
a. _____ back-arc basin; b. _____ strike-slip fault; c. _____ orogen; d. ✓ plateau; e. _____ oceanic ridge system.

7. The Cenozoic Era consists of three periods: the _____, the _____, and the _____.
a. _____ Cretaceous, Eocene, and Pleistocene; b. ✓ Paleogene, Neogene, and Quaternary; c. _____ Triassic, Jurassic, and Cretaceous; d. _____ Proterozoic, Paleocene, and Holocene; e. _____ Nevadan, Sevier, and Laramide.

8. The evidence indicates that the Laramide orogeny ended during _____ time.
a. ✓ Eocene; b. _____ Mesozoic; c. _____ Quaternary; d. _____ Pliocene; e. _____ Permian.

9. Most the sediment on the Atlantic Coastal Plain was eroded from the
a. _____ Rocky Mountains; b. _____ Cascade Range; c. _____ Ozark Plateau; d. _____ Farallon Ridge; e. ✓ Appalachian Mountains.

10. Ongoing volcanism takes place in the
a. _____ Basin and Range Province; b. _____ Gulf Coastal Plain; c. ✓ Cascade Range; d. _____ Himalayan orogeny; e. _____ Colorado Plateau.

11. Twenty years from now you may not remember many of the terms and names of events you learned in this course

unless you become a professional geologist. So how would you explain the evolution of the North American Cordillera in terms that a nongeologist could readily understand?

12. How did the Laramide orogeny differ from other orogenies at convergent plate boundaries?

13. Explain how the San Andreas transform fault formed.

14. Where is the Basin and Range Province, and what events took place there during the Cenozoic Era?

15. What kinds of sedimentary rocks are found on the Great Plains, and what was the source of sediment?

16. A curious high school student wants to know why western North America has many mountain ranges, active volcanoes, and earthquakes, whereas these features or activities are absent or not nearly as common in the eastern part of the continent. How would you explain this disparity, and further, can you think of how this situation might be reversed?

17. Where is the Cascade Range, what kinds of volcanoes are found there, and what accounts for ongoing volcanism there?

18. When and what sequence of events led to the origin of the Himalayas in Asia?

19. What features are shared by the Gulf Coastal Plain and the Atlantic Coastal Plain? Are there any differences between these two areas?

20. The United States uses about 1 billion metric tons of coal each year from a reserve of about 250 billion metric tons. Assuming all this coal could be mined, how long would it last at the current rate of consumption? Is there any reason to think that the current rate of consumption will remain the same, and is it even likely that all of this coal could be mined?

17

Cenozoic Earth History
The Quaternary Period

Glacier Peak, Washington, is in the northern part of the Cascade Range, which stretches from northern California through Oregon and Washington and into southern British Columbia, Canada. It is an imposing composite volcano that last erupted about 300 years ago. Five explosive eruptions have taken place during the last 15,000 years. The volcano is 3,213 m high and supports several glaciers, all of which are now retreating; one glacier on the north side of the mountain disappeared during the 1990s.

Photo by Jim Vallance/USGS

OUTLINE

At the end of this chapter, you will have learned that

- The Pleistocene and the Holocene epochs encompass only the most recent 2.6 million years of Earth history.

- The Pleistocene Epoch, lasting from 2.6 million years to 11,700 years ago, is best known for widespread glaciers but was also a time of continuing orogeny and volcanism.

- Much of our information about Pleistocene climates comes from oxygen isotope ratios, pollen analyses, and the distribution and coiling directions of planktonic foraminifera.

- Pleistocene continental glaciers were present on the Northern Hemisphere continents as well as Antarctica, and thousands of small glaciers were in mountain valleys on all continents.

- Sea level fell and rose during the several Pleistocene advances and retreats of glaciers, depending on how much water from the ocean was frozen on land.

- The tremendous weight of continental glaciers caused Earth's crust to subside into the mantle and to rise again when the glaciers wasted away.

- Many now arid regions far from glaciated areas supported large lakes as a result of greater precipitation and lower evaporation rates during the Pleistocene, and numerous other lakes formed along the margins of glaciers.

- A current widely accepted theory explaining the onset of ice ages relies on irregularities in Earth's rotation and orbit.

- Important Pleistocene mineral resources include sand and gravel, diatomite, peat, and placer deposits of gold.

Introduction

The *Quaternary Period*, consisting of the *Pleistocene* and *Holocene epochs*, is the designation for the most recent 2.6 million years of geologic time (see Figures 1.11 and 16.1). Obviously, the Pleistocene Epoch (2.6 million years ago to 11,700 years ago) makes up most of the Quaternary Period, so our discussion in this chapter focuses on that interval of time. We have mentioned that the Pleistocene was unusual in that it was one of the few times in Earth history when glaciers were widespread. In keeping with that fact, the Pleistocene is commonly called the *Ice Age*. At 2.59 million years long, the Pleistocene is brief when considering Earth's age, but it was nevertheless far longer than we can visualize.

In hindsight, it is difficult to understand why many scientists of the 1830s refused to accept the evidence indicating that widespread glaciers were present on the Northern Hemisphere continents during the recent geologic past. Many critics of this idea invoked the biblical deluge to explain the large boulders throughout Europe far from their sources, whereas others thought that the boulders were rafted to their present positions during vast floods. However, by 1837 the Swiss naturalist Louis Agassiz argued convincingly that these boulders, as well as polished and striated bedrock and U-shaped valleys, resulted from huge masses of ice that moved over the land.

Geologists define a **glacier** as a moving body of ice on land that moves by *plastic flow* (internal deformation of ice in response to pressure) and by *basal slip* (sliding over its underlying surface). **Valley glaciers** are long, narrow tongues of ice confined to mountain valleys, where they flow from higher to lower elevations (▶Figure 17.1a), so both plastic flow and basal slip are important.

Continental glaciers (also called *ice sheets*) (Figure 17.1b), cover at least 50,000 km² and are unconfined by topography, meaning that they flow outward from a center or centers of accumulation. They may also slide over their underlying surface, but plastic flow is much more important than basal slip. An **ice cap** is similar to a continental glacier, but it covers less than 50,000 km² (Figure 17.1c).

We now know that the Pleistocene Epoch was a time of several major episodes of glaciation, each separated by warmer, interglacial intervals. Also, during those times of glacial expansion, more precipitation fell on now arid regions such as the Sahara Desert of North Africa and Death Valley in California, which then supported streams, lakes, and lush vegetation. Indeed, cultures existed in what is now the Sahara Desert as recently as 4,500 years ago. An unresolved question is whether the Ice Age is truly over or if we are simply in an interglacial time that will be followed by renewed glaciation.

Our focus in this chapter is on glaciers because they had such a profound effect on the continents, especially at high latitudes and in high mountains. But even at their maximum extent, glaciers covered no more than 30 percent of Earth's land surface. Of course, the climatic conditions leading to glaciation had worldwide effects, but the processes of tectonics, volcanism, and erosion and deposition continued unabated. From the systems approach that we introduced in Chapter 1, we see that glaciers form a reservoir in the hydrosphere, although some scientists prefer the term *cryosphere* for all of Earth's frozen water, which includes glaciers, sea ice, snow, and even *permafrost* (permanently frozen ground).

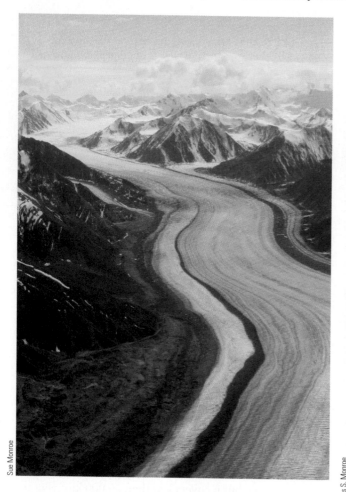

(a) Part of a glacier in Yukon Territory, Canada. As is typical of valley glaciers, it is a narrow tongue of ice confined to a mountain valley. The dark material on the glacier's surface is sediment.

(c) The Penny Ice Cap on Baffin Island, Canada, covers about 6,000 km².

(b) The West and East Antarctic ice sheets merge to form a nearly continuous ice cover that averages 2,160 m thick. The blue lines are lines of equal thickness.

Pleistocene and Holocene Tectonism and Volcanism

The Pleistocene is best known for vast glaciers and their effects, but it was also a time of tectonism and volcanism, processes that continued through the Holocene to the present. Indeed, today plates diverge and converge, and, in places, slide past one another at transform plate boundaries. As a consequence, orogenic activity is ongoing, as is seismic activity and volcanic eruptions.

Tectonism

In Chapter 16, we discussed the continent–continent collision between India and Asia and the convergence of the Pacific plate with South America that formed the Andes. These areas of orogenic activity continue unabated, as

▶ Figure 17.2 **Pleistocene Uplift and Tectonism**

(a) These deformed sedimentary rocks are only a few hundred meters from the San Andreas Fault in southern California.

Marine terraces

(b) Several marine terraces on San Clemente Island, California. Each terrace was at sea level when it formed. The highest is 400 m above sea level.

do those in the Aleutian Islands, Japan, the Philippines, and elsewhere (see Figure 16.4). Interactions between the North America and Pacific plates along the San Andreas transform plate boundary produced folding (▶Figure 17.2a), faulting, and a number of basins and uplifted areas. Marine terraces covered with Pleistocene sediments attest to periodic uplift all along the Pacific Coast of the United States (Figure 17.2b).

Volcanism

Ongoing subduction of remnants of the Farallon plate beneath Central America and the Pacific Northwest accounts for volcanism in these areas. The Cascade Range of California, Oregon, Washington, and British Columbia has a history dating back to the Oligocene, but the large composite volcanoes and Lassen Peak, a large lava dome, formed mostly during the Pleistocene and Holocene (see chapter opening photo and ▶Figure 17.3). Indeed, Lassen Peak in California and Mount St. Helens in Washington State erupted during the 1900s, and in the case of the latter, it showed renewed activity from 2004 to 2005.

Volcanism also took place in several other areas in the western United States, including Arizona, Idaho, and California, and many of the volcanoes in Alaska have erupted in the recent past. Following colossal eruptions, huge calderas formed in the area of Yellowstone National Park, Wyoming. Vast eruptions took place 2.0 and 1.3 million years ago and again 600,000 years ago that left a composite caldera measuring 76 by 45 km. Since its huge eruption, part of the area has risen, presumably from magma below the surface, forming a *resurgent dome*. And finally, between 150,000 and 75,000 years ago, the Yellowstone Tuff erupted and partially filled the caldera.

Elsewhere in the world, volcanoes erupted in South America, the Philippines, Japan, the East Indies, as well

as Iceland, Spitzbergen, and the Azores. So, even though the amount of heat generated within Earth has decreased through time (see Chapter 8), volcanism and the other processes driven by internal heat remain significant.

Pleistocene Stratigraphy

As we noted in the Introduction, the Quaternary Period is made up of two epochs, the Pleistocene and the Holocene. This usage is based on the most recent work by the International Commission on Stratigraphy, which has working committees that continually review and revise the geologic time scale as new information becomes available. The Pleistocene–Holocene boundary at 11,700 years corresponds to warmer conditions concurrent with the melting of the most recent ice sheets. Changes in vegetation, as well as oxygen isotope ratios determined from the shells of marine organisms, provide ample evidence for this climatic change.

Terrestrial Stratigraphy

Not long after Louis Agassiz convinced geologists that widespread glaciation had taken place, research focused on deciphering the history of the Ice Age. This work involved recognizing and mapping terrestrial glacial features and placing them in a stratigraphic sequence. And from these features, including moraines, erratic boulders, and glacial striations, geologists have determined that at their greatest extent, Pleistocene glaciers were up to 3 km thick and covered about three times as much of Earth's surface as they do now, or about 45 million km² (▶Figure 17.4).

Geologists have mapped the distribution of glacial deposits and determined that North America had several

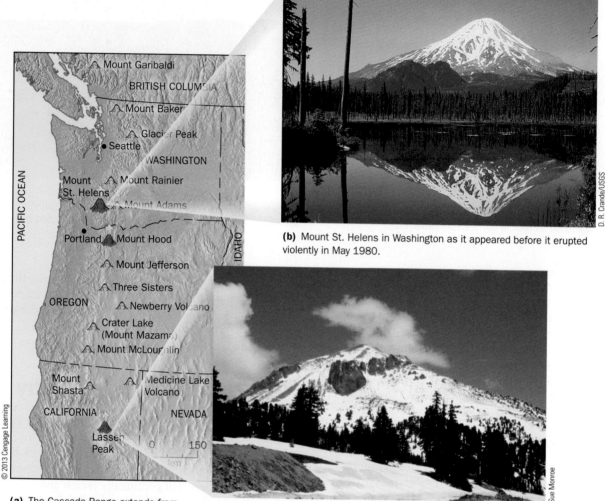

▶ **Figure 17.3** **Pleistocene and Holocene Volcanism in the Cascade Range** Volcanism has occurred in this area since the Oligocene, but the large volcanoes now present developed mostly during the Pleistocene and Holocene.

(b) Mount St. Helens in Washington as it appeared before it erupted violently in May 1980.

(a) The Cascade Range extends from northern California into southern British Columbia, Canada. This map shows the locations of the major volcanoes in the range. All are composite volcanoes with the exceptions of Lassen Peak (lava dome) and two shield volcanoes, Medicine Lake Volcano and Newberry Volcano.

(c) Lassen Peak is a lava dome that formed about 27,000 years ago on the northeast flank of a deeply eroded composite volcano. Lassen Peak erupted numerous times from 1914 to 1917.

glacial stages—that is, times of widespread glaciers—each separated by warmer **interglacial stages**. In the past, scientists recognized four major glacial stages in North America, but now we know that the most recent one (the Wisconsinan stage) and its predecessor (the Illinoian stage) were preceded by 11 episodes of glacial advance and retreat, designated *pre-Illinoian A through K* (▶ Figure 17.5). In short, Pleistocene glaciation was much more complex than formerly accepted.

Other Northern Hemisphere continents also had several episodes of glacial advance and retreat, although the terminology for those events differs from that of North America, and at least 20 warm–cold cycles are indicated by deep-sea sediment cores. During the glacial stages, glaciers were widespread, but comparing these events from different areas is difficult because glacial deposits are typically chaotic mixtures of coarse-grained deposits that are difficult to correlate. In addition, glacial advances and retreats commonly obscure or erode sediments left by previous glacial advances. Even within a single glacial stage, minor advances and retreats of the glacial ice take place; several such fluctuations took place during the Wisconsinan Stage in North America during the last 70,000 years.

Deep-Sea Stratigraphy

Until the 1960s, the traditional view of Pleistocene chronology was based on sequences of glacial sediments on

(a) North America.

(b) Europe and part of Asia.

► Figure 17.5
Geologic Time Scale for the Quaternary Period
This 2011 revision of the geologic time scale is the most recent one proposed by the International Commission on Stratigraphy. In the past, four glacial stages with intervening interglacial stages were recognized, but now there are 11 stages that predate the Illinoian.

Dates are from the 2009 International Stratigraphic Chart. © 2009 by the International Commission of Stratigraphy.

land. However, new evidence from ocean sediment samples indicate numerous climatic fluctuations during the Pleistocene. Evidence for these climatic fluctuations comes from changes in surface ocean temperature recorded in the shells of planktonic foraminifera, which after they die sink to the seafloor and accumulate as sediment.

One way to determine past changes in ocean surface temperatures is to resolve whether planktonic foraminifera were warm- or cold-water species. Many are sensitive to variations in temperature and migrate to different latitudes when the surface water temperature changes. For example, the tropical species *Globorotalia menardii* during periods of cooler climate is found only near the equator, whereas during times of warming its range extends into the higher latitudes.

Some planktonic foraminifera species change the direction in which they coil during growth in response to temperature fluctuations. The Pleistocene species *Globorotalia truncatulinoides* coils predominantly to the right in water temperatures above 10°C but coils mostly to the left in water below 8°–10°C. On the basis of changing coiling ratios, geologists have constructed detailed climatic curves for the Pleistocene and earlier epochs.

Changes in the O^{18}-to-O^{16} ratio in the shells of planktonic foraminifera also provide data about climate. The abundance of these two oxygen isotopes in the calcareous ($CaCO_3$) shells of foraminifera is a function of the oxygen isotope ratio in water molecules and water temperature when the shell forms. The ratio of these isotopes reflects the amount of ocean water stored in glacial ice. Seawater has a higher O^{18}-to-O^{16} ratio than glacial ice, because water containing the lighter O^{16} isotope is more easily evaporated than water containing the O^{18} isotope. Therefore, Pleistocene glacial ice was enriched in O^{16} relative to O^{18}, whereas the heavier O^{18} isotope is concentrated in seawater. The declining percentage of O^{16} and consequent rise of O^{18} in seawater during times of glaciation is preserved in the shells of planktonic foraminifera. Consequently, oxygen isotope fluctuations indicate surface water temperature changes and thus climatic changes.

Unfortunately, geologists have not yet been able to correlate these detailed climatic changes with corresponding changes recorded in the sedimentary record on land. The time lag between the onset of cooling and any resulting glacial advance produces discrepancies between the marine and terrestrial records. Thus, it is unlikely that all the minor climatic fluctuations recorded in deep-sea sediments will ever be correlated with continental deposits.

Onset of the Ice Age

Glacial conditions actually set in about 40 million years ago when surface ocean waters at high southern latitudes rapidly cooled and the water in the deep ocean became much colder than it was previously. The gradual closure of the Tethys Sea during the Oligocene limited the flow of warm water to higher latitudes, and by Middle Miocene time an Antarctic ice sheet had formed, accelerating the formation of very cold oceanic waters. After a brief Pliocene warming trend, continental glaciers began forming in the Northern Hemisphere about 2.6 million years ago—the Pleistocene Ice Age was underway.

Climates of the Pleistocene and Holocene

The climatic conditions leading to Pleistocene glaciation were, as you would expect, worldwide. Contrary to popular belief and depictions in cartoons and movies, Earth was not as cold as commonly portrayed. Rather, evidence of various kinds indicates that the world's climate cooled gradually during the Cenozoic, especially from Eocene through Pleistocene time (▶Figure 17.6). Oxygen isotope ratios (O^{18} to O^{16}) from deep-sea cores reveal that during the last 2 million years Earth has had 20 major warm–cold cycles during which the temperature fluctuated by as much as 10°C (see the section on Deep-Sea Stratigraphy). And studies of glacial deposits attest to several major episodes of glaciation in North America and similar events in Europe.

During glacial growth, those areas covered by or near glaciers experienced short, cool summers and long, wet

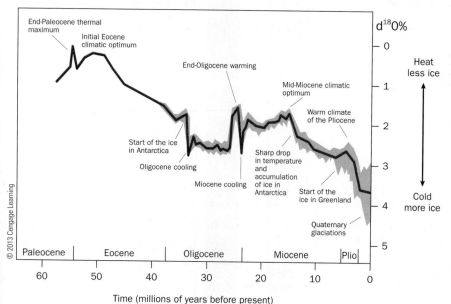

▶ **Figure 17.6 Earth's Cenozoic Climate** Although the climate for the last 65 million years has been complex, the overall trend was cooling that led to Pleistocene glaciation. Notice that glaciers started forming in Antarctica nearly 40 million years ago.

winters. Areas distant from glaciers had varied climates. When glaciers grew and advanced, lower ocean temperatures reduced evaporation rates, so most of the world was drier than now. Some areas now arid were much wetter during the Ice Age. For instance, the expansion of the cold belts at high latitudes compressed the temperate, subtropical, and tropical zones toward the equator. Consequently, the rain that now falls on the Mediterranean then fell farther south, on the Sahara of North Africa, enabling lush forests to grow in what is now desert. In North America, a high-pressure zone over the northern ice sheets deflected storms south, so the arid Southwest was much wetter than today.

Pollen analysis is particularly useful in paleoclimatology (▶Figure 17.7). Pollen grains, produced by the male reproductive bodies of seed plants, have a resistant waxy coating that ensure many will be preserved in the fossil record. Most seed plants disperse pollen by wind, so it settles in streams, lakes, swamps, bogs, and in nearshore marine environments. Once paleontologists recover pollen from sediments, they identify the type of plant it came from, determine the floral composition of the area, and make climatic inferences (Figure 17.7).

Pollen diagrams (Figure 17.7b), tree-ring analysis (see Chapter 4), and studies of the advances and retreats of valley glaciers have yielded a wealth of information about the Northern Hemisphere climate for the last 11,700 years—that is, since the time the last major continental glaciers retreated and disappeared. Data from pollen analysis indicate a continuous trend toward a warmer climate until about 6,000 years ago, although it was warmer only during the summer and in the Northern Hemisphere. Then the climate became cooler and moister, favoring the growth of valley glaciers on the Northern Hemisphere continents. Three episodes of glacial expansion took place during this **neoglaciation**, as it is called. The most recent one, the **Little Ice Age**, a time of cooler temperatures and glacial expansion, occurred between 1500 and the mid- to late-1800s (see Perspective).

Glaciers—How Do They Form?

We defined the terms *continental glacier*, *ice cap*, and *valley glacier* as moving bodies of ice on land (see the Introduction and Figure 17.1). During the Pleistocene, all types of glaciers were much more widespread than now. For example, the only continental glaciers today are the ones in Antarctica and Greenland (▶Figure 17.8), but during the Pleistocene they covered about 30 percent of Earth's land surface, especially on the Northern Hemisphere continents. These continental glaciers formed, advanced, and then retreated several times, forming much of the present topography of the glaciated regions and nearby areas. The Pleistocene was also a time when small valley glaciers were more common in mountain ranges. Indeed, much of the spectacular scenery in such areas as Grand Teton National Park, Wyoming, and Glacier National Park in Montana resulted from erosion by valley glaciers.

The question "How do glaciers form?" is rather easily answered, unlike "What causes the onset of an ice age?" Any area that receives more snow in the cold season than melts in the warm season has a net accumulation over the years. As accumulation takes place, the snow at depth is converted to glacial ice, and when a critical thickness of about 40 m is reached, ice begins to flow in response to pressure. Once a glacier forms, it moves from a *zone of accumulation*, where additions exceed losses, toward its *zone of wastage*, where losses exceed additions. As long as a balance exists between the two, the glacier has a *balanced budget*, but the budget may be *negative* or *positive*, depending on any imbalances that exist in these two zones. Consequently, a glacier's terminus may advance, retreat, or remain stationary, depending on its budget.

▶ Figure 17.7 **Pollen Analyses and Climate**

Courtesy of Vaughn M. Bryant, Jr., Texas A&M University

(a) Scanning electron microscope view of present-day pollen: (1) sunflower, (2) acacia, (3) oak, (4) white mustard, (5) little walnut, (6) agave, and (7) juniper.

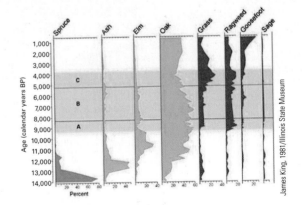

Chatsworth Bog, Illinois

James King, 1981/Illinois State Museum

(b) Pollen diagram for the last 4,000 years of Chatsworth Bog in Illinois. Spruce forest was replaced by ash and elm in a wetter climate. Oak, grasses, and ragweed then increased, indicating prairie development.

The Little Ice Age

Earth has gone through several climatic changes since the end of the Ice Age approximately 11,700 years ago. Around 6,000 years ago, during the Holocene Maximum, summer temperatures in the Northern Hemisphere were slightly warmer than they are now, and some now arid regions, such as the Sahara Desert of North Africa, supported lush vegetation, lakes, and swamps. Then followed a time of cooler temperatures, but from around A.D. 1000 to 1300 was the Medieval Warm Period in Europe, during which wine grapes grew 480 km farther north than they do now. This rather temperate time was followed by another cooling trend, one that led to the Little Ice Age, a time of cooler, wetter summers, colder winters, shorter growing seasons, and the persistence of sea ice at high latitudes for long periods.

During the preceding Medieval Warm Period, the climates of Europe and North America were rather mild, and the North Atlantic Ocean was warmer and more storm free than it is now. Taking advantage of the mild conditions, the Norse, commonly called the Vikings, discovered and settled Iceland, and by A.D. 1200 about 80,000 people were living there. They also sailed to Greenland and North America, where they established two colonies in the former and one in the latter. Unfortunately for these settlers, the climate deteriorated, the North Atlantic became stormier, and sea ice moved farther south and persisted longer each year. The poor sea conditions coupled with political problems in Norway brought a halt to all shipping across the North Atlantic, and the colonies in

Greenland and North America eventually disappeared.

The Little Ice Age was also a time of expansion of small glaciers in mountain valleys (Figure 1). In Europe and Iceland, glaciers moved far down their valleys, reaching their greatest historic extend by the early 1800s. In 1744, one observer in Norway wrote, "When at times [the glacier] pushes forward a great sound is heard, like that of an organ and it pushes in front of it immeasurable soil, grit and rocks bigger than any house could be, which it then crushes small like sand."* A small ice cap formed in Iceland, where none had existed previously, and glaciers in Alaska, Canada, and the mountains of the western United States expanded to their greatest limits during historic time. During the early 1600s, people in the European Alps

Figure 1 Little Ice Age Glaciers in Europe

Reprinted with permission from T. H. Van Andel, *New Views on an Old Planet*, p. 175 (Table 11.1) © 1989 Cambridge University Press.

(a) Samuel Birmann (1793–1847) painted this view in Switzerland, titled *Unterer Grindelwald*, in 1826, at which time the glacier extended onto the valley floor in the foreground.

Sue Monroe

(b) The same glacier today. Its terminus is hidden behind a rock projection at the lower end of its valley.

watched as the Mer de Glace (Sea of Ice Glacier) advanced and overwhelmed farms and villages. Elsewhere glaciers advanced across roadways and pastures, destroyed some villages in Scandinavia, and threatened other villages.

Advancing glaciers caused some problems, but their overall impact on humans was minimal. Far more important from the human perspective was that during much of the Little Ice Age, summers in the Northern Hemisphere were cooler and wetter. Particularly hard hit were the Scandinavian countries and the Norse colonies in Greenland and Iceland, but at times much of Northern Europe was affected as well. The harsher summer weather resulted in shorter growing seasons** reduced crop yields, food shortages, and a number of famines in Iceland, Scotland, and mainland Europe. Iceland's population was down by half, to about 40,000, by 1700. Between 1610 and 1870, sea ice was seen near Iceland for as much as three months a year, and each year the sea ice persisted for long periods. Poor growing seasons and food shortages followed.

Although cool, wet summers caused most of the problems, winter temperatures were colder too, and sea ice was a problem in the North Atlantic. Occasionally Eskimos paddled their kayaks as far south as Scotland, and the canals of Holland froze over during some winters. In 1607 the first "Frost Fair" was held in London on the River Thames, which had begun to freeze over nearly every winter. New York Harbor in the newly founded United States froze over. And 1816 was known as the "year without a summer," when unusually cold temperatures persisted into June and July in New England and northern Europe; both areas experienced frost and even some snow during these months. Eruptions of Tambora in Indonesia in 1815 and Mayon Volcano in the Philippines in 1814 contributed to the cool spring and summer of 1816.

Exactly when the Little Ice Age ended is debatable. Some authorities put the end at 1880, whereas others think it ended as early as 1850. In either case, during the mid to late 1800s, sea ice was retreating northward, glaciers began retreating back from their alleys, and summer weather became more moderate.

*Quoted in C. Officer and J. Page, *Tales of the Earth* (New York: Oxford University Press, 1993), p. 99.

**During the coldest part of the Little Ice Age (1680–1730), the growing season in England was five weeks shorter than it was during the 1900s.

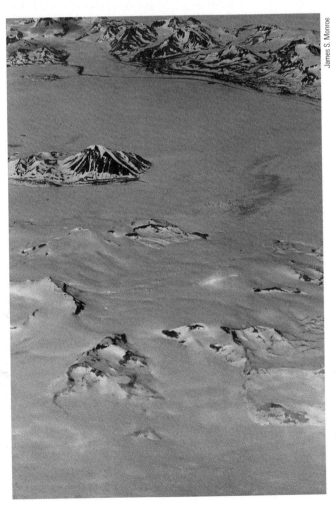

James S. Monroe

▶ Figure 17.8 **The Greenland Ice Sheet** Greenland is mostly covered by a continental glacier that is more than 3,000 m thick. Notice that only a few high mountains are not ice covered. Continental glaciers were much more widespread during the Pleistocene, but today only Greenland and Antarctica have continental glaciers.

Glaciation and Its Effects

Huge glaciers moving over Earth's surface reshaped the previously existing topography and yielded many distinctive glacial landforms. As glaciers formed and wasted away, sea level fell and rose, depending on how much water was frozen on land, and the continental margins were alternately exposed and water-covered. In addition, the climatic changes that initiated glacial growth had effects far beyond the glaciers themselves. Another legacy of the Pleistocene is that areas once covered by thick glaciers are still rising as a result of *isostatic rebound*, discussed later in this chapter.

Glacial Landforms

Continental glaciers have eroded a large part of Canada and parts of some northern states that now have subdued topography, little or no soil, striated and polished bedrock, and poor surface drainage, all features of an **ice-scoured plain**. The erosional power of continental glaciers is also evident from the Finger Lakes of New York, which consists of 11 long, narrow, more or less parallel depressions scoured out by glaciers (▶ Figure 7.9a). Erosion by valley glaciers yields

(a) The Finger Lakes of New York lie in basins scoured out by continental glaciers during the Pleistocene. The bases of some of these depressions extend below sea level.

(b) Sharp, angular peaks and ridges set among broad valleys are typical of mountains eroded by valley glaciers. This view is in the Yukon Territory of Canada.

bowl-shaped depressions on mountainsides called **cirques** and broad valleys known as **U-shaped glacial troughs**, along with several other features (Figure 7.9b).

The deposits of continental and valley glaciers are **moraines**, which are chaotic mixtures of poorly sorted sediment deposited directly by glacial ice, and **outwash** consisting of stream-deposited sand and gravel (▶Figure 17.10). Any moraine deposited at a glacier's terminus is an **end moraine**, but notice from Figure 17.10b that *terminal* and *recessional moraines* are types of end moraines. Terminal moraines and outwash in southern Ohio, Indiana, and Illinois mark the greatest southerly extent of Pleistocene continental glaciers in the midcontinent region. Recessional moraines indicate the positions where the ice front stabilized temporarily during a general retreat to the north (▶Figure 17.11).

Glaciers are, of course, made up of frozen water and thus, constitute an important part of the hydrosphere, one of Earth's major systems. In Chapter 1, we emphasized the systems approach to Earth history, and here we have an excellent opportunity to see interactions among systems at work. Cape Cod, Massachusetts, is a distinctive landform that resembles a human arm extending into the Atlantic Ocean. It and nearby Martha's Vineyard and Nantucket Island owe their existence to deposition by Pleistocene glaciers and modification of these deposits by wind-generated waves and nearshore currents (▶Figure 17.12).

Changes in Sea Level

Today, 28 to 35 million km^3 of water is frozen in glaciers, all of which came from the oceans. During the maximum extent of Pleistocene glaciers, though, more than 70 million km^3 of ice was present on the continents. These huge masses of ice themselves had a tremendous impact on the glaciated areas (see the next section), and they

contained enough frozen water to lower sea level by 130 m. Accordingly, large areas of today's continental shelves were exposed and quickly blanketed by vegetation. In fact, the Bering Strait connected Alaska with Siberia via a broad land bridge across which Native Americans and mammals such as the bison migrated (▶Figure 17.13). The shallow floor of the North Sea was also above sea level, so Great Britain and mainland Europe formed a single landmass. When the glaciers melted, these areas were flooded, drowning the plants and forcing the animals to migrate.

Lower sea level during the several Pleistocene glacial intervals also affected the *base level*, the lowest level to which running water can erode, of rivers and streams flowing into the oceans. As sea level dropped, rivers eroded deeper valleys and extended them across the emergent continental shelves. During times of lower sea level, rivers transported huge quantities of sediment across the exposed continental shelves and onto the continental slopes, where the sediment contributed to the growth of submarine fans. As the glaciers melted, however, sea level rose, and the lower ends of these river valleys along North America's east coast were flooded, whereas those along the west coast formed impressive submarine canyons.

What would happen if the world's glaciers all melted? Obviously, the water stored in them would return to the oceans, and sea level would rise about 70 m. If this were to happen, many of the world's large population centers would be flooded.

Glaciers and Isostasy

In a manner of speaking, Earth's crust floats on the denser mantle below, a phenomenon geologists call **isostasy**. An analogy can help you understand this concept, which is certainly counterintuitive; after all, how can rock float in rock? Consider an iceberg. Ice is slightly less dense than

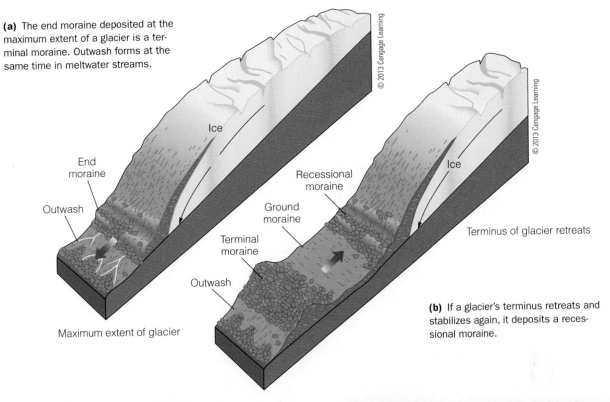

(a) The end moraine deposited at the maximum extent of a glacier is a terminal moraine. Outwash forms at the same time in meltwater streams.

End moraine

Outwash

Maximum extent of glacier

Ice

Recessional moraine

Ground moraine

Terminal moraine

Outwash

Terminus of glacier retreats

© 2013 Cengage Learning

(b) If a glacier's terminus retreats and stabilizes again, it deposits a recessional moraine.

(c) End moraine of the Salmon Glacier in the Coast Mountains of British Columbia, Canada.

(d) This ridge of gravel is a small end moraine that was deposited in 1995 by the Exit Glacier in Kenai Fjords National Park, near Seward, Alaska. The flat-lying gravel deposit is outwash.

water, so an iceberg sinks to its equilibrium position in water with only about 10 percent of its volume above the surface. Earth's crust is more complicated, but it sinks into the mantle, which behaves like a fluid, until it reaches its equilibrium position depending on its thickness and density. Remember, oceanic crust is thinner but denser than continental crust, which varies considerably in thickness.

If the crust has more mass added to it, as when thick layers of sediment accumulate or vast glaciers form, it sinks lower into the mantle until it once again achieves equilibrium. However, if erosion or melting ice reduces the load, the crust slowly rises by **isostatic rebound**. Think of the iceberg again. If some of it above sea level were to melt, it would rise in the water until it regained equilibrium.

No one doubts that Earth's crust subsided from the great weight of glaciers during the Pleistocene or that it has rebounded and continues to do so in some areas. Indeed, the surface in some places was depressed as much as 300 m below preglacial elevations. But as the glaciers melted and wasted away, the downwarped areas gradually rebounded to their former positions. Evidence of isostatic rebound is seen in formerly glaciated areas such as

© 2013 Cengage Learning

▶ **Figure 17.11 Terminal and Recessional Moraines in the Midcontinent Region** These moraines were deposited during the latter part of the Wisconsinan glaciation. The oldest ones, those farthest south, are about 16,000 years old. Only one terminal moraine is present, marking the greatest extent of Wisconsinan glaciers.

Principal glacial deposits in the Great Lakes Region

Drift deposited during middle and late Wisconsinan glaciation

- Till plains
- End moraines
- Outwash plains and valley trains
- Glacial lake deposits
- Undifferentiated drift of earlier glaciations
- Driftless regions

Scandinavia and the North American Great Lakes region. Some coastal cities in Scandinavia have rebounded enough so that docks built only a few centuries ago are now far inland from the shore. And in Canada, as much as 100 m of isostatic rebound has taken place during the last 6,000 years (▶ Figure 17.14).

Pluvial and Proglacial Lakes

During the Wisconsinan glacial stage, many now arid parts of the western United States supported large lakes when glaciers were present far to the north. These **pluvial lakes**, as they are called, existed because of the greater precipitation and overall cooler temperatures, especially during the summer, which lowered the evaporation rate. Wave-cut cliffs, beaches, deltas, and lake deposits along

with fossils of freshwater organisms attest to the presence of these lakes (▶ Figure 17.15).

Death Valley, on the California–Nevada border, is the hottest, driest place in North America, yet during the Wisconsinan it supported Lake Manly, a large pluvial lake (Figure 17.15). It was 145 km long, nearly 180 m deep, and when it dried up, dissolved salts precipitated on the valley floor.

In contrast to pluvial lakes, which are far from areas of glaciation, **proglacial lakes** form where meltwater accumulates along a glacier's margin (Figure 17.15a). Lake Agassiz, named in honor of the French naturalist Louis Agassiz, formed in this manner. It covered about 250,000 km² in North Dakota, Manitoba, Saskatchewan, and Ontario, and persisted until the ice along its northern margin melted, at which time it drained northward into Hudson Bay.

▶ Figure 17.12 **Cape Cod, Massachusetts** Cape Cod and the nearby islands are made up of end moraines that were deposited 23,000 to 14,000 years ago. Nantucket and Martha's Vineyard are deposits of a terminal moraine, whereas the material making up Cape Cod was deposited as a recessional moraine. Both moraines have been modified by waves.

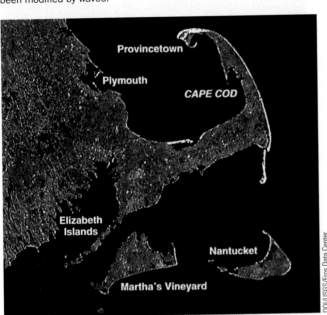

▶ Figure 17.13 **The Bering Land Bridge** During the Pleistocene, sea level was as much as 130 m lower than it is now, and a broad area called the Bering land bridge (Beringia) connected Asia and North America. It was exposed above sea level during times of glacial advances and served as a corridor for the migration of people, animals, and plants.

Bering land bridge

▶ Figure 17.14 **Glaciers and Isostasy**

(a) Isostatic rebound in Scandinavia. The lines show rates of uplift in centimeters per century.

(b) Isostatic rebound in eastern Canada in meters during the last 6,000 years.

(b) The area in this image, taken east of Fallon, Nevada, was covered by Lake Lahontan. In fact, the image was taken from Grimes Point Archaeological Site, where Native Americans lived near the lakeshore.

(a) Pyramid Lake and Great Salt Lake are shrunken remnants of much larger ancient lakes. Of all the lakes shown here, only Lake Columbia and Lake Missoula were proglacial lakes; the others were pluvial lakes. When the 600-m-high ice dam impounding Lake Missoula failed, it drained westward and scoured out the scablands of eastern Washington.

Deposits in lakes adjacent to or near glaciers vary considerably from gravel to mud, but of special interest are the finely laminated mud deposits consisting of alternating dark and light layers. Each dark–light couplet is a **varve** (see Figure 6.13d), representing an annual deposit. The light-colored layer of silt and clay formed during the spring and summer, and the dark layer is made up of smaller particles and organic matter formed during winter when the lake froze over. These varved deposits may also contain gravel-sized particles known as *dropstones*, released from melting ice.

Glacial Lake Missoula In 1923, geologist J Harlan Bretz proposed that a Pleistocene lake in what is now western Montana had periodically burst through its ice dam and flooded a large area in the Pacific Northwest (Figure 17.15a). He further claimed that these huge floods were responsible

for the giant ripple marks and other fluvial features seen in Montana and Idaho as well as the *scablands* of eastern Washington, an area in which the surface deposits were scoured, exposing underlying bedrock (►Figure 17.16a).

Bretz's hypothesis initially met with considerable opposition, but he marshaled his evidence and eventually convinced geologists these huge floods had taken place, the most recent one probably no more than 18,000 to 20,000 years ago. It now is well accepted that Lake Missoula, a large proglacial lake covering about 7,800 km², was impounded by an ice dam in Idaho that periodically failed. In fact, the shorelines of this ancient lake are clearly visible on the mountainsides around Missoula, Montana. When the ice dam failed, the water rushed out at tremendous velocity, accounting for the various fluvial features seen in Montana and Idaho and the scablands in eastern Washington (Figure 17.15a).

Lake Bonneville Lake Bonneville, with a maximum size of about 50,000 km² and a depth of at least 335 m, was a large pluvial lake mostly in what is now Utah, but parts of it extended into eastern Nevada and southern Idaho (Figure 17.15a). About 15,000 years ago, Lake Bonneville flooded catastrophically when it overflowed and rapidly eroded a natural dam at Red Rock Pass in Idaho. The floodwaters followed the course of the Snake River, and, in similar fashion to the Lake Missoula flood, it left abundant evidence of its passing. For example, the Melon Gravels in Idaho consist of rounded basalt boulders up to 3 m in diameter, and the gravel bars are as much as 90 m thick and 2.4 km long (Figure 17.16b). Although the Lake Bonneville flood was a catastrophic one, with an estimated discharge of 1.3 km³/hr, Glacial Lake Missoula flood's discharge was about 30 times as great. The vast salt deposits of the Bonneville Salt Flats west of Salt Lake City, Utah, formed when parts of this ancient lake dried up, and the Great Salt Lake is a shrunken remnant of this once much larger lake.

(a) Glacial Lake Missoula emptied rapidly several times when an ice dam impounding the lake failed. This view shows Palouse Falls in Washington, which is in a canyon eroded by the floodwaters from Glacial Lake Missoula, which was a proglacial lake.

(b) The Melon Gravel, near Hagerman, Idaho, was deposited by the floodwaters from Lake Bonneville, which was a pluvial lake.

A Brief History of the Great Lakes Before the Pleistocene, the Great Lakes region was a rather flat lowland with broad stream valleys. As the continental glaciers advanced southward from Canada, the entire area was ice-covered and deeply eroded. Indeed, four of the five Great Lakes basins were eroded below sea level; glacial erosion is not restricted by base level as erosion by running water is. In any case, the glaciers advanced far to the south, but eventually began retreating north, depositing numerous recessional moraines as they did so (Figure 17.11).

By about 14,000 years ago, parts of the Lake Michigan and Lake Erie basins were ice free, and glacial meltwater began forming proglacial lakes (▶Figure 17.17). As the ice front resumed its retreat northward—although interrupted by minor readvances—the Great Lakes basins eventually became ice free, and the lakes expanded until they reached their present size and shape.

This brief history of the Great Lakes is generally correct, but oversimplified. The minor readvances of the ice front mentioned earlier caused the lakes to fluctuate widely, and as they filled, they overflowed their margins and partly drained. In addition, once the glaciers were gone, isostatic rebound took place, and this too has affected the Great Lakes.

What Caused Pleistocene Glaciation?

We know how glaciers move, erode, transport, and deposit sediment, and we even know the conditions necessary for them to originate—more winter snowfall than melts during the following warmer seasons. But this really does not address the broader question of what caused large-scale glaciation during the Ice Age, and why so few episodes of glaciation have occurred. Geologists, oceanographers, climatologists, and others have tried for more than a century to develop a comprehensive theory explaining all aspects of ice ages but so far have not been completely successful. One reason for their lack of success is that the climatic changes responsible for glaciation, the cyclic occurrence of glacial–interglacial stages, and short-term events such as the Little Ice Age operate on vastly different time scales.

The few periods of glaciation recognized in the geologic record are separated by long intervals of mild climate. Slow geographic changes related to plate tectonic activity are probably responsible for such long-term climatic changes. Plate movements may carry continents into latitudes where glaciers are possible, provided they receive enough snowfall. Long-term climatic changes also take place as plates collide, causing uplift of vast areas far above sea level, and of course the distribution of land and sea has an important influence on oceanic and atmospheric circulation patterns.

One proposed mechanism for the onset of the cooling trend that began following the Mesozoic and culminated with Pleistocene glaciation is decreased levels of carbon dioxide (CO_2) in the atmosphere. Carbon dioxide is a greenhouse gas, so if less were present to trap sunlight, Earth's overall temperature would perhaps be low enough for glaciers to form. The problem is that no hard data exist to demonstrate that such a decrease in CO_2 levels actually occurred, nor do scientists agree on a mechanism to cause a decrease, although uplift of the Himalayas or other mountain ranges has been suggested.

Intermediate climatic changes lasting for a few thousand to a few hundred thousand years, such as the Pleistocene glacial–interglacial stages, have also proved difficult to explain, but the Milankovitch theory, proposed many years ago and discussed next, is now widely accepted.

(a) 13,000 years ago.

(b) 11,500 years ago.

(c) 9,500 years ago.

(d) 6,000 years ago.

Adapted with the permissions of Natural Resources Canada 2011, courtesy of the Geological Survey of Canada (Geology and economic minerals of Canada; 1970; Quaternary geology of Canada by V. K. Prest).

▶ Figure 17.17 **Four Stages in the Evolution of the Great Lakes** As the glacial ice retreated northward, the lake basins began filling with meltwater. The dotted lines indicate the present-day shorelines of the lakes.

that minor irregularities in Earth's rotation and orbit are sufficient to alter the amount of solar radiation received at any given latitude and hence bring about climate changes. Now called the **Milankovitch theory**, his proposal was initially ignored but received renewed interest in the 1970s and presently is widely accepted.

Milankovitch attributed the onset of the Pleistocene Ice Age to variations in three aspects of Earth's orbit. The first is *orbital eccentricity*, which is the degree to which Earth's orbit around the Sun changes over time (▶Figure 17.18a). When the orbit is nearly circular, both the Northern and Southern Hemispheres have similar contrasts between the seasons. However, if the orbit is more elliptic, hot summers and cold winters will occur in one hemisphere, whereas warm summers and cool winters will take place in the other hemisphere. Calculations indicate a roughly 100,000-year cycle between times of maximum eccentricity, which corresponds closely to the 20 warm–cold climatic cycles that took place during the Pleistocene.

Milankovitch also pointed out that the angle between Earth's axis and a line perpendicular to the plane of Earth's orbit shifts about 1.5 degrees from its current value of 23.5 degrees during a 41,000-year cycle (Figure 17.18b). Although changes in this *axial tilt* have little effect on equatorial latitudes, they strongly affect the amount of solar radiation received at high latitudes and the duration of the dark period at and near Earth's poles. Coupled with the third aspect of Earth's orbit, *precession of the equinoxes*, high latitudes might receive as much as 15 percent less solar radiation, certainly enough to affect glacial growth and melting.

This precession of the equinoxes refers to a change in the time of the equinoxes. At present, the equinoxes take place on about March 21 and September 21 when the Sun is directly over the equator. But as Earth rotates on its axis, it also wobbles as its axial tilt varies 1.5 degrees from its current value, thus changing the time of the equinoxes. Taken alone, the time of the equinoxes has little climatic effect, but changes in Earth's axial tilt also change the times of *aphelion* and *perihelion*, which are, respectively, when Earth is farthest from and closest to the Sun during its orbit (Figure 17.18c). Earth is now at perihelion, closest to the Sun, during Northern Hemisphere winters, but in about 11,000 years perihelion will be in July. Accordingly, Earth will be at aphelion, farthest from the Sun, in January and have colder winters.

Continuous variations in Earth's orbit and axial tilt cause the amount of solar heat received at any latitude to vary slightly through time. The total heat received by the planet changes little, but according to Milankovitch, and now many scientists agree, these changes caused complex climatic variations and provided the triggering mechanism for the glacial–interglacial episodes of the Pleistocene.

The Milankovitch Theory

Changes in Earth's orbit as a cause of intermediate-term climatic events was first proposed during the mid 1800s, but the idea was made popular during the 1920s by the Serbian astronomer Milutin Milankovitch. He proposed

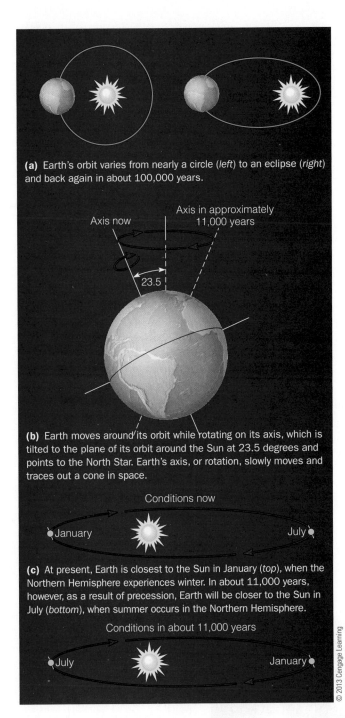

(a) Earth's orbit varies from nearly a circle (*left*) to an eclipse (*right*) and back again in about 100,000 years.

Axis now

Axis in approximately 11,000 years

23.5

(b) Earth moves around its orbit while rotating on its axis, which is tilted to the plane of its orbit around the Sun at 23.5 degrees and points to the North Star. Earth's axis, or rotation, slowly moves and traces out a cone in space.

Conditions now

January   July

(c) At present, Earth is closest to the Sun in January (*top*), when the Northern Hemisphere experiences winter. In about 11,000 years, however, as a result of precession, Earth will be closer to the Sun in July (*bottom*), when summer occurs in the Northern Hemisphere.

Conditions in about 11,000 years

July   January

© 2013 Cengage Learning

Short-Term Climatic Events

Climatic events with durations of several centuries, such as the Little Ice Age, are too short to be accounted for by plate tectonics or Milankovitch cycles. Several hypotheses have been proposed, including variations in solar energy and volcanism.

Variations in solar energy could result from changes within the Sun itself or from anything that would reduce the amount of energy Earth receives from the Sun. The latter could result from the solar system passing through

clouds of interstellar dust and gas or from substances in the atmosphere reflecting solar radiation back into space. Records kept over the past 90 years indicate that during this time the amount of solar radiation has varied only slightly. Although variations in solar energy may influence short-term climatic events, such a correlation has not been demonstrated.

During large volcanic eruptions, tremendous amounts of ash and gases are spewed into the atmosphere, where they reflect incoming solar radiation and thus reduce atmospheric temperatures. Small droplets of sulfur gases remain in the atmosphere for years and can have a significant effect on climate. Several large-scale volcanic events have occurred, such as the 1815 eruption of Tambora, and are known to have had climatic effects. However, no relationship between periods of volcanic activity and periods of glaciation has yet been established.

Glaciers Today

Glaciers today are much more restricted, but they nevertheless remain potent agents of erosion, sediment transport, and deposition. After all, even now they cover about 10 percent of Earth's land surface. Scientists monitor the behavior of glaciers to better understand the dynamics of moving bodies of ice, but they are also interested in glaciers as indicators of climatic change.

No doubt you have heard of *global warming*, a phenomenon of warming of Earth's atmosphere during the last 100 years or so. Many scientists are convinced that the cause of global warming is an increase of greenhouse gases, especially carbon dioxide, in the atmosphere as a result of burning fossil fuels. Others agree that surface temperatures have increased but attribute the increase to normal climatic variation. In either case, glaciers are good indicators of short-term climatic changes.

Any glacier's behavior depends on its budget—that is, its gains versus losses—which in turn is related to temperature and the amount of precipitation. According to one estimate there are about 160,000 valley glaciers and small ice caps outside Antarctica and Greenland, with Alaska alone having several tens of thousands. It is true that not many of these glaciers have been studied, but many of those that have show an alarming trend: they are retreating, ceased moving entirely, or have disappeared.

For example, in 1850 there were about 150 glaciers in Glacier National Park in Montana, but now only about two dozen remain, and nearly all the glaciers in the Cascade Range of the Pacific Northwest are retreating. Glacier Peak in Washington has more than a dozen glaciers, all of which are retreating, and Whitechuck Glacier will soon be inactive. When Mount St. Helens in Washington erupted in May 1980, all 12 of its glaciers were destroyed or considerably diminished. By 1982 the lava dome in the mountain's crater had cooled sufficiently for a new glacier to form; it is now about 190 m thick (▶ Figure 17.19). Remember, though, that Mount St. Helens already had the conditions for glaciers to exist, so the fact that one has become reestablished does not counter the evidence from virtually all other glaciers in the range.

There is one notable exception to shrinking glaciers in the Cascade Range. The seven glaciers on Mount Shasta in California are all growing, probably because of increased precipitation resulting from warming of the Pacific Ocean. Nevertheless, the glaciers are small, and the trend is not likely to continue for long, because as warming continues it will soon overtake the increased snowfall.

The ice sheet in Greenland has lost about 162 km³ of ice during each of the years from 2003 through 2005, and many of the glaciers that flow into the sea from the ice sheet have speeded up markedly. For instance, the Kangerdlugssuag Glacier is moving at about 14 km per year (38.4 m/day), and its terminus retreated 5 km in 2005 alone. The termini of many glaciers in Alaska are also retreating, particularly valley glaciers that flow into the sea (the so-called tidewater glaciers). Two factors account for these phenomena. First, the glaciers are moving faster because more meltwater is present that percolates downward and facilitates basal slip. The other factor is that warmer ocean temperatures melt the glaciers where they flow into the sea.

Most of Antarctica shows no signs of a decreasing volume of ice because the continent is at such high latitudes and so cold that little melting takes place. The greatest concern here is that some of the ice shelves—the parts of vast glaciers that flow into the sea—will collapse and allow the glaciers inland to flow more rapidly. In fact, huge sections of ice shelves have broken off in recent years, allowing land-based glaciers to surge into the ocean. The ice shelves themselves are floating, so when they melt, that does not cause sea level to rise, but when the glacial ice on land flows into the ocean and melts, it does result in a rising sea level.

Quaternary Mineral Resources

Recall that during the Pleistocene, numerous pluvial lakes existed in the southwestern United States. When these lakes evaporated as the climate became more arid, vast salt deposits accumulated in Utah, in Death Valley, and in adjacent basins in southern California. In some areas, halite is mined from these deposits, and borax, one of California's most valuable mineral commodities, was deposited in several lakes. Borax is used in ceramics, fertilizers, glass, solder, and pharmaceuticals.

In many U.S. states, as well as Canadian provinces, the most valuable mineral commodity is sand and gravel used in construction, much of which is recovered from glacial deposits, especially outwash. These same commodities are also recovered from deposits on the continental shelves and from stream deposits unrelated to glaciation. Silica sand is used in the manufacture of glass, and fine-grained glacial lake deposits are used to manufacture bricks and ceramics.

The California Gold Rush of the late 1840s and early 1850s was fueled by the discovery of Pleistocene and Holocene placer deposits of gold in the American River (see

▶ Figure 17.19 **Crater Glacier on Mount St. Helens, Washington** This glacier formed following the eruption of Mount St. Helens in 1980, although eruptions in 2004 modified it. By 2008, the glacier had nearly encircled the lava dome in the volcano's crater.

Photo by John Pallister/USGS

Chapter 14). Most of the $200 million in gold mined in California from 1848 to 1853 came from placer deposits. Discoveries of gold placer deposits in the Yukon Territory of Canada were primarily responsible for settlement of that area.

Peat consisting of semi-carbonized plant material in bogs and swamps is an important resource that has been exploited in Canada and Ireland. It is burned as a fuel in some areas but also finds other uses, as in gardening.

Summary

▶ The most recent part of geologic time is the Quaternary Period (2.6 million years ago to the present), which consists of two epochs, the Pleistocene and the Holocene.

▶ Although the Pleistocene is best known for widespread glaciers, it was also a time of volcanism and tectonism.

▶ Pleistocene glaciers covered about 30 percent of the land surface, and were most widespread on the Northern Hemisphere continents.

▶ Several intervals of extensive Pleistocene glaciation took place in North America, each separated by interglacial stages. Fossils and oxygen isotope data indicate about 20 warm–cold cycles occurred during the Pleistocene.

▶ Areas far beyond the ice were affected by Pleistocene glaciers: climate belts were compressed toward the equator; large pluvial lakes existed in what are now arid regions; and when glaciers were present, sea level was as much as 130 m lower than now.

▶ Moraines, striations, outwash, and various other glacial landforms are found throughout Canada, in the northern

tier of states, and in many mountain ranges where valley glaciers were present.

▶ The tremendous weight of Pleistocene glaciers caused isostatic subsidence of Earth's crust. When the glaciers melted, isostatic rebound began and continues even now in some areas.

▶ Major glacial episodes separated by tens or hundreds of millions of years probably stem from changing positions of plates, which in turn profoundly affects oceanic and atmospheric circulation patterns.

▶ According to the Milankovitch theory, minor changes in Earth's rotation and orbit bring about climatic changes that produce glacial–interglacial intervals.

▶ The causes of short-term climatic changes such as occurred during the Little Ice Age are unknown; two proposed causes are variations in the amount of solar energy and volcanism.

▶ Pleistocene mineral resources include sand and gravel, placer deposits of gold, and some evaporite minerals such as borax.

Important Terms

cirque, p. 354
continental glacier, p. 345
end moraine, p. 354
glacial stage, p. 348
glacier, p. 345
ice cap, p. 345
ice-scoured plain, p. 353

interglacial stage, p. 348
isostasy, p. 354
isostatic rebound, p. 355
Little Ice Age, p. 351
Milankovitch theory, p. 360
moraine, p. 354
neoglaciation, p. 351

outwash, p. 354
pluvial lake, p. 356
pollen analysis, p. 351
proglacial lake, p. 356
U-shaped glacial trough, p. 354
valley glacier, p. 345
varve, p. 358

Review Questions

1. Which one of the following is not a landform resulting from deposition or erosion by valley glaciers?
a. ____ Cirque; b. ____ Moraine; c. ____ Outwash; d. ____ U-shaped glacial trough; e. ____ Pluvial lake.

2. The most recent episode of Pleistocene glaciation was the
a. ____ Illinoian; b. ____ Dakotan; c. ____ Wisconsinan; d. ____ Missoulian; e. ____ Aftonian.

3. The phenomenon in which Earth's crust rises after unloading, as when vast glaciers melt, is called
a. ____ isostatic rebound; b. ____ postglacial adjustment; c. ____ orogenic deformation; d. ____ neoglaciation; e. ____ precession.

4. An important area of Pleistocene and Holocene volcanism in North America is the
a. ____ Great Plains; b. ____ Cascade Range; c. ____ Atlantic Coastal Plain; d. ____ San Andreas Fault; e. ____ Appalachian Mountains.

5. A large pluvial lake that was the forerunner of the Great Salt Lake was
a. ____ Lake Agazziz; b. ____ Lake Ontario; c. ____ Lake Bonneville; d. ____ Lake Champlain; e. ____ Lake Cape Cod.

6. A couplet of light- and dark-colored, laminated, fine-grained sediment deposited in a glacial lake is a
a. ____ moraine; b. ____ ice cap; c. ____ cirque; d. ____ varve; e. ____ ice-scoured plain.

7. An explanation for intermediate-term climatic fluctuations such as the Pleistocene glacial–interglacial stages is the
a. _____ isostasy theory; b. _____ Wegener dictim; c. _____ principle of superposition; d. _____ Milankovitch theory; e. _____ glacial moraine hypothesis.

8. The Pleistocene Epoch began _____ million years ago and ended _____ years ago.
a. _____ 2.6/11,700; b. _____ 66/1.8; c. _____ 40/22,000; d. _____ 250/125,000; e. _____ 15/2,000.

9. Which of the following statements is correct?
a. _____ All of North America was ice-covered during the Pleistocene; b. _____ Continental glaciers are long, narrow tongues of ice in valleys; c. _____ A moraine is a well-sorted deposit of silt and clay; d. _____ Pollen analysis is helpful in determining ancient climates; e. _____ Oxygen isotope ratios are used to determine the thickness of glaciers.

10. Most glaciers move mostly by
a. _____ sliding downhill; b. _____ isostatic rebound; c. _____ plastic flow; d. _____ proglacial advance; e. _____ floating.

11. If you gave a talk on the local geology in Plymouth, Massachusetts, how would you explain how Cape Cod and nearby islands evolved to their present form?

12. What was the Little Ice Age, when did it take place, and what effect did it have on humans?

13. After observing the same glacier for several years, you conclude (1) that the glacier's terminus has retreated 2 km, and yet (2) debris on the glacier's surface has clearly moved at least 1 km toward the terminus. How can you explain these observations?

14. How are oxygen isotope ratios and pollen analysis used to make inferences about ancient climates?

15. Explain how the Milankovitch theory accounts for the onset of glacial ages.

16. What is isostatic rebound, and what evidence indicates that it has occurred?

17. What accounts for the volcanism in the Cascade Range, and why are there no volcanoes along the rest of the U.S. Pacific coast other than in Alaska?

18. Where was Glacial Lake Missoula, and what happened to it?

19. Why is it so difficult to correlate the sequence of warm–cold intervals recorded in seafloor sediments with glacial deposits on land?

20. What are pluvial and proglacial lakes? Give an example of each.

Life of the Cenozoic Era

Restoration of *Titanis walleri*, which was a predatory, flightless bird that stood 2.5 m high and weighed about 150 kg. It was one of 18 species of predatory birds that inhabited South America from the Middle Paleocene to the Pleistocene. *T. walleri*, however, lived only during the Pliocene and Pleistocene, and it is the only one of its family to migrate to North America, where its remains have been found in Texas and Florida.

© Carl Buell

At the end of this chapter, you will have learned that

▶ Survivors of the Mesozoic extinctions evolved and gave rise to the present-day invertebrate marine fauna.

▶ Angiosperms continued to diversify and to dominate land-plant communities, but seedless vascular plants and gymnosperms are still common.

▶ Many of today's families and genera of birds evolved, and large, flightless birds were important Cenozoic predators.

▶ If we could visit the Paleocene, we would not recognize many of the mammals, but more familiar ones evolved during the following epochs.

▶ Small mammals such as rodents, rabbits, insectivores, and bats adapted to the microhabitats unavailable to larger mammals.

▶ Carnivorous mammal fossils are not as common as those of herbivores, but there are enough to show their evolutionary trends and relationships to one another.

▶ The evolutionary histories of odd-toed and even-toed hoofed mammals are well documented by fossils.

▶ Today's giant land mammals (elephants) and giant marine mammals (whales) evolved from small Early Cenozoic ancestors.

▶ Extinctions at the end of the Pleistocene Epoch were most severe in the Americas and Australia, and the animals most affected were large land-dwelling mammals.

▶ As Pangaea continued to fragment during the Cenozoic, intercontinental migrations became increasingly difficult.

▶ A Late Cenozoic land connection formed between North and South America, resulting in migrations in both directions.

Introduction

Remember that Earth and life history for the Cenozoic Era are better known than they are for earlier time intervals, because Cenozoic-age rocks are easily accessible, little altered, and many contain fossils. This most recent episode in life history, at 66 million years long, is comparatively brief, making up only 1.4 percent of all geologic time (see Figure 8.1), but it was certainly long enough for significant changes to take place in the biota.

If we could somehow go back and visit the early part of the Cenozoic Era, we would find many unfamiliar land-dwelling animals, but by the middle of that era more recognizable ones had evolved. In the seas, the invertebrates that survived the Mesozoic extinctions diversified once again and gave rise to today's familiar marine fauna. The flowering plants, or angiosperms, continued their diversification and now constitute more than 90 percent of all plants on land. However, their distribution varied throughout the Cenozoic, depending on changing climates. Birds were present by the Jurassic, but the families now common evolved during the Paleogene and Neogene periods, reached their greatest diversity during the Pleistocene Epoch, and have declined slightly since then.

Recall from Chapter 15 that mammals evolved during the Triassic and were contemporaries of dinosaurs, although most Mesozoic mammals were small, and they were not very diverse. In addition, some of the earliest mammals differed little from their cynodont ancestors, but as they evolved they diversified into numerous lineages that adapted to nearly all terrestrial habitats as well

as aquatic environments—and one group, the bats, became the only fliers among mammals.

We have good fossil records for some mammal lineages, especially for the hoofed mammals such as horses, camels, and rhinoceroses, as well as for elephants, whales, and several others. Indeed, many of the so-called missing links that we discussed in Chapter 7 are not missing at all, a fact that can be easily demonstrated with several lineages of mammals. A casual observer would not realize that the end members of any of these lineages were related, but when the entire sequence is seen, the relationships become clear.

Widespread exposures of Cenozoic-age, fossil-bearing rocks are present in the western part of North America, many of which were deposited in continental environments. Of course, Cenozoic-age fossils are known from the other continents too, and certainly one of the most remarkable concentrations of fossils anywhere is the Eocene Messel fossil site in Germany. Indeed, in addition to the 10,000 or so fossil fish, thousands of insects have been found, as well as birds, amphibians, reptiles, and several types of mammals such as anteaters, pangolins, marsupials, carnivores, bats, rodents, and early hoofed mammals. Fossil mammals are also found in abundance in several areas in North America, where they are featured in some of our state and national parks and monuments, including John Day Fossil Beds National Monument in Oregon, Badlands National Park in South Dakota, and Ashfall Fossil Beds State Historical Park in Nebraska (▶Figure 18.1).

Marine Invertebrates and Phytoplankton

The Cenozoic marine ecosystem was populated mostly by plants, animals, and single-celled organisms that survived the terminal Mesozoic extinction. Gone were the ammonites, rudists, and most of the planktonic foraminifera. Especially prolific Cenozoic invertebrate groups were the foraminifera, radiolarians, corals, bryozoans, mollusks, and echinoids. The marine invertebrate community in general became more provincial during the Cenozoic because of changing ocean currents and latitudinal temperature gradients. In addition, the Cenozoic marine invertebrate faunas became more familiar in appearance to those of today.

Entire families of phytoplankton became extinct at the end of the Cretaceous, with only a few species in each major group surviving into the Paleogene. These species diversified and expanded during the Cenozoic, perhaps because of decreased competitive pressures. Coccolithophores, diatoms, and dinoflagellates all recovered from their Late Cretaceous reduction in numbers to flourish during the Cenozoic. Diatoms were particularly abundant during the Miocene, probably because of increased volcanism during this time. Volcanic ash provided increased dissolved silica in seawater, which diatoms used to construct their skeletons. Massive Miocene diatomite rocks, made up of diatom shells, are present in several western states (▶ Figure 18.2).

The foraminifera were a major component of the Cenozoic marine invertebrate community. Although dominated by relatively small forms (▶ Figure 18.3), it included some exceptionally large forms that lived in the warm waters of the Cenozoic Tethys Sea. Shells of these larger forms accumulated to form thick limestones, some of which the ancient Egyptians used to construct the Sphinx and the Pyramids of Giza.

Corals were perhaps the main beneficiary of the Mesozoic extinctions. Having relinquished their reef-building role to rudists, which are mollusks, during the Mid-Cretaceous, corals again became the dominant reef builders. They formed extensive reefs in the warm waters of the Cenozoic oceans and were especially prolific in the Caribbean and Indo-Pacific regions.

Other suspension feeders such as bryozoans and crinoids were also abundant and successful during the Paleogene and Neogene. Bryozoans, in particular, were very abundant. Perhaps the least important of the Cenozoic marine invertebrates were brachiopods, with fewer than 60 genera surviving today. Brachiopods never recovered from their reduction in diversity at the end of the Paleozoic (see Chapter 12).

Just as during the Mesozoic, bivalves and gastropods were two of the major groups of marine invertebrates during the Cenozoic, and they had a markedly modern appearance (▶ Figures 18.4a and 18.4b).

After the extinction of ammonites and belemnites at the end of the Cretaceous, the Cenozoic cephalopod fauna consisted of nautiloids and shell-less cephalopods such as squids and octopuses.

Echinoids continued their expansion in the infaunal habitat and were very prolific during the Cenozoic. New forms such as sand dollars evolved during this time from biscuit-shaped ancestors (Figure 18.4c).

▸ Figure 18.2 **Miocene Diatomite and Diatoms**

(a) Outcrop of diatomite in the Monterey Formation at Newport Lagoon, California. The diatoms in (b) and (c) are from this formation.

(b) A pinnate diatom.

(c) A centric diatom.

▸ Figure 18.3 **Cenozoic Foraminifera**

(a) *Cibicides americanus* from the Early Miocene of California is a benthonic foraminifera.

(b) *Globigerinoides fistulosus* is a Pleistocene planktonic foraminifera from the South Pacific Ocean.

Cenozoic Vegetation and Climate

Angiosperms continued to diversify during the Cenozoic, as more and more familiar varieties evolved, although seedless vascular plants and gymnosperms were also present in large numbers. In fact, many Paleogene plants would be familiar to us today, but their geographic distribution was not what it is now, because changing climatic conditions along with shifting plant distributions were occurring.

Some plants today are confined to the tropics, whereas others have adapted to drier conditions, and we have every reason to think that climate was a strong control on plant distribution during the past. Furthermore, leaves with entire or smooth margins, many with pointed drip-tips, are dominant in areas with abundant rainfall and high annual temperatures. Smaller leaves with incised margins are more typical of cooler, drier areas (▸ Figure 18.5a). Accordingly, fossil floras with mostly smooth-margined leaves with drip-tips indicate wet, warm conditions, whereas a cool, dry climate is indicated by a predominance of small leaves with incised margins and no drip-tips.

Paleocene rocks in the western interior of North America have fossil ferns and palms, both indicating a warm, subtropical climate. In a Paleocene flora in Colorado with about 100 species of trees, nearly 70 percent of the leaves are smooth margined, and many have drip-tips. The nature of these leaves coupled with the diversity of plants is much like that in today's rain forests. In fact, the Early Oligocene fossil plants at Florissant Fossil Beds National Monument in Colorado indicate that a warm, wet climate persisted then (Figure 18.5c).

Seafloor sediments and geochemical evidence indicate that about 55 million years ago an abrupt warming trend took place. During this time, known as the **Paleocene–Eocene Thermal Maximum** (see Figure 4.15), large-scale

(a) The gastropod *Busycon contrariuim* from Pliocene rocks in Florida.

(c) A Pliocene echinoid (sand dollar) from Mexico.

(b) The bivalve *Chlamya* sp. encrusted with barnacles, from the Miocene of Virginia.

oceanic circulation was disrupted so that heat transfer from equatorial regions to the poles diminished or ceased. As a result, deep oceanic water became warmer, resulting in extinctions of many deep-water foraminifera. Some scientists think that this deep, warm oceanic water released methane from seafloor methane hydrates, contributing a greenhouse gas to the atmosphere and either causing or contributing to the temperature increase at this time.

Subtropical conditions persisted into the Eocene in North America, probably the warmest of all the Cenozoic epochs. Fossil plants in the Eocene John Day Fossil Beds in Oregon include ferns, figs, and laurels, all of which today live in much more humid regions, as in parts of Mexico and Central America. Yellowstone National Park

in Wyoming has a temperate climate now, with warm, dry summers and cold, snowy winters, certainly not an area where you would expect avocado, magnolia, and laurel trees to grow. Yet their presence there during the Eocene indicates the area then had a considerably warmer climate than it does now.

A major climatic change took place at the end of the Eocene, when mean annual temperatures dropped as much as 7°C in about 3 million years (Figure 18.5b). Since the Oligocene, mean annual temperatures have varied somewhat worldwide, but overall they have not changed much in the middle latitudes except during the Pleistocene Epoch.

A general decrease in precipitation during the last 25 million years took place in the midcontinent region of North America. As the climate became drier, the vast forests of the Oligocene gave way first to *savannah* conditions (grasslands with scattered trees) and finally to *steppe* environments (short-grass prairie of the desert margin). Many herbivorous mammals quickly adapted to these new conditions by developing chewing teeth suitable for a diet of grass.

Cenozoic Birds

Birds today are diverse and numerous, making them the most easily observed vertebrates. The first members of many of the living orders, including owls, hawks, ducks, penguins, and vultures, evolved during the Paleogene. Beginning during the Miocene, a marked increase in the variety of songbirds took place, and by 5 to 10 million years ago, many of the existing genera of birds were present. Birds adapted to numerous habitats and continued to

▶ Figure 18.5 **Cenozoic Vegetation and Climate**

University of California Museum of Paleontology. Specimen no. 3827. Photo by Diane M. Erwin

(a) Plants adapted to cool climates typically have small leaves with incised margins (top), but in humid, warm areas they have larger, smooth-margined leaves, many with drip-tips.

(b) Climatic trends for four areas in North America based on the percentages of plant species with smooth-margined leaves.

(c) Fossil leaf from an extinct species of maple from Florissant Fossil Beds National Monument in Colorado. Numerous species of fossil plants and insects indicate that the climate here 35 million years ago was warmer and wetter than it is now.

diversify into the Pleistocene, but since then their diversity has decreased slightly.

Today, birds vary considerably in diet, habitat, adaptations, and size. Nevertheless, their basic skeletal structure remained remarkably constant throughout the Cenozoic. Given that birds evolved from a creature very much like *Archaeopteryx* (see Figure 15.17), this uniformity is not surprising, because adaptations for flying limit variations in structure. Penguins adapted to an aquatic environment, and in some large extinct and living flightless birds, the skeleton became robust and the wings were reduced to vestiges.

Many authorities on prehistoric life are now convinced that birds are so closely related to dinosaurs that they refer to them as *avian dinosaurs* and all the others as *non-avian dinosaurs*. In fact, following the demise of the non-avian dinosaurs at the end of the Mesozoic Era, the dominant large, land-dwelling predators during the Paleocene and well into the Eocene were flightless birds. Among these predators were giants such as *Gastornis*, which stood 2 m tall, had a huge head and beak, toes with large claws, and small, vestigial wings (▶ Figure 18.6). Its massive, short legs indicate that *Gastornis* was not very fast, but neither were the mammals it preyed upon. This extraordinary bird and related genera were widespread in North America and Europe, and in South America they were important predators until they died out 2 million years ago (see the chapter opening photo).

Two of the most notable large, flightless birds were the now-extinct moas of New Zealand and the elephant birds of Madagascar. Moas were up to 3 m tall; elephant birds were shorter but more massive, weighing up to 500 kg. Both are known only from Pleistocene-age deposits and went extinct shortly after humans occupied their respective areas.

Large, flightless birds are truly remarkable creatures, but the real success among birds belongs to the fliers. Even though few skeletal modifications occurred during the Cenozoic, a bewildering array of adaptive types arose. If number of species and habitats occupied is any measure of success, birds have certainly been at least as successful as mammals.

The Age of Mammals Begins

Mammals coexisted with dinosaurs for more than 160 million years, and yet, during this entire time they were not very diverse, and even the largest among them was only about 1 m long. Even at the end of the Cretaceous Period there were only a few families of mammals, a situation that was soon to change. With the demise of non-avian dinosaurs and their relatives, mammals quickly exploited the adaptive opportunities, beginning a diversification that

© 2013 Cengage Learning

James S. Monroe

continued throughout the Cenozoic Era. The Age of Mammals had begun.

We have already mentioned that Cenozoic deposits are easily accessible at or near the surface, and overall they show fewer changes resulting from metamorphism and deformation when compared with older rocks. In addition, because mammals have teeth fully differentiated into various types (see Figure 15.18), they are easier to identify and classify than members of the other classes of vertebrates. In fact, mammal teeth not only differ from front to back of the mouth, but they also differ among various mammalian orders and even among genera and species. This is especially true of chewing teeth, the **premolars** and **molars**; a single chewing tooth is commonly enough to identify the genus from which it came.

Monotremes and Marsupial Mammals

All warm-blooded vertebrates with hair and mammary glands belong to the class Mammalia, which includes the *monotremes* or egg-laying mammals—the platypus and spiny anteater of the Australian region; the *marsupials*, commonly called the "pouched mammals"—kangaroos, opossums, wombats, and so on; and the *placental mammals* (about 18 orders). Female monotremes secrete a milky substance that their young lick from the mother's hair, but both marsupials and placentals have true

mammary glands and milk to nourish their young. Monotremes have the requisite features to be called mammals, but they appear to have had a completely separate evolutionary history from the marsupials and placentals. Unfortunately, they have a very poor fossil record.

When the young of marsupial mammals are born, they are in an immature, almost embryonic state and then undergo further development in the mother's pouch. Marsupials probably migrated to Australia, the only area where they are common now, via Antarctica before Pangaea fragmented completely. They were also common in South America during much of the Cenozoic Era until only a few millions of years ago. However, when a land connection was finally established between the Americas, most of the South American marsupials died out as placental mammals from North America replaced them. Now the only marsupials outside of Australia and nearby islands are species of opossums.

Diversification of Placental Mammals

Like marsupials, placental mammals give birth to live young, but their reproductive method differs in important details. In placentals, the amnion of the amniote egg (see Figure 13.15) has fused with the walls of the uterus, forming a more efficient *placenta* than exists in marsupials. Because nutrients and oxygen flow from mother to embryo through the placenta, the more efficient placenta permits the young to develop more fully before birth than the less efficient one of the marsupials. A measure of the success of placental mammals is related in part to their method of reproduction—more than 90 percent of all mammals, fossil and extinct, are placentals.

In our following discussion of placental mammals, we emphasize the origin and evolution of several of the 18 or so living orders (▶Figure 18.7). During the Paleocene Epoch, several orders of mammals were present, but some were simply holdovers from the Mesozoic or belonged to new but short-lived groups that have no living descendants. These so-called *archaic* mammals, including marsupials, insectivores, and the rodentlike multituburculates, occupied a world with several new mammalian orders (▶Figure 18.8a), such as the first rodents, rabbits, primates, and carnivores, and the ancestors of hoofed mammals.

Most of these Paleocene mammals, even those belonging to orders that still exist, had not yet become clearly differentiated from their ancestors, and the differences between herbivores and carnivores were slight.

The Paleocene mammalian fauna was also made up mostly of small creatures. By Late Paleocene time, though, some rather large mammals were around, although giant terrestrial mammals did not appear until the Eocene. With the evolution of a now-extinct order known as the Dinocerata, better known as uintatheres, and the strange creature known as *Arsinoitherium*, giant mammals of one kind or another have been present ever since (Figure 18.8b). In addition, there were also some large carnivores by Eocene time, such as *Andrewsarchus*, which was a 3-m-long wolf-like predator.

Many mammalian orders that evolved during the Paleocene died out, but of the several that first appeared during the Eocene, only one has become extinct. Thus, by Eocene time many of the mammalian orders existing now were present, yet if we could go back for a visit we would not recognize most of these animals. Surely we would know they were mammals, and some would be at least vaguely familiar, but the ancestors of horses, camels, elephants, whales, and rhinoceroses bore little resemblance to their living descendants.

Warm, humid climates persisted throughout the Paleocene and Eocene of North America, but by Oligocene time drier and cooler conditions prevailed. Most of the archaic Paleocene mammals, as well as several groups that originated during the Eocene, had died out by this time. The large, rhinoceros-like titanotheres died out, and the uintatheres (Figure 18.8b) just mentioned also went extinct. In addition, some other groups of mammals suffered extinctions, including several types of herbivores loosely united as condylarths, carnivorous mammals known as *creodonts*, most of the remaining multituburculates, and some primates. All in all, this was a time of considerable biotic change.

▶ **Figure 18.7 The Diversification of Mammals** Mammals existed during the Mesozoic, but most placental mammals diversified during the Paleocene and Eocene epochs. Among the living orders of mammals, all are placentals except for the monotremes and marsupials. Several extinct orders are not shown. Bold lines indicate actual geologic ranges, whereas the thinner lines indicate the inferred branching of the groups.

© 2013 Cengage Learning

▶ Figure 18.8 **Paleocene and Eocene Mammals**

© Nobumichi Tamura

(a) This squirrel-sized animal is the Paleocene genus *Ptilodus*, a member of a group of mammals known as multituberculates.

© The Field Museum, Chicago. #CK46T

By Oligocene time, most of the existing mammalian orders were present, but they continued to diversify as more and more familiar genera evolved. If we were to encounter some of these animals, we might think them a bit odd, but we would have little difficulty recognizing rhinoceroses (although some were hornless), elephants, horses, rodents, and many others. However, the large, horselike animals known as *chalicotheres*, with claws, and the large piglike entelodonts would be unfamiliar, and others would be found in areas where today we would not expect them—elephants in North America, for instance.

By Miocene, and certainly Pliocene time, most mammals were quite similar to those existing now (▶ Figure 18.9). On close inspection, though, we would see horses with three toes, cats with huge canine teeth, deerlike animals with forked horns on their snouts, and very tall, slender camels. Plus, we would still see a few strange-looking mammals, but overall the fauna would be quite familiar.

Paleogene and Neogene Mammals

Mammals evolved from mammal-like reptiles called *cynodonts* during the Late Triassic and diversified during the Cenozoic, eventually giving rise to the present-day mammalian fauna. Now more than 5,000 species exist, ranging from tiny shrews to giants such as whales and elephants. Nevertheless, when one mentions the term *mammal*, what immediately comes to mind are horses, pigs, cattle, deer, dogs, cats, and so on. Most often we do not think much

(b) Scene from the Eocene showing the rhinoceros-sized animal known as *Unitatherium*, one of the first giants among mammals. It had three pairs of horns and saber-like upper canine teeth.

Chase Studio/Photo Researchers, Inc.

▶ Figure 18.9
Miocene Mammals of Nebraska This scene shows some of the animals that lived in what is now Nebraska 21 million years ago. The animals depicted here include (1) small camels, (2) bear dogs, (3) a chalicothere, (4) shovel-tusked elephants, (5) an entelodont, and (6) piglike oreodonts; (7) a saber-toothed cat in the tree feasts on its prey, and (8) cranes circle overhead.

Paleogene and Neogene Mammals **373**

about small mammals, rodents, shrews, rabbits, and bats. Yet most species of mammals, probably 70 percent, are quite small, weighing less than 1 kg.

Small Mammals—Insectivores, Rodents, Rabbits, and Bats

Insectivores, rodents, rabbits, and bats share a common ancestor, but they have had separate evolutionary histories since they first evolved (Figure 18.7). With the exception of bats, the oldest of which is found in Eocene rocks, these small mammals were present by the Late Mesozoic or Paleocene. The main reason we consider them together is that with few exceptions, they are small enough to have adapted to the microhabitats unavailable to larger mammals. In addition to being small, bats are the only mammals capable of flight.

As you would expect from the name Insectivora, members of this order—today's shrews, moles, and hedgehogs—eat insects. Insectivores have probably not changed much since they appeared during the Late Cretaceous. In fact, an insectivore-like creature very likely lies at the base of the great diversification of placental mammals.

More than 40 percent of all living mammal species are members of the order Rodentia, most of which are very small animals. A few, though, including beavers and the capybara of South America, are sizable animals; the latter is more than 1 m long and weighs 45 kg. One Miocene beaver known as *Paleocastor* was not particularly large, but it constructed some remarkably large burrows (▶ Figure 18.10), and the Miocene rodent found in South America, whimsically called "ratzilla," weighed an estimated 740 kg. Rodents evolved during the Paleocene, diversified rapidly, and adapted to a wide range of habitats. One reason for their phenomenal success is that they can eat almost anything.

Rabbits (order Lagomorpha) superficially resemble rodents but differ from them in several anatomic details. Furthermore, although they arose from a common ancestor during the Paleocene, rabbits and rodents have evolved independently. Like rodents, rabbits are gnawing animals, but details of their gnawing teeth differ. The development of long, powerful hind limbs for speed is the most obvious evolutionary trend in this group. Today's rabbits are not very large animals, rarely weighing more than a few kilograms, but a Pliocene-age fossil rabbit recently found in Spain may have weighed 12 kg.

The oldest fossil bat (order Chiroptera) comes from the Eocene-age Green River Formation of Wyoming, but well-preserved specimens are also known from several other areas. Apart from having forelimbs modified into wings, bats differ little from their immediate ancestors among the insectivores. Indeed, with the exception of wings, they closely resemble living shrews. Unlike pterosaurs and birds, bats use a modification of the hand in which four long fingers support the wings.

▶ Figure 18.10 **Fossil Rodents** *Paleocastor* was a 30-cm-long, Late Oligocene, land-dwelling beaver that made these spiral burrows in Nebraska. The burrows are locally called the Devil's Corkscrew.

Duane Clark/Proctor Museum of Natural Science

© Nobu Tamura

A Brief History of the Primates

The order **Primates** includes the prosimians (tarsiers, lemurs, and lorises), and the monkeys, apes, and humans, collectively referred to as anthropoids. Much of the primate story is more fully told in Chapter 19, where we consider human evolution, so in this chapter we will be brief. Primates may have evolved by Late Cretaceous time, but by the Paleocene they were undoubtedly present.

Small Paleocene primates closely resembled their contemporaries, the shrewlike insectivores. By the Eocene, though, larger primates had evolved, and lemurs and tarsiers that resemble their present descendants lived in Asia and North America. And by Oligocene time, primitive New World and Old World monkeys had developed in South America and Africa, respectively. The Hominoids, the group that includes apes and humans, evolved during the Miocene (see Chapter 19).

The Meat Eaters—Carnivorous Mammals

The order **Carnivora** is extremely varied and includes bears, seals, weasels, skunks, dogs, and cats. All are predators and therefore meat eaters, but their diets vary considerably. For example, cats rarely eat anything but meat, whereas bears, raccoons, and skunks have a varied diet and are thus *omnivorous*. Most carnivores have well-developed, sharp, pointed

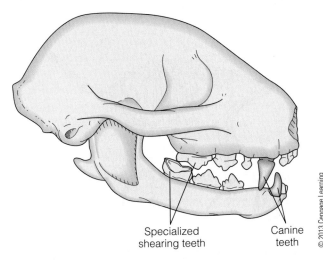

(a) This present-day skull and jaw of a large cat show the specialized, sharp-crested shearing teeth, or carnassials, of carnivorous mammals.

(b) Today's carnivorous mammals evolved from a primitive group known as miacids. This miacid, known as *Tapocyon*, was a coyote-sized animal that lived during the Eocene. All weasels, otters, skunks, badgers, martins, wolverines, seals, sea lions, walruses, bears, raccoons, dogs, hyenas, mongooses, and cats evolved from an ancestor very much like this creature.

canine teeth and specialized shearing teeth known as **carnassials** for slicing meat (▶ Figure 18.11a). Some land-dwelling carnivores depend on speed, agility, and intelligence to chase down prey, but others employ different tactics. For example, badgers dig their prey from burrows, and some small cats depend on stealth and pouncing to catch their meals.

Fossils of carnivorous mammals are not nearly as common as those of many other mammals—but why should this be so? First, in populations of warm-blooded (endothermic) animals, carnivores constitute no more than 5 percent of the total population, usually less. Second, many, but not all, carnivores are solitary animals, so the chance of large numbers of them being preserved together is remote. Nevertheless, fossil carnivores are common enough for us to piece together their overall evolutionary relationships with confidence.

The order Carnivora began to diversify when two distinct lines evolved from *creodonts* and *miacids* during the Paleocene. Both had well-developed canines and carnassials, but they were rather short-limbed and flat-footed. Certainly they were not very fast, but neither was their prey. The creodont branch became extinct by Miocene time, so need not concern us further, but the other branch, evolving from weasel-like miacids, led to all existing carnivorous mammals (Figure 18.11b).

The dogs evolved by the Middle Eocene, and even though they resemble hyenas, the latter are more closely related to cats and viverrids (civits and mongooses), and are another good example of convergent evolution. By Early Oligocene time, cats were present, and by the Miocene some had developed the remarkable enlarged canine teeth of the saber-tooths that existed during most of the Cenozoic Era. Their fossils are especially well known from Pleistocene-age deposits.

Seals, sea lions, and walruses adapted to an aquatic environment so they have a somewhat streamlined body, a layer of blubber for insulation, and limbs modified into paddles. Most eat fish and therefore have rather simple, single-cusped teeth, except walruses, which have flattened teeth for crushing shells. Fossils and studies of living animals clearly indicate that these aquatic carnivores are most closely related to bears.

The Ungulates or Hoofed Mammals

Ungulate is an informal term referring to several groups of living and extinct mammals, particularly the hoofed mammals of the orders **Artiodactyla** and **Perissodactyla**. The artiodactyls, or *even-toed hoofed mammals*, are the most diverse and numerous, with about 170 living species of cattle, goats, sheep, swine, antelope, deer, giraffes, hippopotamuses, camels, and several others. In marked contrast, the perissodactyls, or *odd-toed hoofed mammals*,

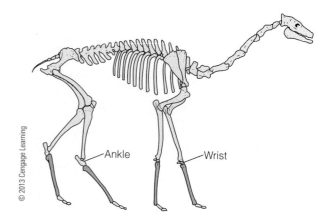

(a) Perissodactyls have one or three functional toes, whereas artio-dactyls have two or four. In perissodactyls, the weight is borne on the third toe, but in artiodactyls, it is borne on toes three and four.

(b) In many hoofed mammals, long, slender limbs evolved as bones between the wrist and toes and the ankle and toes became longer.

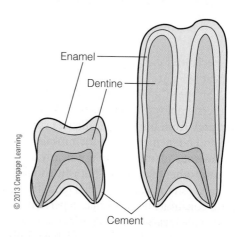

(c) High-crowned, cement-covered chewing teeth (*right*) evolved in hoofed mammals that adapted to a diet of grass. Low-crowned chewing teeth are found in many other mammals, including primates and pigs, both of which have a varied diet.

have only 16 existing species of horses, rhinoceroses, and tapirs. During the Early Cenozoic, though, perissodactyls were more abundant than artiodactyls.

Some defining characteristics of these groups are the number of toes and how the animal's weight is borne on the toes. (Their teeth are also distinctive.) Artiodactyls have either two or four toes, and their weight is borne along an axis that passes between the third and fourth digits (►Figure 18.12a). For those artiodactyls with two toes, such as today's swine and deer, the first, second, and fifth digits have been lost or remain only as vestiges. Perisso-dactyls have one or three toes, although a few fossil spe-cies retained four toes on their forefeet. Nevertheless, their weight is borne on an axis passing through the third toe (Figure 18.12a). Even today's horses have vestigial side toes, and rarely they are born with three toes.

Many hoofed mammals such as antelope and horses depend on speed to escape from predators in their open-grasslands habitat. As a result, they have long, slender limbs, giving them a greater stride length. Notice from Figure 18.12b that the bones of the palm and sole have become very long. In addition, these speedy runners have fewer bones in their feet than their ancestors, mostly because they have fewer toes. Not all hoofed mammals are long-limbed, speedy runners, though. Some are small and dart into heavy vegetation or a hole in the ground when threatened by predators. Size alone is adequate protection in some very large species such as rhinoceroses. In con-trast to the long, slender limbs of horses, antelope, deer, and so on, rhinoceroses and hippopotamuses have devel-oped massive, weight-supporting legs.

All artiodactyls and perissodactyls are herbivorous ani-mals, with their chewing teeth—premolars and molars—modified for a diet of vegetation. One evolutionary trend

in these animals was **molarization**, a change in the pre-molars so that they are more like molars, thus providing a continuous row of grinding teeth. Some ungulates—horses, for example—are characterized as **grazers** because they eat grass, as opposed to **browsers**, which eat the ten-der leaves, twigs, and shoots of trees and shrubs. Grasses are very abrasive, because as they grow through soil they pick up tiny particles of silt and sand that quickly wear teeth down. As a result, once grasses had evolved, many hoofed mammals became grazers and developed high-crowned, abrasion-resistant chewing teeth (Figure 18.12c).

Artiodactyls—Even-Toed Hoofed Mammals The old-est known artiodactyls were Early Eocene rabbit-sized animals that differed little from their ancestors. Yet these small creatures were ancestral to the myriad living and several extinct families of even-toed hoofed mammals (►Figure 18.13). Among the extinct families are the rather

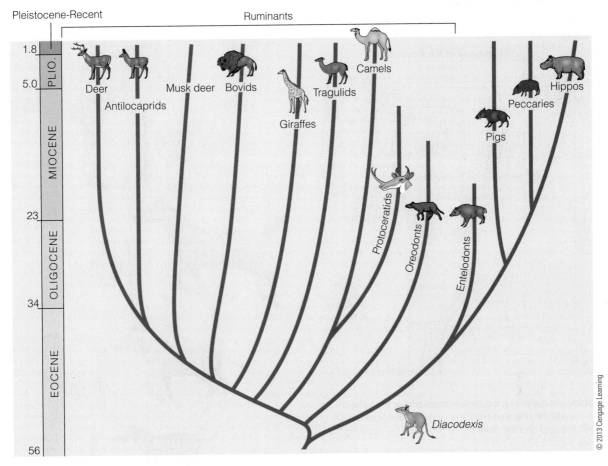

Figure 18.13 Relationships Among Artiodactyls (Even-Toed Hoofed Mammals) Most artiodactyls are ruminants—that is, cud-chewing animals with complex three- or four-chambered stomachs. The bovids are the most diverse and numerous living artiodactyls, whereas some other groups proliferated during the past—the oreodonts, for example. The oldest known artiodactyl, *Diacodexis*, was only about 50 cm long.

piglike oreodonts that were so common in North America until their extinction during the Pliocene and the peculiar genus *Synthetoceras*, with forked horns on their snouts.

During much of the Cenozoic Era, especially in North America, camels of one kind or another were common. The earliest were small four-toed animals, but by Oligocene time all had two toes. We mentioned some of the trends in camel evolution in Chapter 7 (see Figure 7.12b).

Among the various camels in North America were very tall giraffe-like camels; slender, gazelle-like camels; and giants standing 3.5 m high at the shoulder. Most camel evolution took place in North America, but during the Pliocene they migrated to Asia and South America, where the only living species exist now. North American camels went extinct near the end of the Pleistocene Epoch.

Among the artiodactyls, the family Bovidae is by far the most diverse, with dozens of species of cattle, bison, sheep, goats, and antelope. This family did not appear until the Miocene, but most of its diversification took place during Pliocene time on the northern continents. Bovids are now most numerous in Africa and southern Asia. North America still has its share of bovids, such as bighorn sheep

and mountain goats, but the most common ones during the Cenozoic were bison (which migrated from Asia), the pronghorn, and oreodonts, all of which roamed the western interior in vast herds.

Notice from Figure 18.13 that most living artiodactyls are **ruminants**, cud-chewing animals with complex three- or four-chambered stomachs in which food is processed to extract more nutrients. Perissodactyls lack such a complex digestive system. Perhaps the fact that artiodactyls use the same resources more effectively than do perissodactyls explains why artiodactyls have flourished and mostly replaced perissodactyls in the hoofed mammal fauna.

Perissodactyls—Odd-Toed Hoofed Mammals In Table 7.1, we said, "If we examine the fossil record of related organisms such as horses and rhinoceroses, we should find that they were quite similar when they diverged from a common ancestor but became increasingly different as divergence continued." We also discussed how the fossil records for horses, rhinoceroses, tapirs, and their extinct relatives, the chalicotheres and titanotheres, provide precisely this kind of evidence. In

▶ **Figure 18.14** **Evolution of Horses**

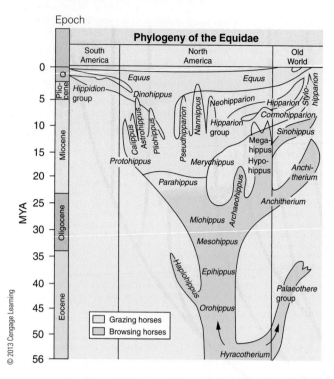

(a) Summary chart showing the relationships among the genera of horses. During the Oligocene two lines emerged, one leading to three-toed browsers and the other to one-toed grazers, including the present-day horse *Equus*.

(b) Simplified diagram showing some trends in horse evolution. Trends include a size increase, a lengthening of the limbs, a reduction in the number of toes, and the development of high-crowned teeth with complex chewing surfaces.

short, when these animals first appeared in the fossil record, they differed slightly in size and the structure of their teeth, but as they evolved, differences between them became more apparent. Perissodactyls evolved from a common ancestor during the Paleocene, reached their greatest diversity during the Oligocene, and have declined markedly since then.

Horse fossils are so common—especially in North America, where most of their evolution took place—that their overall history and evolutionary trends are well known. The earliest member of the horse family (family Equidae) is the fox-sized animal known as **Hyracotherium** (▶ Figure 18.14). This small, forest-dwelling animal had four-toed forefeet and three-toed hind feet, but each toe was covered by a small hoof. Otherwise it possessed few of the features of present-day horses. So how can we be sure it belongs to the family Equidae at all?

Horse evolution was a complex, branching affair, with numerous genera and species existing at various times during the Cenozoic (Figure 18.14a). Nevertheless, their exceptional fossil record clearly shows *Hyracotherium* is linked to the present-day horse, *Equus*, by a series of animals possessing intermediate characteristics. That is, Late Eocene and Early Oligocene horses, followed by more recent ones, show a progressive development of the characteristics found in present-day *Equus* (Table 18.1).

Figure 18.14a shows that horse evolution proceeded along two distinct branches. One led to three-toed browsing horses, all now extinct, and the other led to three-toed grazing horses and finally to one-toed grazers. The appearance of grazing horses, with high-crowned chewing teeth (Figure 18.12c), coincided with the evolution and spread of grasses during the Miocene. Speed was essential in this habitat, and horses' legs became longer and the number of toes was reduced finally to one (Figure 18.14b). Pony-sized *Merychippus* is a good example of the early grazing horses; it had three toes, but its teeth were high-crowned and covered by abrasion-resistant cement.

The other living perissodactyls, rhinoceroses and tapirs, increased in size from Early Cenozoic ancestors, and both became more diverse and widespread than they

A number of horse genera existed during the Cenozoic that evolved differently. For instance, some horses were browsers rather than grazers, never developed high-crowned chewing teeth, and retained three toes.

Trend

1. Size increase.
2. Legs and feet become longer, an adaptation for running.
3. Lateral toes reduced to vestiges. Only the third toe remains functional in *Equus*.
4. Straightening and stiffening of the back.
5. Incisor teeth become wider.
6. Molarization of premolars yields a continuous row of teeth for grinding vegetation.
7. The chewing teeth, molars and premolars, become high-crowned and cement-covered for grinding abrasive grasses.
8. Chewing surfaces of premolars and molars become more complex—also an adaptation for grinding abrasive grasses.
9. Front part of skull and lower jaw become deeper to accommodate high-crowned premolars and molars.
10. Face in front of eye becomes longer to accommodate high-crowned teeth.
11. Larger, more complex brain.

© 2013 Cengage Learning

are now. Most rhinoceroses evolved in the Old World, but North American rhinoceroses were common until they became extinct at the end of the Pleistocene. At more than 5 m high at the shoulder and weighing perhaps 13 or 14 metric tons, a hornless Oligocene–Miocene rhinoceros in Asia was the largest land-dwelling mammal ever.

For the remaining perissodactyls, chalicotheres, and titanotheres, only the latter has a good fossil record. Chalicotheres, although never particularly abundant, are interesting, because the later members of this family, which were the size of large horses, had claws on their feet rather than hooves. The prevailing opinion is that these claws were used to hook and pull down branches. Titanotheres existed only during the Eocene, giving them the distinction of being the shortest-lived perissodactyl family. They evolved from small ancestors to giants standing 2.5 m high at the shoulder (see Figure 7.12a).

Giant Land-Dwelling Mammals— Elephants

We just noted that the largest land-dwelling mammal ever was a hornless rhinoceros, but some of today's elephants, order **Proboscidea**, are also giants. The largest one on record weighed in at nearly 12 metric tons and stood 4.2 m at the shoulder, and some extinct mammoths were equally as large or perhaps slightly larger. In addition to their size, a distinctive feature of elephants is their long snout, or proboscis. During much of the Cenozoic Era, proboscideans of one kind or another were widespread on the northern continents, but now only three species exist, one in southeast Asia and two in Africa. The earliest member of the order was a 100- to 200-kg animal known as *Moeritherium* from the Eocene that possessed few characteristics of elephants. It was probably aquatic.

By Oligocene time, elephants showed the trends toward large size and had developed a long proboscis and large

tusks, which are enlarged incisors. Most elephants developed tusks in the upper jaw only, but a few had them in both jaws, and one, the deinotheres, had only lower tusks (▶Figure 18.15).

The most familiar elephants, other than living ones, are the extinct *mastodons* and *mammoths*. Mastodons evolved in Africa, but from Miocene to Pleistocene time they spread over the Northern Hemisphere continents, and one genus even reached South America. These large, browsing animals died out only a few thousands of years ago. During the Pliocene and Pleistocene, mammoths and living elephants diverged (Figure 18.15). Mammoths were about the size of elephants today, but they had the largest tusks of any elephant. In fact, mammoth tusks are common enough in Siberia that they have been and continue to be a source of ivory. Until their extinction near the end of the Pleistocene, mammoths lived on all Northern Hemisphere continents as well as in India and Africa.

Giant Aquatic Mammals—Whales

Our fascination with huge dinosaurs should not overshadow the fact that by far the largest animal ever is alive today. At more than 30 m long and weighing an estimated 130 metric tons, blue whales greatly exceed the size of any other living thing, except some plants such as redwood trees. But not all whales are large. Consider, for instance, dolphins and porpoises—both are sizable but hardly giants. Nevertheless, an important trend in whale evolution has been increase in body size.

Several kinds of mammals are aquatic or semi-aquatic, but only sea cows (order Sirenia) and whales (order **Cetacea**), are so thoroughly aquatic that they cannot come out onto land. Fossils discovered in Eocene rocks in Southeast Asia indicate that the land-dwelling ancestors of whales were among the small, dog-sized animals (▶Figure 18.16). In fact, fossil evidence has convinced paleontologists that among mammals, whales are

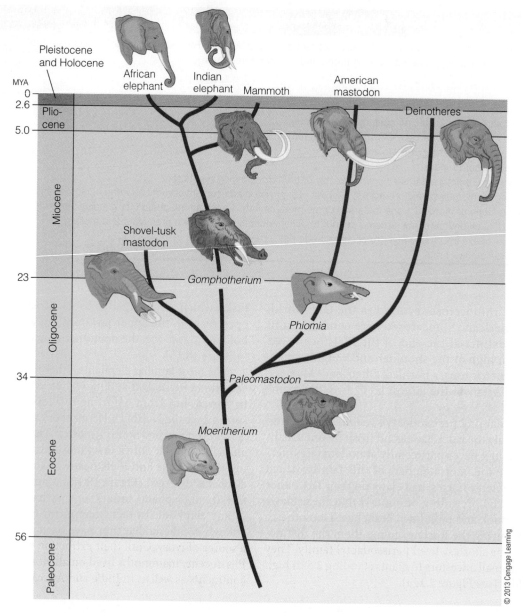

▶ **Figure 18.15 Phylogeny of Elephants and Some of Their Relatives** Large size, a long proboscis, and tusks were some of the evolutionary trends in proboscideans. Proboscideans were more diverse than indicated in this illustration, which excludes several fossil proboscideans.

most closely related to artiodactyls (which are even-toed hoofed mammals) (Figure 18.16). During the transition from land-dwelling animals to aquatic whales, the front limbs modified into paddlelike flippers; the rear limbs were lost; the nostrils migrated to the top of the head; and a large, horizontal tail fluke used for propulsion developed.

We mentioned in Chapter 7 that whales evolved from land-dwelling ancestors in a part of the world where the fossil record was poorly known. Beginning about 20 years ago, though, paleontologists have made some remarkable discoveries that resolved the enigma of how and where whales evolved. For instance, the Early Eocene whale *Ambulocetus* still had limbs capable of supporting

the animal on land, whereas *Basilosaurus*, a 15-m-long whale from the Late Eocene, had only tiny vestigial rear limbs. *Basilosaurus* also retained teeth much like those of the whale ancestors and its nostrils were on its snout. An interesting note on *Basilosaurus* was the so-called "sea serpent" found in Eocene-age rocks of Alabama by Albert Koch during the 1840s (see Perspective).

By Oligocene time, both existing whale groups—baleen whales and toothed whales—had evolved; baleen consists of platelike, filter-feeding devices that differ considerably from teeth. And yet, an Oligocene whale known as *Aetiocetus* had both baleen and teeth, and it also had a blowhole (nostrils) about midway between the snout and the top of the head.

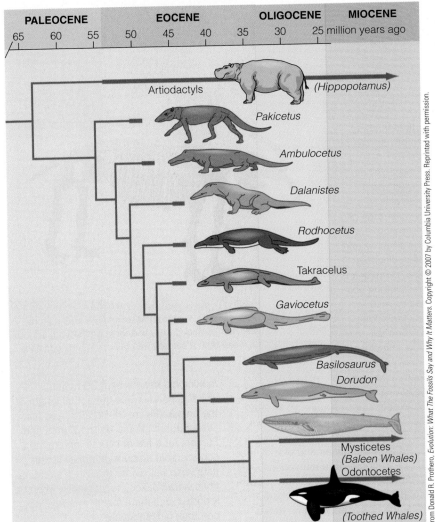

PALEOCENE		EOCENE			OLIGOCENE		MIOCENE
65	60	55	50	45	40	35	30	25 million years ago

Artiodactyls (Hippopotamus)

Pakicetus

Ambulocetus

Dalanistes

Rodhocetus

Takracelus

Gaviocetus

Basilosaurus

Dorudon

Mysticetes (Baleen Whales)

Odontocetes (Toothed Whales)

From Donald R. Prothero, *Evolution: What The Fossils Say and Why It Matters*. Copyright © 2007 by Columbia University Press. Reprinted with permission.

▶ **Figure 18.16 The Evolution of Whales** The land-dwelling ancestors of whales were among the even-toed hoofed mammals, the artiodactyls. Note that *Pakicetus* had well-developed hind limbs, but some of the others, such as *Basilosaurus*, did not. Although *Basilosaurus* was a fully aquatic whale, it differed considerably from today's whales.

Pleistocene Faunas

Unlike the Paleocene fauna, with its archaic mammals, unfamiliar ancestors of living mammalian orders, and large, predatory birds, the fauna of the Pleistocene consists mostly of familiar animals. Even so, their geographic distribution might surprise us, because rhinoceroses, elephants, and camels lived in North America, and a few unusual mammals, such as chalicotheres and the heavily armored glypotodonts, were present. Among the avian fauna, the giant moas and elephant birds were in New Zealand and in Madagascar, respectively.

Ice Age Mammals

The most remarkable aspect of the Pleistocene mammalian fauna is that so many very large species existed. Mastodons, mammoths, giant bison, huge ground sloths, immense camels, and beavers nearly 2 m long were present in North America. South America had its share of giants, too, especially sloths and glyptodonts. Elephants, cave bears, and giant deer known as Irish Elk lived in Europe and Asia (▶ Figure 18.17), and Australia had 3-m-tall kangaroos and wombats the size of rhinoceroses.

Of course, many smaller mammals also existed, but one obvious trend among Pleistocene mammals was large body size. Perhaps this was an adaptation to the cooler conditions that prevailed during that time. Large animals have less surface area compared to their volume and thus retain heat more effectively than do smaller animals.

Some of the world's best-known fossils come from Pleistocene deposits. You have probably heard of the frozen mammals found in Siberia and Alaska, such as mammoths, bison, and a few others. These extraordinary fossils, although very rare, provide much more information than most fossils do (see Figure 5.12b). Contrary to what

Perspective

State Fossils

Most U.S. states and territories have a state gemstone, mineral, and rock, and 42 states have a designated state fossil. Even Washington, D.C., has a fossil dinosaur called *Capitalosaurus*, although it has never been formally described. State fossils include plants, petrified wood, invertebrate animals such as eurypterids, bivalves, trilobites, and brachiopods, as well as dinosaurs and several mammals. California's state fossil, for example, is the saber-toothed cat, whereas Alaska's is the wooly mammoth, and West Virginia's designated fossil is the ground sloth.

All state fossils are interesting in their own right, but perhaps one of the most notable is *Basilosaurus*, a fossil whale that serves as the state fossil for Alabama and Mississippi even though it was first discovered in Louisiana. It was about 18 m long and differed considerably in its proportions and other features from today's whales.

The discovery of a nearly complete specimen of *Basilosaurus* in the 1840s by Albert Koch and his subsequent reconstruction of a huge sea serpent was remarkable. Albert Koch was born in 1804 in Germany, immigrated to the United States when he was 22, and eventually took up residence in St. Louis, Missouri, where he owned a small museum. Koch was certainly a keen observer of geologic phenomena and is now credited with discovering some important

information about Earth history, but he was also a hoaxer. It all began when Koch recovered the remains of a large mastodon and reassembled it with extra vertebrae and wooden blocks to make it even larger; his restoration portrayed an animal nearly 10 m long and 4.6 m tall. In addition, he placed the mastodon's tusks at a peculiar angle, giving his creature an even more unusual appearance (Figure 1). In 1842, he sold his creation, called *Missourium*, to the British Museum, where scientists removed the extra bones and blocks, thus making it a normal-sized mastodon.

Koch attracted visitors to his museum with unusual displays, and the more unusual the better, so when he heard rumors of people finding sea serpent bones in Alabama he went there and eventually discovered a nearly complete specimen of *Basilosaurus*, or what was originally called *Zeuglodon*. In any case, the fossils he collected were real enough, but in his restoration Koch used vertebrae from five different animals, thereby

Figure 1 Koch's mammoth, known as *Missourium*, was a restoration of a real fossil, but he added extra vertebrae and wood blocks to make the animal larger than it actually was. Also, note the odd position of the tusks.

making his sea serpent 35 m long, which is nearly twice the actual length of the animal (Figure 2). He displayed his "sea serpent" in the United States and Europe, for a fee of course, and made a considerable amount of money. Scientists were well aware that Koch's restorations were hoaxes, so, ironically, when he reported legitimate fossil evidence, no one took him seriously.

There have been other hoaxes, as well as honest mistakes, in the history of paleontology, both before and after Koch built his sea serpent. Most of these fooled no one, at least not for very long, but one perpetrated in 1913, called *Piltdown man*, was not revealed as a hoax by scientists for several decades.

Figure 2 This "sea serpent" on display in New York was constructed by Albert Koch from the bones of several fossil whales.

▶ Figure 18.17 **The Irish Elk** Restoration of the giant deer *Megaloceros giganteus*, commonly called the Irish elk. It lived in Europe and Asia during the Pleistocene, and may have persisted in Eastern Europe until only a few thousand years ago. Large males probably weighed 700 kg, about the same as a present-day moose, and had an antler spread of nearly 4 m.

you might hear in the popular press, all these frozen animals were partly decomposed, none were fresh enough to eat, and none were found in blocks of ice or icebergs. All were recovered from permanently frozen ground known as *permafrost.*

Paleontologists have recovered Pleistocene animals from many places in North America; two noteworthy areas are Florida and the La Brea tar pits at Rancho La Brea in southern California. In fact, Florida is one of the few places in the eastern United States where fossils of Cenozoic land-dwelling animals are common (▶Figure 18.18a). At the La Brea tar pits, at least 230 kinds of vertebrate animals have been found trapped in the sticky residue where liquid petroleum seeped out at the surface and then evaporated.

The "tar" is really naturally formed asphalt, not the product manufactured from peat or coal. Many of the fossils are carnivores, especially dire wolves, saber-tooth cats, and vultures, that gathered to dine on mammals that became mired in the "tar" (Figure 18.18b).

Pleistocene Extinctions

During the Pleistocene, the continental interior of North America was teeming with horses, rhinoceroses, camels, mammoths, mastodons, bison, giant ground sloths, glyptodonts, saber-toothed cats, dire wolves, rodents, and

rabbits. Beginning about 14,000 years ago, however, many of these animals became extinct, especially the larger ones. These Pleistocene extinctions were modest compared to those at the end of the Paleozoic and Mesozoic eras, but they were unusual in that they had a profound impact on large, land-dwelling mammals (those weighing more than 44 kg). Particularly hard hit were the mammalian faunas of Australia and the Americas.

In Australia, 15 of the continent's 16 genera of large mammals died out. North America lost 33 of 45 genera of large mammals, and in South America 46 of 58 large mammal genera went extinct. In contrast, Europe had only 7 of 23 large genera die out, whereas Africa south of the Sahara lost only 2 of 44 genera. These data bring up three questions, none of which has been answered completely: (1) What caused Pleistocene extinctions? (2) Why did these extinctions eliminate mostly large mammals? (3) Why were extinctions most severe in Australia and the Americas? Scientists are currently debating two competing hypotheses for these extinctions. One, the *climatic change hypothesis*, holds that rapid changes in climate at the end of the Pleistocene caused extinctions, whereas proponents of *prehistoric overkill* contend that human hunters were responsible.

Rapid changes in climate and vegetation did occur over much of Earth's surface during the Late Pleistocene, as glaciers began retreating. The North American and northern Eurasian open-steppe tundras were replaced by conifer and broadleaf forests as warmer and wetter conditions prevailed. The Arctic region flora changed from a productive herbaceous one that supported a variety of large mammals, to a comparatively barren water-logged tundra that supported a much sparser fauna. The southwestern United States region also changed, from a moist area with numerous lakes, where saber-tooth cats, giant ground sloths, and mammoths roamed, to a semiarid environment unable to support a diverse fauna of large mammals.

Rapid changes in climate and vegetation can certainly affect animal populations, but there are several problems with the climate-change hypothesis. First, why didn't the large mammals migrate to more suitable habitats as the climate and vegetation changed? After all, many animal species did. For example, reindeer and the Arctic fox lived in southern France during the last glaciation and migrated to the Arctic when the climate became warmer. The second argument against the climate hypothesis is the apparent lack of correlation between extinctions and the earlier glacial advances and retreats throughout the Pleistocene Epoch. Previous changes in climate were not marked by episodes of mass extinctions.

Proponents of the prehistoric-overkill hypothesis argue that the mass extinctions in North and South America and Australia coincided closely with the arrival of humans. Perhaps hunters had a tremendous impact on the faunas of North and South America about 11,000 years ago because the animals had no previous experience with humans.

Erika Simons/Florida Museum of Natural History

(a) Among the diverse Pliocene and Pleistocene mammals of Florida were 6-m-long ground sloths and armored glyptodonts that weighed more than 2 metric tons.

TOM MCHUGH/SCIENCE PHOTO

(b) Restoration of a camel trapped in the sticky tar (asphalt) at the La Brea Tar Pits in Los Angeles, California. Vultures and a saber-toothed cat look on. Notice the giant ground sloths in the background.

The same thing happened much earlier in Australia soon after people arrived about 40,000 years ago. No large-scale extinctions occurred in Africa and most of Europe, because animals in those regions had long been familiar with humans.

One problem with the prehistoric-overkill hypothesis is that archaeological evidence indicates the early human inhabitants of North and South America, as well as Australia, probably lived in small, scattered communities, gathering food and hunting. How could a few hunters decimate so many species of large mammals? However, it is true that humans have caused extinctions on oceanic islands. For example, in a period of about 600 years after arriving in New Zealand, humans exterminated several species of the large, flightless birds called moas.

A second problem is that present-day hunters concentrate on smaller, abundant, and less dangerous animals. The remains of horses, reindeer, and other smaller animals

Migrants from
South America

Opossum

Ground sloth

Glyptodon

Armadillo

Caribbean
Sea

Pacific
Ocean

Bear

Horse

Tapir

Squirrel

Camel

Mastodon

Atlantic
Ocean

Migrants from
North America

Cats—including
saber-tooths

Deer

From WICANDER/MONROE, *GEOL, 2e.* Copyright © 2013 Cengage Learning

▶ **Figure 18.19 The Great American Interchange** South America was isolated during much of the Cenozoic, and its mammal fauna consisted of marsupials and placentals unlike those anywhere else. When the Isthmus of Panama formed during the Late Pliocene, many placental mammals migrated south, and many South American mammals went extinct. A few South American mammals migrated north and successfully occupied North America.

are found in many prehistoric sites in Europe, whereas mammoth and woolly rhinoceros remains are scarce.

Finally, few human artifacts are found among the remains of extinct animals in North and South America, and there is usually little evidence that the animals were hunted. Countering this argument is the assertion that the impact on the previously unhunted fauna was so swift as to leave little evidence.

The reason for the extinctions of large Pleistocene mammals is unresolved and probably will be for some time. It may turn out that the extinctions resulted from a combination of different circumstances. Populations that were already under stress from climate changes were perhaps more vulnerable to hunting, especially if small females and young animals were the preferred targets.

Intercontinental Migrations

Intercontinental migrations account for the many similarities of North American, Asian, and European mammalian faunas throughout the Cenozoic. Even today, Asia and

North America are only narrowly separated at the Bering Strait, which at several times during the Cenozoic formed a land corridor across which mammals migrated (see Figure 17.13). During the Early Cenozoic, a land connection between Europe and North America allowed mammals to roam across all the northern continents. Many did; camels and horses are only two examples.

In contrast, the southern continents were largely separate island continents during much of the Cenozoic. Africa, however, remained fairly close to Eurasia, and at times faunal interchange between those two continents was possible. For example, elephants first evolved in Africa, but they migrated to all the northern continents.

South America was isolated from all other landmasses from Late Cretaceous until a land connection with North America formed about 5 million years ago. Before the connection was established, the South American fauna was made up of marsupials and several orders of placental mammals that lived nowhere else. These animals thrived in isolation and showed remarkable convergence with North American placental mammals (see Figure 7.11a). When the Isthmus of Panama formed, migrants from North America soon replaced many of the indigenous South American mammals, whereas fewer migrants from the south were successful in North America (▶ Figure 18.19). As a result of this *great American interchange*, today about 50 percent of South America's mammalian fauna came from the north, but in North America only 20 percent of its mammals came from the south. Even today, the coyote (*Canis latrans*) is extending its range from the north through Central America.

Most living species of marsupials are restricted to the Australian region. Recall from Chapter 15 that marsupials occupied Australia before its separation from Gondwana, but apparently placentals, other than bats and a few rodents, never got there until they were introduced by humans. So, unlike South America, which now has a connection with another continent, Australia has remained isolated, and its fauna is unique.

Summary

▶ The marine invertebrate groups that survived the Mesozoic extinctions diversified throughout the Cenozoic. Bivalves, gastropods, corals, and several kinds of phytoplankton such as foraminifera proliferated.

▶ During much of the Early Cenozoic, North America was covered by subtropical and tropical forests, but the climate became drier by Oligocene and Miocene time, especially in the midcontinent region.

▶ Birds belonging to the living orders and families evolved during the Paleogene Period. Large, flightless, predatory birds of the Paleogene were eventually replaced by mammalian predators.

▶ Evolutionary history is better known for mammals than for other classes of vertebrates, because mammals have a good fossil record, their teeth are so distinctive, and Cenozoic deposits are easily accessible.

▶ Egg-laying mammals (monotremes) and marsupials exist mostly in the Australian region. The placental mammals—by far the most common mammals—owe their success to their method of reproduction.

▶ All placental and marsupial mammals descended from shrewlike ancestors that existed from Late Cretaceous to Paleogene time.

▶ Small mammals such as insectivores, rodents, and rabbits occupy the microhabitats unavailable to larger mammals. Bats, the only flying mammals, have forelimbs modified into wings but otherwise differ little from their ancestors.

▶ Most carnivorous mammals have well-developed canines and specialized shearing teeth, although some aquatic carnivores such as seals have peglike teeth.

▶ The most common ungulates are the even-toed hoofed mammals (artiodactyls) and odd-toed hoofed mammals (perissodactyls), both of which evolved during the Eocene. Many ungulates show evolutionary trends such as molarization of the premolars as well as lengthening of the legs for speed.

▶ During the Paleogene, perissodactyls were more common than artiodactyls, but now their 16 living species constitute less than 10 percent of the world's hoofed mammal fauna.

▶ Although present-day *Equus* differs considerably from the oldest known member of the horse family, *Hyracotherium*, an excellent fossil record shows a continuous series of animals linking the two.

▶ Even though horses, rhinoceroses, and tapirs, as well as the extinct titanotheres and chalicotheres, do not closely resemble one another, fossils show they diverged from a common ancestor during the Eocene.

▶ The fossil record for whales verifies that they evolved from land-dwelling ancestors during the Eocene.

▶ Elephants evolved from rather small ancestors, became quite diverse and abundant, especially on the Northern Hemisphere continents, and then dwindled to only three living species.

▶ Horses, camels, elephants, and other mammals spread across the northern continents during the Cenozoic because land connections existed between those landmasses at various times.

▶ During most of the Cenozoic, South America was isolated, and its mammal fauna was unique. A land connection was established between the Americas during the Late Cenozoic, and migrations in both directions took place.

▶ One important evolutionary trend in Pleistocene mammals and some birds was toward giantism. Many of these large species died out, beginning about 40,000 years ago.

▶ Changes in climate and prehistoric overkill are the two hypotheses explaining Pleistocene extinctions.

Important Terms

Review Questions

1. The Artiodacyla is an order of mammals that includes the
   a. ____ carnivores; b. ____ elephants and whales; c. ____ even-toed hoofed mammals; d. ____ rodents, rabbits, and bats; e. ____ primates.

2. Cenozoic bryozoans were particularly abundant and successful
   a. ____ suspension feeders; b. ____ predatory birds; c. ____ marine reptiles; d. ____ meat-eating mammals; e. ____ bony fish.

3. One of the defining features of the carnivorous mammals is
   a. ____ short, massive limbs; b. ____ molarization of the premolars; c. ____ teeth adapted to grazing; d. ____ four-chambered stomach; e. ____ carnassial teeth.

4. Which one of the following statements is incorrect?
   a. ____ The oldest known member of the horse family is *Hyracotherium*; b. ____ *Ungulate* is a designation for all hoofed mammals; c. ____ Browsers have high-crowned chewing teeth; d. ____ Multituberculates were rodent-like mammals that survived from the Mesozoic into the Cenozoic; e. ____ Flightless, predatory birds were common in South America during the Cenozoic.

5. The three geologic periods, from oldest to youngest, for the Cenozoic Era are
   a. ____ Quaternary, Paleogene, Neogene; b. ____ Neogene, Miocene, Pleistocene; c. ____ Mesozoic, Eocene, Holocene; d. ____ Paleogene, Neogene, Quaternary; e. ____ Cretaceous, Pliocene, Oligocene.

6. The monotremes, one of the three basic groups of mammals, is the only one that
   a. ____ lays eggs; b. ____ bears live young; c. ____ is warm-blooded; d. ____ has vestigial wings; e. ____ has a fairly good fossil record.

7. One of the main components of the Cenozoic marine fauna was
   a. ____ foraminifera; b. ____ cetaceans; c. ____ ammonites; d. ____ trilobites; e. ____ ungulates.

8. Which one of the following has been proposed as a cause of Pleistocene extinctions?
   a. ____ Meteorite impact; b. ____ Climate change; c. ____ Evolution of grasses; d. ____ Migration of mammals from Australia; e. ____ Increased energy from the Sun.

9. One indication of a cool climate is
   a. ____ an increase in diversity of odd-toed hoofed mammals; b. ____ extinction of coccolithophores; c. ____ small leaves with incised margins; d. ____ mammals with nostrils on the top of the head; e. ____ evolution of large, flightless birds.

10. During the Cenozoic Era, Earth's temperatures were highest during the _____ epoch.
    a. ____ Eocene; b. ____ Holocene; c. ____ Miocene; d. ____ Pleistocene; e. ____ Oligocene.

11. You are a science teacher who receives numerous unlabeled mammal and plant fossils. You are not too concerned with identifying genera and species, but you do want to show your students mammal adaptations for diet and speed. What features of the skulls, teeth, and limb bones would allow you to infer which animals were herbivores (grazers versus browsers), which were carnivores, and which ones were speedy runners? Also, could you use the plant fossils to make inferences about ancient climates?

12. Dogs and hyenas are similar in their appearance and in their social structures, but they are only very distantly related. Explain how this came to be so.

13. What are the main evolutionary trends that took place in hoofed mammals that adapted to grazing in an open plains habitat?

14. Summarize the major events that took place in the evolution from fish to mammals without using technical language. In short, how would you explain the overall evolution of vertebrates to an interested but uninformed audience?

15. Why are so many fossil land-dwelling mammals found in the western United States, but so few in the eastern part of the country?

16. How do you explain the fact that so much more is known about Cenozoic Earth and life history than about the previous intervals of geologic time?

17. How do fossil leaves and the composition of paleofloras give clues about ancient climates?

18. Explain how and why the South American Cenozoic mammal fauna differed so much from faunas elsewhere, and how that fauna changed.

19. What are the hypotheses for the extinctions that occurred at the end of the Pleistocene Epoch? Does any evidence support either or both hypotheses?

20. Discuss three evolutionary trends seen in whales.

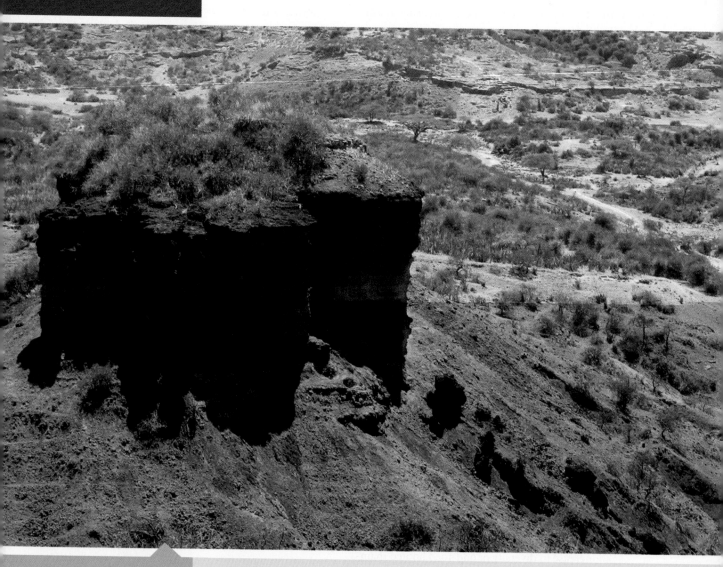

19 Primate and Human Evolution

Olduvai Gorge, on the eastern Serengeti Plain, Northern Tanzania, is often referred to as the "Cradle of Mankind" because of the many important hominid discoveries made there. The gorge, part of the East African Rift Valley, is 48 km long and 9 m deep and formed as a result of the tectonic forces shaping East Africa.

Courtesy of Persis Sturgis

At the end of this chapter, you will have learned that

▶ Primates are difficult to characterize as an order because they lack strong specializations found in most other mammalian orders.

▶ Primates are divided into two suborders: the prosimians, which include lemurs, lorises, tarsiers, and tree shrews, and the anthropoids, which include monkeys, apes, and humans.

▶ The hominids include present-day humans and their extinct ancestors, which have a fossil record extending back almost 7 million years.

▶ Hominid, and therefore human evolution, is very complex and in a constant state of flux owing to various new discoveries.

▶ The most famous of all fossil humans are the Neanderthals, who inhabited Europe and the Middle East beginning about 200,000 years ago, and were succeeded by the Cro-Magnons about 30,000 years ago.

Introduction

Who are we? Where did we come from? What is the human genealogy? These are basic questions that we probably have all asked ourselves at some time or another. Just as many people enjoy tracing their own family history as far back as they can, paleoanthropologists are discovering that the human family tree goes back much farther than we thought.

So where does this leave us, evolutionarily speaking? It leaves us at a very exciting time as we seek to unravel the history of our species. Our understanding of our own phylogeny is presently in flux, and each new fossil hominid discovery sheds more light on our ancestry. Although some may find it frustrating, human evolution is just like that of other groups in that we have followed an uncertain evolutionary path. As new species evolved, they filled ecological niches and either gave rise to descendants better adapted to the changing environment, or they became extinct. So it should not surprise us that our own evolutionary history has many "dead-end" side branches.

In this chapter, we examine the various primate groups, in particular the origin and evolution of the hominids, the group that includes our ancestors. However, we must point out that recent discoveries of fossil hominids, as well as innovative techniques for scientific analysis, are leading to new hypotheses about our ancestry. By the time you read this chapter, it is very likely that new discoveries may change some of the conclusions stated here. Such is the nature of paleoanthropology—and one reason why the study of hominids is so fascinating and exciting.

What Are Primates?

Primates are difficult to characterize as an order because they lack the strong specializations found in most other mammalian orders. We can, however, point to several trends in their evolution that help define primates and are related to their *arboreal,* or tree-dwelling, ancestry. These include changes in the skeleton and mode of locomotion; an increase in brain size; a shift toward smaller, fewer, and less specialized teeth; stereoscopic vision, and the evolution of a grasping hand with an opposable thumb. Not all of these trends took place in every primate group, nor did they evolve at the same rate in each group. In fact, some primates have retained certain primitive features, whereas others show all or most of these trends.

The Primate order is divided into two suborders, the Prosimii and Anthropoidea (▶Figure 19.1, Table 19.1). The *prosimians* include the lemurs, lorises, tarsiers, and tree shrews, whereas the *anthropoids* include monkeys, apes, and humans.

Prosimians

Prosimians are generally small, ranging from species the size of a mouse up to those as large as a house cat. They are arboreal, have five digits on each hand and foot with

TABLE 19.1 Classification of the Primates

Order Primates: Lemurs, lorises, tarsiers, tree shrews, monkeys, apes, humans

    Suborder Prosimii: Lemurs, lorises, tarsiers, tree shrews

    Suborder Anthropoidea: Monkeys, apes, humans

        Superfamily Cercopithecoidea: Macaque, baboon, proboscis monkey (Old World monkeys)

        Superfamily Ceboidea: Howler, spider, and squirrel monkeys (New World monkeys)

        Superfamily Hominoidea: Apes, humans

            Family Pongidae: Chimpanzees, orangutans, gorillas

            Family Hylobatidae: Gibbons, siamangs

            Family Hominidae: Humans

(a) Tarsier.

(b) Baboon.

▶ Figure 19.1 **Primates** Primates are divided into two suborders: the prosimians (a) and the anthropoids (b–d), which are further subdivided into three superfamilies: Old World monkeys (b), New World monkeys (c), and hominoids, which include the apes (d) and humans.

(c) Spider monkey.

(d) Chimpanzee.

either claws or nails, and are typically omnivorous. They have large, forwardly directed eyes specialized for night vision—hence, most are nocturnal (Figure 19.1a). As their name implies (*pro* means "before," and *simian* means "ape"), they are the oldest primate lineage, with a fossil record extending back to the Paleocene.

During the Eocene, prosimians were abundant, diversified, and widespread in North America, Europe, and Asia (▶Figure 19.2). As the continents moved northward during the Cenozoic, and the climate changed from warm-tropical to cooler midlatitude conditions, the prosimian population decreased in both abundance and diversity.

By the Oligocene, hardly any prosimians were left in the northern continents as the once widespread Eocene populations migrated south to the warmer latitudes of Africa, Asia, and Southeast Asia. Presently, prosimians are found only in the tropical regions of Asia, India, Africa, and Madagascar.

Anthropoids

Anthropoids evolved from a prosimian lineage sometime during the Late Eocene, and by the Oligocene, they were a well-established group. Much of our knowledge about the

early evolutionary history of anthropoids comes from fossils found in the Fayum district, a small desert area southwest of Cairo, Egypt. During the Late Eocene and Oligocene, this region of Africa was a lush, tropical rainforest that supported a diverse and abundant fauna and flora. Within this forest lived many different arboreal anthropoids as well as various prosimians. In fact, several thousand fossil specimens representing more than 20 primate

▶ Figure 19.2 **Eocene Prosimian** *Notharctus*, a primitive Eocene prosimian from North America.

species have been recovered from rocks of this region. One of the earliest anthropoids was *Aegyptopithecus,* a small, Late Eocene, fruit-eating, arboreal primate that weighed about 5 kg (▶ Figure 19.3). *Aegyptopithecus* had not only monkey characteristics, but also features that were more like those of apes as well. As such, it is presently the closest link we have to the Old World primates.

Scientists divide anthropoids into three superfamilies: Cercopithecoidea (Old World monkeys), Ceboidea (New World monkeys), and Hominoidea (apes and humans) (Table 19.1). *Old World monkeys* (superfamily Cercopithecoidea) include the macaque, baboon, and proboscis monkey, and are characterized by close-set, downward-directed nostrils (like those of apes and humans), grasping hands, and a nonprehensile tail (Figure 19.1b). Present-day Old World monkeys are distributed throughout the tropical regions of Africa and Asia and are thought to have evolved from a primitive anthropoid ancestor, like *Aegyptopithecus,* sometime during the Oligocene.

New World monkeys (superfamily Ceboidea) are found only in Central and South America. They are characterized by a prehensile tail, flattish face, and widely separated nostrils, and include the howler, spider, and squirrel monkeys (Figure 19.1c). New World monkeys probably evolved from African monkeys that migrated across the widening Atlantic sometime during the Early Oligocene, and they have continued evolving in isolation to this day. No evidence exists of any prosimian or other primitive primates in Central or South America, or of any contact with Old World monkeys after the initial immigration from Africa.

Hominoids (superfamily Hominoidea) consist of three families: the *great apes* (family Pongidae), which include chimpanzees, orangutans, and gorillas (Figure 19.1d); the *lesser apes* (family Hylobatidae), which are gibbons and siamangs; and the *hominids* (family Hominidae), which are humans and their extinct ancestors. The hominoid lineage diverged from Old World monkeys sometime before the Miocene, but exactly when is still being debated. It is

generally accepted, however, that hominoids evolved in Africa, probably from the ancestral group that included *Aegyptopithecus.*

Recall that beginning in the Late Eocene the northward movement of the continents resulted in pronounced climatic shifts. In Africa, Europe, Asia, and elsewhere, a major cooling trend began (see Figure 18.5b), and the tropical and subtropical rain forests slowly began to change to a variety of mixed forests separated by savannas and open grasslands as temperatures and rainfall decreased. As the climate changed, the primate populations also changed. Prosimians and monkeys became rare, whereas hominoids diversified in the newly forming environments and became abundant. Ape populations became reproductively isolated from each other within the various forests, leading to adaptive radiation and increased diversity among the hominoids. During the Miocene, Africa collided with Eurasia, producing additional changes in the climate, as well as providing opportunities for migration of animals between the two landmasses.

Two apelike groups evolved during the Miocene that ultimately gave rise to present-day hominoids. Although there is still no agreement on the early evolutionary relationships among the hominoids, fossil evidence and molecular DNA similarities between present-day hominoid families is providing a clearer picture of the evolutionary pathways and relationships among the hominoids.

The first group, the *dryopithecines,* evolved in Africa during the Miocene and subsequently spread to Eurasia following the collision between the two continents. The dryopithecines were a group of hominoids that varied in size, skeletal features, and lifestyle. The best known of all later hominoids is *Proconsul,* an apelike, fruit-eating animal that led a quadrupedal arboreal existence with limited activity on the ground (▶ Figure 19.4). The dryopithecines were very abundant and diverse during the Miocene and Pliocene, particularly in Africa.

▶ Figure 19.4 **Proconsul** Probable appearance of *Proconsul*, a dryopithecine.

The second group, the *sivapithecids*, evolved in Africa during the Miocene, then spread throughout Eurasia. The fossil remains of sivapithecids are plentiful and consist mostly of skulls, jaws, and isolated teeth. Body or limb bones are rare, limiting our knowledge about what they looked like and how they moved around. We do know that sivapithecids had powerful jaws and thick-enameled teeth with flat chewing surfaces, suggesting a diet of hard and coarse foods, including nuts.

It is clear from the fossil evidence that sivapithecids were not involved in the evolutionary branch leading to humans, but they were probably the ancestral stock from which present-day orangutans evolved. In fact, one genus, *Gigantopithecus*, was a contemporary of early *Homo* in Eastern Asia.

Although many pieces are still missing, particularly during critical intervals in the African hominoid fossil record, molecular DNA as well as fossil evidence indicates the dryopithecines, African apes, and hominids form a closely related lineage. The sivapithecids and orangutans, as discussed, form a different lineage, one that did not lead to humans.

Hominids

The **hominids** (family Hominidae), the primate family that includes present-day humans and their extinct ancestors (Table 19.1), have a fossil record extending back almost 7 million years. Several features distinguish them from other hominoids. Hominids are bipedal; that is, they have an upright posture, which is indicated by several modifications in their skeleton (▶Figure 19.5a). In addition, they show a trend toward a large and internally reorganized brain (Figure 19.5b). Other features include a reduced face and reduced canine teeth, omnivorous feeding, increased manual dexterity, and the ability to use sophisticated tools.

Many anthropologists think that these hominid features evolved in response to major climatic changes that began during the Miocene and continued into the Pliocene. During this time, vast savannas replaced the African tropical rain forests where the prosimians and Old World monkeys had been so abundant. As the savannas and grasslands continued to expand, the hominids made the transition from true forest dwelling to life in an environment of mixed forests and grasslands.

At present, there is no clear consensus on the evolutionary history of the hominid lineage. This is due, in part, to the incomplete fossil record of hominids, as well as new discoveries, and also because some species are known only from partial specimens or fragments of bone. Because of these factors, there is even disagreement on the total number of hominid species. A complete discussion of all the proposed hominid species and the various competing schemes of hominid evolution is beyond the scope of this chapter. However, we will briefly discuss the generally accepted taxa (▶Figure 19.6) and present some of the current theories of hominid evolution.

Remember that although the fossil record of hominid evolution is not complete, what exists is well documented. However, the interpretation of that fossil record precipitates the often vigorous and sometimes acrimonious debates concerning our evolutionary history.

Discovered in northern Chad's Djurab Desert in July 2002, the nearly 7-million-year-old skull and dental remains of *Sahelanthropus tchadensis* (but nicknamed *Tourmaï*, which means "hope of life" in the local Goran language) make it the oldest known hominid yet unearthed and one that existed at or very near to the time when humans and chimpanzees diverged from a common ancestor (▶Figure 19.7). Currently, most paleoanthropologists accept that the human–chimpanzee stock separated from gorillas approximately 8 million years ago and humans separated from chimpanzees about 5 million years ago.

Besides being the oldest hominid, *Sahelanthropus tchadensis* shows a mosaic of primitive and advanced features that has both excited and puzzled paleoanthropologists. Its small brain case and most of its teeth (except the canines) are chimplike. However, the nose, which is fairly flat, and the prominent brow ridges are features seen only, until now, in the human genus *Homo*. It is hypothesized that *Sahelanthropus tchadensis* was probably bipedal in its walking habits, but until bones from its legs and feet are found, that supposition remains conjecture.

The next oldest hominid is *Orrorin tugenensis*, whose fossils have been dated at 6 million years old and consist of bits of jaw, isolated teeth, and finger, arm, and partial upper leg bones (Figure 19.6). At this time, there is still debate as to exactly where *Orrorin tugenensis* fits in the hominid lineage.

Sometime between 5.8 and 5.2 million years ago, another hominid, *Ardipithecus kadabba*, was present in eastern Africa (Figure 19.6). Recently, the discovery of a second species of *Ardipithecus* was announced. *Ardipithecus ramidus* consists of a skeleton and skull of a female, 1.2 m tall, weighing approximately 50 kg, and nicknamed "Ardi" (▶Figure 19.8). Ardi, who lived approximately 4.4 million years ago, displays an interesting mosaic of evolutionary characteristics that include a dexterous hand for grasping and a foot with an opposable big toe but which lacked the flexibility of an ape's foot, which is used for climbing and maneuvering in trees. Based on these and other features, scientists have concluded that Ardi was able to walk upright on the ground, while retaining the ability to climb and maneuver in trees.

The discovery of Ardi has caused paleoanthropologists to once again rethink the human evolutionary lineage. But, as we've pointed out repeatedly throughout this book, this is how science works! With the discovery of new hominid fossils, some questions are answered, and yet other new ones arise, thus either confirming our previously held notions or forcing us to form new hypotheses.

▶ **Figure 19.5** **Comparison of Gorilla and Human Locomotion and Anthropoid Brain Size**

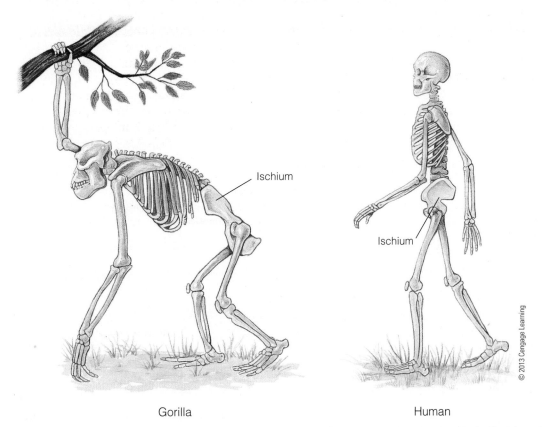

Gorilla   Human

(a) In gorillas, the ischium bone is long, and the entire pelvis is tilted toward the horizontal. In humans, the ischium bone is much shorter, and the pelvis is vertical.

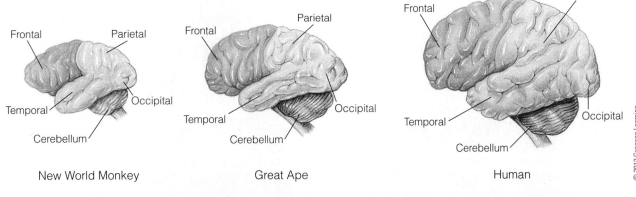

New World Monkey   Great Ape   Human

(b) An increase in brain size and organization is apparent in comparing the brains of a New World monkey (Superfamily Ceboidea), a great ape (Superfamily Hominoidea; Family Pongidae), and a present-day human (Superfamily Hominoidea; Family Hominidae).

Australopithecines

Australopithecine is a collective term for all members of the genus *Australopithecus.* Currently, five species are generally recognized: *A. anamensis, A. afarensis, A. africanus, A. robustus,* and *A. boisei.* However, a recently discovered sixth species may result in a new interpretation of the origin of our own genus, *Homo* (see Perspective). Notwithstanding this new discovery, many paleoanthropologists accept the evolutionary scheme in which *A. anamensis,* the

oldest known australopithecine, is ancestral to *A. afarensis,* who in turn is ancestral to *A. africanus* and the genus *Homo,* as well as the side branch of australopithecines represented by *A. robustus* and *A. boisei.*

The oldest known australopithecine is *Australopithecus anamensis.* Discovered at Kanapoi, a site near Lake Turkana, Kenya, by Meave Leakey, of the National Museums of Kenya, and her colleagues, this 4.2-million-year-old bipedal species has many features in common with its

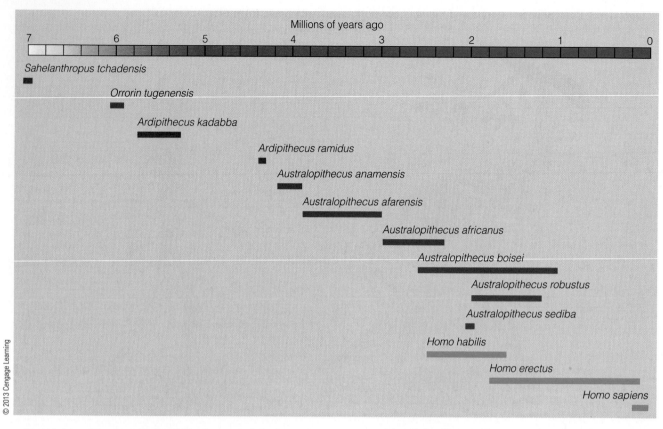

© 2013 Cengage Learning

▶ Figure 19.6 **The Stratigraphic Record of Hominids** The geologic ranges for the commonly accepted species of hominids (the branch of primates that includes present-day humans and their extinct ancestors).

© M.P.F.T.

▶ Figure 19.7 **Sahelanthropus tchadensis** Discovered in Chad in 2002, and dated at nearly 7 million years, this skull of *Sahelanthropus tchadensis* is presently the oldest known hominid.

HO/AFP/Getty Images/NEWSCOM

▶ Figure 19.8
Ardipithecus ramidus
Frontal view of a female *Ardipithecus ramidus*, known as "Ardi." *A. ramidus*, who lived approximately 4.4 million years ago in what is now Ethiopia, shows an interesting variety of features, such as a dexterous hand for grasping and a foot with an opposable big toe (clearly shown here) that allowed it to walk upright, yet also climb and maneuver in trees.

Perspective

Australopithecus sediba—The Ancestor of *Homo*?

At a site called Malapa, approximately 40 km northwest of Johannesburg, South Africa, fossils from an eroded limestone cave contain the bones of hominid individuals that might represent the oldest known species of the human lineage. Currently named *Australopithecus sediba* (Figure 1), it appears to be an intermediate form between the youngest australopithecines and *Homo* (Figure 19.6).

Owing to the geography and geology of the region, the Malapa cave site has preserved the bones of numerous animals, such as hyenas, antelope, and saber-toothed cats, as well as several hominid individuals that lived in the area approximately 2 million years ago. At that time, the landscape, which consisted of rolling hills and wooded valleys, was underlain by a limestone aquifer riddled with caves, caverns, and passageways. These were connected to the surface by various fractures and tunnels that varied in size.

During the wet periods, the water table in the region rose and water seeped out at the surface, forming pools that animals in the area drank from. When drier times prevailed, these ponds became open pits, leading to the caves below. Some of the shafts, which were nearly vertical and connected the caves to the surface openings, were as much as 50 m long, meaning that any animals that were unlucky enough to fall in them were unable to climb back out. Consequently, the animals, including hominids, that fell into these caves and caverns were quickly buried by sediments and thus preserved until later unearthed by paleoanthropologists.

What makes *A. sediba* such an exciting discovery, besides the nearly complete skeletons so far recovered, is that it contains a combination of primitive and advanced features, placing it squarely between *Australopithecus anemensis* and our oldest known ancestor, *Homo habilis* (Figure 19.14). One of the primitive traits that *A. sediba* shares with australopithicines is a small brain, 420 cc in size, comparable to that of other australopithicines, yet smaller that the larger-brained *H. habilis* (700 cc average). Other primitive features include long upper arms; long, high cheekbones; primitive molar cusps; a primitive heel bone; and an overall small size (1.3 m tall).

On the other hand, the shape of *A. sediba*'s brain, as well as such facial features as smaller teeth and chewing muscles, are advanced traits that are not characteristic of earlier australopithicines. Other advanced features are a humanlike pelvis, long legs, and a hand with the capability to bring its thumb and fingers together to form a precision grip, a feature necessary for the manipulation of objects and the making and using of tools (Figure 2).

At this time, the team that discovered *A. sediba* is proposing that it is the australopithicine species most likely to be the ancestor to our own genus, *Homo*. Furthermore, they also are proposing that rather than evolving into *H. habilis*, *A. sediba* evolved directly to *Homo erectus*, thus making *H. habilis* a side branch in the *Homo* lineage.

Just as has happened with earlier hominid discoveries, this proposed hypothesis has generated a tremendous amount of debate and consternation among paleoanthropologists. And, as with previous new hominid finds, *A. sediba*'s place in the human phylogenetic tree will eventually be determined by the bones themselves.

Figure 1 The skull of *Australopithecus sediba* contains both primitive and advanced features. One of the primitive features is a small brain, comparable in size to that of other australopithicines. On the other hand, the skull displays such advanced features as relatively small premolars and molars, as well as facial features that are more similar to *Homo*.

Figure 2 This image shows the skeleton of the right hand of *Australopithecus sediba* against a modern human hand. The hand lacks three wrist bones and four terminal phalanges (fingertip bones), but is otherwise complete. An important and advanced feature of the hand of *A. sediba* is that the thumb and fingers can be brought together to grip items, a feature necessary for the manipulation of objects.

David L. Brill

▶ **Figure 19.9** **Skeleton of Lucy** A reconstruction of Lucy's skeleton (*Australopithecus afarensis*) by Owen Lovejoy and his students at Kent State University, Ohio. Lucy, whose fossil remains were discovered by Donald Johanson, is an approximately 3.5-million-year-old *Australopithecus afarensis* individual. This reconstruction illustrates how adaptations in Lucy's hip, leg, and foot allowed a fully bipedal means of locomotion.

younger relative, *Australopithecus afarensis,* yet is more primitive in other characteristics, such as its teeth and skull. *A. anamensis* is estimated to have been between 1.3 and 1.5 m tall and weighed 33 to 50 kg.

A discovery of fossils of *Australopithecus anamensis* from the Middle Awash area in northeastern Ethiopia has shed new light on the transition between *Ardipithecus* and *Australopithecus*. Prior to this discovery, the origin of *Australopithecus* has been hampered by a sparse fossil record. The discovery of *Ardipithecus* in the same region of Africa and at the same time as the earliest *Australopithecus* provides strong evidence that *Ardipithecus* evolved into *Australopithecus* and links these two genera in the evolutionary lineage leading to humans.

Australopithecus afarensis (▶Figure 19.9), who lived 3.9–3.0 million years ago, was fully bipedal and exhibited great variability in size and weight. Members of this species ranged from just over 1 m to about 1.5 m tall and weighed between 29 and 45 kg. They had a brain size of 380–450 cubic centimeters (cc), larger than the 300–400 cc of a chimpanzee, but much smaller than that of present-day humans (1,350 cc average).

The skull of *A. afarensis* retained many apelike features, including massive brow ridges and a forward-jutting jaw, but its teeth were intermediate between those of apes and humans. The heavily enameled molars were probably an adaptation to chewing fruits, seeds, and roots (▶Figure 19.10).

Recall that one of the features that distinguish hominids from the rest of the hominoids is the fact that they are bipedal. *Australopithecus afarensis* was certainly fully bipedal, as indicated by its fossil bones (Figure 19.9). However, what is the fossil record of hominid footprints, and what can they tell us about when hominids walked upright? Although fossil trackways of various animals extend as far back as the Paleozoic (see Chapter 13 opening photo), the oldest hominid footprints come from an

Darwen and Vally Hennings

▶ **Figure 19.10** **African Pliocene Landscape** Re-creation of a Pliocene landscape showing members of *Australopithecus afarensis* gathering and eating various fruits and seeds.

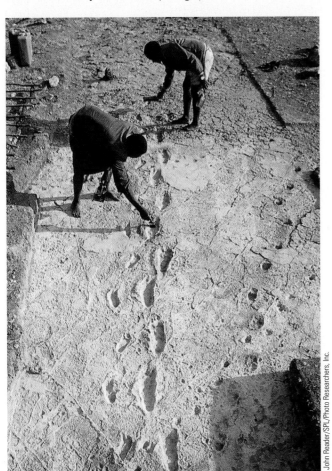

somewhat larger brain. Furthermore, it appears the limbs of *A. africanus* may not have been as well adapted for bipedalism as those of *A. afarensis*.

A. afarensis and *A. africanus* both differ markedly from the so-called robust australopithicine species, *A. boisei* (2.6–1.0 million years ago) and *A. robustus* (2.0–1.2 million years ago). Both robust species were about the same size (1.1–1.4 m tall) and weight (32–49 kg) and had similar morphologic features. For example, *A. robustus* had a flat face, and the crown of its skull had an elevated bony crest that provided additional area for the attachment of strong jaw muscles (▶ Figure 19.13). In addition, its broad, flat molars indicated that it was a vegetarian. Most scientists accept the idea that the robust australopithecines form a separate lineage from the other australopithecines and went extinct around 1 million years ago.

The Human Lineage

Homo habilis The earliest member of our own genus, **Homo**, is *Homo habilis*, who lived 2.5–1.6 million years ago. Its remains were first found at Olduvai Gorge in Tanzania (see chapter opening photo) by Mary and Louis Leakey, but it is also known from Kenya, Ethiopia, and South Africa. *H. habilis* evolved from the *A. afarensis* and *A. africanus* lineage and coexisted with *A. africanus* for approximately 200,000 years (Figure 19.6). *H. habilis* had a larger brain (700 cc average) than its australopithecine

ash bed radiometrically dated at 3.8–3.4 million years old (▶ Figure 19.11).

Discovered in 1978 at Laetoli in northern Tanzania by Mary Leakey, these 54 individual footprints comprise a trackway 27 m long. Dubbed the Footprint Tuff, this volcanic ash layer reveals the footprints of three individuals, two adults and perhaps a child (Figure 19.11). What makes this find so exciting and scientifically valuable is that the footprints prove that early hominids were fully bipedal and had an erect posture long before the advent of stone toolmaking, or an increase in the size of the brain. Furthermore, the footprints showed that early hominids walked like modern humans by placing the full weight of the body on the ball of the heel.

A. afarensis was stratigraphically succeeded by *Australopithecus africanus*, who lived 3.0–2.3 million years ago (▶ Figure 19.12). The differences between the two species are relatively minor. They were both about the same size and weight, but *A. africanus* had a flatter face and

▶ Figure 19.13 **Australopithecus robustus** The skull of *Australopithecus robustus* had a massive jaw, powerful chewing muscles, and large, broad, flat chewing teeth apparently used for grinding up coarse plant food.

▶ Figure 19.14 **Homo habilis** *Homo habilis* is the earliest species of the *Homo* lineage. Shown is an approximately 1.9-million-year-old skull from Kenya.

Replica courtesy of Carolina Biological Supply. Photo by Sue Monroe

© The Bridgeman Art Library/Getty Images

▶ Figure 19.15 **Homo erectus** A reconstruction of the skull of *Homo erectus*, a widely distributed species whose remains have been found in Africa, Europe, India, China, and Indonesia.

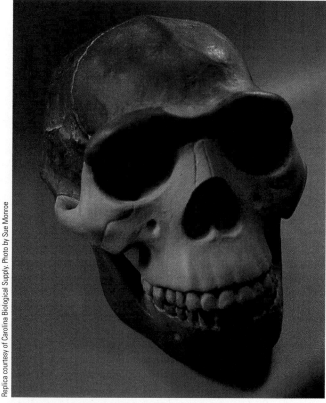

Replica courtesy of Carolina Biological Supply. Photo by Sue Monroe

ancestors, but smaller teeth (▶ Figure 19.14). It was between 1.2 and 1.3 m tall, had disproportionately long arms compared with modern humans, and weighed only 32 to 37 kg.

The evolutionary transition from *H. habilis* to *Homo erectus* appears to have occurred in a short period of time, between 1.8 and 1.6 million years ago. However, based on findings published in 2007, which suggest that *H. habilis* and *H. erectus* apparently coexisted for approximately 500,000 years, some scientists think that *H. habilis* and *H. erectus* may have evolved from a common ancestor and thus represent separate lineages of *Homo* rather than the traditional linear view of *H. erectus* evolving from *H. habilis*.

Homo erectus In contrast to the australopithecines and *H. habilis,* which are unknown outside Africa, *Homo erectus* was a widely distributed species, having migrated from Africa during the Pleistocene. Specimens have been found not only in Africa, where it evolved 1.8 million years ago, but also in Europe, India, China ("Peking Man"), and Indonesia ("Java Man"). By 1 million years ago, *H. erectus* was present in southeastern and eastern Asia, where it survived until about 100,000 years ago.

Although *H. erectus* developed regional variations in form, the species differed from modern humans in several ways. Its brain size of 800–1,300 cc, although much larger than that of *H. habilis*, was still less than the average for *Homo sapiens* (1,350 cc). The skull of *H. erectus* was thick-walled, its face was massive, it had prominent brow ridges, and its teeth were slightly larger than those of present-day humans (▶ Figure 19.15). *H. erectus* was comparable in size to modern humans, standing between 1.6 and 1.8 m tall and weighing between 53 and 63 kg.

The archaeological record indicates that *H. erectus* was a toolmaker. Furthermore, some sites show evidence that its members used fire and lived in caves, an advantage for those living in more northerly climates (▶ Figure 19.16).

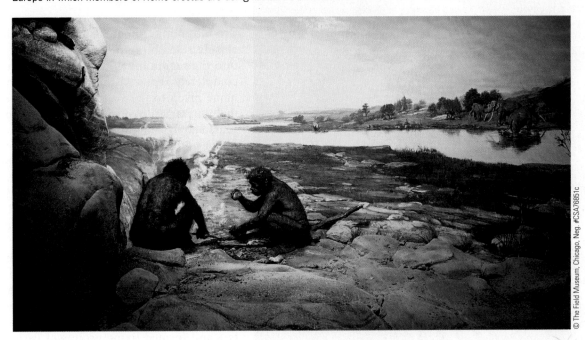

Debate still surrounds the transition from *H. erectus* to our own species, *Homo sapiens*. Paleoanthropologists are split into two camps. On the one side are those who support the "out of Africa" view. According to this view, early modern humans evolved from a single woman in Africa, whose offspring then migrated, perhaps as recently as 100,000 years ago, and populated Europe and Asia, driving the earlier hominid populations there to extinction.

The alternative explanation, the "multiregional" view, maintains that early modern humans did not have an isolated origin in Africa, but rather that they established separate populations throughout Eurasia. Occasional contact and interbreeding between these populations enabled our species to maintain its overall cohesiveness while still preserving the regional differences in people we see today. Regardless of which theory turns out to be correct, our species, *H. sapiens*, most certainly evolved from *H. erectus*.

Neanderthals Perhaps the most famous of all fossil humans are the **Neanderthals**, who inhabited Europe and the Near East from about 200,000 to 30,000 years ago and, according to the best estimates, never exceeded 15,000 individuals in western Europe. Some paleoanthropologists regard the Neanderthals as a variety or subspecies (*Homo sapiens neanderthalensis*), whereas others consider them as a separate species (*Homo neanderthalensis*). In any case, their name comes from the first specimens found in 1856 in the Neander Valley near Düsseldorf, Germany.

The most notable difference between Neanderthals and present-day humans is in the skull. Neanderthal skulls were long and low with heavy brow ridges, a projecting mouth, and a weak, receding chin (▶Figure 19.17). Their brain was slightly larger, on average, than our own and somewhat differently shaped. The Neanderthal body was more massive and heavily muscled than ours, with a flaring rib cage and rather short lower limbs, much like those of other cold-adapted people of today. Neanderthal males averaged between 1.6 and 1.7 m in height and weighed about 83 kg.

In 2007, scientists announced that they had isolated a pigmentation gene from a segment of Neanderthal DNA that indicated that at least some Neanderthals had red hair and light skin. Furthermore, the gene is different from that of modern red-haired people, suggesting that perhaps Neanderthals and present-day humans developed the trait independently, in response to similar higher-northern-latitude environmental pressures.

Based on specimens from more than 100 sites, we now know that Neanderthals were not much different from us, only more robust. Europe's Neanderthals were the first humans to move into truly cold climates, enduring miserably long winters and short summers as they pushed north into tundra country (▶Figure 19.18). Their remains are found chiefly in caves and hut-like rock shelters, which also contain a variety of specialized stone tools and weapons. Furthermore, archaeological evidence indicates that Neanderthals commonly took care of their injured and buried their dead, frequently with such grave items as tools, food, and perhaps even flowers.

As more fossil discoveries are made, and increasingly sophisticated techniques of DNA extraction and analysis are carried out, our view of Neanderthals and their

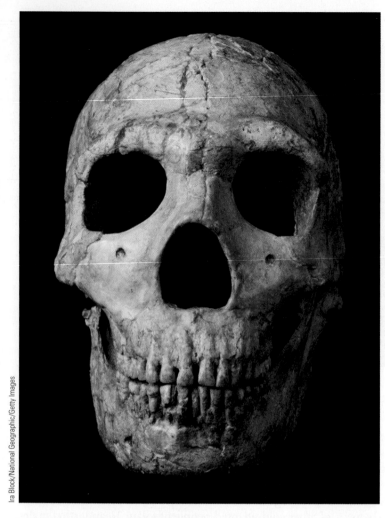

▶ **Figure 19.17 Neanderthal Skull** The Neanderthals were characterized by prominent heavy brow ridges, a projecting mouth, and a weak chin as seen in this Neanderthal skull from Wadl Amud, Israel. In addition, the Neanderthal brain was slightly larger, on average, than that of modern humans.

society, as well as their place in human evolution, will continue to change.

Cro-Magnons About 30,000 years ago, humans closely resembling modern Europeans moved into the region inhabited by Neanderthals and completely replaced them. **Cro-Magnons**, the name given to the successors of the Neanderthals in France, lived from about 35,000 to 10,000 years ago. During this period, the development of art and technology far exceeded anything the world had seen before. Using paints made from manganese and iron oxides, Cro-Magnon people painted hundreds of scenes on the ceilings and walls of caves in France and Spain, where many of them are still preserved today (▶Figure 19.19).

Cro-Magnons were also skilled nomadic hunters, following the herds in their seasonal migrations and using a variety of specialized tools in their hunts, including perhaps the bow and arrow. They sought refuge in caves and rock shelters and formed living groups of various sizes (▶Figure 19.20).

With the appearance of Cro-Magnons, human evolution has become almost entirely cultural rather than biologic. Since the evolution of the Neanderthals approximately 200,000 years ago, humans have gone from a stone culture to a technology that has allowed us to visit other planets with space probes and land astronauts on the Moon. It remains to be seen how we will use this technology in the future and whether we will continue as a species, evolve into another species, or become extinct as many groups have before us.

▶ **Figure 19.18 Pleistocene Cave Setting with Neanderthals** Archaeological evidence indicates that Neanderthals lived in caves and participated in ritual burials, as depicted in this painting of a burial ceremony such as occurred approximately 60,000 years ago at Shanidar Cave, Iraq.

Robert Harding Picture Library

► Figure 19.20 **Pleistocene Cro-Magnon Camp in Europe** Cro-Magnons were highly skilled hunters who formed living groups of various sizes.

© 2013 Cengage Learning

Summary

- Primates evolved during the Paleocene. Several trends help characterize primates and differentiate them from other mammalian orders, including a change in overall skeletal structure and mode of locomotion, an increase in brain size, stereoscopic vision, and evolution of a grasping hand with opposable thumb.

- The primates are divided into two suborders: prosimians and anthropoids. The prosimians are the oldest primate lineage and include lemurs, lorises, tarsiers, and tree shrews. The anthropoids include the New and Old World monkeys, apes, and hominids, which are humans and their extinct ancestors.

- The oldest known hominid is *Sahelanthropus tchadensis*, dated at nearly 7 million years. It was followed by *Orrorin tugenensis* at 6 million years, then two species of *Ardipithecus* at 5.8 and 4.4 million years, respectively. These early hominids were succeeded by the australopithecines, a fully bipedal group that evolved in Africa 4.2 million years ago. Recent discoveries indicate that *Ardipithecus* evolved into various species of *Australopithecus*, of which five species are generally recognized.

- The human lineage began approximately 2.5 million years ago in Africa with the evolution of *Homo habilis*, which survived as a species until about 1.6 million years ago.

- *Homo erectus* evolved from *H. habilis* sometime between 1.8 million and 1.6 million years ago and was the first hominid to migrate out of Africa, spreading to Europe, India, China, and Indonesia between 1.8 and 1 million years ago.

- The transition from *H. erectus* to *H. sapiens* is still unresolved because there is presently insufficient evidence to determine which hypothesis—the "out of Africa" or the "multiregional"—is correct. Nonetheless, we know that *H. erectus* used fire, made tools, and lived in caves.

▶ Neanderthals inhabited Europe and the Near East between 200,000 and 30,000 years ago and were not much different from present-day humans. They were, however, more robust and had differently shaped skulls. They also made specialized tools and weapons, apparently took care of their injured, and buried their dead.

▶ The Cro-Magnons were the successors of the Neanderthals and lived from about 35,000 to 10,000 years ago. They were highly skilled nomadic hunters, formed living groups of various sizes, and were also skilled cave painters.

▶ Modern humans succeeded the Cro-Magnons about 10,000 years ago and have spread throughout the world, as well as having set foot on the Moon.

Important Terms

anthropoid, p. 390
australopithecine, p. 393
Cro-Magnon, p. 400

hominid, p. 392
hominoid, p. 391
Homo, p. 397

Neanderthal, p. 399
Primates, p. 389
prosimian, p. 389

Review Questions

1. Which of the following is a hominid?
   a. _____ Chimpanzee; b. _____ Gibbon; c. _____ Prosimian; d. _✓_ Australopithecine; e. _____ Gorilla.
2. Which of the following evolutionary trends characterize primates?
   a. _____ Change in overall skeletal structure; b. _____ Grasping hand with opposable thumb; c. _____ Increase in brain size; d. _____ Stereoscopic vision; e. _✓_ All of the previous answers.
3. The oldest currently known hominid is
   a. _✓_ *Sahelanthropus tchadensis*; b. _____ *Orrorin tugenensis*; c. _____ *Ardipithecus ramidus*; d. _____ *Australopithecus anamensis*; e. _____ *Homo erectus*.
4. Which of the following features distinguish hominids from other hominoids?
   a. _____ A large and internally reorganized brain; b. _____ A reduced face and reduced canine teeth; c. _____ Bipedalism; d. _____ Use of sophisticated tools; e. _✓_ All of the previous answers.
5. When did primates evolve?
   a. _✓_ Paleocene; b. _____ Eocene; c. _____ Oligocene; d. _____ Miocene; e. _____ Pliocene.
6. Which extinct lineage of humans were skilled hunters and cave painters?
   a. _____ Acheuleans; b. _____ Archaic; c. _____ Neanderthals; d. _✓_ Cro-Magnons; e. _____ None of the previous answers.
7. According to archaeological evidence, which were the first hominids to use fire?
   a. _____ *Australopithecus robustus*; b. _____ *Homo sapiens*; c. _✓_ *Homo erectus*; d. _____ *Homo habilis*; e. _____ *Australopithecus boisei*.
8. The first hominids to migrate out of Africa and from which we evolved were
   a. _____ *Australopithecus robustus*; b. _✓_ *Homo erectus*; c. _____ *Homo sapiens*; d. _____ *Ardipithecus ramidus*; e. _____ *Homo habilis*.

9. Which is the oldest primate lineage?
   a. _____ Anthropoids; b. _✓_ Prosimians; c. _____ Insectivores; d. _____ Omnivores; e. _____ Hominids.
10. The human lineage began with the evolution of which species?
    a. _____ *Orrorin tugenensis*; b. _____ *Ardipithecus ramidus*; c. _____ *Sahelanthropus tchadensis*; d. _✓_ *Homo habilis*; e. _____ *Australopithecus boisei*.
11. Discuss the importance of the discovery of *Ardipithecus ramidus* ("Ardi") in terms of hominid evolution.
12. What factors do you think will influence the future course of human evolution? Do you think it is possible that we can control the direction that evolution takes?
13. Discuss what is currently known about the evolutionary history of hominids.
14. How has the discovery of *Australopithecus sediba* changed the currently held hypothesis of the origin of *Homo*?
15. What major evolutionary trends characterize the primates and set them apart from the other orders of mammals?
16. Discuss how climate changes that began in the Miocene might have influenced the evolution of hominids.
17. What are the main differences between the Neanderthals and Cro-Magnons?
18. Discuss the merits of the "out of Africa" and "multiregional" views in regard to the transition between *Homo erectus* and *Homo sapiens*.
19. Discuss the evolutionary history of the genus *Homo*.
20. Discuss the differences between the prosimians and anthropoids.

Epilogue

OUTLINE

Overpopulation is considered one of the greatest environmental problems facing the world today. Not only are we covering the land with our structures, such as seen here in the urban sprawl of Mexico City, Mexico, but we are contributing to the pollution of the atmosphere and possibly influencing climate change.

Guido Alberto Rossi/TIPS IMAGES/age fotostock

403

Introduction

Throughout this book, we have emphasized that Earth is a complex, dynamic planet that has changed continuously since its origin some 4.6 billion years ago. These changes, and the present-day features we observe, are the result of interactions between the various interrelated internal and external Earth systems, subsystems, and cycles. In addition, these interactions have influenced the evolution of the biosphere, as is obvious from an examination of the fossil record. Perhaps the most important lesson to be learned from your study of historical geology is that Earth has continued evolving throughout its history.

The *rock cycle* (see Figure 2.7), with its recycling of Earth materials to form the three major rock groups, illustrates the interrelationships between Earth's internal and external processes. The *hydrologic cycle* explains the continuous recycling of water from the oceans, to the atmosphere, to the land, and eventually back to the oceans. Changes within this cycle can have profound effects on Earth's topography as well as its biota. For example, a rise in global temperature will cause the ice caps to melt, contributing to rising sea level, which will greatly affect coastal areas, where many of the world's large population centers are presently located (see Chapter 17). We have seen the effect of changing sea level on continents in the past—that is, large-scale transgressions and regressions (see Chapters 10, 11, and 14). Some of these were caused by growing and shrinking continental ice caps when landmasses moved over the South Pole as a result of plate movements (see Chapter 11).

On a larger scale, the movement of plates has had a profound effect on the formation of landscapes, the distribution of mineral resources, and atmospheric and oceanic circulation patterns, as well as the evolution and diversification of life.

The launching in 1957 of *Sputnik 1*, the world's first artificial satellite, ushered in a new global consciousness in terms of how we view Earth and our place in the global ecosystem. Satellites have provided us with the ability to view not only the beauty of our planet, but also the fragility of Earth's biosphere and thus to realize the role humans play in shaping and modifying the environment. The pollution of the atmosphere, oceans, and many of our lakes and streams; the denudation of huge areas of tropical forests; the scars from strip mining; the depletion of the ozone layer—all are visible in the satellite images beamed back from space and attest to the impact humans have had on the ecosystem (see Chapter 4 Perspective).

Accordingly, we must understand that changes we make in the global ecosystem can have wide-ranging effects of which we may not even be aware. Furthermore, we must remember that humans are part of the global ecosystem, and, like all other life-forms, our presence alone affects the ecosystem. We must therefore act in a responsible manner, based on sound scientific knowledge, so that future generations will inherit a habitable environment. An understanding of geology, and science in general, is of paramount importance if we are to be good stewards of Earth.

One of the most important objectives of this book, as well as much of your secondary education, is to develop your critical thinking skills. As opposed to simple disagreement, critical thinking involves evaluating the supporting evidence for a particular point of view. Although your exposure to geology is probably limited, you do have the fundamental knowledge needed to appraise why geologists accept plate tectonic theory, why they think that Earth is 4.6 billion years old, and why scientists are convinced that the theory of evolution is well supported by evidence. Your abilities as a critical thinker will probably help you more effectively evaluate the arguments about global warming, ozone depletion, groundwater contamination, and many other environmental issues.

When such environmental issues as acid rain, the depletion of the ozone layer, and the greenhouse effect and global warming are discussed and debated (see Chapter 4), it is important to remember that they are not isolated topics but are part of a larger system that involves the entire Earth (see Figure 1.1). In addition, you should keep in mind that Earth goes through cycles of much longer duration than the human perspective of time (see Figure 1.13). Although they may have disastrous effects on the human species, global warming and cooling are part of a larger cycle that has resulted in numerous glacial advances and retreats during the past 2.6 million years (see Chapter 17).

For this very reason—the long periods of time involved in such issues—geologists can make important contributions to the debate on global warming because of their geologic perspective. They can study long-term trends by analyzing deep-sea sediments, ice cores, changes in sea level during the geologic past, and the distribution of plants and animals through time (see Figure 4.14).

Interestingly, most of the major environmental issues currently being discussed and debated, such as acid rain, ozone depletion, and global warming, seemingly have as a common denominator the effects of human actions (see chapter opening photo). The role that humans play in many of these issues, as well as the degree to which they play that role, is being examined and debated not only by scientists, but by economists, sociologists, and politicians, to name a few. It is therefore more important than ever that citizens become versed in what science can and cannot do, as well as becoming critical thinkers in the arguments regarding the impact of humans on the global ecosystem (see Chapter 4 Perspective).

Recognizing that humans are part of the global ecosystem, as well as one species in an evolving biota, we can now examine the interrelated topics of acid rain, ozone depletion, and global warming from the perspective of historical geology.

Acid Rain

One result of industrialization is atmospheric pollution, which causes smog, possible disruption of the ozone layer, global warming, and acid rain (▶ Figure E.1). Acidity, a measure of hydrogen ion concentration, is measured on the pH scale (▶ Figure E.2a). A pH value of 7 is neutral, whereas acidic conditions correspond to values less than 7, and values greater than 7 denote alkaline, or basic, conditions. Normal rain has a pH value of slightly less than 6.0. Some areas experience acid snow and even acid fog with a pH as low as 1.7.

▶ Figure E.1 **The Causes and Effects of Acid Rain**

(a) Pollution from the smoke stacks of steel mills in Pittsburgh, Pennsylvania, during the 1970s contributed to the formation of acid rain in the eastern United States and Canada.

Yoram Lehman/Photo Researchers, Inc.

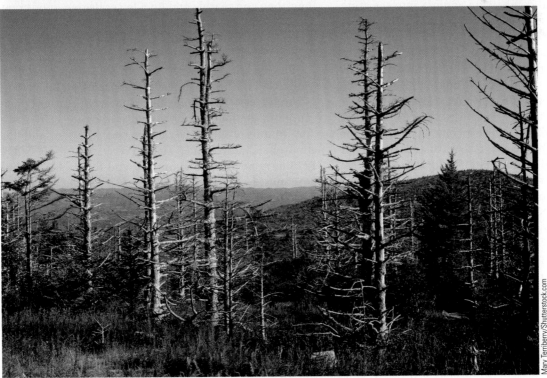

(b) The effects of acid rain can be seen in these dead trees in Grayson Highlands State Park, Virginia.

Mary Terriberry/Shutterstock.com

▶ Figure E.2 **Acid Rain**

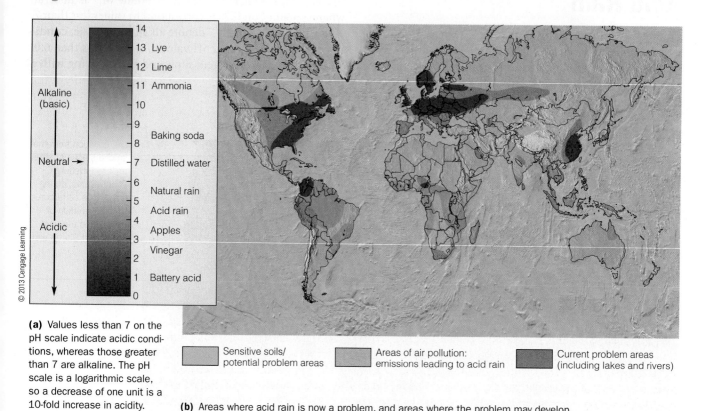

(a) Values less than 7 on the pH scale indicate acidic conditions, whereas those greater than 7 are alkaline. The pH scale is a logarithmic scale, so a decrease of one unit is a 10-fold increase in acidity.

Sensitive soils/potential problem areas

Areas of air pollution: emissions leading to acid rain

Current problem areas (including lakes and rivers)

(b) Areas where acid rain is now a problem, and areas where the problem may develop.

Several natural processes, including soil-bacteria metabolism and volcanism, release gases into the atmosphere that contribute to acid rain. Human activities also produce added atmospheric stress, especially the burning of fossil fuels, which release carbon dioxide and nitrogen oxide from internal combustion engines into the atmosphere. Both of these gases add to acid rain, but the greatest culprit is sulfur dioxide, released mostly by burning coal. Coal contains sulfur, which oxidizes to form sulfur dioxide (SO_2). As sulfur dioxide rises into the atmosphere, it reacts with oxygen and water droplets to form sulfuric acid (H_2SO_4), the main component of acid rain.

Robert Angus Smith first recognized acid rain in England in 1872, but not until 1961 did it become an environmental concern. At that time, scientists realized that acid rain is corrosive and irritating, kills vegetation, and has a detrimental effect on surface waters. Since then, the effects of acid rain are apparent in Europe (especially in Eastern Europe) and the eastern part of North America, where the problem has been getting worse for the last three decades (Figure E.2b).

The areas affected by acid rain invariably lie downwind from plants that emit sulfur gases, but the effects of acid rain in these areas may be modified by local conditions. For instance, if the area is underlain by limestone or alkaline soils, acid rain tends to be neutralized; however, granite has little or no modifying effect. Small lakes lose their ability to neutralize acid rain and become more and more acidic until some types of organisms disappear, and in some cases, all life-forms eventually die. However, as

discussed in Chapter 7, some organisms are able to adapt to new, and seemingly hostile, conditions, such as certain plants that have adapted to contaminated soils caused by coal mining operations.

Acid rain also causes increased chemical weathering of limestone and marble and, to a lesser degree, sandstone. The effects are especially evident on buildings, monuments, and tombstones, as in Gettysburg National Military Park in Pennsylvania.

The devastation caused by sulfur gases on vegetation near coal-burning power plants is apparent, and many forests in the eastern United States show signs of stress that cannot be attributed to other causes (Figure E.1b). Millions of tons of sulfur dioxide are released yearly into the atmosphere in the United States. Power plants built before 1975 have no emission controls, and the problems they pose must be addressed if emissions are to be reduced to an acceptable level. Various methods can be implemented to significantly reduce sulfur dioxide emissions by both older power plants and more recently constructed ones. However, these methods are costly, and in some cases, it is simply too expensive to upgrade older power plants to reduce their emissions.

Acid rain, like global warming, is a worldwide problem that knows no national boundaries. Wind may blow pollutants from the source in one country to another where the effects are felt. For instance, much of the acid rain in eastern Canada actually comes from sources in the United States.

Ozone Depletion

Earth supports life because of (1) its distance from the Sun, (2) its abundance of liquid water, and (3) its oxygen-rich atmosphere. An ozone layer (O_3) in the stratosphere (10 to 48 km above the surface) protects Earth because it blocks out most of the harmful ultraviolet radiation that bombards our planet (see Chapter 8). During the early 1980s, scientists discovered an ozone hole over Antarctica that has continued to grow. In fact, depletion of the ozone layer is now also recognized over the Arctic region and elsewhere. Any ozone depletion is viewed with alarm because it allows more dangerous radiation to reach the surface, increasing the risk of skin cancer, among other results (see Chapter 13 Perspective).

This discovery unleashed a public debate about the primary cause of ozone depletion and how best to combat the problem. Scientists proposed that one cause of ozone depletion is chlorofluorocarbons (CFCs), which are used in several consumer products—for instance, in aerosol cans. According to this theory, CFCs rise into the upper atmosphere where reactions with ultraviolet radiation liberate chlorine, which in turn reacts with and depletes ozone (▶Figure E.3). As a result of this view, an international agreement called the *Montreal Protocol* was reached in 1983, limiting the production of CFCs, along with other ozone-depleting substances.

However, during the 1990s this view was challenged. Those opposed to the idea that CFCs were the cause of ozone depletion proposed an alternative idea that natural causes rather than commercial products (such as CFCs) were the main culprits. They pointed out that volcanoes release copious quantities of hydrogen chloride (HCl) gas that rises into the stratosphere and could be responsible for ozone depletion. Furthermore, they claimed that because CFCs are heavier than air, they would not rise into the stratosphere.

It is true that volcanoes release HCl gas, as well as several other gases, some of which are quite dangerous. Nevertheless, most eruptions are too weak to inject gases of any kind high into the stratosphere. Even when it is released, HCl gas from volcanoes is very soluble and quickly removed from the atmosphere by rain and even by steam (water vapor) from the same eruption that released the HCl gas in the first place. Measurements of chlorine concentrations in the stratosphere show that only temporary increases occur following huge eruptions. For example, the largest volcanic outburst since 1912, the eruption of Mount Pinatubo in 1991, caused little increase in upper atmospheric chlorine. The impact of volcanic eruptions is certainly not enough to produce the average rate of ozone depletion taking place each year.

Conversely, as mentioned in the Chapter 13 Perspective, a study using a two-dimensional, atmospheric chemistry-transport model examined the possible role that the Siberian

▶ **Figure E.3 Ozone Depletion** Ozone is destroyed by chlorofluorocarbons (CFCs). Chlorine atoms are continually regenerated, so one chlorine atom can destroy many ozone molecules.

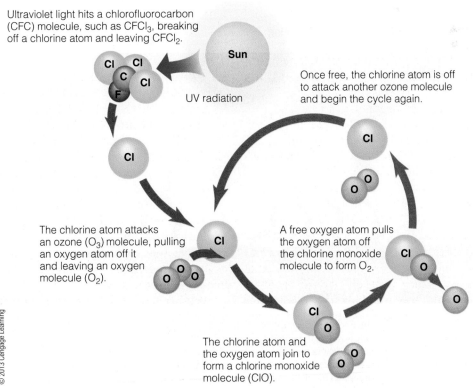

Ultraviolet light hits a chlorofluorocarbon (CFC) molecule, such as $CFCl_3$, breaking off a chlorine atom and leaving $CFCl_2$.

Sun

UV radiation

Once free, the chlorine atom is off to attack another ozone molecule and begin the cycle again.

The chlorine atom attacks an ozone (O_3) molecule, pulling an oxygen atom off it and leaving an oxygen molecule (O_2).

A free oxygen atom pulls the oxygen atom off the chlorine monoxide molecule to form O_2.

The chlorine atom and the oxygen atom join to form a chlorine monoxide molecule (ClO).

© 2013 Cengage Learning

Traps flood basalt eruptions at the end of the Permian might have played in atmospheric ozone depletion. It found that such an eruption was far greater than that of Mount Pinatubo.

Moreover, although it is true that CFCs are heavier than air, this does not mean that they cannot rise into the stratosphere. Earth's surface heats differentially, meaning that more heat may be absorbed in one area than in an adjacent one. The heated air above a warmer area becomes less dense, rises by convection, and carries with it CFCs and other substances that are actually denser than air. Once in the stratosphere, ultraviolet radiation, which is usually absorbed by ozone, breaks up CFC molecules and releases chlorine that reacts with ozone. Indeed, a single chlorine atom can destroy 100,000 ozone molecules (Figure E.3). In contrast to the HCl gas produced by volcanoes, CFCs are absolutely insoluble; it is the fact that they are inert that made them so desirable for commercial products. Because a CFC molecule can last for decades, any increase in CFCs is a long-term threat to the ozone layer.

Another indication that CFCs are responsible for the Antarctic ozone hole is that the rate of ozone depletion has decreased since 1989, when an international agreement (the Montreal Protocol mentioned earlier) was implemented to reduce suspected ozone-depleting substances. A sound understanding of the science behind these atmospheric processes helped world leaders act quickly to address this issue. Scientists hope that with continued compliance with the protocol, the ozone layer will recover by the middle of this century.

Global Warming

As discussed throughout this book, global warming and cooling are part of a larger cycle that has resulted in numerous glacial advances and retreats throughout Earth history (see Figure 1.13). However, from a human perspective, one of the concerns is the role that humans have played in contributing to the greenhouse effect. The issue is not whether we have a greenhouse effect, because we do, but to what degree human activity, such as the burning of fossil fuels, has contributed to global warming.

Carbon dioxide is produced as a by-product of respiration and the burning of organic material. As such, it is a component of the global ecosystem and is constantly being recycled as part of the carbon cycle. The recycling of carbon dioxide between the crust and atmosphere is an important climatic regulator because carbon dioxide—as well as other gases such as methane, nitrous oxide, chlorofluorocarbons, and water vapor—allow for sunlight to pass through them but trap the heat reflected back from Earth's surface. This retention of heat is called the *greenhouse effect* (▶ Figure E.4). It results in an increase in the temperature of Earth's surface and, more importantly, its atmosphere, thus producing global warming.

Because of the increase in human-produced greenhouse gases during the last 200 years, many scientists worry that a global warming trend has already begun and will result in severe global climatic shifts. Presently, most climate researchers use a range of scenarios for greenhouse gas emissions when making predictions for future warming rates. Based on state-of-the-art climate model simulations,

▶ Figure E.4 **The Greenhouse Effect and Global Warming**

(a) Short-wavelength radiation from the Sun that is not reflected back into space penetrates the atmosphere and warms Earth's surface.

(b) Earth's surface radiates heat in the form of long-wavelength radiation back into the atmosphere, where some of it escapes into space. The rest is absorbed by greenhouse gases and water vapor and reradiated back toward Earth.

(c) Increased concentrations of greenhouse gases trap more heat near Earth's surface, causing a general increase in surface and atmospheric temperatures, which leads to global warming.

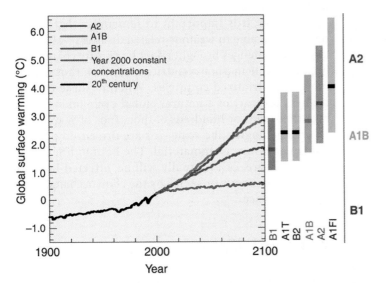

(a) Solid lines are multimodel global averages of surface warming (relative to 1980–1999) for the 2000 IPCC Special Report on Emissions Scenarios (SRES) A2, A1B, and B1, shown as continuations of the 20th century simulations. The orange line is for the experiment where concentrations were held constant at year 2000 values. The bars in the middle of the figure indicate the best estimate (solid line within each bar) and the likely range assessed for the six SRES marker scenarios at 2090–2099 relative to 1980–1999. The assessment of the best estimate and likely ranges in the bars includes the Atmosphere-Ocean General Circulation Models (AOGCMs) in the left part of the figure, as well as results from a hierarchy of independent models and observational constraints.

(b) Projected surface temperature changes for the early and late 21st century relative to the period 1980–1999. The panels show the multi-AOGCM average projections for the A2 (top), A1B (middle), and B1 (bottom) SRES scenarios averaged over decades 2020–2029 (left) and 2090–2099 (right).

the fourth Intergovernmental Panel on Climate Change (IPCC) report, issued in 2007, showed a predicted increase in global average temperature from year 2000 to year 2100 of 1 to 3°C under the best conditions, to a 2.5 to 6.5°C rise under "business-as-usual" conditions (▶Figure E.5).

These predicted increases in temperatures are based on various scenarios that explore different global development pathways. They are grouped into four scenario families (A1, A2, B1, and B2) that cover a wide range of economic, technological, and demographic possibilities and their resultant greenhouse gas emissions. The A1FI scenario (Figure E.5a) is based on a "business-as-usual" outlook in which the world experiences very rapid economic growth, global population peaks in the mid-century with declines thereafter, and the world continues a fossil-fuel intensive energy consumption strategy. At the opposite end of the scenario spectrum is the B1 scenario (Figure E.5a), which also assumes a global population peaking in mid-century and declining thereafter, a shift toward a service and information economy, and an emphasis on reducing materials usage with the introduction of clean and resource-efficient technologies. The A1B combination (Figure E.5a) is the "middle-of-the-road" scenario in which global population trends follow that for the A1FI and B1 scenarios, but there is a balance across all energy sources and technologies used.

Regardless of which scenario is followed, the global temperature change will be uneven, with the greatest warming occurring in the higher latitudes of the northern hemisphere. Furthermore, there will be greater warming for the continents than for the oceanic regions because land areas tend to heat and cool more rapidly than oceans. Thus, there will be greater warming in the more land-dominated northern hemisphere than in the ocean-dominated southern hemisphere (Figure E.5b). As a consequence of this warming, rainfall patterns will shift dramatically. This change in rainfall patterns will have a major effect on the largest grain-producing areas of the world, such as the American Midwest. Drier and hotter conditions will intensify the severity and frequency of droughts, leading to more crop failures and higher food prices (▶Figure E.6). With such shifts in climate, Earth may experience an increase in the expansion of deserts, which will remove valuable crop and grazing lands.

We cannot leave the subject of global warming without pointing out that some scientists are still not convinced that the global warming trend is the direct result of increased human activity related to industrialization. They say that although the level of greenhouse gases has increased, we are still uncertain about their rate of generation and rate of removal, and about whether the rise in global temperatures during the past century resulted from

▸ Figure E.6 **Withered Corn Crop as a Result of Drought Conditions**

normal climatic variations through time or from human activity. Furthermore, they conclude that even if a general global warming occurs during the next hundred years, it is not certain that the dire predictions made by proponents of global warming will come true.

Earth, as we know, is a remarkably complex system, with many feedback mechanisms and interconnections throughout its various subsystems. It is very difficult to predict all of the consequences that global warming would have for atmospheric and oceanic circulation patterns and its ultimate effect on Earth's biota.

However, it is important to remember that everyone is vulnerable to weather-related disasters—although large-scale changes brought about by climate change will impact people in poor countries more than those in the more industrialized countries. Whether these climate changes are part of a natural global cycle taking place over thousands or hundreds of thousands of years—that is, on a geologic time scale—or are driven, in part, by human activities is immaterial. The bottom line is that we already are, or eventually will be, affected in some way, be it economic or social, by the climate changes that are taking place.

A Final Word

In conclusion, the most important lesson to be learned from the study of historical geology is that Earth is an extremely complex and ever-changing planet. Interactions among its various systems and subsystems have resulted in changes in the atmosphere, lithosphere, and biosphere through time. By studying how Earth has evolved in the past, we can better understand how the different Earth systems and subsystems work and interact with one another, and, more importantly, how our actions affect the delicate balance between these systems and subsystems.

We tend to view our planet from the perspective of a human lifetime and commonly overlook the fact that Earth has changed markedly in the context of geologic time and that it continues to do so. As you now realize and can appreciate, historical geology is not a static science, but one that, like the dynamic Earth it seeks to understand, is constantly evolving as new information and methods of investigation become available. It is thus incumbent on all of us to apply the geologic lessons we have learned to make Earth a more sustainable planet for both current and future generations.

Appendix A
English-Metric Conversion Chart

English Unit	Conversion Factor	Metric Unit	Conversion Factor	English Unit
Length				
Inches (in)	2.54	Centimeters (cm)	0.39	Inches (in)
Feet (ft)	0.305	Meters (m)	3.28	Feet (ft)
Miles (mi)	1.61	Kilometers (km)	0.62	Miles (mi)
Area				
Square inches (in^2)	6.45	Square centimeters (cm^2)	0.16	Square inches (in^2)
Square feet (ft^2)	0.093	Square meters (m^2)	10.8	Square feet (ft^2)
Square miles (mi^2)	2.59	Square kilometers (km^2)	0.39	Square miles (mi^2)
Volume				
Cubic inches (in^3)	16.4	Cubic centimeters (cm^3)	0.061	Cubic inches (in^3)
Cubic feet (ft^3)	0.028	Cubic meters (m^3)	35.3	Cubic feet (ft^3)
Cubic miles (mi^3)	4.17	Cubic kilometers (km^3)	0.24	Cubic miles (mi^3)
Weight				
Ounces (oz)	28.3	Grams (g)	0.035	Ounces (oz)
Pounds (lb)	0.45	Kilograms (kg)	2.20	Pounds (lb)
Short tons (st)	0.91	Metric tons (t)	1.10	Short tons (st)
Temperature				
Degrees Fahrenheit (°F)	−32° × 0.56	Degrees Centigrade (Celsius) (°C)	× 1.80 + 32°	Degrees Fahrenheit (°F)

Examples: 10 inches = 25.4 centimeters; 10 centimeters = 3.9 inches
100 square feet = 9.3 square meters; 100 square meters = 1,080 square feet
50°F = 10.08°C; 50°C = 122°F

Appendix B
Classification of Organisms

Any classification is an attempt to make order out of disorder and to group similar items into the same categories. All classifications are schemes that attempt to relate items to each other based on current knowledge and therefore are progress reports on the current state of knowledge for the items classified. Because classifications are to some extent subjective, classification of organisms may vary among different texts.

The classification that follows is based on the five-kingdom system of classification of Margulis and Schwartz.* We have not attempted to include all known life-forms, but rather the major categories of both living and fossil groups.

Kingdom Monera

Prokaryotes
    Phylum Archaebacteria—(Archean–Recent)
    Phylum Cyanobacteria—Blue-green algae or blue-green bacteria (Archean–Recent)

Kingdom Protoctista

Solitary or colonial unicellular eukaryotes
    Phylum Acritarcha—Organic-walled unicellular algae of unknown affinity (Proterozoic–Recent)
    Phylum Bacillariophyta—Diatoms (Jurassic–Recent)
    Phylum Charophyta—Stoneworts (Silurian–Recent)
    Phylum Chlorophyta—Green algae (Proterozoic–Recent)
    Phylum Chrysophyta—Golden-brown algae, silicoflagellates and coccolithophorids (Jurassic–Recent)
    Phylum Euglenophyta—Euglenids (Cretaceous–Recent)
    Phylum Myxomycophyta—Slime molds (Proterozoic–Recent)
    Phylum Phaeophyta—Brown algae, multicellular, kelp, seaweed (Proterozoic–Recent)
    Phylum Protozoa—Unicellular heterotrophs (Cambrian–Recent)
        Class Sarcodina—Forms with pseudopodia for locomotion (Cambrian–Recent)
            Order Foraminifera—Benthonic and planktonic sarcodinids most commonly with calcareous tests (Cambrian–Recent)
            Order Radiolaria—Planktonic sarcodinids with siliceous tests (Cambrian–Recent)

Phylum Pyrrophyta—Dinoflagellates (Silurian?, Permian–Recent)
Phylum Rhodophyta—Red algae (Proterozoic–Recent)
Phylum Xanthophyta—Yellow-green algae (Miocene–Recent)

Kingdom Fungi

    Phylum Zygomycota—Fungi that lack cross walls (Proterozoic–Recent)
    Phylum Basidiomycota—Mushrooms (Pennsylvanian–Recent)
    Phylum Ascomycota—Yeasts, bread molds, morels (Mississippian–Recent)

Kingdom Plantae

Photosynthetic eukaryotes
    Division* Bryophyta—Liverworts, mosses, hornworts (Devonian–Recent)
    Division Psilophyta—Small, primitive vascular plants with no true roots or leaves (Silurian–Recent)
    Division Lycopodophyta—Club mosses, simple vascular systems, true roots and small leaves, including scale trees of Paleozoic Era (lycopsids) (Devonian–Recent)
    Division Sphenophyta—Horsetails (scouring rushes) and sphenopsids such as the Carboniferous *Calamites* (Devonian–Recent)
    Division Pteridophyta—Ferns (Devonian–Recent)
    Division Pteridospermophyta—Seed ferns (Devonian–Jurassic)
    Division Coniferophyta—Conifers or cone-bearing gymnosperms (Carboniferous–Recent)
    Division Cycadophyta—Cycads (Triassic–Recent)
    Division Ginkgophyta—Maidenhair tree (Triassic–Recent)
    Division Angiospermophyta—Flowering plants and trees (Cretaceous–Recent)

Kingdom Animalia

Nonphotosynthetic multicellular eukaryotes (Proterozoic–Recent)
    Phylum Porifera—Sponges (Cambrian–Recent)
        Order Stromatoporoida—Extinct group of reef-building organisms (Cambrian–Oligocene)

*Margulis, L., and K. V. S. Schwartz (1982). *Five Kingdoms.* New York: W. H. Freeman and Co.

*In botany, division is the equivalent to phylum.

Phylum Archaeocyatha—Extinct spongelike organisms (Cambrian)

Phylum Cnidaria—Hydrozoans, jellyfish, sea anemones, corals (Cambrian–Recent)

Class Hydrozoa—Hydrozoans (Cambrian–Recent)

Class Scyphozoa—Jellyfish (Proterozoic–Recent)

Class Anthozoa—Sea anemones and corals (Cambrian–Recent)

Order Tabulata—Exclusively colonial corals with reduced to nonexistent septa (Ordovician–Permian)

Order Rugosa—Solitary and colonial corals with fourfold symmetry (Ordovician–Permian)

Order Scleractinia—Solitary and colonial corals with six-fold symmetry. Most colonial forms have symbiotic dinoflagellates in their tissues. Important reef builders today (Triassic–Recent)

Phylum Bryozoa—Exclusively colonial suspension-feeding marine animals that are useful for correlation and ecological interpretations (Ordovician–Recent)

Phylum Brachiopoda—Marine suspension-feeding animals with two unequal-sized valves. Each valve is bilaterally symmetrical (Cambrian–Recent)

Class Inarticulata—Primitive chitinophosphatic or calcareous brachiopods that lack a hinging structure. They open and close their valves by means of complex muscles (Cambrian–Recent)

Class Articulata—Advanced brachiopods with calcareous valves that are hinged (Cambrian–Recent)

Phylum Mollusca—A highly diverse group of invertebrates (Cambrian–Recent)

Class Monoplacophora—Segmented, bilaterally symmetrical crawling animals with cap-shaped shells (Cambrian–Recent)

Class Amphineura—Chitons. Marine crawling forms, typically with eight separate calcareous plates (Cambrian–Recent)

Class Scaphopoda—Curved, tusk-shaped shells that are open at both ends (Ordovician–Recent)

Class Gastropoda—Single-shelled, generally coiled crawling forms. Found in marine, brackish, and fresh water as well as terrestrial environments (Cambrian–Recent)

Class Bivalvia—Mollusks with two valves that are mirror images of each other. Typically known as clams and oysters (Cambrian–Recent)

Class Cephalopoda—Highly evolved swimming animals. Includes shelled sutured forms as well as non-shelled types such as octopus and squid (Cambrian–Recent)

Order Nautiloidea—Forms in which the chamber partitions are connected to the wall along simple, slightly curved lines (Cambrian–Recent)

Order Ammonoidea—Forms in which the chamber partitions are connected to the wall along wavy lines (Devonian–Cretaceous)

Order Coleoidea—Forms in which the shell is reduced or lacking. Includes octopus, squid, and the extinct belemnoids (Mississippian–Recent)

Phylum Annelida—Segmented worms. Responsible for many of the Phanerozoic burrows and trail trace fossils (Proterozoic–Recent)

Phylum Arthropoda—The largest invertebrate group, comprising about 80 percent of all known animals. Characterized by a segmented body and jointed appendages (Cambrian–Recent)

Class Trilobita—Earliest appearing arthropod class. Trilobites had a head, body, and tail and were bilaterally symmetrical (Cambrian–Permian)

Class Crustacea—Diverse class characterized by a fused head and body and an abdomen. Included are barnacles, copepods, crabs, ostracodes, and shrimp (Cambrian–Recent)

Class Insecta—Most diverse and common of all living invertebrates, but rare as fossils (Silurian–Recent)

Class Merostomata—Characterized by four pairs of appendages and a more flexible exoskeleton than crustaceans. Includes the extinct eurypterids, horseshoe crabs, scorpions, and spiders (Cambrian–Recent)

Phylum Echinodermata—Exclusively marine animals with fivefold radial symmetry and a unique water vascular system (Cambrian–Recent)

Subphylum Crinozoa—Forms attached by a calcareous jointed stem (Cambrian–Recent)

Class Crinoidea—Most important class of Paleozoic echinoderms. Suspension-feeding forms that are either free-living or attached to seafloor by a stem (Cambrian–Recent)

Class Blastoidea—Small class of Paleozoic suspension-feeding, sessile forms with short stems (Ordovician–Permian)

Class Cystoidea—Globular to pear-shaped suspension-feeding, benthonic, sessile forms with very short stems (Ordovician–Devonian)

Subphylum Homalozoa—A small group with flattened, asymmetrical bodies with no stems. Also called carpoids (Cambrian–Devonian)

Subphylum Echinozoa—Globose, predominantly benthonic mobile echinoderms (Cambrian–Recent)

Class Helioplacophora—Benthonic, mobile forms, shaped like a top with plates arranged in a helical spiral (Early Cambrian)

Class Edrioasteroidea—Benthonic, sessile or mobile, discoidal, globular or cylindrical forms with five straight or curved feeding areas shaped like a starfish (Cambrian–Pennsylvanian)

Class Holothuroidea—Sea cucumbers. Sediment feeders having calcareous spicules embedded in a tough skin (Ordovician–Recent)

Class Echinoidea—Largest group of echinoderms. Globe- or disk-shaped, with movable spines. Predominantly grazers or sediment feeders. Epifaunal and infaunal (Ordovician–Recent)

Subphylum Asterozoa—Stemless, benthonic, mobile forms (Ordovician–Recent)

Class Asteroidea—Starfish. Arms merge into body (Ordovician–Recent)

Class Ophiuroidea—Brittle star. Distinct central body (Ordovician–Recent)

Phylum Hemichordata—Characterized by a notochord sometime during their life history. Modern acorn worms and extinct graptolites (Cambrian–Recent)

Class Graptolithina—Colonial marine hemichordates having a chitinous exoskeleton. Predominantly planktonic (Cambrian–Mississippian)

Phylum Chordata—Animals with notochord, hollow dorsal nerve cord, and pharyngeal pouches during at least part of their life cycle (Cambrian–Recent)

Subphylum Urochordata—Sea squirts, tunicates. Larval forms with notochord in tail region (Cambrian–Recent)

Subphylum Cephalochordata—Small marine animals with notochords and small fishlike bodies (Cambrian–Recent)

Subphylum Vertebrata—Animals with a backbone of vertebrae (Cambrian–Recent)

Class Agnatha—Jawless fish. Includes the living lampreys and hagfish as well as extinct armored ostracoderms (Cambrian–Recent)

Class Acanthodii—Primitive jawed fish with numerous spiny fins (Silurian–Permian)

Class Placodermii—Primitive armored jawed fish (Silurian–Permian)

Class Chondrichthyes—Cartilaginous fish such as sharks and rays (Devonian–Recent)

Class Osteichthyes—Bony fish (Devonian–Recent)

Subclass Actinopterygii—Ray-finned fish (Devonian–Recent)

Subclass Sarcopterygii—Lobe-finned, air-breathing fish (Devonian–Recent)

Order Coelacanthimorpha—Lobe-finned fish *Latimeria* (Devonian–Recent)

Order Crossoptergii—Lobe-finned fish that were ancestral to amphibians (Devonian–Permian)

Order Dipnoi—Lungfish (Devonian–Recent)

Class Amphibia—Amphibians. The first terrestrial vertebrates (Devonian–Recent)

Subclass Labyrinthodontia—Earliest amphibians. Solid skulls and complex tooth pattern (Devonian–Triassic)

Subclass Salientia—Frogs, toads, and their relatives (Triassic–Recent)

Subclass Condata—Salamanders and their relatives (Triassic–Recent)

Class Reptilia—Reptiles. A large and varied vertebrate group characterized by having scales and laying an amniote egg (Mississippian–Recent)

Subclass Anapsida—Reptiles whose skull has a solid roof with no openings (Mississippian–Recent)

Order Cotylosauria—One of the earliest reptile groups (Pennsylvanian–Triassic)

Order Chelonia—Turtles (Triassic–Recent)

Subclass Euryapsida—Reptiles with one opening high on the side of the skull behind the eye. Mostly marine (Permian–Cretaceous)

Order Protorosauria—Land-living ancestral euryapsids (Permian–Cretaceous)

Order Placodontia—Placodonts. Bulky, paddle-limbed marine reptiles with rounded teeth for crushing mollusks (Triassic)

Order Ichthyosauria—Ichthyosaurs. Dolphin-shaped swimming reptiles (Triassic–Cretaceous)

Subclass Diapsida—Most diverse reptile class. Characterized by two openings in the skull behind the eye. Includes lizards, snakes, crocodiles, thecodonts, dinosaurs, and pterosaurs (Permian–Recent)

Infraclass Lepidosauria—Primitive diapsids including snakes, lizards, and the mosasaurs, a large Cretaceous marine reptile group (Permian–Recent)

Order Mosasauria—Mosasaurs (Cretaceous)

Order Plesiosauria—Plesiosaurs (Triassic–Cretaceous)

Order Squamata—Lizards and snakes (Triassic–Recent)

Order Rhynchocephalia—The living tuatara *Sphenodon* and its extinct relatives (Jurassic–Recent)

Infraclass Archosauria—Advanced diapsids (Triassic–Recent)

Order Thecodontia—Thecodontians were a diverse group that was ancestral to the crocodilians, pterosaurs, and dinosaurs (Permian–Triassic)

Order Crocodilia—Crocodiles, alligators, and gavials (Triassic–Recent)

Order Pterosauria—Flying and gliding reptiles called pterosaurs (Triassic–Cretaceous)

Infraclass Dinosauria—Dinosaurs (Triassic–Cretaceous)

Order Saurischia—Lizard-hipped dinosaurs (Triassic–Cretaceous)

Suborder Theropoda—Bipedal carnivores (Triassic–Cretaceous)

Suborder Sauropoda—Quadrupedal herbivores, including the largest known land animals (Jurassic–Cretaceous)

Order Ornithischia—Bird-hipped dinosaurs (Triassic–Cretaceous)

Suborder Ornithopoda—Bipedal herbivores, including the duck-billed dinosaurs (Triassic–Cretaceous)

Suborder Stegosauria—Quadrupedal herbivores with bony spikes on their tails and bony plates on their backs (Jurassic–Cretaceous)

Suborder Pachycephalosauria—Bipedal herbivores with thickened bones of the skull roof (Cretaceous)

Suborder Ceratopsia—Quadrupedal herbivores, typically with horns or a bony frill over the top of the neck (Cretaceous)

Suborder Ankylosauria—Heavily armored quadrupedal herbivores (Cretaceous)

Subclass Synapsida—Mammal-like reptiles with one opening low on the side of the skull behind the eye (Pennsylvanian–Triassic)

Order Pelycosauria—Early mammal-like reptiles including those forms in which the vertebral spines were extended to support a "sail" (Pennsylvanian–Permian)

Order Therapsida—Advanced mammal-like reptiles with legs positioned beneath the body and the lower jaw formed largely of a single bone. Many therapsids may have been endothermic (Permian–Triassic)

Class Aves—Birds. Endothermic and feathered (Jurassic–Recent)

Class Mammalia—Mammals. Endothermic animals with hair (Triassic–Recent)

Subclass Prototheria—Egg-laying mammals (Triassic–Recent)

Order Docodonta—Small, primitive mammals (Triassic)

Order Triconodonta—Small, primitive mammals with specialized teeth (Triassic–Cretaceous)

Order Monotremata—Duck-billed platypus, spiny anteater (Cretaceous–Recent)

Subclass Allotheria—Small, extinct early mammals with complex teeth (Jurassic–Eocene)

Order Multituberculata—The first mammalian herbivores and the most diverse of Mesozoic mammals (Jurassic–Eocene)

Subclass Theria—Mammals that give birth to live young (Jurassic–Recent)

Order Symmetrodonta—Small, primitive Mesozoic therian mammals (Jurassic–Cretaceous)

Order Upantotheria—Trituber-
culates (Jurassic–Cretaceous)
Order Creodonta—Extinct
ancient carnivores
(Cretaceous–Paleocene)
Order Condylartha—
Extinct ancestral hoofed
placentals (ungulates)
(Cretaceous–Oligocene)
Order Marsupialia—
Pouched mammals. Opos-
sum, kangaroo, koala
(Cretaceous–Recent)
Order Insectivora—Primi-
tive insect-eating mammals.
Shrew, mole, hedgehog
(Cretaceous–Recent)
Order Xenungulata—Large South
American mammals that
broadly resemble pantodonts
and uintatheres (Paleocene)
Order Taeniodonta—Includes
some of the most highly
specialized terrestrial pla-
centals of the Late Paleocene and Early Eocene
(Paleocene–Eocene)
Order Tillodontia—Large,
massive placentals with
clawed, five-toed feet
(Paleocene–Eocene)
Order Dinocerata—Uintatheres;
large herbivores with bony
protuberances on the skull
and greatly elongated canine
teeth (Paleocene–Eocene)
Order Pantodonta—North
American forms are the size
of large sheep to rhinoceroses;
Asian forms are as small as a
rat (Paleocene–Eocene)
Order Astropotheria—Large
placental mammals with slender rear legs, stout forelimbs,
and elongate canine teeth
(Paleocene–Miocene)
Order Notoungulata—Larg-
est assemblage of South
American Ungulates, with
a wide range of body forms
(Paleocene–Pleistocene)
Order Liptoterna—
Extinct South Ameri-
can hoofed mammals
(Paleocene–Pleistocene)
Order Rodentia—Squirrel,
mouse, rat, beaver, porcupine,
gopher (Paleocene–Recent)

Order Lagomorpha—Hare, rab-
bit, pika (Paleocene–Recent)
Order Primate—Lemur, tar-
sier, loris, monkey, human
(Paleocene–Recent)
Order Edentata—Anteater,
sloth, armadillo, glyptodont
(Paleocene–Recent)
Order Carnivora—Modern
carnivorous placentals.
Dog, cat, bear, skunk, seal,
weasel, hyena, raccoon,
panda, sea lion, walrus
(Paleocene–Recent)
Order Pyrotheria—Large
mammals with long bodies
and short, columnar limbs
(Eocene–Oligocene)
Order Chiroptera—Bats
(Eocene–Recent)
Order Dermoptera—Flying
lemur (Eocene–Recent)
Order Cetacea—Whale,
dolphin, porpoise
(Eocene–Recent)
Order Tubulidentata—Aardvark
(Eocene–Recent)
Order Perissodactyla—Odd-
toed ungulates (hoofed placentals). Horse, rhinoceros,
tapir, titanothere, chalicothere
(Eocene–Recent)
Order Artiodactyla—Even-toed
ungulates. Pig, hippo, camel,
deer, elk, bison, cattle, sheep,
antelope, entelodont, oreo-
dont (Eocene–Recent)
Order Proboscidea—Elephant,
mammoth, mastodon
(Eocene–Recent)
Order Sirenia—Sea cow, mana-
tee, dugong (Eocene–Recent)
Order Embrithopodoa—Known
primarily from a single locality in Egypt. Large mammals
with two gigantic bony processes arising from the nose
area (Oligocene)
Order Desmostyla—Amphibi-
ous or seal-like in habit.
Front and hind limbs well
developed, but hands and
feet somewhat specialized as
paddles (Oligocene–Miocene)
Order Hyracoidea—Hyrax
(Oligocene–Recent)
Order Pholidota—Scaly anteater
(Oligocene–Recent)

Appendix C
Mineral Identification

To identify most common minerals, geologists use physical properties such as color, luster, crystal form, hardness, cleavage, specific gravity, and several others (Tables C1 and C2). Notice that the Mineral Identification Table (C3) is arranged with minerals having a metallic luster grouped separately from those with a nonmetallic luster. After determining luster, ascertain hardness and note that each part of the table is arranged with minerals in order of increasing hardness. Thus, if you have a nonmetallic mineral with a hardness of 6, it must be augite, hornblende, plagioclase, or one of the two potassium feldspars (orthoclase or microcline). If this hypothetical mineral is dark green or black, it must be augite or hornblende. Use other properties to make a final determination.

TABLE C1 Physical Properties Used to Identify Minerals

Mineral Property	Comment
Luster	Appearance in reflected light; if it has the appearance of a metal, the luster is metallic; those with a nonmetallic luster do not look like metals
Color	Rather constant in minerals with metallic luster; varies in minerals with nonmetallic luster
Streak	Powdered mineral on an unglazed porcelain plate (streak plate) is more typical of a mineral's true color
Crystal form	Useful if crystals visible (see Figure 2.5)
Cleavage	Minerals with cleavage tend to break along a smooth plane or planes of weakness
Hardness	A mineral's resistance to abrasion (see Table C2)
Specific gravity	Ratio of a mineral's weight to an equal volume of water
Reaction with HCl (hydrochloric acid)	Calcite reacts vigorously, but dolomite reacts only when powdered
Other properties	Talc has a soapy feel; graphite writes on paper; magnetite is magnetic; closely spaced, parallel lines visible on plagioclase; halite tastes salty

© 2013 Cengage Learning

TABLE C2 Moh's Hardness Scale

Austrian geologist Frederich Mohs devised this relative hardness scale for ten minerals. He assigned a value of 10 to diamond, the hardest mineral known, and lesser values to the other minerals. You can determine the relative hardness of minerals by scratching one mineral with another or by using objects of known hardness.

Hardness	Mineral	Hardness of Some Common Objects
10	Diamond	
9	Corundum	
8	Topaz	
7	Quartz	
		Steel file (6.5)
6	Orthoclase	
		Glass (5.5–6)
5	Apatite	
4	Fluorite	
3	Calcite	Copper penny (3)
		Fingernail (2.5)
2	Gypsum	
1	Talc	

© 2013 Cengage Learning

TABLE C3 Mineral Identification Tables

Metallic Luster

Mineral	Chemical Composition	Color	Hardness Specific Gravity	Other Features	Comments
Graphite	C	Black	1–2 2.09–2.33	Greasy feel; writes on paper; one direction of cleavage	Used for pencil "leads." Mostly in metamorphic rocks.
Galena	PbS	Lead gray	2.5 7.6	Cubic crystals; three cleavages at right angles	The ore of lead. Mostly in hydrothermal rocks.
Chalcopyrite	$CuFeS_2$	Brassy yellow	3.5–4 4.1–4.3	Usually massive; greenish black streak; iridescent tarnish	The most common copper mineral. Mostly in hydrothermal rocks.
Magnetite	Fe_3O_4	Black	5.5–6.5 5.2	Strong magnetism	An ore of iron. An accessory mineral in many rocks.
Hematite	Fe_2O_3	Red brown	6 4.8–5.3	Usually granular or massive; reddish brown streak	Important iron ore. An accessory mineral in many rocks.
Pyrite	FeS_2	Brassy yellow	6.5 5.0	Cubic and octahedral crystals	Found in some igneous and hydrothermal rocks and in sedimentary rocks associated with coal.

Nonmetallic Luster

Mineral	Chemical Composition	Color	Hardness Specific Gravity	Other Features	Comments
Talc	$Mg_3Si_4O_{10}(OH)_2$	White, green	1 2.82	One cleavage direction; usually in compact masses; soapy feel	Formed by the alteration of magnesium silicates. Mostly in metamorphic rocks.
Clay minerals	Varies	Gray, buff, white	1–2 2.5–2.9	Earthy masses; particles too small to observe properties	Found in soils, mudrocks, slate, phyllite.
Chlorite	$(Mg,Fe)_3(Si,Al)_4O_{10}$ $(Mg,Fe)_3(OH)_6$	Green	2 2.6–3.4	One cleavage; occurs in scaly masses	Common in low-grade metamorphic rocks such as slate.
Gypsum	$CaSO_4 \cdot 2H_2O$	Colorless, white	2 2.32	Elongate crystals; fibrous and earthy masses	The most common sulfate mineral. Found mostly in evaporite deposits.
Muscovite (Mica)	$KAl_2Si_3O_{10}(OH)_2$	Colorless	2–2.5 2.7–2.9	One direction of cleavage; cleaves into thin sheets	Common in felsic igneous rocks, metamorphic rocks, and some sedimentary rocks.
Biotite (Mica)	$K(Mg,Fe)_3AlSi_3O_{10}$ $(OH)_2$	Black, brown	2.5 2.9–3.4	One cleavage direction; cleaves into thin sheets	Occurs in both felsic and mafic igneous rocks, in metamorphic rocks, and in some sedimentary rocks.
Calcite	$CaCO_3$	Colorless, white	3 2.71	Three cleavages at oblique angles; cleaves into rhombs; reacts with dilute HCl	The most common carbonate mineral. Main component of limestone and marble.
Anhydrite	$CaSO_4$	White, gray	3.5 2.9–3.0	Crystals with two cleavages; usually in granular masses	Found in limestones, evaporite deposits, and the cap rock of salt domes.

Nonmetallic Luster

Mineral	Chemical Composition	Color	Hardness Specific Gravity	Other Features	Comments
Halite	$NaCl$	Colorless, white	3–4 2.2	Three cleavages at right angles; cleaves into cubes; cubic crystals; salty taste	Occurs in evaporite deposits.
Dolomite	$CaMg(CO_3)_2$	White, yellow, gray, pink	3.5–4 2.85	Cleavage as in calcite; reacts with dilute hydrochloric acid when powdered	The main constituent of dolostone. Also found associated with calcite in some limestones and marble.
Fluorite	CaF_2	Colorless, purple, green, brown	4 3.18	Four cleavage directions; cubic and octahedral crystals	Occurs mostly in hydrothermal rocks and in some limestones and dolostones.
Augite	$Ca(Mg,Fe,Al)(Al,Si)_2O_6$	Black, dark green	6 3.25– 3.55	Short eight-sided crystals; two cleavages; cleavages nearly at right angles	The most common pyroxene mineral. Found mostly in mafic igneous rocks.
Hornblende	$NaCa_2(Mg,Fe,Al)_5$ $(Si,Al)_8O_{22}(OH)_2$	Green, black	6 3.0–3.4	Elongate, six-sided crystals; two cleavages intersecting at 56° and 124°	A common rock-forming amphibole mineral in igneous and metamorphic rocks.
Plagioclase feldspars	Varies from $CaAl_2Si_2O_8$ to $NaAlSi_3O_8$	White, gray, brown	6 2.56	Two cleavages at right angles	Common in igneous rocks and a variety of metamorphic rocks. Also in some arkoses.
Potassium feldspars — Microcline	$KAlSi_3O_8$	White, pink, green	6 2.56	Two cleavages at right angles	Common in felsic igneous rocks, some metamorphic rocks, and arkoses.
Potassium feldspars — Orthoclase	$KAlSi_3O_8$	White, pink	6 2.56	Two cleavages at right angles	
Olivine	$(Fe,Mg)_2SiO_4$	Olive green	6.5 3.3–3.6	Small mineral grains in granular masses; conchoidal fracture	Common in mafic igneous rocks.
Quartz	SiO_2	Colorless, white, gray, pink, green	7 2.67	Six-sided crystals; no cleavage; conchoidal fracture	A common rock-forming mineral in all rock groups and hydrothermal rocks. Also occurs in varieties known as chert, flint, agate, and chalcedony.
Garnet	$Fe_3Al_2(SiO_4)_3$	Dark red, green	7–7.5 4.32	Twelve-sided crystals common; uneven fracture	Found mostly in gneiss and schist.
Zircon	Zr_2SiO_4	Brown, gray	7.5 3.9–4.7	Four-sided, elongate crystals	Most common as an accessory in granitic rocks.
Topaz	$Al_2SiO_4(OH,F)$	Colorless, white, yellow, blue	8 3.5–3.6	High specific gravity; one cleavage direction	Found in pegmatites, granites, and hydrothermal rocks.
Corundum	Al_2O_3	Gray, blue, pink, brown	9 4.0	Six-sided crystals and great hardness are distinctive	An accessory mineral in some igneous and metamorphic rocks.

Glossary

abiogenesis The origin of life from non-living matter.

Absaroka Sequence A widespread succession of Pennsylvanian and Permian sedimentary rocks bounded above and below by unconformities; deposited during a transgressive–regressive cycle of the Absaroka Sea.

absolute dating Assigning an age in years to geologic events before the present; absolute dates are determined by radioactive-decay dating techniques. Also called *numerical dating*. (See *relative dating*.)

Acadian orogeny A Devonian episode of mountain building in the northern Appalachian mobile belt resulting from a collision of Baltica with Laurentia.

Alleghenian orogeny Pennsylvanian to Permian mountain building in the Appalachian mobile belt from New York to Alabama.

allele A variant form of a single gene. (See *gene*.)

allopatric speciation Model for the origin of a new species from a small population that became isolated from its parent population.

alluvial fan A cone-shaped accumulation of mostly sand and gravel where a stream flows from a mountain valley onto an adjacent lowland.

Alpine–Himalayan orogenic belt A linear zone of deformation extending from the Atlantic eastward across southern Europe and North Africa, through the Middle East and into Southeast Asia. (See *circum-Pacific orogenic belt*.)

Alpine orogeny A Late Mesozoic–Early Cenozoic episode of mountain building affecting southern Europe and North Africa.

amniote Any vertebrate animal that produces amniote eggs; includes reptiles, birds, and mammals.

amniote egg An egg in which an embryo develops in a liquid-filled cavity (the amnion); a waste sac is present as well as a yolk sac for nourishment.

anaerobic Refers to organisms that do not depend on oxygen for respiration.

analogous structure Body parts, such as wings of insects and birds, that serve the same function but differ in structure and development. (See *homologous structure*.)

Ancestral Rockies Late Paleozoic uplift in the southwestern part of the North American craton.

angiosperm Vascular plants having flowers and seeds; the flowering plants.

angular unconformity An unconformity below which strata dip at a steeper angle than those above. (See *disconformity*, *nonconformity*, and *unconformity*.)

anthropoid Any member of the primate suborder Anthropoidea; includes New World and Old World monkeys, apes, and humans.

Antler orogeny A Late Devonian to Mississippian episode of mountain building that affected the Cordilleran mobile belt from Nevada to Alberta, Canada.

Appalachian mobile belt A long, narrow region of tectonic activity along the eastern margin of the North American craton extending from Newfoundland to Georgia; probably continuous to the southwest with the Ouachita mobile belt.

Archaeopteryx The oldest positively identified fossil bird; it had feathers but retained many reptile characteristics; from Jurassic-age rocks in Germany.

archosaur A term referring to the ruling reptiles—dinosaurs, pterosaurs, crocodiles, and birds.

artificial selection The practice of selectively breeding plants and animals with desirable traits.

Artiodactyla The mammalian order whose members have two or four toes; the even-toed hoofed mammals such as deer, goats, sheep, antelope, bison, swine, and camels.

asthenosphere Part of the upper mantle over which the lithosphere moves; it behaves as a plastic and flows.

Atlantic Coastal Plain The broad, low-relief area of eastern North America extending from the Appalachian Mountains to the Atlantic shoreline.

atom The smallest unit of matter that retains the characteristics of an element.

atomic mass number The total number of protons and neutrons in an atom's nucleus.

atomic number The number of protons in an atom's nucleus.

australopithecine A collective term for all species of the extinct genus *Australopithecus* that existed in South Africa during the Pliocene and Pleistocene.

autotrophic Describes organisms that synthesize their organic nutrients from inorganic raw materials; photosynthesizing bacteria and plants are autotrophs. (See *heterotrophic* and *primary producer*.)

back-arc basin A marine basin, such as the Sea of Japan, lying between a volcanic island arc and a continent; probably forms by back-arc spreading.

Baltica One of six major Paleozoic continents; composed of Russia west of the Ural Mountains, Scandinavia, Poland, and northern Germany.

banded iron formation (BIF) Sedimentary rocks made up of alternating thin layers of chert and iron minerals, mostly the iron oxides hematite and magnetite.

barrier island A long sand body more or less parallel with a shoreline but separated from it by a lagoon.

Basin and Range Province An area of Cenozoic block-faulting centered on Nevada but extending into adjacent states and northern Mexico.

benthos All bottom-dwelling marine organisms that live on the seafloor or within seafloor sediments.

Big Bang A theory for the evolution of the universe from a dense, hot state followed by expansion, cooling, and a less dense state.

biogenic sedimentary structure Any feature such as tracks, trails, and burrows in sedimentary rocks produced by the activities of organisms. (See *trace fossil*.)

biogeography The study of the geographic distribution of organisms and communities of organisms.

biostratigraphic unit A unit of sedimentary rock defined solely by its fossil content.

bioturbation The churning of sediment by organisms that burrow through it.

biozone A general term referring to all biostratigraphic units such as range zones and concurrent range zones.

bipedal Walking on two legs as a means of locomotion, as in birds and humans. (See *quadrupedal*.)

black smoker A submarine hydrothermal vent that emits a plume of black water colored by dissolved minerals. (See *submarine hydrothermal vent*.)

body fossil The shells, teeth, bones, or (rarely) the soft parts of organisms preserved in the fossil record. (See *fossil* and *trace fossil*.)

bonding The processes whereby atoms join with other atoms.

bony fish Members of the class Osteichthyes that evolved during the Devonian; characterized by a bony internal skeleton; includes the ray-finned fishes and the lobe-finned fishes.

braided stream A stream with an intricate network of dividing and rejoining channels.

browser An animal that eats tender shoots, twigs, and leaves. (See *grazer*.)

Caledonian orogeny A Silurian–Devonian episode of mountain building that took place along the northwestern margin of Baltica, resulting from the collision of Baltica with Laurentia.

Canadian shield The Precambrian shield in North America; mostly in Canada but also exposed in Minnesota, Wisconsin, Michigan, and New York.

carbon-14 dating An absolute-dating technique relying on the ratio of C^{14} to C^{12} in organic substances; useful back to about 70,000 years ago.

carbonate mineral Any mineral with the negatively charged carbonate ion $(CO_3)^{-2}$ (e.g., calcite $[CaCO_3]$ and dolomite $[CaMg(CO_3)_2]$).

carbonate rock Any rock composed mostly of carbonate minerals (such as limestone and dolostone).

carnassials A pair of specialized shearing teeth in members of the mammal order Carnivora.

Carnivora An order of mammals consisting of meat eaters such as dogs, cats, bears, weasels, and seals.

carnivore-scavenger Any animal that eats other animals, living or dead, as a source of nutrients.

cartilaginous fish Fish, such as sharks and their living and extinct relatives, that have an internal skeleton of cartilage.

Cascade Range A mountain range made up of volcanic rock stretching from northern California through Oregon and Washington and into British Columbia, Canada.

Catskill Delta A Devonian clastic wedge deposited adjacent to the highlands that formed during the Acadian orogeny.

Cetacea The mammal order that includes whales, porpoises, and dolphins.

chemical sedimentary rock Rock formed of minerals derived from materials dissolved during weathering.

China One of six major Paleozoic continents; composed of all of Southeast Asia, including China, Indochina, part of Thailand, and the Malay Peninsula.

chordate Any member of the phylum Chordata, all of which have a notochord, dorsal hollow nerve cord, and pharyngeal pouches at some time during their life cycle.

chromosome Complex, double-stranded, helical molecule of deoxyribonucleic acid (DNA); specific segments of chromosome are genes.

circum-Pacific orogenic belt One of two major Mesozoic–Cenozoic areas of large-scale deformation and the origin of mountains; includes orogens in South and Central America, the North American Cordillera, and the Aleutian, Japanese, and Philippine arcs. (See *Alpine–Himalayan orogenic belt*.)

cirque A steep-walled, bowl-shaped depression formed on a mountainside by glacial erosion.

cladistics A type of analysis of organisms in which they are grouped together on the basis of derived versus primitive characteristics.

cladogram A diagram showing the relationships among members of a clade, including their most recent common ancestor.

clastic wedge An extensive accumulation of mostly detrital sedimentary rocks eroded from, and deposited adjacent to, an area of uplift, as in the Catskill Delta or Queenston Delta.

Colorado Plateau A vast upland area in Colorado, Utah, Arizona, and New Mexico with only slightly deformed Phanerozoic rocks, deep canyons, and volcanic mountains.

compound A substance made up of different atoms bonded together (such as water $[H_2O]$ and quartz $[SiO_2]$).

concurrent range zone A biozone established by plotting the overlapping geologic ranges of fossils.

conformable Refers to a sequence of sedimentary rocks deposited one after the other with no discontinuities, or only minor ones.

contact metamorphism Metamorphism taking place adjacent to a body of magma (a pluton) or beneath a lava flow from heat and chemically active fluids.

continental accretion The process whereby continents grow by additions of Earth materials along their margins.

continental–continental plate boundary A convergent plate boundary along which two continental lithospheric plates collide, such as the collision of India with Asia. (See *convergent plate boundary*, *oceanic–continental plate boundary*, and *oceanic–oceanic plate boundary*.)

continental drift The hypothesis proposed by Alfred Wegener that all continents were once joined into a single landmass that broke apart, with the various fragments (continents) moving with respect to one another.

continental glacier A glacier covering at least 50,000 km^2 and unconfined by topography. Also called an *ice sheet*.

continental interior An area in North America made up of the Great Plains and the Central Lowlands, bounded by the Rocky Mountains, the Canadian shield, the Appalachian Mountains, and parts of the Gulf Coastal Plain.

continental red bed Red-colored rock, especially mudrock and sandstone, on the continents; iron oxides account for their color.

continental rise The gently sloping part of the seafloor lying between the base of the continental slope and the deep seafloor.

continental shelf The area where the seafloor slopes gently seaward between a shoreline and the continental slope.

continental slope The relatively steep part of the seafloor between the continental shelf and continental rise or an oceanic trench.

convergent evolution The origin of similar features in distantly related organisms as they adapt in comparable ways, such as ichthyosaurs and porpoises. (See *parallel evolution*.)

convergent plate boundary The boundary between two plates that move toward one another. (See *continental–continental plate boundary*, *oceanic–continental plate boundary*, and *oceanic–oceanic plate boundary*.)

Cordilleran mobile belt An area of extensive deformation in western North America bounded by the Pacific Ocean and the Great Plains; it extends north–south from Alaska into central Mexico.

Cordilleran orogeny A period of deformation affecting the western part of North America from Jurassic to Early Cenozoic time; divided into three phases known as the Nevadan, Sevier, and Laramide orogenies.

core The inner part of Earth from a depth of about 2,900 km, consisting of a liquid outer part and a solid inner part; probably composed mostly of iron and nickel.

correlation Demonstration of the physical continuity of stratigraphic units over an area; also refers to matching up time-equivalent events in different areas.

craton Name applied to a stable nucleus of a continent, consisting of a Precambrian shield and a platform of buried ancient rocks.

cratonic sequence A widespread association of sedimentary rocks bounded above and below by unconformities that were deposited during a transgressive–regressive cycle of an epeiric sea; for example, the Sauk Sequence.

Cretaceous Interior Seaway A Late Cretaceous arm of the sea that effectively divided North America into two large landmasses.

Cro-Magnon A race of *Homo sapiens* that lived mostly in Europe from 35,000 to 10,000 years ago.

cross-bedding A type of bedding in which individual layers are deposited at an angle to the surface on which they accumulate, as in sand dunes.

crossopterygian A specific type of lobe-finned fish that had lungs.

crust The upper part of Earth's lithosphere, which is separated from the mantle by the Moho; consists of continental crust with an overall granitic composition and thinner, denser oceanic crust made up of basalt and gabbro.

crystalline solid A solid with its atoms arranged in a regular three-dimensional framework.

Curie point The temperature at which iron-bearing minerals in a cooling magma attain their magnetism.

cyclothem A sequence of cyclically repeated sedimentary rocks resulting from alternating periods of marine and nonmarine deposition; commonly contains a coal bed.

cynodont A type of therapsid (advanced mammal-like reptile); ancestors of mammals are among the cynodonts.

delta A deposit of sediment where a stream or river enters a lake or the ocean.

deoxyribonucleic acid (DNA) The chemical substance of which chromosomes are composed.

depositional environment Any area where sediment is deposited; a depositional site where physical, chemical, and biological processes operate to yield a distinctive kind of deposit.

detrital sedimentary rock Rock made up of the solid particles derived from preexisting rocks, as in sandstone.

dinosaur Any of the Mesozoic reptiles belonging to the orders Saurischia and Ornithischia.

disconformity A type of unconformity above and below which the strata are parallel. (See *angular unconformity, nonconformity,* and *unconformity.*)

divergent evolution The diversification of a species into two or more descendant species.

divergent plate boundary The boundary between two plates that move apart; characterized by seismicity, volcanism, and the origin of new oceanic lithosphere.

drift A collective term for all sediment deposited by glacial activity; includes till deposited directly by ice, and outwash deposited by streams discharging from glaciers. (See *outwash.*)

dynamic metamorphism Metamorphism in fault zones where rocks are subjected to high differential pressure.

ectotherm A cold-blooded vertebrate, such as an amphibian or reptile, that depends on external heat to regulate body temperature. (See *endotherm.*)

Ediacaran fauna Name for all Late Proterozoic faunas with animal fossils similar to those of the Ediacara fauna of Australia.

element A substance composed of only one kind of atom (such as calcium [Ca] or silicon [Si]).

end moraine A pile or ridge of rubble deposited at the terminus of a glacier.

endosymbiosis A type of mutually beneficial symbiosis in which one symbiont lives within the other.

endotherm A warm-blooded vertebrate, such as a bird or mammal, that maintains its body temperature, within narrow limits, by internal processes. (See *ectotherm.*)

epeiric sea A broad, shallow sea that covers part of a continent; six epeiric seas were present in North America during the Phanerozoic Eon, such as the Sauk Sea.

eukaryotic cell A cell with an internal, membrane-bounded nucleus, containing chromosomes and other internal structures such as mitochondria, that are not present in prokaryotic cells. (See *prokaryotic cell.*)

evaporite Sedimentary rock formed by inorganic chemical precipitation from evaporating water (for example, rock salt and rock gypsum).

extrusive igneous rock An igneous rock that forms as lava cools and crystallizes or when pyroclastic materials are consolidated. (See *volcanic rock.*)

Farallon plate A Late Mesozoic–Cenozoic oceanic plate that was largely subducted beneath North America; the Cocos and Juan de Fuca plates are remnants.

fission-track dating The dating process in which small linear tracks (fission tracks) resulting from alpha decay are counted in mineral crystals.

fluvial Relating to streams and rivers and their deposits.

formation The basic lithostratigraphic unit; a mappable unit of strata with distinctive upper and lower boundaries.

fossil Remains or traces of prehistoric organisms preserved in rocks. (See *body fossil* and *trace fossil.*)

gene A specific segment of a chromosome constituting the basic unit of heredity. (See *allele.*)

geologic column A diagram showing a composite column of rocks arranged with the oldest at the bottom, followed upward by progressively younger rocks. (See *geologic time scale.*)

geologic record The record of prehistoric physical and biologic events preserved in rocks.

geologic time scale A chart arranged so that the designation for the earliest part of geologic time appears at the bottom, followed upward by progressively younger time designations. (See *geologic column.*)

geology The science concerned with the study of Earth; includes study of Earth materials, internal and surface processes, and Earth and life history.

glacial stage A time of extensive glaciation that occurred several times during the Pleistocene, especially on the Northern Hemisphere continents.

glacier A mass of ice on land that moves by plastic flow and basal slip.

***Glossopteris* flora** A Late Paleozoic association of plants found only on the Southern Hemisphere continents and India; named after its best-known genus, *Glossopteris.*

Gondwana One of six major Paleozoic continents; composed of South America, Africa, Australia, India, and parts of Southern Europe, Arabia, and Florida.

graded bedding A sediment layer in which grain size decreases from the bottom up.

granite-gneiss complex One of the two main rock associations found in areas of Archean rocks.

grazer An animal that eats low-growing vegetation, especially grasses. (See *browser.*)

greenstone belt A linear or podlike association of rocks particularly common in Archaen terranes; typically synclinal and consists of lower and middle volcanic units and an upper sedimentary unit.

Grenville orogeny An episode of deformation that took place in the eastern United States and Canada during the Neoproterozoic.

guide fossil Any easily identified fossil with a wide geographic distribution and short geologic range; useful for determining relative ages of strata in different areas.

Gulf Coastal Plain The broad, low-relief area along the Gulf Coast of the United States.

gymnosperm A flowerless, seed-bearing plant.

half-life The time necessary for one-half of the original number of radioactive atoms of an element to decay to a stable daughter product; for example, the half-life of potassium 40 is 1.3 billion years.

herbivore An animal dependent on vegetation as a source of nutrients.

Hercynian (Variscan)–Alleghenian orogeny Pennsylvanian to Permian deformation in the Hercynian mobile belt of Europe, and the Appalachian and Ouachita mobile belts of North America.

heterotrophic Organism such as an animal that depends on preformed organic molecules from its environment for nutrients. (See *autotrophic* and *primary producer.*)

Himalayan orogeny A Cenozoic episode of mountain building that began 50 to 40 million years ago (Eocene) when the Indian plate collided with Asia.

hominid Abbreviated term for Hominidae, the family that includes bipedal primates such as *Australopithecus* and *Homo.* (See *hominoid.*)

hominoid Abbreviated term for Hominoidea, the superfamily that includes apes and humans. (See *hominid.*)

Homo The genus of hominids consisting of *Homo sapiens* and their ancestors, *Homo erectus* and *Homo habilis.*

homologous structure Body part in different organisms that has a similar structure, similar relationships to other organs, and similar development, but does not necessarily serve the same function, such as forelimbs in whales, bats, and dogs. (See *analogous structure.*)

hot spot Localized zone of melting below the lithosphere; detected by volcanism at the surface.

hypothesis A provisional explanation for observations that is subject to continual testing and modification if necessary; if well supported by evidence, hypotheses may become theories.

Hyracotherium A small Early Eocene mammal that was ancestral to today's horses.

Iapetus Ocean A Paleozoic ocean between North America and Europe; it eventually closed as North America and Europe moved toward one another and collided during the Late Paleozoic.

ice cap A dome-shaped mass of glacial ice covering less than 50,000 km².

ice-scoured plain An area eroded by glaciers resulting in low-relief, extensive bedrock exposures with glacial polish, striations, and little soil.

ichthyosaur Any of the porpoise-like, Mesozoic marine reptiles.

igneous rock Rock formed when magma or lava cools and crystallizes and when pyroclastic materials become consolidated.

inheritance of acquired characteristics Jean-Baptiste de Lamarck's mechanism for evolution; holds that characteristics acquired during an individual's lifetime can be inherited by descendants.

interglacial stage A time of warmer temperatures between episodes of widespread glaciation.

intrusive igneous rock Igneous rock that cools and crystallizes from magma intruded into or formed within the crust. (See *plutonic rock*.)

iridium anomaly The occurence of a higher than usual concentration of the element iridium at the Cretaceous–Palogene boundary.

isostasy The concept of Earth's crust "floating" on the more dense underlying mantle. As a result of isostasy, thicker, less dense continental crust stands higher than oceanic crust.

isostatic rebound The phenomenon in which unloading of the crust causes it to rise, as when extensive glaciers melt, until it attains equilibrium.

Jovian planet Any planet with a low mean density that resembles Jupiter (Jupiter, Saturn, Uranus, and Neptune); the Jovian planets, or gas giants, are composed largely of hydrogen, helium, and frozen compounds such as methane and ammonia. (See *terrestrial planet*.)

Kaskaskia Sequence A widespread association of Devonian and Mississippian sedimentary rocks bounded above and below by unconformities; deposited during a transgressive–regressive cycle of the Kaskaskia Sea.

Kazakhstania One of six major Paleozoic continents; a triangular-shaped continent centered on Kazakhstan.

labyrinthodont Any of the Devonian to Triassic amphibians characterized by complex folding in the enamel of the teeth.

Laramide orogeny Late Cretaceous to Early Paleogene phase of the Cordilleran orogeny; responsible for many of the structural features in the present-day Rocky Mountains.

Laurasia A Late Paleozoic, Northern Hemisphere continent made up of North America, Greenland, Europe, and Asia.

Laurentia A Proterozoic continent composed mostly of North America and Greenland, parts of Scotland, and perhaps parts of the Baltic shield of Scandinavia.

lava Magma that reaches the surface.

lithification The process of converting sediment into sedimentary rock.

lithosphere The outer, rigid part of Earth, consisting of the upper mantle, oceanic crust, and continental crust; lies above the asthenosphere.

lithostratigraphic unit A body of sedimentary rock, such as a formation, defined solely by its physical attributes.

Little Ice Age An interval from about 1500 to the mid- to late-1800s during which glaciers expanded to their greatest historic extent.

living fossil An existing organism that has descended from ancient ancestors with little apparent change.

lobe-finned fish Fish with limbs containing a fleshy shaft and a series of articulating bones; one of the two main groups of bony fish.

macroevolution Evolutionary changes that account for the origin of new species, genera, orders, and so on. (See *microevolution*.)

magma Molten rock material below the surface.

magnetic anomaly Any change, such as the average strength, in Earth's magnetic field.

magnetic field The area in which magnetic substances are affected by lines of magnetic force emanating from Earth.

magnetic reversal The phenomenon involving the complete reversal of the north and south magnetic poles.

magnetism A physical phenomenon resulting from moving electricity and the spin of electrons in some solids, in which magnetic substances are attracted toward one another.

mantle The inner part of Earth surrounding the core, accounting for about 85 percent of the planet's volume; probably composed of peridotite.

marine regression Withdrawal of the sea from a continent or coastal area caused by emergence of the land with a resulting seaward migration of the shoreline.

marine transgression Invasion of a coastal area or much of a continent by the sea as sea level rises, resulting in a landward migration of the shoreline.

marsupial mammal The pouched mammals, such as wombats and kangaroos, that give birth to their young in a very immature state.

mass extinction Greatly accelerated extinction rates resulting in marked decrease in biodiversity, such as the mass extinction at the end of the Cretaceous.

meandering stream A stream with a single, sinuous channel with broadly looping curves.

meiosis Cell division yielding sex cells (sperm and eggs in animals, and pollen and ovules in plants), in which the number of chromosomes is reduced by half. (See *mitosis*.)

metamorphic rock Any rock altered in the solid state from preexisting rocks by any combination of heat, pressure, and chemically active fluids.

microevolution Evolutionary changes within a species. (See *macroevolution*.)

Midcontinent Rift A Mesoproterozoic intracontinental rift in Lauentia in which volcanic and sedimentary rocks accumulated.

Milankovitch theory A theory that explains cyclic variations in climate and the onset of glacial episodes triggered by irregularities in Earth's rotation and orbit.

mineral A naturally occurring, inorganic, crystalline solid having characteristic physical properties and a narrowly defined chemical composition.

mitosis Cell division resulting in two cells with the same number of chromosomes as the parent cell; takes place in all cells except sex cells. (See *meiosis*.)

mobile belt Elongated area of deformation generally at the margins of a craton, such as the Appalachian mobile belt.

modern synthesis A combination of ideas of various scientists yielding a view of evolution that includes the chromosome theory of inheritance, mutations as a source of variation, and gradualism; it also rejects inheritance of acquired characteristics.

molar Any of a mammal's teeth that are used for grinding and chewing.

molarization An evolutionary trend in hoofed mammals in which the premolars become more like molars, giving the animals a continuous series of grinding teeth.

monomer A comparatively simple organic molecule, such as an amino acid, that can link with other monomers to form more complex polymers such as proteins. (See *polymer*.)

monotremes The egg-laying mammals; including only the platypus and spiny anteater of the Australian region.

moraine A ridge or mound of unsorted, unstratified debris deposited by a glacier.

mosaic evolution The concept that not all parts of an organism evolve at the same rate, thus accounting for organisms with features retained from the ancestral condition as well as more recently evolved features.

mosasaur A term referring to a group of Mesozoic marine lizards.

mud crack A crack in clay-rich sediment that forms in response to drying and shrinkage.

multicelled organism An organism made up of many cells as opposed to a single cell; possesses cells specialized to perform specific functions.

mutation Any change in the genes of organisms; yields some of the variation on which natural selection acts.

natural selection A mechanism accounting for differential survival and reproduction among members of a species; the mechanism proposed by Charles Darwin and Alfred Wallace to account for evolution.

Neanderthal A type of human that inhabited the Near East and Europe from 200,000 to 30,000 years ago; may be a subspecies (*Homo sapiens neanderthalensis*) of *Homo* or a separate species (*Homo neanderthalensis*).

nekton Actively swimming organisms such as fish, whales, and squid. (See *plankton*.)

neoglaciation An episode in Earth history from about 6,000 years ago until the mid to late 1800s during which three periods of glacial expansion took place.

Nevadan orogeny Late Jurassic to Cretaceous phase of the Cordilleran orogeny; most strongly affected the western part of the Cordilleran mobile belt.

nonconformity An unconformity in which stratified sedimentary rocks overlie an erosion surface cut into igneous or metamorphic rocks. (See *angular unconformity, disconformity,* and *unconformity*.)

North American Cordillera A complex mountainous region in western North America extending from Alaska into central Mexico.

oceanic–continental plate boundary A convergent plate boundary along which oceanic and continental lithosphere collide; characterized by subduction of the oceanic plate, seismicity, and volcanism. (See *continental–continental plate boundary, convergent plate boundary,* and *oceanic–oceanic plate boundary*.)

oceanic–oceanic plate boundary A convergent plate boundary along which oceanic lithosphere collides with oceanic lithosphere; characterized by subduction of one of the oceanic plates, seismicity, and volcanism. (See *continental–continental plate boundary, convergent plate boundary,* and *oceanic–continental plate boundary*.)

Old Red Sandstone A Devonian-age clastic wedge that grew eastward from the Caledonian highlands onto the Baltica craton.

organic evolution See *theory of evolution*.

organic reef A wave-resistant limestone structure with a framework of animal skeletons, such as a coral reef or stromatoporoid reef.

Ornithischia One of the two orders of dinosaurs; characterized by a birdlike pelvis; includes ornithopods, stegosaurs, ankylosaurs, pachycephalosaurs, and ceratopsians. (See *Saurischia*.)

orogen A linear part of Earth's crust that was or is being deformed during an orogeny; part of an orogenic belt.

orogeny An episode of mountain building involving deformation, usually accompanied by igneous activity, metamorphism, and crustal thickening.

ostracoderm The "bony-skinned" fish, characterized by bony armor but no jaws or teeth; appeared during the Late Cambrian, making them the oldest known vertebrates.

Ouachita mobile belt An area of deformation along the southern margin of the North American craton; probably continuous to the northeast with the Appalachian mobile belt.

Ouachita orogeny A period of mountain building that took place in the Ouachita mobile belt during the Pennsylvanian and Permian periods.

outgassing The process whereby gases released from Earth's interior by volcanism formed an atmosphere.

outwash All sediment deposited by streams that issue from glaciers. (See *drift*.)

Pacific–Farallon Ridge A spreading ridge that was located off the coast of western North America during part of the Cenozoic Era.

Paleocene–Eocene Thermal Maximum A warming trend that began abruptly about 55 million years ago.

paleogeography The study of Earth's ancient geography on a global as well as a regional scale.

paleontology The use of fossils to study life history and relationships among organisms.

Pangaea Alfred Wegener's name for a Late Paleozoic supercontinent made up of most of Earth's landmasses.

Pannotia A supercontinent that existed during the Neoproterozoic.

Panthalassa Ocean A Late Paleozoic ocean that surrounded Pangaea.

parallel evolution Evolution of similar features in two separate but closely related lines of descent as a result of comparable adaptations. (See *convergent evolution*.)

pelycosaur Pennsylvanian to Permian reptile that possessed some mammal characteristics; many species had large fins on their back.

period The fundamental unit in the hierarchy of time units; part of geologic time during which the rocks of a system were deposited.

Perissodactyla The order of odd-toed hoofed mammals; consists of present-day horses, tapirs, and rhinoceroses.

photochemical dissociation A process whereby water molecules in the upper atmosphere are disrupted by ultraviolet radiation, yielding oxygen (O_2) and hydrogen (H).

photosynthesis The metabolic process in which organic molecules are synthesized from water and carbon dioxide (CO_2) using the radiant energy of the Sun as it is captured by chlorophyll-containing cells.

phyletic gradualism The concept that a species evolves gradually and continuously as it gives rise to new species. (See *punctuated equilibrium*.)

placental mammal All mammals with a well-developed placenta to nourish the developing embryo, as opposed to egg-laying mammals (monotremes) and pouched mammals (marsupials), which have a less efficient placenta.

placoderm Late Silurian through Permian "plate-skinned" fish with jaws and bony armor, especially in the head-shoulder region.

plankton Animals and plants that float passively, such as phytoplankton and zooplankton. (See *nekton*.)

plate A segment of Earth's crust and upper mantle (lithosphere) varying from 50 to 250 km thick.

plate tectonic theory Theory holding that lithospheric plates move with respect to one another at divergent, convergent, and transform plate boundaries.

platform The buried extension of a Precambrian shield, which together with the shield makes up a craton.

playa lake A temporary lake in an arid region.

plesiosaur A type of Mesozoic marine reptile; both short-necked and long-necked plesiosaurs existed.

plutonic rock Igneous rock that cools and crystallizes from magma intruded into or formed within the crust. (See *igneous rock*.)

pluvial lake Any lake that formed in non-glaciated areas during the Pleistocene as a result of increased precipitation and reduced evaporation rates during that time.

pollen analysis Identification and statistical analysis of pollen from sedimentary rocks; provides information about ancient floras and climates.

polymer A comparatively complex organic molecule, such as nucleic acids and proteins, formed by monomers linking together. (See *monomer*.)

Precambrian shield An area in which a continent's ancient craton is exposed, as in the Canadian shield.

premolar Any of a mammal's teeth between the canines and the molars; premolars and molars together are a mammal's chewing teeth.

primary producer Organism in a food chain, such as bacteria and green plants, that manufacture their own organic molecules and on which all other members of the food chain depend for sustenance. (See *autotrophic*.)

Primates The order of mammals that includes prosimians (lemurs and tarsiers), and anthropoids (monkeys, apes, and humans).

principle of cross-cutting relationships A principle holding that an igneous intrusion or fault must be younger than the rocks it intrudes or cuts across.

principle of fossil succession A principle holding that fossils, especially groups or assemblages of fossils, succeed one another through time in a regular and determinable order.

principle of inclusions A principle holding that inclusions or fragments in a rock unit are older than the rock itself—for example, granite inclusions in sandstone are older than the sandstone.

principle of lateral continuity A principle holding that rock layers extend outward in all directions until they terminate.

principle of original horizontality A principle holding that sediments are deposited in horizontal or nearly horizontal layers.

principle of superposition A principle holding that sedimentary rocks in a vertical sequence formed one on top of the other so that the oldest layer is at the bottom of the sequence whereas the youngest is at the top.

principle of uniformitarianism A principle holding that we can interpret past events by understanding present-day processes, based on the idea that natural processes have always operated as they do now.

Proboscidea The order of mammals that includes elephants and their extinct relatives.

proglacial lake A lake formed of meltwater accumulating along the margin of a glacier.

progradation The seaward (or lakeward) migration of a shoreline as a result of nearshore sedimentation.

prokaryotic cell A cell lacking a nucleus and organelles such as mitochondria and plastids; the cells of bacteria and archaea are prokaryotic. (See *eukaryotic cell*.)

prosimian A subgroup of the order Primates, including tree shrews, lemurs, lorises, and tarsiers.

protorothyrid A loosely grouped category of small, lizardlike reptiles.

pterosaur Any of the Mesozoic flying reptiles that had a long finger to support a wing.

punctuated equilibrium A concept holding that new species evolve rapidly, in perhaps a few thousand years, then remain much the same during their several million years of existence. (See *phyletic gradualism*.)

pyroclastic materials Fragmental materials, such as ash, explosively erupted from volcanoes.

quadrupedal A term referring to locomotion on all four limbs, as in dogs and horses. (See *bipedal*.)

Queenston Delta A clastic wedge resulting from deposition of sediment eroded from the highland formed during the Taconic orogeny.

radioactive decay The spontaneous change in an atom by emission of a particle from its nucleus (alpha and beta decay) or by electron capture, thus changing the atom to a different element.

range zone A biostratigraphic unit defined by the occurrence of a single type of organism such as a species or a genus.

regional metamorphism Metamorphism taking place over a large but usually elongated area resulting from heat, pressure, and chemically active fluids.

relative dating The process of placing geologic events in their proper chronological order with no regard to when the events took place in terms of the number of years ago. (See *absolute dating*.)

relative geologic time scale The geologic time scale as deduced from the geologic column, showing only relative time; that is, Silurian rocks are younger than those of the Ordovician but older than those designated Devonian.

Rio Grande rift A linear depression made up of several interconnected basins extending from Colorado into Mexico.

ripple mark Wavelike structure on a bedding plane, especially in sand, formed by (1) unidirectional flow of air or water currents or (2) by oscillating currents as in waves.

rock An aggregate of one or more minerals, as in granite (feldspars and quartz) and limestone (calcite), but also includes rocklike materials such as natural glass (obsidian) and altered organic material (coal).

rock cycle A sequence of processes through which Earth materials may pass as they are transformed from one rock type to another.

rock-forming mineral Any of about two dozen minerals common enough in rocks to be important for their identification and classification.

Rodinia The name of a Neoproterozoic supercontinent.

rounding The process involving abrasion of sedimentary particles during transport so that their sharp edges and corners are smoothed off.

ruminant Any cud-chewing placental mammal with a complex three- or four-chambered stomach, such as deer, cattle, antelope, and camels.

San Andreas transform fault A major transform fault extending from the Gulf of Mexico through part of California to its termination in the Pacific Ocean off the north coast of California. (See *transform fault*.)

sand dune A ridge or mound of wind-deposited sand.

sandstone-carbonate-shale assemblage An association of sedimentary rocks typically found on passive continental margins.

Sauk Sequence A widespread association of sedimentary rocks bounded above and below by unconformities, deposited during a Neoproterozoic to Early Ordovician transgressive–regressive cycle of the Sauk Sea.

Saurischia An order of dinosaurs; characterized by a lizardlike pelvis; includes theropods, prosauropods, and sauropods. (See *Ornithischia*.)

scientific method A logical, orderly approach involving data gathering, formulating and testing hypotheses, and proposing theories.

seafloor spreading The phenomenon involving the origin of new oceanic crust at spreading ridges that then moves away from ridges and is eventually consumed at subduction zones.

sedimentary facies Any aspect of sediment or sedimentary rocks that makes them recognizably different from adjacent rocks of about the same age, such as a sandstone facies.

sedimentary rock Any rock composed of (1) particles of preexisting rocks, (2) minerals derived from solution by inorganic chemical processes or by the activities of organisms, or (3) masses of altered organic matter as in coal.

sedimentary structure All features in sedimentary rocks, such as ripple marks, cross-beds, and burrows, that formed as a result of physical or biologic processes that operated in a depositional environment.

sediment-deposit feeder Animal that ingests sediment and extracts nutrients from it.

seedless vascular plant Plant with specialized tissues for transporting fluids and nutrients and that reproduces by spores rather than seeds, such as ferns and horsetail rushes.

sequence stratigraphy The study of rock relationships within a time-stratigraphic framework of related facies bounded by widespread unconformities.

Sevier orogeny Cretaceous phase of the Cordilleran orogeny that affected the continental shelf and slope areas of the Cordilleran mobile belt.

Siberia One of six major Paleozoic continents; composed of Russia east of the Ural Mountains and Asia north of Kazakhstan and south of Mongolia.

silicate A mineral containing silica, a combination of silicon and oxygen, and usually one or more other elements.

solar nebula theory An explanation for the origin and evolution of the solar system from a rotating cloud of gases.

Sonoma orogeny A Permian–Triassic orogeny caused by the collision of an island arc with the southwestern margin of North America.

sorting The process whereby sedimentary particles are selected by size during transport; deposits are poorly sorted to well sorted, depending on the range of particle sizes present.

species A population of similar individuals that in nature can interbreed and produce fertile offspring.

stratigraphy The branch of geology concerned with the composition, origin, areal extent, and age relationships of stratified (layered) rocks; concerned with all rock types but especially sediments and sedimentary rocks.

stratification (bedding) The layering in sedimentary rocks; layers less than 1 cm thick are laminations, whereas beds are thicker.

stromatolite A biogenic sedimentary structure, especially in limestone, produced by entrapment of sediment grains on sticky mats of photosynthesizing bacteria.

submarine hydrothermal vent A crack or fissure in the seafloor through which superheated water issues. (See *black smoker*.)

Sundance Sea A wide seaway that existed in western North America during the Middle Jurassic Period.

supercontinent A landmass consisting of most of Earth's continents (such as Pangaea).

suspension feeder Animal that consumes microscopic plants and animals or dissolved nutrients from water.

system The fundamental unit in the hierarchy of time-stratigraphic units, such as the Devonian System; also a combination of related parts that interact in an organized manner; Earth's systems include the atmosphere, hydrosphere, and biosphere, as well as the lithosphere, mantle, and core.

Taconic orogeny An Ordovician episode of mountain building resulting in deformation of the Appalachian mobile belt.

Tejas epeiric sea A Cenozoic sea largely restricted to the Gulf and Atlantic Coastal Plains, coastal California, and the Mississippi Valley.

terrane A small, lithospheric block with characteristics quite different from those of surrounding rocks; probably consists of seamounts, oceanic rises, and other seafloor features accreted to continents during orogenies.

terrestrial planet Any of the four small inner planets (Mercury, Venus, Earth, and Mars) similar to Earth (*Terra*); all have high mean densities, indicating they are composed of rock. (See *Jovian planet*.)

theory An explanation for some natural phenomenon with a large body of supporting evidence; theories must be testable by experiments and/or observations, as is the case with plate tectonic theory.

theory of evolution The theory holding that all living things are related and that they descended with modification from organisms that lived during the past.

therapsid Permian to Triassic mammal-like reptiles; the ancestors of mammals are among one group of therapsids known as cynodonts.

thermal convection cell A type of circulation of material in the asthenosphere during which hot material rises, moves laterally, cools and sinks, then is reheated and continues the cycle.

tidal flat A broad, extensive area along a coastline that is alternately water-covered at high tide and exposed at low tide.

till Sediment deposited directly by glacial ice, as in an end moraine.

time-stratigraphic unit A body of strata that was deposited during a specific interval of geologic time; for example, the Devonian System, a time-stratigraphic unit, was deposited during that part of geologic time designated the Devonian Period.

time unit Any of the units such as eon, era, period, epoch, and age referring to specific intervals of geologic time.

Tippecanoe Sequence A widespread body of sedimentary rocks bounded above and below by unconformities; deposited during an Ordovician to Early Devonian transgressive–regressive cycle of the Tippecanoe Sea.

trace fossil Any indication of prehistoric organic activity such as tracks, trails, burrows, and nests. (See *biogenic sedimentary structure*, *body fossil*, and *fossil*.)

Transcontinental Arch Area extending from Minnesota to New Mexico that stood above sea level as several large islands during the Cambrian transgression of the Sauk Sea.

transform fault A type of fault that changes one kind of motion between plates into another type of motion; recognized on land as a strike-slip fault. (See *San Andreas transform fault*.)

transform plate boundary Plate boundary along which adjacent plates slide past one another and crust is neither produced nor destroyed.

tree-ring dating The process of determining the age of a tree or wood in a structure by counting the number of annual growth rings.

unconformity A break or gap in the geologic record resulting from erosion or nondeposition or both; the surface separating younger from older rocks where a break in the geologic record is present. (See *angular unconformity*, *nonconformity*, and *disconformity*.)

ungulate An informal term referring to a variety of mammals but especially the hoofed mammals of the orders Artiodactyla and Perissodactyla.

U-shaped glacial trough A valley with steep or nearly vertical walls and a broad, concave, or rather flat floor; formed by movement of a glacier through a stream valley.

valley glacier A glacier confined to a mountain valley.

varve A dark-light couplet of sedimentary laminations representing an annual deposit in a glacial lake.

vascular plant A plant with specialized tissues for transporting fluids.

vertebrate Any animal possessing a segmented vertebral column, as in fish, amphibians, reptiles, birds, and mammals; members of the subphylum Vertebrata.

vestigial structure Any structure in an organism that no longer serves any function or serves only a limited or different function; examples are dewclaws in dogs, wisdom teeth in humans, and middle ear bones in mammals.

volcanic rock An igneous rock that forms as lava cools and crystallizes or when pyroclastic materials are consolidated. (See *extrusive igneous rock*.)

Walther's law A concept holding that the facies in a conformable vertical sequence will be found laterally to one another.

Wilson cycle The relationship between mountain building (orogeny) and the opening and closing of ocean basins.

Zuni epeiric sea A widespread sea that was present in North America mostly during the Cretaceous, but persisted into the Paleogene.

Answers to Multiple-Choice Review Questions

Chapter 1
1. d; 2. d; 3. c; 4. b; 5. d; 6. e; 7. d; 8. a; 9. d; 10. e

Chapter 2
1. b; 2. d; 3. a; 4. e; 5. c; 6. a; 7. b; 8. d; 9. e; 10. b

Chapter 3
1. c; 2. d; 3. a; 4. e; 5. d; 6. a; 7. e; 8. c; 9. a; 10. b

Chapter 4
1. a; 2. b; 3. e; 4. d; 5. a; 6. d; 7. b; 8. a; 9. b; 10. b

Chapter 5
1. b; 2. e; 3. a; 4. a; 5. c; 6. c; 7. a; 8. e; 9. c; 10. b

Chapter 6
1. c; 2. c; 3. a; 4. e; 5. b; 6. a; 7. d; 8. c; 9. b; 10. d

Chapter 7
1. b; 2. d; 3. a; 4. b; 5. e; 6. c; 7. a; 8. d; 9. e; 10. b

Chapter 8
1. b; 2. e; 3. c; 4. a; 5. a; 6. c; 7. b; 8. c; 9. e; 10. d

Chapter 9
1. b; 2. d; 3. c; 4. e; 5. b; 6. a; 7. b; 8. b; 9. b; 10. d

Chapter 10
1. e; 2. d; 3. b; 4. e; 5. b; 6. c; 7. b; 8. b; 9. b; 10. b; 11. a; 12. d; 13. d

Chapter 11
1. e; 2. a; 3. e; 4. b; 5. d; 6. c; 7. d; 8. b; 9. d; 10. c

Chapter 12
1. a; 2. c; 3. e; 4. c; 5. a; 6. b; 7. e; 8. e; 9. c; 10. e; 11. c; 12. d

Chapter 13
1. c; 2. e; 3. a; 4. c; 5. e; 6. d; 7. c; 8. b; 9. c; 10. b; 11. e; 12. d

Chapter 14
1. d; 2. e; 3. c; 4. a; 5. a; 6. e; 7. e; 8. e; 9. e; 10. e

Chapter 15
1. d; 2. b; 3. c; 4. e; 5. d; 6. a; 7. c; 8. a; 9. c; 10. b

Chapter 16
1. c; 2. a; 3. d; 4. d; 5. d; 6. a; 7. b; 8. a; 9. e; 10. c

Chapter 17
1. e; 2. c; 3. a; 4. b; 5. c; 6. d; 7. d; 8. a; 9. d; 10. c

Chapter 18
1. c; 2. a; 3. e; 4. a; 5. d; 6. a; 7. a; 8. b; 9. c; 10. a

Chapter 19
1. d; 2. e; 3. a; 4. e; 5. a; 6. d; 7. c; 8. b; 9. b; 10. d

Index